从心灵到神经

FROM MIND TO NERVOUS SYSTEM

认知神经科学文献选编

SELECTED ARTICLES
ON COGNITIVE NEUROSCIENCE

燕 燕 李福华 白 玉 主编

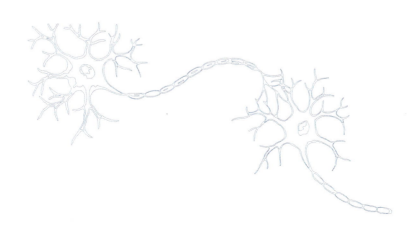

社会科学文献出版社
SOCIAL SCIENCES ACADEMIC PRESS (CHINA)

序　言

　　心灵是西方哲学、心理学领域里的一个古老的话题。作为哲学思辨的对象，心灵首先是一种假定存在的实体，它使有形可见的表现出一种秩序与动力。对心灵存在的哲学反思既指向物理世界，又反观人性自身。如果人是心灵与身体的复合体，那么作为实体的心灵是怎样与物性的身体相结合的呢？又在哪里结合呢？当亚里士多德把心灵理解为身体的形式，心身关系是形式与质料的关系时，亚里士多德却无法回答形式的心灵究竟出自我们的骨骼还是我们的肉，以及它又为什么能够从身体之肉中示现出来。心灵作为哲学思辨的结果，是思想中的一个对象，即它是一个概念性的存在。概念性的心灵一直延续到思潮涌起、世殊事异的 18 世纪。这是一个见证了科学兴盛、实证快然的世纪。物理科学及其研究模式的显圣成为那个世纪的人们理解自然与人性的鸿规。相应的，对心灵的反思就从传统哲学的经验性的思辨转向基于生理实证科学的思辨。心灵俯就物质性的神经之轭。这就是神经哲学家查尔斯·伯奈特（Charles Bonnet）的开拓性工作。

　　查尔斯·伯奈特把意识现象与神经活动联系起来，把心理活动与生理活动相对应，认为大脑、神经系统是心灵活动的座架。查尔斯·伯奈特把神经系统理解为由神经纤维组成的机器，这或多或少钩沉了笛卡尔的影子。但不同于笛卡尔把感觉、运动与意志视为独立心灵的不同形式，查尔斯·伯奈特把意识活动解释为神经系统传导运动的不同形式。当然，笛卡尔也从大脑论述人的情感与意识，正如同柏拉图也曾把高级的心灵活动归结于大脑，但把意识与神经系统结合起来并进行系统性阐释的第一位哲学家正是查尔斯·伯奈特。从此以后，思辨的心灵就演变为生物性的精灵。它在中空的导管里由感官向脑室流入并从脑室流出。没有神经就没有感知，感知与情绪的多样性正是神经振动方式的多样性。

查尔斯·伯奈特的新径既举要 18 世纪科学思想之纲维，又凿开哲学反思新范式之户牖。对心灵、人性的认识与反观不再拘囿于传统哲学疏阔寡要的思辨，而是转向以实证表窥可测量的枢区与神经。18 世纪神经矩步心灵的思绪在 19 世纪兴发如潮，并塑造了学科间的渗透与通融。在脑科学领域里，通过神经透视心灵的卓越者当数英国神经生理学家、哲学家查尔斯·谢灵顿（Sir Charles Scott Sherrington），在心理学领域则是师心独见的威廉·詹姆斯（William James）。两者破理之见皆是认识到了身体的智慧，即心灵是身体的意识。当然，谢灵顿的洞悉也并非别出机杼的创见，早于他一个世纪的英国生理学家查尔斯·贝尔（Sir Charles Bell）就已思有出辙并流声后代，比如查尔斯·达尔文（Charles Robert Darwin）的情绪理论。不同的是，贝尔过多地关注了情绪的身体系统，而谢灵顿则借助神经系统窥探心灵。20 世纪早期，顺流谢灵顿的思想并义尚光大者是加拿大一流神经生理学家威尔德·彭菲尔德（Wilder Penfield）。作为谢灵顿的高足，彭菲尔德在脑科学、神经科学领域对心灵的标识，以及对语言、情景记忆、动作在大脑皮层分区的描绘，事实上已经标明了一门交叉学科的诞生。尽管某些行为有直接对应的脑区在 19 世纪中期已有实证的研究，比如布洛卡语言区、维尔尼克语言区等，然而，正是在彭菲尔德手执电极的操作中，心灵成为可触的，思维成为可视的。更为重要的是，传统哲学所信奉的那个高级、独立、自持、理性的心灵在彭菲尔德的探针下却是平行结构的。我们并不是欲望的身体与理智的心灵的耦合。相反，心灵是身体的不同结构与形式。记忆、想象、阅读、推理等学习以及认知活动是神经系统、皮质以及皮下组织的机制。可是，幽深处的机制究竟是怎样的呢？钩深取极彭菲尔德神经规式的认知原理成全了唐纳德·赫布（Donald Hebb）定律，即突触连结的学习原理。曾在彭菲尔德实验室里学习的赫布发现，神经元与神经元之间的反射性特性增强了神经元之间的连结。换言之，新行为的习得是神经元的可塑性。

脑科学、神经科学、生理医学等关于人性的实证科学与传统心灵哲学在 20 世纪精密地并轨交织使神经科学规式下的心灵探幽之风飞动腾跃。然而，神经科学家并非仅仅聚焦在脑与认知的关系上。从脑科学、神经科学解开情绪的隐机构成了另一支流。成就斐然者当属美国认知神经科学家安东尼奥·达玛西奥（Antonio Damasio）。正是 18 世纪从心灵到神经的思绪初发，19 世纪紧扣其端的局促与推送，才有 20 世纪振绪播风又壮采时逢的认知与神经科学的结合。当然，在走向学科融合的进路中，还有流源于詹姆士身体哲学的行为主义的强势崛起，以及反行为主义的认知科学的标立。种种流派、理论的枝附并推陈出新铺垫了 20 世纪 70 年代 "认知神经科学" 的赋名显义，一门交叉学科赫然耸立。

正名立范之后的认知神经科学气势奋发。从脑皮层、神经系统到镜像神经元，从大脑的自组织结构到神经系统的 theta 正弦活动方式，从长程的神经协同作业到大脑的 3-D 表征等，认知神经科学正在用它的探极将隐蔽幽暗的机制带入我们的视域中。而这些内容也构成了本编著改弦易辙的上编。

然而，在人性的环境里发展、发达的神经系统从来就不是生理学意义上的实在论，也不是解剖学意义上的物质性，而是文化、文明膏润的体性，个性历史浸身的风骨。从视觉到行动，从评价到决策，从情绪到理性，脑皮层是社会化的定势，神经系统是文化化了的情采。虽然心灵并不能还原为突触、神经元，也不是它们的集合，心灵也不是可触的大脑皮层或皮下组织，可正是借助认知神经科学，我们认识到心灵是这具肉身的精灵，是具身的结构。从另一个方面来说，虽然我们都是身体与心灵的统一体，虽然我们的脑结构在解剖学意义上来说是相同的（尽管爱因斯坦的大脑的解剖结构具有特异性），但正是文化、文明以及社会化把我们相互区分。所以，我们不能说是我们的大脑在看，也不能说我们的大脑或自我是突触的连结。相反，个体是语言、历史、文化、经验等交织的个体，是体现着特定文化样式并负载传承这种样式的责任与生活意义的生命体。认知神经科学虽然有着短暂的历史，但它脱胎于其中的实证科学却有着漫长的过去。在历史与当下的渗透中，认知神经科学也将带着幽远的反思探赜更为隐匿的人性。正如哲学家 Walter Glannon 的结括之语：我们每个人是 "大脑、身体与心灵——一张人脸的神经伦理学"。

正因为我们不仅是这有形的大脑皮层交织交错的皮下组织，也不只是可见的物性的肉身，而是种种认知形式可从中间出的、活性的系统。这活性的系统正好可以通过勘测结构性组织的脑区与神经系统来例证。所以，认知神经科学的影响迅速渗透到教育学、心理学、认知科学、康复学等领域。它对教育学的启发是：教育、教学的实施再也不是基于经验的堆砌而是基于科学的实证，即大脑的损伤会导致认知障碍，而设计、运用针对性的教育、教学项目可以改善损伤处。同时，了解了大脑的运作原理与机制，我们也就可以更为高效地创设学习环境、工作环境以提高学习、认知的能力。换言之，可基于脑皮层、神经系统的活动机制与原理来提高脑的工作效率以及社会化的能力。这种效率与能力既是情感的又是认知的。关于认知神经科学，学界更强调神经系统、大脑皮层以及皮下组织的认知功能。然而，如果我们的神经元具有镜像效应，具有空间的表征能力等，如果我们的皮质或皮下组织能够让我们视听以及体验痛苦与欢乐等，那么，我们就不能说神经系统或脑组织是认知的。

恰恰相反，它们是感性的，或情感的。种种的认知形式正是它们的感性形式。认知神经科学指括结要的是认知，是情本的晶体。这也是编者编著此书的用意与目的。

本编著从 *Annual Review of Neuroscience* 期刊 17 年间的研究成果中选编了 18 篇文章，既包含经典神经元的研究，比如镜像神经元，又兼顾知觉、行为等文献综述，旨在突出西方学者的研究进路、内容以及方法。虽然这是认知神经科学文献的选编，但其中的篇章对理解中国哲学,比如庄子哲学的"官知止而神欲行"中的"神"与"道"提示了路径。加拿大汉学家 Edward Slingerland 说，中国哲学的一些概念只有在当下实证科学的帮助下才能得到更好的理解，颇切肯綮。仔细审读第一部分的第 1 篇与第二部分的第 1 篇，我们就有所知会庄子用中国文化的"神"与"道"等概念来表示的正是习惯性动作的神经机制。也正是在这里，庄子的思想与美国心理学家威廉·詹姆斯钩沉的身体的意识契合遥应。再者,心理学关于行为的形成一直有模仿说。可模仿为什么能够发生呢？这可能就要归功于我们的镜像神经元。文学、心理学领域里有通感之说。钱钟书先生曾把通感归结于心理的联想，然而认知神经科学却揭示它是视、听、触等知觉特异化的神经通道共享一种结构模式的表征。同理，格式塔心理学描述的知觉完形也是不同脑区的神经环路协同共振的神经现象，且是文化世界里的经验性的神经现象。第一部分的第 4、5 篇也证实了我们的时间意识内具在神经元应节的活动频率中，是神经系统的存在方式。换言之，时间是我们存在的维度。而这种揭示又是威廉·詹姆斯在 *The Principles of Psychology* 一书中时间观的余绪。当然，我们并不是还原论者。这就是文化的意义。第二部分的第 4 篇告诉我们的正是超越还原论。法国著名的物理学家、数学家、科学哲学家彭加勒曾说，我们不能在二维的平面上画出三维的图像。悖论的是，在绘画中我们的确看到了深度。如此，第三维的深度知觉究竟来自哪里呢？现象学家梅洛－庞蒂解释说，对深度的知觉来自肉身内在的动作。这也是第一部分第 7 篇文章的实证研究。现象学家又说我们的肉身是拓扑的结构。认知神经科学家用实证性的语言来图示化现象学家的灼见。进一步的，神经生理学的研究追溯到单细胞的源头向我们揭示了肉身的拓扑结构是怎样肇始于细胞的分化以及拓扑地自组织过程。这就是第一部分的第 3 篇文章的贡献。仅此几例以发凡。

苦于版权受限，对前文中生理学、神经科学、脑科学等实证研究的背景知识以及历史路径梳理涉及的 12 部经典书稿的节选以及 6 篇美国科学院的相关文章未能拾掇汇集。尽管如此，编者仍然要感谢美国波士顿大学 Steven Katz 教授以及 Rika 夫人不计繁缛地为编者争取版权。虽然编者在选读文献过程中呕心吐胆，劳形损精，但终感劳深绩寡，飔焰欠缺。编者感谢教育学院王家云院长始终如一地支持本编著的出版，编者同样感谢李晓岩博士。从 PDF 格式转成 Word 文档后的校对过程同样是劳深力多的繁苛。李晓岩博士欣然应允编者的个人求助，带领研究生挤出时间，无私沉默地奉献。编者向她以及研究生们敬致谢忱。学界素知认知神经科学是典型的跨学科专业，然而，编者在编辑的过程中更深切地体味到，对这一博广又精深领域之优秀成果的编撰同样需要跨学科的团队合作。回望编著历程，虽然不短亦不长，但感受却深切骨髓，那就是学识并体能上的势单力薄。虽然编者的专业领域是现象学身体哲学，始于 2009 年学习认知神经科学，虽然认知神经科学是现象学尤其是梅洛－庞蒂现象学身体哲学的循体而成势，但终非编者的术业专攻。挈瓶之智，难以见全。编者伫候专家同行赐教斧正。

燕　燕

2019 年 3 月 18 日于滨湖

Contents

I

Lived Nervous System: From Mirror-Neuron to the Structured Nervous System

Intentional Maps in Posterior Parietal Cortex   / 3

The Mirror-Neuron System   / 41

The Role of Organizers in Patterning the Nervous System   / 72

Mechanisms and Functions of Theta Rhythms   / 100

Neural Basis of the Perception and Estimation of Time   / 124

Long-range Neural Synchrony in Behavior   / 155

3-D Maps and Compasses in the Brain   / 186

Neuronal Mechanisms of Visual Categorization: An Abstract
    View on Decision Making   / 219

The Role of the Lateral Intraparietal Area in (the Study of) Decision
    Making   / 243

II

Socialization of Nervous System: From Neuroculture to Socialized Cognition

Habits, Rituals, and the Evaluative Brain   / 277

Brain Plasticity Through the Life Span: Learning to Learn and Action
    Video Games   / 313

Neural Basis of Reinforcement Learning and Decision Making　/ 347

Social Control of the Brain　/ 375

Cortical Control of Arm Movements: A Dynamical Systems
　Perspective　/ 400

Emotion and Decision Making: Multiple Modulatory
　Neural Circuits　/ 431

Embodied Cognition and Mirror Neurons: A Critical Assessment　/ 464

Establishing Wiring Specificity in Visual System Circuits: From the
　Retina to the Brain　/ 484

Neural Mechanisms of Social Cognition in Primates　/ 522

目　录

一、活的神经系统：从镜像神经元到结构化的神经系统

后顶叶皮质的意向图示 ························· 3

镜像神经元系统 ····························· 41

组织者在神经系统形成中的作用 ················· 72

θ 波的机制与功能 ·························· 100

知觉和时间估计的神经基础 ··················· 124

行为的远程神经同步 ························ 155

大脑的 3D 图示与方向感 ···················· 186

视觉分类的神经机制：关于决策的抽象观点 ·········· 219

侧顶叶在决策（研究）中的作用 ················· 243

二、神经系统的社会化：从神经文化到社会化认知

习惯、习俗与评价性的大脑 ··················· 277

生命全程的大脑可塑性：学会学习与动作视频游戏 ······ 313

强化学习和决策的神经基础 ··················· 347

大脑的社会控制 ··························· 375

手臂运动的皮质控制：动力学系统视角 …………………………… 400

情绪与决策：多通道神经回路 …………………………… 431

具身认知与镜像神经元：批判性评价 …………………………… 464

在视觉系统回路中建立联结特异性：从视网膜到大脑 ………… 484

灵长类动物社会认知的神经机制 …………………………… 522

I

Lived Nervous System: From Mirror-Neuron to the Structured Nervous System

活的神经系统：从镜像神经元到结构化的神经系统

Intentional Maps in Posterior Parietal Cortex

Richard A. Andersen and Christopher A. Buneo

Division of Biology, California Institute of Technology, Mail Code 216-76, Pasadena, California 91125; email: andersen@vis.caltech.edu; chris@vis.caltech.edu

Key Words

eye movements, arm movements, optic flow, spatial representations, neural prosthetics

Abstract

The posterior parietal cortex (PPC), historically believed to be a sensory structure, is now viewed as an area important for sensory-motor integration. Among its functions is the forming of intentions, that is, high-level cognitive plans for movement. There is a map of intentions within the PPC, with different subregions dedicated to the planning of eye movements, reaching movements, and grasping movements. These areas appear to be specialized for the multisensory integration and coordinate transformations required to convert sensory input to motor output. In several subregions of the PPC, these operations are facilitated by the use of a common distributed space representation that is independent of both sensory input and motor output. Attention and learning effects are also evident in the PPC. However, these effects may be general to cortex and operate in the PPC in the context of sensory-motor transformations.

INTRODUCTION

The posterior parietal cortex (PPC) has traditionally been viewed as a sensory "association" area, associating different modalities and having higher-level sensory functions such as spatial attention and spatial awareness. In this review, we highlight a new view of the PPC that is emerging. It is proposed that the PPC, rather than serving a purely sensory or motor role, subserves higher-level cognitive functions related to action. Among these higher cognitive functions is the formation of intentions, or early plans for movement. These intentions are anatomically segregated within the PPC, with regions specialized for the planning of saccades, reaches, and grasps. Moreover, these intentions are highly abstract and are evident in the discharge of single neurons even when a specific intention is not carried out.

The different intention-related regions of the PPC appear to participate in operations critical to the earliest stages of movement planning: multisensory integration and coordinate transformations. These functions are facilitated by employing a rather unique, distributed code. The response fields of neurons in at least two regions are in retinal coordinates, independent of both the sensory modality used to cue target locations (i.e., audition vs. vision) and the action that will ultimately be performed (i.e., reaches vs. saccades). However, these retinal fields are also gain modulated by eye, head, and limb positions. As a result, groups of parietal cells do not generally represent space in a single, defined spatial reference frame. Rather, they code locations in a distributed manner, which can be read out by other groups of neurons in a variety of reference frames.

We describe a potential medical application that utilizes the finding that the PPC encodes movement intentions. The intention-related activity in the PPC can, in principle, be used to operate a neural prosthesis for paralyzed patients. Such a neural prosthesis would consist of recording the activity of PPC neurons, interpreting the movement intentions of the subject with computer algorithms, and using these predictions of the subject's intentions to operate external devices such as a robot limb or a computer. We describe preliminary investigations in healthy monkeys that estimate the number of parietal cells needed to operate such a prosthesis (Meeker et al. 2001, Shenoy et al. 1999b). We also describe a recent finding that monkeys can use this intended movement activity to position a cursor on a computer screen just by thinking about a reach movement, without actually generating a reach (D. Meeker, S. Cao, J. W. Burdick & R. A. Andersen, unpublished observations). This result was obtained without extensive training and strongly suggests that we are in fact tapping into the highly abstract neural signals that represent the earliest plans for movement.

THE PPC SUBSERVES COGNITIVE FUNCTIONS RELATED TO ACTION

Many of the deficits observed following lesions of the PPC are consistent with the area playing a high-level, cognitive role in sensory-motor integration. Patients with PPC lesions do not have primary sensory or motor deficits. However, when they attempt to connect these functions, for instance during sensory guided movements, then defects become apparent. Patients with PPC lesions often suffer from optic ataxia; that is, difficulty in estimating the location of stimuli in 3D space, as indicated by pronounced errors in reaching movements (Balint 1909, Rondot et al. 1977). Patients with PPC lesions can also suffer from one or more of the apraxias, a class of deficits characterized by the inability to plan movements (Geshwind & Damasio 1985). These can range from a complete inability to follow verbal commands for simple movements, to difficulty in performing sequences of movements. Patients with parietal lobe damage also have difficulty correctly shaping their hands as they prepare to grasp objects, which again points to a disconnection between the visual sensory apparatus that registers the shape of objects and the motor systems that shape the configuration of the hand (Goodale & Milner 1992, Perenin & Vighetto 1988).

Neglect is another deficit commonly attributed to lesions of the PPC, although there is currently some debate about whether it is damage to the PPC or to the nearby superior temporal gyrus that is the source of this defect (Critchley 1953, Karnath et al. 2001). The hallmark of neglect is the lack of awareness within the personal and extrapersonal space contralateral to the lesioned hemisphere, with the most profound deficits seen with right hemisphere lesions in right-handed humans.

These clinical results are extremely informative and useful, and have helped guide much of the neurophysiological investigation of the PPC. However, to understand the neural mechanisms and circuits within the PPC that are involved in sensory-motor integration requires that the investigator, rather than relying on the happenstance of medical defects, be able to control the parameters of the experiments. Moreover, refined techniques need to be applied. In the case of humans, this has generally taken the form of fMRI studies, and in the case of monkeys, electrophysiological recording and anatomical studies. The monkey has proven to be a good model for the study of the PPC, since sophisticated motor behaviors such as hand-eye coordination are similar in the two species of primates, and there is extensive evidence to suggest that the PPC in both species performs similar functions (Connolly et al. 2000, DeSouza et al. 2000, Rushworth et al. 2001b). Evidence from these studies provides additional support for the concept that the PPC is neither strictly sensory nor motor but rather is involved in high-level cognitive functions related to action (Mountcastle et al. 1975, Andersen 1987, Goodale & Milner 1992). These functions include early-movement planning, particularly the coordinate transformations required for sensory-guided movement. The activity of PPC may also be influenced by spatial attention and learning. However, these functions are general to cortex, and in the PPC appear to operate in the more specific context of sensory-motor operations.

INTENTION

Intention is an early plan for a movement. It specifies the goal of a movement and the type of movement. For instance, "I wish to pick up the coffee cup" specifies both the goal and type of movement. An intention is high level and abstract. For instance, we can have intentions without actually acting upon them. Moreover, a neural correlate of intention does not necessarily contain information about the details of a movement, for instance the joint angles, torques, and muscle activations required to make a movement. As discussed below, intentions are initially coded in visual coordinates in at least some of the cortical areas within the PPC. This encoding is consistent with a more cognitive representation of intentions, specifying the goals of movements rather than the exact muscle activations required to execute the movement.

An intention is also a broad category of cortical functions, which include decision making (Gold & Shadlen 2001) and "motor attention" (Rushworth et al. 2001a). For instance, decision making can be considered a competition between potential movement intentions (Platt & Glimcher 1999). It may also be the case that the earliest intentions sit atop a sequence of increasingly more specific movement plans. In the example above, the earliest intention may reflect the desire to grasp the cup, with further specifications including which limb (right or left), the trajectory of the movement to avoid obstacles, the coordination of eye and hand movements, the speed of the movement, etc. Only further research will be able to resolve which parameters of a movement are coded at which stages in the sensory-motor pathway.

Distinguishing Intention from Attention

The issue of intention versus attention has been most prominent in the study of the PPC, which is perhaps not surprising considering this area is at the interface between sensory and motor systems. Mountcastle and colleagues (1975) first noted neural activity in the PPC related to the behaviors of monkeys. Robinson and colleagues (1978) later argued that these effects could be due to sensory stimulation and attention during movement. In experiments designed to tease apart sensory and movement components of activity, Andersen et al. (1987) found both, which is consistent with a role for this area in sensory-motor transformations.

One common method of separating sensory from motor components is the so-called memory task (Hikosaka & Wurtz 1983) in which an animal is cued as to the location for a movement by a briefly flashed stimulus but must withhold the response until a go signal. Typically, PPC neurons show bursts of activity to the cue and the movement, indicating both sensory- and motor-related activity. However, during the memory period the cells in many parietal areas have persistent activity, even in the dark (Gnadt & Andersen 1988, Snyder et al. 1997). This persistent activity by and large does not represent the sensory memory of the target. This can be demonstrated using tasks in which animals memorize the locations of two stimuli and subsequently make movements to both locations. For eye and arm movements the persistent activity in the delay period for nearly all neurons in the PPC is only present for the next planned movement (Batista & Andersen 2001, Mazzoni et al. 1996a), even though the animals must hold in memory two cued locations. This result indicates that the sensory memory of the target locations is either contained in a very small subset of neurons within the PPC, or in areas outside the PPC, perhaps in the frontal lobe (Tian et al. 2000).

The results of the double movement tasks rule out the coding of a sensory memory in the delay period activity. However, this activity could reflect either the direction of a movement plan or the direction of attention. Experimentally it has been very difficult to distinguish movement planning or preparation from spatial attention. Most studies of attention in monkeys use experimental paradigms that require animals to make eye or limb movements as part of the experimental design, or have the potential artifact of the animal covertly planning these movements. This fact is reason for concern in studies of the dorsal, sensory-motor pathway since there is extensive overlap of circuitry concerned with attention and eye movements, as demonstrated by fMRI experiments in humans (Corbetta et al. 1998). The finding that the locus of spatial attention can affect the metrics of saccades electrically evoked from the superior colliculus (SC) further argues for a very tight coupling of spatial attention and eye movements (Kustov & Robinson 1996). These and other results have led Rizzolatti and colleagues to argue for a motor theory of spatial attention (1994). They propose that spatial attention is an early form of motor preparation, at least for eye movements.

Some investigators have used antisaccade and antireach tasks to separate sensory from movement processing. In these paradigms, animals are trained to make movements in the opposite direction from flashed visual targets. For the case of reaches, activity in the medial intraparietal area (MIP) has been reported to code mostly the direction of the movement, and not the location of the stimulus (Eskandar & Assad 1999, Kalaska 1996). Gottlieb & Goldberg (1999) have reported that the reverse is true in the lateral intraparietal area (LIP) for eye movements, i.e., that cells respond to the stimulus and not the direction of planned movement. However, a recent report by Zhang & Barash (2000) indicates that, after a brief transient linked to the stimulus, most cells in LIP code the direction of the planned eye movement. Moreover, a smaller class of cells encode both the location of the stimulus and the movement plan, which suggests that LIP is involved in the intermediate stages of the sensory-motor transformations required for the antisaccade task. Overall, these antisaccade and antireach results reinforce the idea that PPC cells have both sensory- and movement-related responses, and occupy an intermediate stage in the sensory-motor transformation process.

We recently conducted an experiment specifically designed to separate the effects of spatial attention from those of intention (Snyder et al. 1997). In this experiment, animals attended to a flashed target and planned a movement to it during a delay period, but in one case they were instructed to plan a saccade and in the other a reach (see Figure 1*a*). The only difference in the task during the memory period was the movement the animals were planning to make. We reasoned that if PPC activity reflected a sensory memory or attention, it should be the same in the two conditions, but if it reflected the movement plan it should be different.

Figure 1 shows two intention-specific neurons, one from area LIP (*b*) and one from an area we refer to as the parietal reach region (PRR) (*c*). In this task the monkey plans an eye or an arm movement to the same location in space. The activity of the LIP neuron illustrated in panel (*b*) shows a transient response to the onset of the briefly flashed target. This is followed by activity during the delay period if the animal is planning an eye movement (left histogram), but not if he is planning an arm movement to the same location (right histogram). In contrast, the cell in panel (*c*) shows no activity above baseline in the delay period when the animal is planning an eye movement, but strong activity when he is planning an arm movement. Such results were typical in the PPC: In general, we found that during eye movement planning area LIP was much more active, and during limb movement planning PRR was more active. PRR included MIP, 7a, and the dorsal aspect of the parieto-occipital (PO) area, though MIP was found to have the highest concentration of reach-related neurons. The results from both LIP and PRR argue strongly for a role of the PPC in movement planning.

A subsequent experiment showed that activity in the PPC is also related to the shifting of movement plans, when spatial attention is held constant (Snyder et al. 1998a). Cells with a particular movement preference (reach or saccade) showed greater activity if a plan was changed from the nonpreferred to the preferred movement (for the same target location), compared to simply reaffirming the preferred plan. This result is reminiscent of proposals that the PPC plays a role in shifting attention (Steinmetz & Constantinidis 1995), but in this case it is the intended movement that shifts, and not the spatial locus of attention.

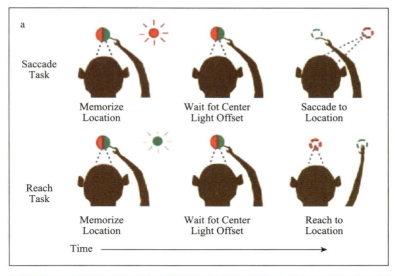

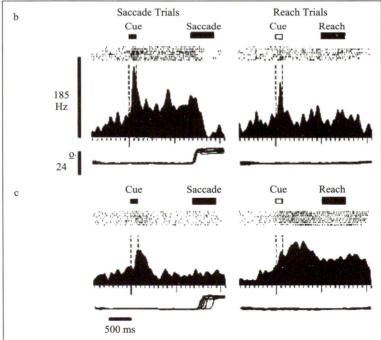

Figure 1

(*a*) Tasks used in Snyder et al. (1997) to separate the effects of spatial attention from those of intention. Animals made either a saccade (*top*) or a reach (*bottom*) to the remembered location of a flashed visual target (red flash: saccade, green flash: reach). Movements were made in complete darkness, after a delay period. (*b*) A lateral intraparietal area cell showing elevated delay period activity before a saccade (*left*) but not before a reach (*right*). *Vertical dashed lines and short horizontal bars* indicate the timing of target ("Cue") presentation (red flashes: filled bars, green flashes: open bars) and *long horizontal bars* indicate the timing of the motor response ("Saccade" or "Reach"). Each panel shows eight rasters of tick marks corresponding to every third action potential recorded during each of eight trials. Below each set of rasters is a spike density histogram representing the average rate of action potential firing over all trials (generated by convolution with a triangular kernel) that is aligned on cue presentation. *Thin horizontal lines* below each histogram represent the animal's vertical eye position on each trial. During the delay interval (150–600 msec after target extinction) firing depended specifically on motor intent. For illustration purposes, data for this cell were collected using a fixed delay interval. (*c*) A PRR cell showing reach rather than saccade specificity during the delay interval (Modified from Snyder et al. 1997).

Default Plans

The experiments by Snyder et al. (1997) were not the first to attempt to separate attention from intended movement activity. Bushnell and colleagues (1981) trained animals to either reach or saccade to a target while recording from PPC neurons. They reasoned that if the PPC was involved in attention then they should see the same level of activity regardless of the motor output, and this was what they reported. However, they recorded from only nine cells, and inspection of their Figure 1 suggests that the animal may have looked to the stimulus after the reach. Thus the animal may have been planning an eye movement as well as an arm movement during the task.

This potential problem of covert planning of eye movements is a general problem for experiments examining attention to targets placed away from the fixation point. While the formation of covert plans is unlikely to be critical for studies of attention in the ventral visual pathway, which current evidence suggests is largely involved in visual recognition, it is certainly a problem when studying the dorsal pathway, which is involved in movement planning. The issue of covert planning was directly addressed in Snyder et al. (1997). In the population of cells from which we recorded, 68% were significantly modulated in the delay period by one movement plan (reach or saccade) but not the other. Interestingly, even during the cue period 44% showed this specificity. We reasoned that the remaining cells showing significant activity for both movement plans might reflect covert plans for movement, since it is very natural to look to where you reach. To control for this possibility, we had the animals also perform a "dissociation" task in which they simultaneously planned an eye and an arm movement in different directions, with one movement into the response field and the other outside.

Figure 2 shows an example of a neuron that had activity for both eye and arm movements in the single-movement task. In the top row of histograms, the monkey performed only saccades, making eye movements into the response field (left histogram) or in the opposite direction (right histogram). In the middle row, the animal made reaches instead of saccades, and a similar level of activity is seen when the animal reaches into the response field. The bottom row of histograms shows activity from the same neuron while the animal was performing the dissociation task. In the histogram on the left, the animal was simultaneously planning an arm movement into the response field and an eye movement out of the response field. In this case the cell was very active.

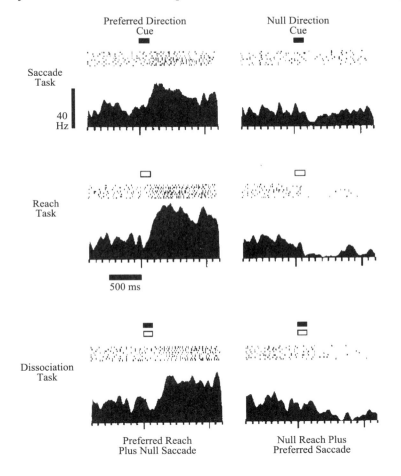

Figure 2

A posterior parietal cortex neuron whose motor specificity was revealed by a dissociation task. In saccade (*top*) and reach (*middle*) tasks, delay period activity was greater before movements in the preferred direction (*left*) compared to the null direction (*right*). Thus, in single-movement tasks, this neuron appeared to code remembered target location independent of movement intent. However, firing was vigorous in the delay period preceding a reach in the preferred direction when this reach was combined with a saccade in the null direction (*bottom left*), but firing was nearly absent before a saccade in the preferred direction combined with a null reach. Thus, when both a reach and a saccade were planned, delay-period activity reflected the intended reach and not the intended saccade. Panel formats similar to Figure 1 except that every other action potential is shown in the rasters (Modified from Snyder et al. 1997.).

In the histogram on the right, the animal was planning an eye movement into the response field, but an arm movement out of the field. Although this is the same eye movement plan that evoked activity in the single-movement case, now there is no activity or even a slight suppression. The pattern of activity of this neuron can be explained if the animal is forming a covert plan in the single-movement cases, in this example a covert arm movement plan. Of the neurons in the population, 62% that were not specific for single movements were specific in the dissociation task, bringing to 84% the number of cells that showed movement planning specificity in the delay period. Interestingly, more cells also revealed specificity for the cue response in the dissociation task, with a total of 45% being specific for reaches and 62% for saccades.

Covert planning may explain activity that is seen in go/no-go tasks. In these tasks a stimulus appears in the response field and the animal is later cued whether to make a movement to it or not. Activity in area 5 for reaches (Kalaska & Crammond 1995), and in LIP for saccades (Pare & Wurtz 1997) continues when the animal is cued not to move. This result is not consistent with attention or intention activity, since the target is no longer important to the animal's behavior (Pare & Wurtz 1997). However, it is consistent with a covert or default plan, which remains if no new movement plans are being formed. Evidence for this alternative explanation comes from experiments in which the plan is cancelled, and a new movement plan is put in place (Bracewell et al. 1996). An example is shown in Figure 3a and 3b, in which the monkey changed the movement plan three times before a saccade, alternating between planning into, out of, and into the response field (*a*), or out, in, and out (*b*). A similar result is found even when the type of movement plan is changed, but not the direction (Snyder et al. 1998a). In Figure 3c, the dark histogram shows activity for a reach-specific neuron when the first cue presented in the response field instructs a reach (R1), and the second stimulus appears at the same location instructing a change in plan to a saccade (S2). Note that although no movement is made during the time shown in the histogram, the activity turns off when the animal changes to the nonpreferred plan for a movement to the same location. The lighter histogram shows activity for the same cell when the monkey plans a saccade first (S1), and then changes the plan to a reach (R2). Again the activity is consistent with the cell's activity expressing the intent of the animal, with baseline activity after the cue transient when the animal is planning a saccade, and high activity during the delay when he changes his plan to a reach. Taken together, the data from various labs suggest that default plans are formed in parietal areas to stimuli of behavioral significance in the case of no alternative plans, but are erased if alternative plans are formed.

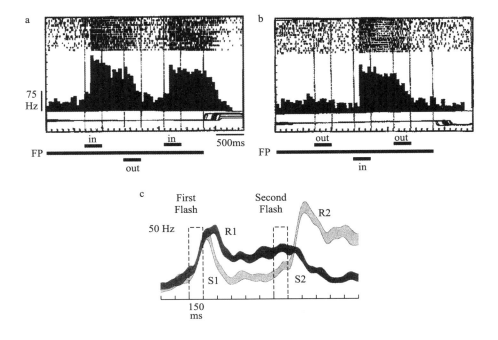

Figure 3

Responses of PPC neurons when movement plans are changed. (*a*) Activity of an LIP neuron when saccades are planned into, out of, and then into the receptive field. The long horizontal bar indicates the onset and offset of a visual fixation point. Short horizontal bars represent the timings of three successive target presentations. Spike rasters show every action potential recorded in a given trial. Spike-density histograms were constructed using 50 ms binwidths. Thin horizontal lines below each histogram represent the animal's vertical eye position. Importantly, no saccades were initiated until after the fixation point was extinguished; thus changes in activity during the time of fixation correspond to changes in the animal's plans or intentions (Modified from Bracewell et al. 1996). (*b*) Activity of the same neuron when saccades are planned out of, into, and then out of the cell's receptive field (Modified from Bracewell et al. 1996). (*c*) Activity of a neuron in the parietal reach region when the type of movement plan, but not its direction, is changed. Activity resulting from an instruction to plan a reach (R1) was abolished when a second flash changed the plan to a saccade (S2). An initial instruction to plan a saccade elicited only a transient response (S1) but when the plan changed to a reach activity increased (R2). Each ribbon represents the mean response of 8–12 trials +/−1 SE. Data were smoothed with a 121-point digital low-pass filter, transition band 20–32 Hz. Dashed rectangles indicate the timings target flashes (Modified from Snyder et al. 1998a).

Dynamic Evolution of Intention-Related Activity

Several studies point to a dynamic evolution in the relation of PPC activity to task requirements, changing from sensory to cognitive to motor as the demands of the task change. For instance, we recently examined the activity of PRR neurons when monkeys plan reaches to auditory versus visual targets in a memory-reach task. We found that at cue onset activity for visually cued trials carried more information about spatial location than activity for auditory cued trials. However, as the trials progressed and the animal was preparing a movement, the amount of spatial location information increased for the auditory cued trials so that by the time of the reach movement, it was not significantly different from the visually cued trials (Y. C. Cohen & R. A. Andersen, unpublished observations).

In another study, we trained animals to make saccades to a specific location cued on an object, but after the cue and before the saccade the object was rotated. Early in the task area LIP cells carried information about the location of the cue and the orientation of the object, both pieces of information being important for solving the task. However, near the time of the eye movement many of these same neurons predominately coded just the direction of the intended movement (Breznen et al. 1999).

Platt & Glimcher (1999) showed in a delayed eye movement task that the early activity of LIP neurons varied as a function of the expected probability that a stimulus was a target for a saccade, as well as the amount of reward previously associated with the target. However, during later periods of the trial the cells coded only the direction of the planned eye movement. A similar evolution has been shown in LIP and dorsal prefrontal cortex in eye movement tasks instructed by motion signals. The strength of the motion signal is an important determinant of activity in the beginning of the trial, but at the end of the trial the activity codes the decision or movement plan of the animal (Leon & Shadlen 1999, Shadlen & Newsome 1996). These studies emphasize the fact that the circuits involved in sensory-motor transformations are distributed in nature, involving parietal, frontal, and prefrontal areas (Chaffee & Goldman-Rakic 1998). Moreover, activity in these circuits can evolve dynamically to reflect sensory, cognitive, and movement components of behavior.

Intentional Maps

The above studies point to a map of intentions within the PPC (Figure 4). Area LIP is more specialized for saccade planning, and area MIP for reaching. Work by other investigators implicates areas 5,PO,7m,and Pec as additional reaching-related regions within the posterior parietal cortex (Battaglia-Mayer et al. 2000, Ferraina et al. 2001, Ferraina et al. 1997, Kalaska 1996). Recent studies by Sakata and colleagues (1995, 1997) point to the anterior intraparietal area (AIP) as specialized for grasping. Cells in this area respond to the shapes of objects and the configuration of the hand for grasping the objects. Reversible inactivations of AIP produce deficits in shaping the hand prior to grasping in monkeys. This deficit is reminiscent of problems in shaping the hands prior to grasping found in humans with parietal lobe damage (Perenin & Vighetto 1988). The medial superior temporal area (MST) appears to play a specialized role in smooth-pursuit eye movements. Cells in this area are active for pursuit, even during brief periods when the pursuit target is extinguished (Newsome et al. 1988). Inactivations of this area produce pursuit deficits that are not a result of sensory deficits (Dursteler & Wurtz 1988).

Experiments using fMRI in humans are consistent with the monkey results. Rushworth and colleagues (2001b) found that peripheral attention tasks activated the lateral bank of the intraparietal sulcus, whereas planning manual movements activated the medial bank. They concluded that their results were consistent with the monkey studies, with the medial bank specialized for manual movements and the lateral bank for attention and eye movements. A similar result has recently been reported by Connolly et al. (2000) using event-related fMRI and an eye and hand movement task similar to the one employed by Snyder et al. (1997). An area specialized for grasping has also been identified in the anterior aspect of the intraparietal sulcus in humans (Binkofski et al. 1998). This area may be homologous to monkey AIP.

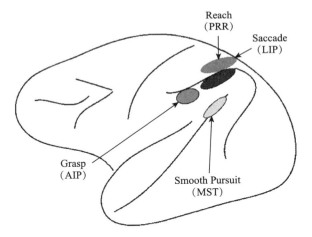

Figure 4
Anatomical map of intentions in the PPC.

Multisensory Integration and Coordinate Transformations

Producing a movement in response to a sensory stimulus requires that a host of problems be solved. From the sensory side, different sensory modalities are coded in different reference frames, vision in retinal or eye-centered coordinates, sound in head-centered coordinates, and touch in body-centered coordinates. These different coordinate frames need to be resolved in some way, since a particular movement might need to be directed to a visual, auditory, or somatosensory stimulus or any combination of these. From the motor side, the locations of these stimuli must ultimately be transformed into the natural coordinates of the muscles in order to make movements.

Lesions to the PPC can produce optic ataxia, where patients mislocalize stimuli to which they are reaching. These mislocalizations are apparent in all three dimensions. Since these effects are often present with no primary sensory or motor defects, they suggest that the PPC is important for multisensory integration as well as for the coordinate transformations required for making sensory-guided movements. In this section, we discuss electrophysiological experiments supporting this view. In particular, we provide evidence that spatial locations are represented in a common coordinate frame in at least some parts of the PPC, independent of sensory input or motor output.

AREA LIP: SACCADE PLANNING IN EYE-CENTERED COORDINATES We can easily make eye movements to visual or auditory targets. If area LIP is involved in making eye movements, then cells in this area should respond when an animal is planning an eye movement, regardless of the sensory modality of the stimulus. Recently we have found this to be the case (Grunewald et al. 1999, Linden et al. 1999, Mazzoni et al. 1996b). However, this observation raises the question, do these two modalities share a common reference frame, and if so, what is it? Cells in the intermediate layers of the SC use a common, eye-position-dependent reference frame for representing saccades to visual, auditory, or somatosensory stimuli (Groh & Sparks 1996a, Groh & Sparks 1996b, Jay & Sparks 1987a, Jay & Sparks 1987b). This is not surprising given that the SC is near the final motor output stage for saccades and that motor error is expressed in eye-centered coordinates. However, area LIP is intermediate between sensory and motor areas; thus it is not immediately apparent what reference frame should be used to represent visual and auditory targets in this region.

In experiments in which monkeys made saccades to auditory targets, we found that a majority of the neurons coded these targets in eye-centered coordinates, although some also coded auditory targets in head-centered coordinates, or in a reference frame intermediate between the eye and head reference frames (Stricanne et al. 1996). Moreover, many of the response fields of LIP neurons were gain modulated by eye position. These data suggest that area LIP may be one of the sites involved in the transformation of auditory signals from head- to eye-centered coordinates. Recent experiments examining cells in the temporo-parietal cortex (TpT) (Wu & Andersen 2001), an auditory association area that projects into the PPC (Pandya & Kuypers 1969), and the inferior colliculus (Groh et al. 2001), indicate that cells antecedent to LIP code auditory locations in head-centered coordinates, with many neurons also gain modulated by eye position. These results support a model in which head-centered auditory signals are gain modulated by eye position and are then read out at subsequent levels in eye-centered coordinates (Xing & Andersen 2000b).

PRR: REACH PLANNING IN EYE-CENTERED COORDINATES The LIP results suggest that this area encodes sensory stimuli as motor error for saccades. If this is the case, then one might predict that PRR would code sensory stimuli as motor error as well, i.e., in limb coordinates. We tested this prediction by training monkeys to reach to targets from two different initial arm positions while fixating their gaze in two different directions. As illustrated for one PRR neuron in Figure 5, the response field did not vary with changes in limb position (*a,b*), but shifted with gaze direction (*c,d*). This result indicates that PRR codes limb movements in eye-centered coordinates. This result, as well as those obtained in LIP, indicates that the PPC is capable of encoding intended movements in eye-centered coordinates independent of the type of movement to be made, i.e., saccades (LIP) and reaches (PRR) (see Figure 6).

The finding that area LIP encodes intended saccades in eye-centered coordinates for both visual and auditory stimuli, as well as the finding that area PRR encodes reaches in eye-centered coordinates, led us to an unusual prediction: that PRR would code reaches to auditory stimuli in eye-centered coordinates. This prediction is based on the assumption that the PPC may use a common reference frame for movement planning, independent of sensory input or motor output. Such a result would be quite surprising since sounds, which are initially coded in head-centered coordinates, could simply be converted to body- and then limb-centered coordinates—a transformation to eye-centered coordinates is not required. In a study in which monkeys planned reaches to sounds in complete darkness, we found that this prediction was correct. Under these conditions, many cells in PRR encoded the intended movement in eye-centered coordinates (Cohen & Andersen 2000).

EYE-CENTERED CODING IN OTHER AREAS Recent stimulation studies suggest that the SC, rather than coding desired gaze displacement or gaze direction in space, encodes the desired gaze direction in retinal coordinates (Klier et al. 2001). Electrophysiological studies of the SC have provided evidence for an eye-centered coding of limb movements in this structure as well (Stuphorn et al. 2000). The ventral premotor cortex also appears to contain neurons that code the location of reach targets in eye-centered coordinates (Mushiake et al. 1997), though these cells may coexist with others having more arm-centered properties (Graziano et al. 1994, 1997). These results, as well as those obtained in the PPC, support the existence of a distributed network devoted to eye-hand coordination that uses a common eye-centered reference frame for representing the spatial aspects of eye and arm movements (Figure 6).

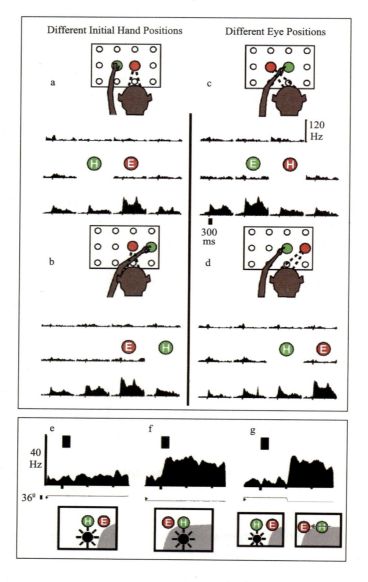

Figure 5

(*a–d*) A PRR neuron that codes target locations in eye-centered coordinates. Icons depict the four possible behavioral conditions at the beginning of a trial; initial hand position and the point of visual fixation are represented by *green and red circles,* respectively. *Open circles* represent target locations on a vertically oriented board of push buttons. Below each icon, spike density histograms (aligned at cue onset and smoothed as in Figure 1) are plotted at positions corresponding to the target locations on the board (11 locations in *a, b,* and *d*; 10 locations in c). *Short horizontal bar* below the histograms in (*c*) represents the timing of the cue. The response field of this neuron did not vary with changes in limb position (*a, b*) but shifted with gaze direction (*c, d*) (Modified from Batista et al. 1999). (*e–g*) Activity of a PRR neuron in an intervening saccade experiment. Each spike density histogram shows the response of the cell for the experimental conditions illustrated in the corresponding icon (*below*). *Shaded region* represents the spatial extent of the cell's response field. (*e*) Activity when the target is presented outside of the response field. (*f*) Response when the target is in the response field. (*g*) Activity when an eye movement carries the reach goal into the neuron's response field. This cell compensated for the saccade to maintain the correct coding of the reach target in eye coordinates. The H and E in the icons below the histograms indicate the position of the hand and eye on the reach board, the *black target* indicates the location of a flashed reached target, and the *shaded area* indicates the spatial extent of the response field of a PRR neuron. The *squares* above the histograms show the time of the flashed target, and the *traces* below the histograms are the recorded eye positions (Modified from Batista et al. 1999).

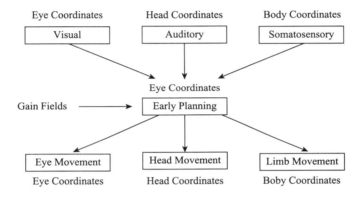

Figure 6
Theory of multisensory integration and coordinate transformations subserved by the PPC.

COMPENSATION FOR EYE MOVEMENTS If saccade and reach plans are coded in eye-centered coordinates, then problems can arise in situations where a movement plan is formed and an intervening saccade is made before the movement is executed. The problem is particularly acute in cases of movements planned to remembered locations in the dark. Mays & Sparks (1980) found that under these circumstances, activity shifts within the eye movement map of the SC to compensate for the intervening saccade and to still code the correct motor vector. Gnadt & Andersen (1988) reported a similar result in area LIP. Duhamel et al. (1992) extended these results by showing that it was not necessary to make an eye movement for this updating to take place. They interpreted this updated activity as sensory and a mechanism for maintaining perceptual stability across eye movements. The results of Snyder et al. (1997, 1998a) provide an alternative explanation: that this activity reflects a default plan for an eye movement. Whether this shift also accounts for perceptual stability remains a possibility and requires additional investigation.

Accounting for eye movements is also a problem when reach plans are coded in eye-centered coordinates. Imagine that an animal plans an arm movement in eye-centered coordinates to the remembered location of a stimulus in the dark and then makes an intervening eye movement before the arm movement takes place. If the reach plan is not adjusted to take into account the retinal position of the stimulus after the eye movement, then areas downstream of PRR will use the previous retinal position of the stimulus to calculate the motor error. This will result in an error corresponding to the size and direction of the intervening saccade.

We directly tested the effect of intervening saccades on intended reach activity in PRR. Figure 5*e-g* shows the design of the experiment and results from one cell. When the flash occurred outside the response field there was no response (*e*), and when it fell within the response field there was robust planning activity (*f*). Note that the histograms demonstrate planning activity; the actual arm movement occurs at a time later than that shown. In (*g*), the task began with the same configuration of eye, hand, and stimulus as in (*e*). However, after the stimulus was extinguished, the animal was instructed to make a saccade to a new location on the board, bringing the response field over the location on the board where the animal was planning the reach. The activity shifted in PRR such that the cell was now active, coding the correct location of the planned reach in eye coordinates even though the reach cue never appeared in the response field. Thus the cell compensated for the saccade to maintain the correct coding of the reach target in eye coordinates. All 34 PRR cells tested with this paradigm showed such compensation for saccades. A remapping of reach plans in eye coordinates has been demonstrated in psychophysical experiments in humans (Henriques et al. 1998), consistent with this physiological finding.

Another possible example of this type of compensation for eye movements, which must still be experimentally verified, is the compensation for smooth-pursuit eye movements that must occur for self-motion perception. During forward locomotion, self-motion perception is estimated from the focus of expansion of the visual field. However, when subjects make smooth eye movements during forward locomotion, as would occur when tracking an object on the ground, these eye movements introduce an additional, laminar motion on the retinas that disrupts the focus of expansion, generally shifting it in the direction of the eye movement. Cells in the dorsal subdivision of the medial superior temporal area (MSTd) are thought to play a role in self-motion perception because they are sensitive to optic flow stimuli and because they are tuned to the spatial location of the focus of expansion (Duffy & Wurtz 1995). In experiments from our laboratory, we found that these focus tuning curves shift to compensate for smooth-pursuit gaze movements. This compensation appears to depend on both efference copies of commands to move the eye or head and the visual information in the optic flow pattern (Bradley et al. 1996, Shenoy et al. 1999a). To guide locomotion, this signal would eventually need to be coded in body- or world-centered coordinates; however, it is currently not known in what reference frame MSTd neurons code focus-position signals. It would be consistent with the data from LIP and PRR if the MSTd cells compensated for the eye movements to maintain the correct heading direction in eye-centered coordinates.

GAIN FIELDS The common representation of space in the PPC is embodied not only in eye-centered response fields, but also in the gain modulation of these fields by body-position signals. These gain field effects are found throughout the PPC and include modulation of retinotopic fields by eye-, head-, body-, and limb-position signals (Andersen et al. 1993). Computational studies have shown that these gain effects can be the mechanism for transforming between coordinate frames (Salinas & Abbott 1995, Zipser & Andersen 1988). Moreover, groups of neurons with retinal response fields, modulated by various body part–position signals, can conceivably be read out in multiple frames of reference (Pouget & Snyder 2000, Xing & Andersen 2000b) as would be needed to direct movements of the eyes, head, or hand. Thus the representation of space in the PPC is distributed and is comprised of eye-centered response fields with gain modulation.

A potential problem with this type of representation is the "curse of dimensionality." For example, if it takes 10 cells to tile each dimension in visual space, and 10 for each dimension of eye position, head position, etc., the number of cells required to represent all possible combinations of such signals quickly exceeds the number of neurons in the brain. One method the PPC employs to avoid this combinatorial explosion is to code only a limited number of variables in each of its subdivisions (Snyder et al. 1998b). We have found that area LIP and area 7a both carry information about head position that is used to gain modulate the cells in this area. In area LIP this information is derived from neck proprioceptive signals indicating the orientation of the head on the body, whereas in area 7a the information is derived from vestibular signals and indicates the orientation of the head in the world. Thus, activity in LIP can be read out in body-centered coordinates, while activity in 7a can be read out in world-centered coordinates. One possible reason for this paucity of dimensionality is that area LIP may be concerned primarily with representing space for gaze shifts and eye-head coordination, whereas area 7a may be more related to representing space for navigation. In other words, these areas and possibly other cortical areas may only represent as many dimensions as are needed for the particular functions they perform. Knowledge of those dimensions may provide clues to the function of a particular area.

GAIN FIELDS AND REMAPPING As mentioned above, eye movements are compensated for within PPC representations by shifting activity within eye-centered maps. Such a remapping of activity is necessary if coordinate transformations are to be accurately achieved using a gain mechanism. For instance, if the eyes move, the new location of a stimulus or planned movement must be adjusted in eye coordinates to correctly read out the head-centered location of the target. An important question is how this remapping is achieved. It could be accomplished using an eye displacement signal, or it could be accomplished using an eye position signal. Both eye displacement and eye position–related signals are found in the PPC (Mountcastle et al. 1975).

The experimental protocol typically used for examining remapping is the "double saccade" paradigm, in which an animal remembers two sequentially flashed targets and makes eye movements to the remembered locations of the targets in the order of their appearance. Activity in LIP appears for the next impending movement and disappears for the previous movement (Gnadt & Andersen 1988, Mazzoni et al. 1996a). More importantly, the activity for the second saccade specifies the direction and amplitude of the planned saccade, not the location on the retina in which the second flash occurred prior to the first eye movement. This compensation requires taking into account either the eye displacement for the first saccade or the new eye position after the first saccade. Patients with PPC lesions performing double saccades can make the first eye movement into the unhealthy visual field, but are not able to generate an accurate second saccade (Heide et al. 1995). Although it has been argued that this proves that an eye-displacement signal mediates remapping, eye displacement and eye position are in fact confounded in this task. The deficit was seen when the displacement of the eyes was in the direction of the unhealthy field, but also when the eye position after the first movement was in the unhealthy field.

In other experiments, area LIP was reversibly inactivated in monkeys in experiments designed to directly examine whether eye-displacement or eye-position signals are used for remapping in double-saccade experiments (Li & Andersen 2001). Both initial eye position and the direction of eye movements were varied in individual trials in order to tease apart eye-position and eye-displacement contributions. It was found that the largest deficits were seen when the animal made the first eye movement into the unhealthy visual field, largely independent of the direction of the eye movement. This result suggests that eye-position signals play a large role in the compensation for intervening saccades.

A recent computational study illustrates that dynamic neural networks can be trained to perform the double saccade task using eye-position signals (Xing & Andersen 2000a). These networks show activity similar to that recorded from LIP when monkeys perform the same task. These include eye-centered response fields that are gain modulated by eye position, and activity that shifts within an eye-centered map of visual space to correct for intervening saccades. Thus, the gain field mechanism can account for dynamic compensation for intervening eye movements in eye-centered coordinates.

GAIN FIELDS: OTHER USES Since their discovery in areas of the PPC, gain effects have been identified throughout the brain. This suggests that multiplicative and additive interactions between different inputs to neurons may reflect a general method of neural computation. Although the role of gain fields in coordinate transformations has been highlighted in this review, gain fields appear to play a role in many other functions, including attention, navigation, decision making, and object recognition. Some examples are discussed briefly below (see also Salinas & Thier 2000).

The direction of attention can modulate the activity of V4 neurons (McAdams & Maunsell 2000, Reynolds et al. 2000), and this effect has been proposed to play a role in the binding of features in objects (Salinas & Abbott 1997). In addition, although smooth pursuit shifts the focus tuning of many MSTd neurons, as mentioned above, other MSTd neurons do not shift their focus tuning but are gain modulated by the pursuit signal (Bradley et al. 1996, Shenoy et al. 1999a). This gain modulation is consistent with an intermediate step toward the production of shifting focus-tuning curves, and thus may play a role in the perception of self-motion for navigation. Monkeys and humans have been shown to choose between two targets for a reach depending on eye position, essentially choosing targets that tend to center the reach with respect to the head (Scherberger et al. 1999). Eye-position gain effects have been shown in PRR and may bias the decision of animals to choose targets based on eye position (Scherberger & Andersen 2001).

We have recently trained monkeys to make object-based saccades by cueing a location on an object, extinguishing the object, and then presenting the object again at a different orientation. In this task, the animals must saccade to the previously cued location on the object to obtain their reward (Breznen et al. 1999). We find that area LIP does not code the cued location in an explicit object-centered reference frame in this task, even though it requires the animal to code the target in such a reference frame. Rather, cells in area LIP carry information about the cued location, movement vector, and orientation of the object, all in retinal coordinates. Some cells show a gain modulation of the cue or movement vector activity by object orientation. This result is surprising given the finding that lesions to the PPC in humans often produce deficits in object-centered coordinates (Arguin & Bub 1993, Driver & Mattingley 1998). However, computational studies show that it is not necessary to use cells with object-centered response fields to solve object-based tasks; rather, distributed coding using retinal response fields for target locations and object orientations, and gain modulations between the two, is sufficient (Pouget & Sejnowski 1997). This distributed representation can be used to form response fields in object-centered coordinates, as has been reported in the supplementary eye fields (SEF) (Olson 2001). However, the SEF results may also be explained as a result of gain modulation of retinal response fields by object position (Pouget & Sejnowski 1997), and more thorough mapping of the response fields will be required to distinguish between the two possibilities.

A COMMON DISTRIBUTED CODE FOR INTENDED MOVEMENTS IN AREAS LIP AND PRR
The above results, summarized in Figure 6, suggest that LIP and PRR use a common space representation in which response fields are represented in eye-centered coordinates. This representation exists independent of whether the targets are visual or auditory. Likewise, this representation is used regardless of whether the output is to move the limb or make an eye movement. This general scheme generated a nonintuitive, but correct, prediction that auditory targets for reach would be coded in eye-centered coordinates in PRR. Currently, we do not know if somatosensory stimuli, such as proprioceptive signals coding the position of the hand, are coded in eye coordinates in PRR and LIP. This would be an interesting question for future experimentation, and if true, would provide further evidence for the generality of this model. In both LIP and PRR, the eye-centered response fields are gain modulated by eye-, head-, and limb-position signals. This gain modulation may provide the mechanism for converting stimuli in various reference frames into eye-centered coordinates. Likewise, these gain modulations may allow other areas to read out signals from LIP and PRR in different coordinate frames, including eye-, head-, body-, and limb-centered coordinates.

Why use a common coordinate frame for PRR and LIP? One possibility is to facilitate hand-eye coordination. Presumably the orchestration of these movements would be facilitated if they used a common reference frame (Battaglia-Mayer et al. 2000). A second reason may be that vision is the most accurate spatial sense in primates. This dominance of vision may explain certain illusions, such as the ventriloquist effect, in which the spatial locations of sounds are referred to seen objects.

COORDINATE TRANSFORMATIONS FOR REACH—DIRECT TRANSFORMATION To make visually guided reaching movements, the location of the target for the reach must be converted from eye- to limb-centered coordinates. There have been two general schemes for this transformation in the literature. One we refer to as the sequential method, shown in Figure 7*a* (Flanders et al. 1992; McIntyre et al. 1997, 1998). In this scheme, visual signals in retinal coordinates are combined with eye-position signals to represent targets in head-centered coordinates. Next, head position is combined with the representation of target location in head-centered coordinates to form a representation of the target in body-centered coordinates. Finally, the current location of the limb (in body-centered coordinates) is subtracted from the location of the target (in body-centered coordinates) to generate the motor vector, in limb-centered coordinates. There are two drawbacks to this method. One is that it requires a number of stages and separate computations, which would likely require a large number of neurons and cortical areas. The second is that, although there are some reports of cells in the PPC coding targets in head-centered coordinates (Duhamel et al. 1997, Galletti et al. 1993), the vast majority of PPC cells code visual targets in eye-centered coordinates and not head- or body-centered coordinates.

A second scheme is referred to as the combinatorial method. As shown in Figure 7*b*, retinal target location, eye, head, and limb position are all combined at once, and the target location in limb-centered coordinates is then read out from this representation (Battaglia-Mayer et al. 2000). As mentioned above, a drawback to this approach is the "curse of dimensionality."

A third scheme we refer to as the direct method. Figure 7*c* shows that this approach subtracts the current position of the hand (in eye coordinates) from the position of the target (in eye coordinates) to directly generate the motor vector in limb coordinates. An advantage of this approach over the sequential method is that it requires fewer computational stages. In addition the computation is restricted to only dimensions in eye coordinates and does not suffer the curse of dimensionality of the combinatorial approach.

Recently, we have provided evidence for the direct transformation scheme. Single cells in area 5, a somatomotor cortical area within the PPC, have been found to code target locations simultaneously in eye- and limb-centered coordinates (Buneo et al. 2002). This result suggests that the PPC transforms target locations directly between these two frames of reference. Moreover, cells in PRR code the target location in eye-centered coordinates, but the initial hand position introduces a gain on this response that is also eye centered. These two findings, taken together, suggest that a simple gain field mechanism underlies the transformation from eye- to limb-centered coordinates. A convergence of input from cells in PRR onto area 5 neurons can perform this transformation directly (see Figure 7*d*) without having to resort to intermediate coordinate frames or a large combination of retina-, eye-, head-, and limb-position signals.

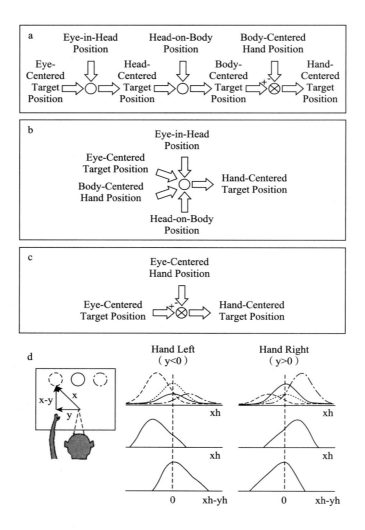

Figure 7

(*a–c*) Schemes for transforming target position from eye-centered to hand-centered coordinates. (*a*) Sequential method. (*b*) Combinatorial method. (*c*) Direct method. (*d*) Illustration showing how convergent input from neurons encoding target position (x) and initial hand position (y) in eye coordinates can drive downstream responses that encode target position in hand-centered coordinates (x-y), as in the direct method. *Top row of curves* represents the responses of four idealized neurons at two initial hand positions. *Middle and bottom rows* show the responses of a downstream neuron, derived as a weighted sum of the responses in the top row minus a constant (Salinas & Abbott 1995, Salinas & Thier 2000). When initial hand position varies, the peak response of the downstream neuron shifts in eye-centered coordinates (*middle row*), but remains fixed in hand-centered coordinates (*bottom row*). For simplicity, only the horizontal components of responses are shown.

Psychophysical evidence supporting a sequential scheme has been provided by Flanders et al. (1992), Henriques et al. (1998), and McIntyre et al. (1997, 1998). These results, as well as our own physiological studies supporting an alternative direct scheme, may reflect an underlying context dependence in the coordinate transformations that subserves visually guided reaching (Carrozzo et al. 1999). For example, direct transformations may be the preferred scheme when both target location and the current hand position are simultaneously visible, even for a brief instant. In contrast, a sequential scheme may be used when visual information about the current position of the hand is unavailable.

MOVEMENT DECISIONS

Experiments in LIP by Platt & Glimcher (1999) and by Shadlen and colleagues (Kim & Shadlen 1999, Shadlen & Newsome 1996) have found activity related to the decision of a monkey to make eye movements. Both the prior probability and amount of reward influence the effectiveness of visual stimuli in LIP, consistent with a role for this area in decision making. As monkeys accumulate sensory information to make a movement plan, activity increases for neurons in LIP and the prefrontal cortex (Kim & Shadlen 1999, Leon & Shadlen 1999, Shadlen & Newsome 1996). These results are consistent with these areas weighting decision variables for the purpose of planning eye movements (Gold & Shadlen 2001). The fact that these effects appear in multiple brain areas suggests that decision making is a distributed function that includes the PPC.

ATTENTION

The PPC has been classically thought to play a central, perhaps controlling role in attention. Strong evidence for this idea is the finding of neglect, an inability to attend to the contralateral visual field, after PPC lesions in humans (Critchley 1953). However, many of the processes involved in visual-motor transformations, for example the shaping of the hand for grasping, appear to operate unconsciously (Goodale & Milner 1992). In fact, lesions to the PPC in monkeys produce visual-motor deficits and not neglect (Faugier-Grimaud et al. 1978, Lamotte & Acuna 1978). Rather, it has been reported that lesions to the superior temporal gyrus produce neglect similar to that found in humans (Watson et al. 1994). Interestingly, a recent report by Karnath and colleagues (2001) suggests that the superior temporal gyrus damage may also be the source of neglect seen in humans. Thus, the locus of cortical lesions that produce neglect is still an open question that will likely be resolved with further research.

Although there have been several studies reporting attentional effects on neural activity in the PPC, these experiments have been performed in conjunction with eye or limb movements, or in peripheral attention paradigms where animals are likely to form covert plans for eye movements. As yet no experiments have been performed similar to those of Snyder et al. (1997, 1998a). In those experiments, intention was isolated from attentional effects; similar experimental designs are needed to isolate attentional effects from intentional effects.

It has been argued that the lower degree of activation of LIP neurons when monkeys reach rather than saccade to targets is due to less attention being required for reaching (Colby & Goldberg 1999). This reasoning would predict less activity in PRR as well, but in fact the reverse is true. Figure 8 shows the population activity from one monkey for recordings obtained in LIP and PRR. When this monkey planned, a saccade activity was high in LIP and low in PRR. On the other hand, when reaches were planned the reverse was true. This figure also shows that when covert planning was controlled in dual-movement trials the separation for saccades and reaches was even greater. This double dissociation between saccades and reaches for LIP and PRR shows that the effects are due to planning, and not a general reduction in attention when the animal reaches.

In a recent study, Powell & Goldberg (2000) flashed stimuli around the time of eye movements and found responses for LIP neurons even while the animals were planning eye movements outside of the cells' response fields. They argued that this demonstrated that LIP is more involved in registering the salience of visual stimuli than in planning eye movements. This interpretation is at odds with that of Mazzoni et al. (1996a), who showed that, when monkeys were planning eye movements outside of the response field of LIP neurons, a flash in the center of their response field produced only a very brief transient before the activity was suppressed. A closer inspection of the data in Powell & Goldberg shows a similar effect, with activity dying out prior to and during the eye movement. A similar sensory transient can be seen when a reach target is flashed within an LIP response field, which quickly dies away in this case hundreds of msec before the reach movement (see Figure 1*b*, right histogram). Interestingly, units in simulated networks programmed to hold a movement plan for one location while a distractor stimulus is briefly flashed at another location [conditions similar to the experiment of Powell & Goldberg (2000)] also exhibit input transients (Xing & Andersen 2000a). However, lateral inhibitory connections in these networks quickly suppressed the activity due to the distractor, similar to the suppression seen in LIP experiments.

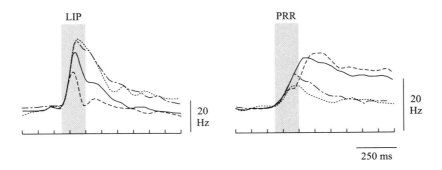

Figure 8

Population response from one monkey for areas LIP (*left*) and PRR (*right*). Cells had significant activity during the delay period of either the reach or saccade task of Snyder et al. (1997). Solid gray traces represent the average activity of the population of cells for saccades into the response field. Solid black traces represent activity for reaches into the response field. Dashed lines represent activity in the dual-movement task, with gray traces representing saccades into the response field and black traces representing reaches. Histograms were smoothed with a 181-point digital low-pass filter with a −3 dB point at 9 Hz. (From Snyder et al. 2000.)

In conclusion, it is likely that pure attentional effects will be found in the PPC. However, because this area is specialized for sensory-motor integration, there is the additional challenge of designing paradigms that rule out movement planning as a source of activity. This concern is less problematic in areas of the visual cortex that are involved in recognition, where there is ample evidence of attentional effects. Attentional effects in PPC may be general, since attention-related activity has been reported throughout the cerebral cortex. Thus attention effects in PPC would be related to planning movements, much like attention effects in inferotemporal cortex would be related to visual recognition. Interestingly, recent studies suggest PPC and prefrontal structures may regulate spatial aspects of attention in the ventral, recognition pathway (Kastner & Ungerleider 2000). Whether PPC is specialized for attention and is the controller for attention throughout cortex is an important question.

LEARNING AND ADAPTATION

Like attention, learning is a distributed function and like atttentional effects, learning effects in the PPC tend to be most apparent in the context of sensory-motor operations, for example prism adaptation. As first demonstrated by Held & Hein (1963), when human subjects reach to visual targets while wearing displacing prisms, they initially miss-reach in the direction of target displacement but gradually recover and reach correctly if provided with appropriate feedback about their errors. Using positron emission tomography (PET) to monitor changes in cerebral blood flow, Clower and colleagues (1996) showed that the prism adaptation process results in selective activation of the PPC contralateral to the reaching arm, when confounding sensory, motor, and cognitive effects are ruled out. Similarly, Rosetti and colleagues (1998) found that hemispatial neglect resulting from damage to the right hemisphere can be at least partially ameliorated by first having affected patients make reaching movements in the presence of a prismatic shift, then removing the prisms. They interpreted these effects as resulting from the stimulation of neural structures responsible for sensorimotor transformations, including the PPC as well as the cerebellum. A recent electrophysiological study employing a prism adaptation paradigm suggests that the ventral premotor cortex plays a role in this process as well (Kurata & Hoshi 1999).

An example of the effects of learning in the PPC was revealed in a recent electrophysiological study of LIP (Grunewald et al. 1999). The responses of LIP neurons to auditory stimuli in a passive fixation task were examined before and after animals were trained to make saccades to auditory targets. Before such training, the number of cells responding to auditory stimuli in LIP was statistically insignificant. After training, however, 12% showed significant responses to auditory stimuli. This indicates that at least some LIP neurons become active for auditory stimuli only after an animal has learned that these stimuli are important for oculomotor behavior. As with the learning effects discussed above, effects of this nature have been reported in other areas of cortex, e.g., area 3a for tactile discrimination (Recanzone et al. 1992), premotor cortex for arbitrary associations (Mitz et al. 1991), and the frontal eye fields (FEF) for visual search training (Bichot et al. 1996), highlighting the distributed nature of learning in cortex.

READING OUT INTENTIONS—THE PPC AND NEURAL PROSTHETICS

The experimental results reviewed above indicate that activity related to an animal's intentions to make reaches or saccades is strong and robust in the PPC. These experiments are typical of neurophysiological experiments in demonstrating correlations between behaviors or perception and neural activity. It would be powerful to be able to test these proposed linkages more directly. One method to achieve a more direct demonstration of intention coding is to perform experiments that close the loop. This type of experiment is demonstrated in Figure 9.

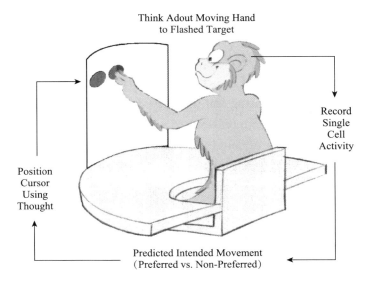

Think Aobut Moving Hand
to Flashed Target

Record
Single
Cell
Activity

Position
Cursor
Using
Thought

Predicted Intended Movement
(Preferred vs. Non-Preferred)

Figure 9

Schematic of an experiment in "closing the loop." In this experiment, the activity of a single-PRR neuron is first isolated and a database of the neuron's responses for reaches in the preferred and nonpreferred direction is constructed. In the relevant trials, the animal fixates and touches a central cursor on a display screen and a target is then presented either in the preferred or nonpreferred direction of the cell. Based on the response of the neuron for this one trial, and the statistics of the previously constructed database, we then predict whether this animal is intending to reach in either the preferred or nonpreferred direction. Importantly, the animal never actually makes the intended movement on these trials. A cursor moves to the location predicted from the cell activity and the animal is rewarded if the prediction corresponds to the cued location.

Rather than simply recording the monkeys' reach intentions from PRR and then rewarding them for making a reach, we record their intentions and use them to move a cursor to the location on the display screen where we predict, based on the neural activity, that they are intending to reach. This cursor provides feedback to the animals, indicating where the reach is predicted to end up, based purely on the animals' thoughts, i.e., without the animals making any actual reaches.

In preliminary studies we have used PRR activity to predict one out of two possible reach directions in real time (D. Meeker, S. Cao, J. W. Burdick & R. A. Andersen, unpublished observations). In other words, the animals perform a task in which they locate a cursor on the computer screen by using intended movement activity without making an actual reach. In other studies, offline analysis of single-cell recordings from PRR indicates that reliably predicting one out of eight directions could require only a small number of simultaneously recorded neurons, perhaps in the range of 10 to 15 (Meeker et al. 2001, Shenoy et al. 1999b).

The high-level planning activity observed in PRR could be used in the control of a neural prosthetic for paralyzed patients. Patients with paralysis due to peripheral neuropathies, trauma, and stroke can often still think about making movements but cannot execute them. The idea of a cortical prosthetic is to record these intentions to move, interpret the intentions using real-time decode algorithms running on computers, and then convert these decoded intentions to control signals that can operate external devices. These external devices could include stimulators imbedded in the patient's muscles that would allow the patient to move his/her own body, a robot limb, or a computer interface for communication.

Research on neural prosthetics is a burgeoning and young field. Several groups are working toward using motor cortex for such a prosthesis (Chapin et al. 1999, Isaacs et al. 2000, Wessberg et al. 2000, Kennedy et al. 2000, Maynard et al. 1999), which makes sense because it is the area of cortex closest to the motor output. There may also be advantages for using higher, cognitive areas of the sensory-motor system such as PRR for the control of prosthetics. Because this is a field that is still at its infancy, it is not clear if one area is optimal for prosthetic control. Moreover, because different areas of the sensory-motor pathway no doubt provide different, useful information, it may turn out to be most optimal to develop multiarea prosthetics that can read out and decode these different signals. Below are listed some potential attributes of PRR for prosthetic control.

1.) Motor cortex is known to undergo degradation as a result of paralysis. For instance with spinal cord lesions, cortico-spinal neurons are destroyed and the somatosensory reafferent signals to this area of the cortex are also lost. However, PRR may undergo less degradation after paralysis as it is more closely tied to the visual system.

2.) Learning is an important aspect of neural prosthetic success. For instance, cochlear prosthetics do not produce natural auditory sensations with electrical stimulation, but patients learn to interpret these stimulus-induced percepts, for instance in understanding speech. Implanted arrays of electrodes in humans will likely only sample a part of the workspace of the subject. Although other parts of space not well sampled can be inferred from the activity of cells in sampled regions, the resolution cannot be as good as would result from a more even sampling. Thus, neural plasticity would be an important advantage for a cortical prosthetic and, as mentioned above, there is evidence that the PPC does play a role in adjusting the registration of sensory-motor representations for accurate behaviors.

3.) To successfully close the loop, patients require feedback to the cortex regarding the success of the movement. Whereas these feedback signals are lost with paralysis in motor areas, they are largely intact in PRR, as the reafference to this region is largely visual. Moreover, the evidence presented above, that the PPC performs early coordinate transformations required for reaching in retinal coordinates, suggests that vision can be used by the patients for correcting motor error computations.

4.) The fact that the intended movement signal is a high-level, cognitive signal may have advantages. For instance, if PRR is coding intentions in abstract or general terms, it may require fewer neurons to control devices that are dissimilar to the human limb. In addition, the fact that this planning activity can be sustained for long periods of time may help in the decoding of the movement plan by providing prolonged, stable signals related to the intentions of the patient.

5.) We have recently found that, during the delays when monkeys are planning a movement, there are broadband gamma oscillations in the local field potentials (LFPs) in both LIP and PRR. The LFPs in both areas are tuned to the direction of planned movements and change their strength with behavioral state, becoming much larger in amplitude when the animal plans a movement and rapidly decreasing during the movement execution. Although it is somewhat difficult to record single-cell activity over long periods of time with chronically implanted electrodes, LFPs are stable and relatively easy to record with chronic electrodes, since they reflect the activity of cortical columns rather than single cells. Thus, using LFPs in LIP and PRR may be an important breakthrough for obtaining long-term recordings with existing technology.

CONCLUSION

The PPC is important for sensory-motor integration, particularly the forming of intentions or high-level cognitive plans for movement. There is a map of intentions within the PPC, with subregions for saccades, reaching, and grasping. These regions appear to be specialized for multisensory integration and coordinate transformations. Within each subregion is a map of the working space. In PRR and LIP these maps are in eye-centered coordinates, regardless of sensory input or movement plan. Within these maps, activity is gain modulated by eye-, head-, and limb-position signals. This distributed, abstract representation is consistent with intentions in this area being high level and cognitive.

These ideas perhaps raise more questions than they answer. For instance, what other features of intended movements are coded in the activity of PPC neurons? Are the trajectories, distance of reaching, or the dynamics of the planned movement coded in PRR activity? Is the transition from a high-level, cognitive intention to the motor output of the cortex one of multiple stages and continual refinement of the plan, or is the intention to move converted to an executable plan in one step of convergence onto motor cortical areas? Are there additional cortical areas in PPC for intending other types of movement, such as leg movements and head movements?

There are a number of outstanding questions regarding the finding of a common coordinate frame for spatial location in LIP and PRR. For instance, do other parietal areas, such as the grasp area, also code in retinal coordinates with gain modulation by eye- and other body part–position signals? Are somatosensory signals also represented in eye coordinates in these areas? Does this common reference frame also support intended head movements? Are there tasks in which the intentions to move do not use this common, distributed reference frame? Experiments designed to answer these questions will help to determine how generally the concepts of cognitive intentions and common reference frames can be applied to the PPC.

ACKNOWLEDGMENTS

We wish to acknowledge the generous support of the James G. Boswell Foundation, the National Institutes of Health (NIH), the Defense Advanced Research Projects Agency (DARPA), Sloan-Swartz Center for Theoretical Neurobiology, the Office of Naval Research (ONR), and the Christopher Reeves Foundation. We also thank Dr. Paul Glimcher for helpful comments.

LITERATURE CITED

Andersen RA. 1987. *The Role of the Inferior Parietal Lobule in Spatial Perception and Visual-Motor Integration.* Bethesda, MD: Am. Physiol. Soc., pp. 483–518

Andersen RA, Essick GK, Siegel RM. 1987. Neurons of area 7a activated by both visual stimuli and oculomotor behavior. *Exp. Brain Res.* 67:316–22

Andersen RA, Snyder LH, Li CS, Stricanne B. 1993. Coordinate transformations in the representation of spatial information. *Curr. Opin. Neurobiol.* 3:171–76

Arguin M, Bub DN. 1993. Evidence for an independent stimulus-centered spatial reference frame from a case of visual hemineglect. *Cor- tex* 29:349–57

Balint R. 1909. Seelenlahmung des "Schau-ens," optische Ataxie, raumliche Storung der Aufmerksamkeit. *Monatsschr. Psychiatr. Neurol.* 25:51–81

Batista AP, Buneo CA, Snyder LH, Andersen RA. 1999. Reach plans in eye-centered coordinates. *Science* 285:257–60

Batista AP, Andersen RA. 2001. The parietal reach region codes the next planned movement in a sequential reach task. *J. Neurophysiol.* 85:539–44

Battaglia-Mayer A, Ferraina S, Mitsuda T, Marconi B, Genovesio A, et al. 2000. Early coding of reaching in the parietooccipital cortex. *J. Neurophysiol.* 83:2374– 91

Bichot NP, Schall JD, Thompson KG. 1996. Visual feature selectivity in frontal eye fields induced by experience in mature macaques. *Nature* 381:697–99

Binkofski F, Dohle C, Posse S, Stephan KM, Hefter H, et al. 1998. Human anterior intraparietal area subserves prehension: a combined lesion and functional MRI activation study. *Neurology* 50:1253–59

Bracewell RM, Mazzoni P, Barash S, Andersen RA. 1996. Motor intention activity in the macaque's lateral intraparietal area. II. Changes of motor plan. *J. Neurophysiol.* 76:1457–64

Bradley DC, Maxwell M, Andersen RA, Banks MS, Shenoy KV. 1996. (Neural) mechanisms of heading perception in primate visual cortex. *Science* 273:1544–47

Breznen B, Sabes PN, Andersen RA. 1999. Parietal coding of object-based saccades: reference frames. *Soc. Neurosci. (Abstr.)* 25: 1547

Buneo CA, Jarvis MR, Batista AP, Andersen RA. 2002. Direct visuomotor transformations for reaching. *Nature.* In press

Bushnell MC, Goldberg ME, Robinson DL. 1981. Behavioral enhancement of visual responses in monkey cerebral cortex. I. Modulation in posterior parietal cortex related to selective visual attention. *J. Neurophysiol.* 46:755–74

Carrozzo M, McIntyre J, Zago M, Lacquaniti F. 1999. Viewer-centered and body-centered frames of reference in direct visuomotor transformations. *Exp. Brain Res.* 129:201– 10

Chaffee MV, Goldman-Rakic PS. 1998. Matching patterns of activity in primate prefrontal area 8a and parietal area 7ip during a spatial working memory task. *J. Neurophysiol.* 79:2919–40

Chapin JK, Moxon KA, Markowitz RS, Nicolelis MA. 1999. Real-time control of a robot arm using simultaneously recorded neurons in the motor cortex. *Nat. Neurosci.* 2:664–70

Clower DM, Hoffman JM, Votaw JR, Faber TL, Woods RP, Alexander GE. 1996. Role of posterior parietal cortex in the recalibration of visually guided reaching. *Nature* 383:618– 21

Cohen YE, Andersen RA. 2000. Reaches to sounds encoded in an eye-centered reference frame. *Neuron* 27:647–52

Colby CL, Goldberg ME. 1999. Space and attention in parietal cortex. *Annu. Rev. Neurosci.* 22:319–49

Connolly JD, Menon RS, Goodale MA. 2000. Human frontoparietal areas active during a pointing but not a saccade delay. *Soc. Neurosci. (Abstr.)* 26:1329

Corbetta M, Akbudak E, Conturo TE, Snyder AZ, Ollinger JM, et al. 1998. A common network of functional areas for attention and eye movements. *Neuron* 21:761–73

Critchley M. 1953. *The Parietal Lobes.* London: Arnold

DeSouza JF, Dukelow SP, Gati JS, Menon RS, Andersen RA, Vilis T. 2000. Eye position signal modulates a human parietal pointing region during memory-guided movements. *J. Neurosci.* 20:5835–40

Driver J, Mattingley JB. 1998. Parietal neglect and visual awareness. *Nat. Neurosci.* 1:17– 22

Duffy CJ, Wurtz RH. 1995. Response of monkey MSTd neurons to optic flow stimuli with shifted centers of motion. *J. Neurosci.* 15:5192–208

Duhamel JR, Colby CL, Goldberg ME. 1992. The updating of the representation of visual space in parietal cortex by intended eye movements. *Science* 255:90–92

Duhamel JR, Bremmer F, Ben Hamed S, Grof W. 1997. Spatial invariance of visual receptive fields in parietal cortex neurons. *Nature* 389:845–48

Dursteler MR, Wurtz RH. 1988. Pursuit and optokinetic deficits following chemical lesions of cortical areas Mt and Mst. *J. Neurophysiol.* 60:940–65

Eskandar EN, Assad JA. 1999. Dissociation of visual, motor and predictive signals in parietal cortex during visual guidance. *Nat. Neurosci.* 2:88–93

Faugier-Grimaud S, Frenois C, Stein DG. 1978. Effects of posterior parietal lesions on visually guided behavior in monkeys. *Neuropsy- chologia* 16:151–68

Ferraina S, Battaglia-Mayer A, Genovesio A, Marconi B, Onorati P, Caminiti R. 2001. Early coding of visuomanual coordination during reaching in parietal area PEc. *J. Neurophysiol.* 85:462–67

Ferraina S, Garasto MR, Battaglia-Mayer A, Ferraresi P, Johnson PB, et al. 1997. Visual control of hand-reaching movement: activity in parietal area 7m. *Eur. J. Neurosci.* 9:1090– 95

Flanders M, Helms-Tillery SI, Soechting JF.1992. Early stages in a sensorimotor transformation. *Behav. Brain Sci.* 15:309–62

Galletti C, Battaglini PP, Fattori P. 1993. Parietal neurons encoding spatial locations in craniotopic coordinates. *Exp. Brain Res.* 96: 221–29

Geshwind N, Damasio AR. 1985. Apraxia. In *Handbook of Clinical Neurology,* ed. PJ Vinken, GW Bruyn, HL Klawans, pp. 423–32. Amsterdam: Elsevier

Gnadt JW, Andersen RA. 1988. Memory related motor planning activity in posterior parietal cortex of macaque. *Exp. Brain Res.* 70:216–20

Gold JI, Shadlen MN. 2001. Neural computations that underlie decisions about sensory stimuli. *Trends Cogn. Sci.* 5:10–16

Goodale MA, Milner AD. 1992. Separate visual pathways for perception and action. *Trends Neurosci.* 15:20–25

Gottlieb J, Goldberg ME. 1999. Activity of neurons in the lateral intraparietal area of the monkey during an antisaccade task. *Nat. Neurosci.* 2:906–12

Graziano MS, Hu XT, Gross CG. 1997. Visuospatial properties of ventral premotor cortex. *J. Neurophysiol.* 77:2268–92

Graziano MS, Yap GS, Gross CG. 1994. Coding of visual space by premotor neurons. *Science* 266:1054–57

Groh JM, Sparks DL. 1996a. Saccades to somatosensory targets. 2. Motor convergence in primate superior colliculus. *J. Neurophysiol.* 75:428–38

Groh JM, Sparks DL. 1996b. Saccades to somatosensory targets. 3. Eye-positiondependent somatosensory activity in primate superior colliculus. *J. Neurophysiol.* 75:439–53

Groh JM, Trause AS, Underhill AM, Clark KR, Inati S. 2001. Eye position influences auditory responses in primate inferior colliculus. *Neuron* 29:509–18

Grunewald A, Linden JF, Andersen RA. 1999. Responses to auditory stimuli in macaque lateral intraparietal area. I. Effects of training. *J. Neurophysiol.* 82:330–42

Heide W, Blankenburg M, Zimmermann E, Kompf D. 1995. Cortical control of double-step saccades—implications for spatial orientation. *Ann. Neurol.* 38:739–48

Held R, Hein A. 1963. Movement produced stimulation in the development of visually guided behavior. *J. Comp. Physiol. Psychol.* 53:236–41

Henriques DY, Klier EM, Smith MA, Lowy D, Crawford JD. 1998. Gaze-centered remapping of remembered visual space in an open- loop pointing task. *J. Neurosci.* 18:1583–94

Hikosaka O, Wurtz RH. 1983. Visual and oculomotor functions of monkey substantia nigra pars reticulata.

III. Memory-contingent visual and saccade responses. *J. Neurophys-iol.* 49:1268–84

Isaacs RE, Weber DJ, Schwartz AB. 2000. Work toward real-time control of a cortical neural prothesis. *IEEE Trans. Rehabil. Eng.* 8:196–98

Jay MF, Sparks DL. 1987a. Sensorimotor integration in the primate superior colliculus. 1. Motor convergence. *J. Neurophysiol.* 57:22–34

Jay MF, Sparks DL. 1987b. Sensorimotor integration in the primate superior colliculus. 2. Coordinates of auditory signals. *J. Neurophysiol.* 57:35–55

Kalaska JF. 1996. Parietal cortex area 5 and visuomotor behavior. *Can. J. Physiol. Pharmacol.* 74:483–98

Kalaska JF, Crammond DJ. 1995. Deciding not to GO: neural correlates of response selection in a GO/NOGO task in primate premotor and parietal cortex. *Cerebr. Cortex* 5:410–28

Karnath H, Ferber S, Himmelbach M. 2001. Spatial awareness is a function of the temporal not the posterior parietal lobe. *Nature* 411:950–53

Kastner S, Ungerleider LG. 2000. Mechanisms of visual attention in the human cortex. *Annu. Rev. Neurosci.* 23:315–41

Kennedy PR, Bakay RAE, Moore MM, Adams K, Goldwaithe J. 2000. Direct control of a computer from the human central nervous system. *IEEE Trans. Rehabil. Eng.* 8:198–202

Kim JN, Shadlen MN. 1999. Neural correlates of a decision in the dorsolateral prefrontal cortex of the macaque. *Nat. Neurosci.* 2:176–85

Klier EM, Wang H, Crawford JD. 2001. The superior colliculus encodes gaze commands in retinal coordinates. *Nat. Neurosci.* 4:627–32

Kurata K, Hoshi E. 1999. Reacquisition deficits in prism adaptation after muscimol microinjection into the ventral premotor cortex of monkeys. *J. Neurophysiol.* 81:1927–38

Kustov AA, Robinson DL. 1996. Shared neural control of attentional shifts and eye movements. *Nature* 384:74–77

Lamotte RH, Acuna C. 1978. Defects in accuracy of reaching after removal of posterior parietal cortex in monkeys. *Brain Res.* 139:309–26

Leon MI, Shadlen MN. 1999. Effect of expected reward magnitude on the response of neurons in the dorsolateral prefrontal cortex of the macaque. *Neuron* 24:415–25

Li CS, Andersen RA. 2001. Inactivation of macaque lateral intraparietal area delays initiation of the second saccade predominantly from contralesional eye positions in a double-saccade task. *Exp. Brain Res.* 137:45–57

Linden JF, Grunewald A, Andersen RA. 1999. Responses to auditory stimuli in macaque lateral intraparietal area. II. Behavioral modulation. *J. Neurophysiol.* 82:343–58

Mays LE, Sparks DL. 1980. Saccades are spatially, not retinocentrically, coded. *Science* 208:1163–65

Maynard EM, Hatsopoulos NG, Ojakangas CL, Acuna BD, Sanes JN, et al. 1999. Neuronal interactions improve cortical population coding of movement direction. *J. Neurosci.* 19: 8083–93

Mazzoni P, Bracewell RM, Barash S, Andersen RA. 1996a. Motor intention activity in the macaque's lateral intraparietal area. I. Dissociation of motor plan from sensory memory. *J. Neurophysiol.* 76:1439–56

Mazzoni P, Bracewell RM, Barash S, Andersen RA. 1996b. Spatially tuned auditory responses in area LIP of macaques performing delayed memory saccades to acoustic targets. *J. Neurophysiol.* 75:1233–41

McAdams CJ, Maunsell JHR. 2000. Attention to both space and feature modulates neuronal responses in macaque area V4. *J. Neurophysiol.* 83:1751–55

McIntyre J, Stratta F, Lacquaniti F. 1997. Viewer-centered frame of reference for pointing to memorized targets in three-dimensional space. *J. Neurophysiol.* 78:1601– 18

McIntyre J, Stratta F, Lacquaniti F. 1998. Short-term memory for reaching to visual targets: psychophysical evidence for body-centered reference frames. *J. Neurosci.* 18:8423– 35

Meeker D, Shenoy KV, Cao S, Pesaran B, Scherberger H, et al. 2001. Cognitive control signals for prosthetic systems. *Soc. Neurosci. (Abstr.)* 27:63.6

Mitz AR, Godschalk M, Wise SP. 1991. Learning-dependent neuronal-activity in the premotor cortex-activity during the acquisition of conditional motor associations. *J. Neurosci.* 11:1855–72

Mountcastle VB, Lynch JC, Georgopoulos A, Sakata H, Acuna C. 1975. Posterior parietal association cortex of the monkey: command functions for operations within extrapersonal space. *J. Neurophysiol.* 38:871–908

Mushiake H, Tanatsugu Y, Tanji J. 1997. Neuronal activity in the ventral part of premotor cortex during target-reach movement is modulated by direction of gaze. *J. Neurophysiol.* 78:567–71

Newsome WT, Wurtz RH, Komatsu H. 1988. Relation of cortical areas MT and MST to pursuit eye movements. II. Differentiation of retinal from extraretinal inputs. *J. Neuro- physiol.* 60:604–20

Olson CR. 2001. Object-based vision and attention in primates. *Curr. Opin. Neurobiol.* 11:171–79

Pandya DN, Kuypers HG. 1969. Cortico-cortical connections in the rhesus monkey. *Brain Res.* 13:13–36

Pare M, Wurtz RH. 1997. Monkey posterior parietal cortex neurons antidromically activated from superior colliculus. *J. Neurophysiol.* 78:3493–97

Perenin MT, Vighetto A. 1988. Optic ataxia: a specific disruption in visuomotor mechanisms. I. Different aspects of the deficit in reaching for objects. *Brain* 111:643–74

Platt ML, Glimcher PW. 1999. Neural corre- lates of decision variables in parietal cortex. *Nature* 400:233–38

Pouget A, Sejnowski TJ. 1997. A new view of hemineglect based on the response properties of parietal neurones. *Philos. Trans. R. Soc. Lond. B. Biol. Sci.* 352:1449–59

Pouget A, Snyder LH. 2000. Computational approaches to sensorimotor transformations. *Nat. Neurosci.* 3:1193–98

Powell KD, Goldberg ME. 2000. Response of neurons in the lateral intraparietal area to a distractor flashed during the delay period of a memory-guided saccade. *J. Neurophysiol.* 84:301–10

Recanzone GH, Merzenich MM, Jenkins WM. 1992. Frequency discrimination training engaging a restricted skin surface results in an emergence of a cutaneous response zone in cortical area 3a. *J. Neurophysiol.* 67:1057–70

Reynolds JH, Pasternak T, Desimone R. 2000. Attention increases sensitivity of V4 neurons. *Neuron* 26:703–14

Rizzolatti G, Riggio L, Sheliga B. 1994. Space and selective attention. In *Attention and Performance*, ed. C Umilta, M Moscovitch, pp. 231–65. Cambridge, MA: MIT Press

Robinson DL, Goldberg ME, Stanton GB. 1978. Parietal association cortex in the primate: sensory mechanisms and behavioral modulations. *J. Neurophysiol.* 41:910–32

Rondot P, Recondo J, de Ribadeau Dumas J. 1977. Visuomotor ataxia. *Brain* 100:355–76

Rossetti Y, Rode G, Pisella L, Farne A, Li L, et al. 1998. Prism adaptation to a rightward optical deviation rehabilitates left hemispatial neglect. *Nature* 395:166–69

Rushworth MFS, Ellison A, Walsh V. 2001a. Complementary localization and lateralization or orienting and motor attention. *Nat. Neurosci.* 4:656–61

Rushworth MFS, Paus T, Sipila PK. 2001b. Attention systems and the organization of the human parietal cortex. *J. Neurosci.* 21:5262– 71

Sakata H, Taira M, Murata A, Mine S. 1995. Neural mechanisms of visual guidance of hand action in the parietal cortex of the monkey. *Cereb. Cortex* 5:429–38

Sakata H, Taira M, Kusunoki M, Murata A, Tanaka Y. 1997. The TINS lecture. The parietal association cortex in depth perception and visual control of hand action. *Trends Neurosci.* 20:350–57

Salinas E, Abbott LF. 1995. Transfer of coded information from sensory to motor networks. *J. Neurosci.* 15:6461–74

Salinas E, Abbott LF. 1997. Invariant visual responses from attentional gain fields. *J. Neurophysiol.* 77:3267–72

Salinas E, Thier P. 2000. Gain modulation: a major computational principle of the central nervous system. *Neuron* 27:15–21

Scherberger H, Andersen RA. 2001. Neural activity in the posterior parietal cortex during decision processes for generating visually- guided eye and arm movements in the monkey. *Soc. Neurosci. (Abstr.)* 27:237.7

Scherberger H, Goodale MA, Andersen RA. 1999. Reaching and saccadic target selection both follow a head-rather than trunk-centered reference frame during visual double simultaneous stimulation in the monkey. *Soc. Neurosci. (Abstr.)* 25:2189

Shadlen MN, Newsome WT. 1996. Motion perception: seeing and deciding. *Proc. Natl. Acad. Sci. USA* 93:628–33

Shenoy KV, Bradley DC, Andersen RA. 1999a. Influence of gaze rotation on the visual response of primate MSTd neurons. *J. Neuro- physiol.* 81:2764–86

Shenoy KV, Kureshi SA, Meeker D, Gillikan BL, Dubowitz DJ, et al. 1999b. Toward prosthetic systems controlled by parietal cortex. *Soc. Neurosci. (Abstr.)* 25:383

Snyder LH, Batista AP, Andersen RA. 1997. Coding of intention in the posterior parietal cortex. *Nature* 386:167–70

Snyder LH, Batista AP, Andersen RA. 1998a. Change in motor plan, without a change in the spatial locus of attention, modulates activity in posterior parietal cortex. *J. Neurophysiol.* 79:2814–19

Snyder LH, Grieve KL, Brotchie P, Andersen RA. 1998b. Separate bodyand worldreferenced representations of visual space in parietal cortex. *Nature* 394:887–91

Snyder LH, Batista AP, Andersen RA. 2002. Intention-related activity in the posterior parietal cortex: a review. *Vision Res.* 40: 1433–41

Steinmetz MA, Constantinidis C. 1995. Neuro-physiological evidence for a role of posterior parietal cortex in redirecting visual attention. *Cereb. Cortex* 5:448–56

Stricanne B, Andersen RA, Mazzoni P. 1996. Eye-centered, head-centered, and intermediate coding of remembered sound locations in area LIP. *J. Neurophysiol.* 76:2071–76

Stuphorn V, Bauswein E, Hoffman K-P. 2000. Neurons in the primate superior colliculus coding for arm movements in gaze-related coordinates. *J. Neurophysiol.* 83:1283– 99

Tian J, Schlag J, Schlag-Rey M. 2000. Testing quasi-visual neurons in the monkey's frontal eye field with the triple-step paradigm. *Exp. Brain Res.* 130:433–40

Watson RT, Valenstein E, Day A, Heilman KM. 1994. Posterior neocortical systems subserving awareness and neglect—neglect associated with superior temporal sulcus but not area-7 lesions. *Arch. Neurol.* 51:1014–21

Wessberg J, Stambaugh CR, Kralik JD, Beck PD, Laubach M, et al. 2000. Real-time prediction of hand trajectory by ensembles of cortical neurons in primates. *Nature* 408:361–65

Wu S, Andersen RA. 2001. The representation of auditory space in temporo-parietal cortex. *Soc. Neurosci. (Abstr.)* 27:166.15

Xing J, Andersen RA. 2000a. Memory activity of LIP neurons for sequential eye movements simulated with neural networks. *J. Neurophysiol.* 84:651–65

Xing J, Andersen RA. 2000b. Models of the posterior parietal cortex which perform multimodal integration and represent space in several coordinate frames. *J. Cogn. Neurosci.* 12:601–14

Zhang M, Barash S. 2000. Neuronal switching of sensorimotor transformations for antisaccades. *Nature* 408:971–75

Zipser D, Andersen RA. 1988. A back-propagation programmed network that simulates response properties of a subset of posterior parietal neurons. *Nature* 331:679–84

The Mirror-Neuron System

Giacomo Rizzolatti[1] *and Laila Craighero*[2]

[1]Dipartimento di Neuroscienze, Sezione di Fisiologia, via Volturno, 3, Universita di Parma, 43100, Parma, Italy; email: giacomo.rizzolatti@unipr.it
[2]Dipartimento SBTA, Sezione di Fisiologia Umana, via Fossato di Mortara, 17/19, Universita di Ferrara, 44100 Ferrara, Italy; email: crh@unife.it

Key Words

mirror neurons, action understanding, imitation, language, motor cognition

Abstract

A category of stimuli of great importance for primates, humans in particular, is that formed by actions done by other individuals. If we want to survive, we must understand the actions of others. Furthermore, without action understanding, social organization is impossible. In the case of humans, there is another faculty that depends on the observation of others' actions: imitation learning. Unlike most species, we are able to learn by imitation, and this faculty is at the basis of human culture. In this review we present data on a neurophysiological mechanism—the mirror-neuron mechanism—that appears to play a fundamental role in both action understanding and imitation. We describe first the functional properties of mirror neurons in monkeys. We review next the characteristics of the mirror-neuron system in humans. We stress, in particular, those properties specific to the human mirror-neuron system that might explain the human capacity to learn by imitation. We conclude by discussing the relationship between the mirror-neuron system and language.

INTRODUCTION

Mirror neurons are a particular class of visuomotor neurons, originally discovered in area F5 of the monkey premotor cortex, that discharge both when the monkey does a particular action and when it observes another individual(monkey or human) doing a similar action (Di Pellegrino et al. 1992, Gallese et al. 1996, Rizzolatti et al. 1996a). A lateral view of the monkey brain showing the location of area F5 is presented in Figure 1.

The aim of this review is to provide an updated account of the functional properties of the system formed by mirror neurons. The review is divided into four sections. In the first section we present the basic functional properties of mirror neurons in the monkey, and we discuss their functional roles in action understanding. In the second section, we present evidence that a mirror-neuron system similar to that of the monkey exists in humans. The third section shows that in humans, in addition to action understanding, the mirror-neuron system plays a fundamental role in action imitation. The last section is more speculative. We present there a theory of language evolution, and we discuss a series of data supporting the notion of a strict link between language and the mirror-neuron system (Rizzolatti & Arbib 1998).

THE MIRROR-NEURON SYSTEMIN MONKEYS

F5 Mirror Neurons: Basic Properties

There are two classes of visuomotor neuronsinmonkey area F5: canonical neurons, which respond to the presentation of an object, and mirror neurons, which respond when the monkey sees object-directed action(Rizzolatti & Luppino2001). In order to be triggered by visual stimuli, mirror neurons require an interaction between a biological effector (hand or mouth) and an object. The sight of an object alone, of an agent mimicking an action, or of an individual making intransitive (nonobject-directed) gestures are all ineffective. The object significance for the monkey has no obvious influence on the mirror-neuron response. Grasping a piece of food or a geometric solid produces responses of the same intensity.

Mirror neurons show a large degree of generalization. Presenting widely different visual stimuli, but which all represent the same action, is equally effective. For example, the same grasping mirror neuron that responds to a human hand grasping an object responds also when the grasping hand is that of a monkey. Similarly, the response is typically not affected if the action is done near or far from the monkey, in spite of the fact that the size of the observed hand is obviously different in the two conditions.

It is also of little importance for neuron activation if the observed action is eventually rewarded. The discharge is of the same intensity if the experimenter grasps the food and gives it to the recorded monkey or to another monkey introduced in the experimental room.

An important functional aspect of mirror neurons is the relation between their visual and motor properties. Virtually all mirror neurons show congruence between the visual actions they respond to and the motor responses they code. According to the type of congruence they exhibit, mirror neurons have been subdivided into "strictly congruent" and "broadly congruent" neurons (Gallese et al. 1996).

Mirror neurons in which the effective observed and effective executed actions correspond in terms of goal (e.g., grasping) and means for reaching the goal (e.g., precision grip) have been classed as "strictly congruent." They represent about one third of F5 mirror neurons. Mirror neurons that, in order to be triggered, do not require the observation of exactly the same action that they code motorically have been classed as "broadly congruent." They represent about two thirds of F5 mirror neurons.

F5 Mouth Mirror Neurons

The early studies of mirror neurons concerned essentially the upper sector of F5 where Hand action are mostly represented. Recently, a study was carried out on the properties of neurons located in the lateral part of F5 (Ferrari et al. 2003), where, in contrast, most neurons are related to mouth actions.

The results showed that about 25% of studied neurons have mirror properties. According to the visual stimuli effective in triggering the neurons, two classes of mouth mirror neurons were distinguished: ingestive and communicative mirror neurons.

Ingestive mirror neurons respond to the observation of actions related to ingestive functions, such as grasping food with the mouth, breaking it, or sucking. Neurons of this class form about 80% of the total amount of the recorded mouth mirror neurons. Virtually all ingestive mirror neurons show a good correspondence between the effective observed and the effective executed action. In about one third of them, the effective observed and executed actions are virtually identical(strictly congruent neurons); in the remaining, the effective observed and executed actions are similar or functionally related (broadly congruent neurons).

More intriguing are the properties of the communicative mirror neurons. The most effective observed action for them is a communicative gesture such as lip smacking, for example. However, from a motor point of view they behave as the ingestive mirror neurons, strongly discharging when the monkey actively performs an ingestive action.

This discrepancy between the effective visual input (communicative) and the effective active action (ingestive) is rather puzzling. Yet, there is evidence suggesting that communicative gestures, or at least some of them, derived from ingestive actions in evolution (MacNeilage 1998, Van Hoof 1967). From this perspective one may argue that the communicative mouth mirror neurons found in F5 reflect a process of corticalization of communicative functions not yet freed from their original ingestive basis.

The Mirror-Neuron Circuit

Neurons responding to the observation of actions done by others are present not only in area F5. A region in which neurons with these properties have been described is the cortex of the superior temporal sulcus (STS; Figure 1) (Perrett et al. 1989, 1990; Jellema et al. 2000; see Jellema et al. 2002). Movements effective in eliciting neuron responses in this region are walking, turning the head, bending the torso, and moving the arms. A small set of STS neurons discharge also during the observation of goal-directed hand movements (Perrett et al. 1990).

If one compares the functional properties of STS and F5 neurons, two points emerge. First, STS appears to code a much larger number of movements than F5. This may be ascribed, however, to the fact that STS output reaches, albeit indirectly (see below), the whole ventral premotor region and not only F5. Second, STS neurons do not appear to be endowed with motor properties.

Another cortical area where there are neurons that respond to the observation of actions done by other individuals is area 7b or PF of Von Economo(1929)(Fogassi et al. 1998, Gallese et al. 2002). This area (see Figure1) forms the rostral part of the inferior parietal lobule. It receives input from STS and sends an important output to the ventral premotor cortex including area F5.

PF neurons are functionally heterogeneous. Most of them (about 90%) respond to sensory stimuli, but about 50% of them also have motor properties discharging when the monkey performs specific movements or actions (Fogassi et al. 1998, Gallese et al. 2002, Hyvarinen 1982).

PF neurons responding to sensory stimuli have been subdivided into "somatosensory neurons" (33%), "visual neurons" (11%), and "bimodal (somatosensory and visual) neurons" (56%). About 40% of the visually responsive neurons respond specifically to action observation and of them about two thirds have mirror properties (Gallese et al. 2002).

In conclusion, the cortical mirror neuron circuit is formed by two main regions: the rostral part of the inferior parietal lobule and the ventral premotor cortex. STS is strictly related to it but, lacking motor properties, cannot be considered part of it.

Function of the Mirror Neuron in the Monkey: Action Understanding

Two main hypotheses have been advanced on what might be the functional role of mirror neurons. The first is that mirror-neuron activity mediates imitation (see Jeannerod 1994); the second is that mirror neurons are at the basis of action understanding (see Rizzolatti et al. 2001).

Both these hypotheses are most likely correct. However, two points should be specified. First, although we are fully convinced(for evidence see next section)that the mirror neuron mechanism is a mechanism of great evolutionary importance through which primates understand actions done by their conspecifics, we cannot claim that this is the only mechanism through which actions done by others may be understood (see Rizzolatti et al. 2001). Second, as is shown below, the mirror-neuron system is the system at the basis of imitation in humans. Although laymen are often convinced that imitation is a very primitive cognitive function, they are wrong. There is vast agreement among ethologists that imitation, the capacity to learn to do an action from seeing it done (Thorndyke 1898), is present among primates, only in humans, and (probably) in apes (see Byrne 1995, Galef 1988, Tomasello & Call 1997, Visalberghi & Fragaszy 2001, Whiten & Ham 1992). Therefore, the primary function of mirror neurons cannot be action imitation.

How do mirror neurons mediate understanding of actions done by others? The proposed mechanism is rather simple. Each time an individual sees an action done by another individual, neurons that represent that action are activated in the observer's premotor cortex. This automatically induced, motor representation of the observed action corresponds to that which is spontaneously generated during active action and whose outcome is known to the acting individual. Thus, the mirror system transforms visual information into knowledge (see Rizzolatti et al. 2001).

Evidence in Favor of the Mirror Mechanism in Action Understanding

At first glance, the simplest, and most direct, way to prove that the mirror-neuron system underlies action understanding is to destroy it and examine the lesion effect on the monkey's capacity to recognize actions made by other monkeys. In practice, this is not so. First, the mirror-neuron system is bilateral and includes, as shown above, large portions of the parietal and premotor cortex. Second, there are other mechanisms that may mediate action recognition (see Rizzolatti et al. 2001). Third, vast lesions as those required to destroy the mirror neuron system may produce more general cognitive deficits that would render difficult the interpretation of the results.

An alternative way to test the hypothesis that mirror neurons play a role in action understanding is to assess the activity of mirror neurons in conditions in which the monkey understands the meaning of the occurring action but has no access to the visual features that activate mirror neurons. If mirror neurons mediate action understanding, their activity should reflect the meaning of the observed action, not its visual features.

Prompted by these considerations, two series of experiments were carried out. The first tested whether F5 mirror neurons are able to recognize actions from their sound (Kohler et al. 2002), the second whether the mental representation of an action triggers their activity (Umiltà et al. 2001).

Kohler et al. (2002) recorded F5 mirror neuron activity while the monkey was observing a noisy action (e.g., ripping a piece of paper) or was presented with the same noise without seeing it. The results showed that about 15% of mirror neurons responsive to presentation of actions accompanied by sounds also responded to the presentation of the sound alone. The response to action sounds did not depend on unspecific factors such as arousal or emotional content of the stimuli. Neurons responding specifically to action sounds were dubbed "audio-visual" mirror neurons. Neurons were also tested in an experimental design in which two noisy actions were randomly presented in vision-and-sound, sound-only, vision-only, and motor conditions. In the motor condition, the monkeys performed the object-directed action that they observed or heard in the sensory conditions. Out of 33 studied neurons, 29 showed auditory selectivity for one of the two hand actions. The selectivity in visual and auditory modality was the same and matched the preferred motor action.

The rationale of the experiment by Umiltà et al. (2001) was the following. If mirror neurons are involved in action understanding, they should discharge also in conditions in which monkey does not see the occurring action but has sufficient clues to create a mental representation of what the experimenter does. The neurons were tested in two basic conditions. In one, the monkey was shown a fully visible action directed toward an object ("full vision" condition). In the other, the monkey saw the same action but with its final, critical part hidden ("hidden" condition). Before each trial, the experimenter placed a piece of food behind the screen so that the monkey knew there was an object there. Only those mirror neurons were studied that discharged to the observation of the final part of a grasping movement and/or to holding.

Figure 2 shows the main result of the experiment. The neuron illustrated in the figure responded to the observation of grasping and holding (*A*, full vision). The neuron discharged also when the stimulus-triggering features (a hand approaching the stimulus and subsequently holding it) were hidden from monkey's vision (*B*, hidden condition). As is the case for most mirror neurons, the observation of a mimed action did not activate the neuron (*C*, full vision, and *D*, hidden condition). Note that from a physical point of view B and D are identical. It was therefore the understanding of the meaning of the observed actions that determined the discharge in the hidden condition.

More than half of the tested neurons discharged in the hidden condition. Out of them, about half did not show any difference in the response strength between the hidden- and full-vision conditions. The other half responded more strongly in the full-vision condition. One neuron showed a more pronounced response in the hidden condition than in full vision.

In conclusion, both the experiments showed that the activity of mirror neurons correlates with action understanding. The visual features of the observed actions are fundamental to trigger mirror neurons only insomuch as they allow the understanding of the observed actions. If action comprehension is possible on another basis (e.g., action sound), mirror neurons signal the action, even in the absence of visual stimuli.

THE MIRROR-NEURON SYSTEM IN HUMANS

There are no studies in which single neurons were recorded from the putative mirror-neuron areas in humans. Thus, direct evidence for the existence of mirror neurons in humans is lacking. There is, however, a rich amount of data proving, indirectly, that a mirror-neuron system does exist in humans. Evidence of this comes from neurophysiological and brain-imaging experiments.

Neurophysiological Evidence

Neurophysiological experiments demonstrate that when individuals observe an action done by another individual their motor cortex becomes active, in the absence of any overt motor activity. A first evidence in this sense was already provided in the 1950s by Gastaut and his coworkers (Cohen-Seat et al. 1954, Gastaut & Bert 1954). They observed that the desynchronization of an EEG rhythm recorded from central derivations (the so-called mu rhythm) occurs not only during active movements of studied subjects, but also when the subjects observed actions done by others.

This observation was confirmed by Cochin et al.(1998, 1999)and by Altschuler et al.(1997, 2000) using EEG recording, and by Hari et al.(1998) using magnetoencephalographic (MEG) technique. This last study showed that the desynchronization during action observation includes rhythms originating from the cortex inside the central sulcus (Hari & Salmelin 1997, Salmelin & Hari 1994).

More direct evidence that the motor system in humans has mirror properties was provided by transcranial magnetic stimulation (TMS) studies. TMS is a noninvasive technique for electrical stimulation of the nervous system. When TMS is applied to the motor cortex, at appropriate stimulation intensity, motor-evoked potentials (MEPs) can be recorded from contralateral extremity muscles. The amplitude of these potentials is modulated by the behavioral context. The modulation of MEPs' amplitude can be used to assess the central effects of various experimental conditions. This approach has been used to study the mirror neuron system.

Fadiga et al. (1995) recorded MEPs, elicited by stimulation of the left motor cortex, from the right hand and arm muscles in volunteers required to observe an experimenter grasping objects (transitive hand actions) or performing meaningless arm gestures (intransitive arm movements). Detection of the dimming of a small spot of light and presentation of 3-D objects were used as control conditions. The results showed that the observation of both transitive and intransitive actions determined an increase of the recorded MEPs with respect to the control conditions. The increase concerned selectively those muscles that the participants use for producing the observed movements.

Facilitation of the MEPs during movement observation may result from a facilitation of the primary motor cortex owing to mirror activity of the premotor areas, to a direct facilitatory input to the spinal cord originating from the same areas, or from both. Support for the cortical hypothesis (see also below, Brain Imaging Experiments) came from a study by Strafella & Paus (2000). By using a double-pulse TMS technique, they demonstrated that the duration of intracortical recurrent inhibition, occurring during action observation, closely corresponds to that occurring during action execution.

Does the observation of actions done by others influence the spinal cord excitability? Baldissera et al. (2001) investigated this issue by measuring the size of the H-reflex evoked in the flexor and extensor muscles of normal volunteers during the observation of hand opening and closure done by another individual. The results showed that the size of H-reflex recorded from the flexors increased during the observation of hand opening, while it was depressed during the observation of hand closing. The converse was found in the extensors. Thus, while the cortical excitability varies in accordance with the seen movements, the spinal cord excitability changes in the opposite direction. These findings indicate that, in the spinal cord, there is an inhibitory mechanism that prevents the execution of an observed action, thus leaving the cortical motor system free to "react" to that action without the risk of overt movement generation.

In a study of the effect of hand orientation on cortical excitability, Maeda et al. (2002) confirmed (see Fadiga et al. 1995) the important finding that, in humans, intransitive movements, and not only goal-directed actions, determine motor resonance. Another important property of the human mirror-neuron system, demonstrated with TMS technique, is that the time course of cortical facilitation during action observation follows that of movement execution. Gangitano et al. (2001) recorded MEPs from the hand muscles of normal volunteers while they were observing grasping movements made by another individual. The MEPs were recorded at different intervals following the movement onset. The results showed that the motor cortical excitability faithfully followed the grasping movement phases of the observed action.

In conclusion, TMS studies indicate that a mirror-neuron system (a motor resonance system) exists in humans and that it possesses important properties not observed in monkeys. First, intransitive meaningless movements produce mirror-neuron system activation in humans (Fadigaetal. 1995, Maedaetal. 2002, Patuzzo et al. 2003), whereas they do not activate mirror neurons in monkeys. Second, the temporal characteristics of cortical excitability, during action observation, suggest that human mirror-neuron systems code also for the movements forming an action and not only for action as monkey mirror-neuron systems do. These properties of the human mirror-neuron system should play an important role in determining the humans' capacity to imitate others' action.

Brain Imaging Studies: The Anatomy of the Mirror System

A large number of studies showed that the observation of actions done by others activates in humans a complex network formed by occipital, temporal, and parietal visual areas, and two cortical regions whose function is fundamentally or predominantly motor (e.g., Buccino et al. 2001; Decety et al. 2002; Grafton et al. 1996; Grèzes et al. 1998; Grèzes et al. 2001; Grèzes et al. 2003; Iacoboni et al. 1999, 2001; Koski et al. 2002, 2003; Manthey et al. 2003; Nishitani & Hari 2000, 2002; Perani et al. 2001; Rizzolatti et al. 1996b). These two last regions are the rostral part of the inferior parietal lobule and the lower part of the precentral gyrus plus the posterior part of the inferior frontal gyrus (IFG). These regions form the core of the human mirror-neuron system.

Which are the cytoarchitectonic areas that form these regions? Interpretation of the brain-imaging activations in cytoarchitectonic terms is always risky. Yet, in the case of the inferior parietal region, it is very plausible that the mirror activation corresponds to areas PF and PFG, where neurons with mirror properties are found in the monkeys (see above).

More complex is the situation for the frontal regions. A first issue concerns the location of the border between the two main sectors of the premotor cortex: the ventral premotor cortex (PMv) and the dorsal premotor cortex (PMd). In nonhuman primates the two sectors differ anatomically (Petrides & Pandya 1984, Tanné-Gariepy et al. 2002) and functionally (see Rizzolatti et al. 1998). Of them, PMv only has (direct or indirect) anatomical connections with the areas where there is visual coding of action made by others (PF/PFG and indirectly STS) and, thus, where there is the necessary information for the formation of mirror neurons (Rizzolatti & Matelli 2003).

On the basis of embryological considerations, the border between human PMd and PMv should be located, approximately, at Z level 50 in Talairach coordinates (Rizzolatti & Arbib 1998, Rizzolatti et al. 2002). This location derives from the view that the superior frontal sulcus (plus the superior precentral sulcus) represents the human homologue of the superior branch of the monkey arcuate sulcus. Because the border of monkey PMv and PMd corresponds approximately to the caudal continuation of this branch, the analogous border should, in humans, lie slightly ventral to the superior frontal sulcus.

The location of human frontal eye field (FEF) supports this hypothesis (Corbetta 1998, Kimmingetal. 2001, Paus 1996, Petitetal. 1996).In monkeys, FEF lies in the anterior bank of the arcuate sulcus, bordering posteriorly the sector of PMv where arm and head movements are represented (area F4). If one accepts the location of the border between PMv and PMd suggested above, FEF is located in a similar position in the two species. In both of them, the location is just anterior to the upper part of PMv and the lowest part of PMd.

The other issue concerns IFG areas. There is a deeply rooted prejudice that these areas are radically different from those of the precentral gyrus and that they are exclusively related to speech (e.g., Grèzes & Decety 2001).This is not so. Already at the beginning of the last century, Campbell (1905) noted clear anatomical similarities between the areas of posterior IFG and those of the precentral gyrus. This author classed both the *pars opercularis* and the *pars triangularis* of IFG together with the precentral areas and referred to them collectively as the "intermediate precentral" cortex. Modern comparative studies indicate that the *pars opercularis* of IFG (basically corresponding to area 44) is the human homologue of area F5 (Von Bonin & Bailey 1947, Petrides & Pandya 1997). Furthermore, from a functional perspective, clear evidence has been accumulating in recent years that human area 44, in addition to speech representation, contains (as does monkey area F5) a motor representation of hand movements (Binkofski et al. 1999, Ehrsson et al. 2000, Gerardin et al. 2000, Iacoboni et al. 1999, Krams et al.1998). Taken together, these data strongly suggest that human PMv is the homologue of monkey area F4, and human area 44 is the homologue of monkey area F5. The descending branch of the inferior precentral sulcus (homologue to the monkey inferior precentral dimple) should form the approximate border between the two areas (for individual variations of location and extension area 44, see Amunts et al. 1999 and Tomaiuolo et al. 1999).

If the homology just described is correct, one should expect that the observation of neck and proximal arm movements would activate predominantly PMv, whereas hand and mouth movements would activate area 44. Buccino et al.(2001)addressed this issue in an fMRI experiment. Normal volunteers were presented with video clips showing actions performed with the mouth, hand/arm, and foot/leg. Both transitive (actions directed toward an object) and intransitive actions were shown. Action observation was contrasted with the observation of a static face, hand, and foot (frozen pictures of the video clips), respectively.

Observation of object-related mouth movements determined activation of the lower part of the precentral gyrus and of the *pars opercularis* of the inferior frontal gyrus (IFG), bilaterally. In addition, two activation foci were found in the parietal lobe. One was located in the rostral part of the inferior parietal lobule (most likely area PF), whereas the other was located in the posterior part of the same lobule. The observation of intransitive actions activated the same premotor areas, but there was no parietal lobe activation.

Observation of object-related hand/arm movements determined two areas of activation in the frontal lobe, one corresponding to the *pars opercularis* of IFG and the other located in the precentral gyrus. The latter activation was more dorsally located than that found during the observation of mouth movements. As for mouth movements, there were two activation foci in the parietal lobe. The rostral focus was, as in the case of mouth actions, in the rostral part of the inferior parietal lobule, but more posteriorly located, whereas the caudal focus was essentially in the same location as that for mouth actions. During the observation of intransitive movements the premotor activations were present, but the parietal ones were not.

Finally, the observation of object-related foot/leg actions determined an activation of a dorsal sector of the precentral gyrus and an activation of the posterior parietal lobe, in part overlapping with those seen during mouth and hand actions, in part extending more dorsally. Intransitive foot actions produced premotor, but not parietal, activation.

A weakness of the data by Buccino et al. (2001) is that they come from a group study. Data from single individuals are badly needed for a more precise somatotopic map. Yet, they clearly show that both the frontal and the parietal "mirror" regions are somatotopically organized. The somatotopy found in the inferior parietal lobule is the same as that found in the monkey. As far as the frontal lobe is concerned, the data appear to confirm the predictions based on the proposed homology. The activation of the *pars opercularis* of IFG should reflect the observation of distal hand actions and mouth actions, whereas that of the precentral cortex activation should reflect that of proximal arm actions and of neck movements.

It is important to note that the observation of transitive actions activated both the parietal and the frontal node of the mirror-neuron system, whereas the intransitive actions activated the frontal node only. This observation is in accord with the lack of inferior parietal lobule activation found in other studies in which intransitive actions were used (e.g., finger movements; Iacoboni et al. 1999, 2001; Koski et al. 2002, 2003). Considering that the premotor areas receive visual information from the inferior parietal lobule, it is hard to believe that the inferior parietal lobule was not activated during the observation of intransitive actions. It is more likely, therefore, that when an object is present, the inferior parietal activation is stronger than when the object is lacking, and the activation, in the latter case, does not reach statistical significance.

Brain Imaging Studies: Mirror-Neuron System Properties

As discussed above, the mirror-neuron system is involved in action understanding. An interesting issue is whether this is true also for actions done by individuals belonging to other species. Is the understanding by humans of actions done by monkeys based on the mirror-neuron system? And what about more distant species, like dogs?

Recently, an fMRI experiment addressed these questions (Buccino et al. 2004). Video clips showing silent mouth actions performed by humans, monkeys, and dogs were presented to normal volunteers. Two types of actions were shown: biting and oral communicative actions (speech reading, lip smacking, barking). As a control, static images of the same actions were presented.

The results showed that the observation of biting, regardless of whether it was performed by a man, a monkey, or a dog, determined the same two activation foci in the inferior parietal lobule discussed above and activation in the *pars opercularis* of the IFG and the adjacent precentral gyrus (Figure 3). The left rostral parietal focus and the left premotor focus were virtually identical for all three species, whereas the right side foci were stronger during the observation of actions made by a human being than by an individual of another species. Different results were obtained with communicative actions. Speech reading activated the left *pars opercularis* of IFG; observation of lip smacking, a monkey communicative gesture, activated a small focus in the right and left *pars opercularis* of IFG; observation of barking did not produce any frontal lobe activation (Figure 4).

These results indicated that actions made by other individuals could be recognized through different mechanisms. Actions belonging to the motor repertoire of the observer are mapped on his/her motor system. Actions that do not belong to this repertoire do not excite the motor system of the observer and appear to be recognized essentially on a visual basis without motor involvement. It is likely that these two different ways of recognizing actions have two different psychological counterparts. In the first case the motor "resonance" translates the visual experience into an internal "personal knowledge" (see Merleau-Ponty 1962), whereas this is lacking in the second case.

One may speculate that the absence of the activation of the frontal mirror area reported in some experiments might be due to the fact that the stimuli used (e.g., light point stimuli, Grèzes et al. 2001) were insufficient to elicit this "personal" knowledge of the observed action.

An interesting issue was addressed by Johnson Frey et al. (2003). Using event-related fMRI, they investigated whether the frontal mirror activation requires the observation of a dynamic action or if the understanding of the action goal is sufficient. Volunteers were presented with static pictures of the same objects being grasped or touched. The results showed that the observation of the goals of hand-object interactions was sufficient to activate selectively the frontal mirror region.

In this experiment, *pars triangularis* of IFG has been found active in several subjects (see also Rizzolatti etal. 1996b, Grafton et al. 1996). In speech, this sector appears to be mostly related to syntax (Bookheimer 2002). Although one may be tempted to speculate that this area may code also the syntactic aspect of action (see Greenfield 1991), there is at present no experimental evidence in support of this proposal. Therefore, the presence of activation of *pars triangularis* lacks, at the moment, a clear explanation.

Schubotz & Von Cramon (2001, 2002a,b) tested whether the frontal mirror region is important not only for the understanding of goal-directed actions, but also for recognizing predictable visual patterns of change. They used serial prediction tasks, which tested the participants' performance in a sequential perceptual task without sequential motor responses. Results showed that serial prediction caused activation in premotor and parietal cortices, particularly within the right hemisphere. The authors interpreted these findings as supporting the notion that sequential perceptual events can be represented independent of preparing an intended action toward the stimulus. According to these authors, the frontal mirror-neuron system node plays, in humans, a crucial role also in the representation of sequential information, regardless of whether it is perceptual or action related.

MIRROR-NEURON SYSTEM AND IMITATION

Imitation of Actions Present in the Observer's Repertoire

Psychological experiments strongly suggest that, in the cognitive system, stimuli and responses are represented in a commensurable format (Brass et al. 2000, Craighero et al. 2002, Wohlschlager & Bekkering 2002; see Prinz 2002). When observers see a motor event that shares features with a similar motor event present in their motor repertoire, they are primed to repeat it. The greater the similarity between the observed event and the motor event, the stronger the priming is (Prinz 2002).

These findings, and the discovery of mirror neurons, prompted a series of experiments aimed at finding the neural substrate of this phenomenon(Iacoboni et al. 1999, 2001; Nishitani & Hari 2000, 2002).

Using fMRI, Iacoboni et al. (1999) studied normal human volunteers in two conditions: observation-only and observation-execution. In the "observation-only" condition, subjects were shown a moving finger, a cross on a stationary finger, or a cross on an empty background. The instruction was to observe the stimuli. In the "observation-execution" condition, the same stimuli were presented, but this time the instruction was to lift the right finger, as fast as possible, in response to them.

The most interesting statistical contrast was that between the trials in which the volunteers made the movement in response to an observed action (imitation) and the trials in which the movement was triggered by the cross. The results showed that the activity was stronger during imitation trials than during the other motor trials in four areas: the left *pars opercularis* of the IFG, the right anterior parietal region, the right parietal operculum, and the right STS region (see for this last activation Iacoboni et al. 2001). Further experiments by Koski et al. (2002) confirmed the importance of Broca's area, in particular when the action to be imitated had a specific goal. Grèzes et al. (2003) obtained similar results, but only when participants had to imitate pantomimes. The imitation of object-directed actions surprisingly activated PMd.

Nishitani & Hari (2000,2002) performed two studies in which they investigated imitation of grasping actions and of facial movements, respectively. The event-related MEG was used. The first study confirmed the importance of the left IFG (Broca'sarea) in imitation. In the second study (Nishitani & Hari2002),the authors asked volunteers to observe still pictures of verbal and nonverbal (grimaces) lip forms, to imitate them immediately after having seen them, or to make similar lip forms spontaneously. During lip form observation, cortical activation progressed from the occipital cortex to the superior temporal region, the inferior parietal lobule, IFG (Broca'sarea), and finally to the primary motor cortex. The activation sequence during imitation of both verbal and nonverbal lip forms was the same as during observation. Instead, when the volunteers executed the lip forms spontaneously, only Broca's area and the motor cortex were activated.

Taken together, these data clearly show that the basic circuit underlying imitation coincides with that which is active during action observation. They also indicate that, in the posterior part of IFG, a direct mapping of the observed action and its motor representation takes place.

The studies of Iacoboni et al. (1999, 2001) showed also activations—superior parietal lobule, parietal operculum, and STS region—that most likely do not reflect a mirror mechanism. The activation of the superior parietal lobule is typically not present when subjects are instructed to observe actions without the instruction to imitate them (e.g., Buccino et al. 2001). Thus, a possible interpretation of this activation is that the request to imitate produces, through backward projections, sensory copies of the intended actions. In the monkey, superior parietal lobule and especially its rostral part (area PE) contains neurons that are active in response to proprioceptive stimuli as well as during active arm movements (Kalaska et al. 1983, Lacquaniti et al. 1995, Mountcastle et al. 1975). It is possible, therefore, that the observed superior parietal activation represents a kinesthetic copy of the intended movements. This interpretation fits well previous findings by Grèzesetal. (1998), who, in agreement with Iacoboni et al. (1999), showed a strong activation of superior parietal lobule when subjects' tasks were to observe actions in order to repeat them later.

An explanation in terms of sensory copies of the intended actions may also account for the activations observed in the parietal operculum and STS. The first corresponds to the location of somatosensory areas hidden in the sylvian sulcus (Disbrow et al. 2000), whereas the other corresponds to higher-order visual areas of the STS region (see above). Thus, these two activations might reflect somatosensory and visual copies of the intended action, respectively.

The importance of the *pars opercularis* of IFG in imitation was further demonstrated using repetitive TMS (rTMS), a technique that transiently disrupts the functions of the stimulated area (Heiser et al. 2003). The task used in the study was, essentially, the same as that of Iacoboni et al. (1999). The results showed that following stimulation of both left and right Broca's area, there was significant impairment in imitation of finger movements. The effect was absent when finger movements were done in response to spatial cues.

Imitation Learning

Broadly speaking, there are two types of newly acquired behaviors based on imitation learning. One is substitution, for the motor pattern spontaneously used by the observer in response to a given stimulus, of another motor pattern that is more adequate to fulfill a given task. The second is the capacity to learn a motor sequence useful to achieve a specific goal (Rizzolatti 2004).

The neural basis of the capacity to form a new motor pattern on the basis of action observation was recently studied by Buccino et al. (G. Buccino, S. Vogt, A. Ritzl, G.R. Fink, K. Zilles, H.J. Freund & G. Rizzolatti, submitted manuscript), using an event-related fMRI paradigm. The basic task was the imitation, by naive participants, of guitar chords played by an expert guitarist. By using an event-related paradigm, cortical activation was mapped during the following events: (*a*) action observation, (*b*) pause(new motor pattern formation and consolidation), (*c*)chord execution, and (*d*) rest. In addition to imitation condition, there were three control conditions: observation without any motor request, observation followed by execution of a nonrelated action (e.g., scratching the guitar neck), and free execution of guitar chords.

The results showed that during the event observation-to-imitate there was activation of a cortical network that coincided with that which is active during observation-without-instruction-to-imitate and during observation in order not to imitate. The strength of the activation was, however, much stronger in the first condition. The areas forming this common network were the inferior parietal lobule, the dorsal part of PMv, and the *pars opercularis* of IFG. Furthermore, during the event observation-to-imitate, but not during observation-without-further-motor-action, there was activation of the superior parietal lobule, anterior mesial areas plus a modest activation of the middle frontal gyrus.

The activation during the pause event in imitation condition involved the same basic circuit as in event observation-of-the-same-condition, but with some important differences: increase of the superior parietal lobule activation, activation of PMd, and, most interestingly, a dramatic increase in extension and strength of the middle frontal cortex activation (area 46) and of the areas of the anterior mesial wall. Finally, during the execution event, not surprisingly, the activation concerned mostly the sensorimotor cortex contralateral to the acting hand.

These data show that the nodal centers for new motor pattern formation coincide with the nodal mirror-neuron regions. Although fMRI experiments cannot give information on the mechanism involved, it is plausible (see the neurophysiological sections) that during learning of new motor patterns by imitation the observed actions are decomposed into elementary motor acts that activate, via mirror mechanism, the corresponding motor representations in PF and in PMv and in the *pars opercularis* of IFG. Once these motor representations are activated, they are recombined, according to the observed model by the prefrontal cortex. This recombination occurs inside the mirror-neuron circuit with area 46 playing a fundamental orchestrating role.

To our knowledge, there are no brain-imaging experiments that studied the acquisition of new sequences by imitation from the perspective of mirror neurons. Theoretical aspect of sequential learning by imitation and its possible neural basis have been discussed by Arbib (2002), Byrne (2002), and Rizzolatti (2004). The interested reader can find there an exhaustive discussion of this issue.

MIRROR-NEURON SYSTEM AND COMMUNICATION

Gestural Communication

Mirror neurons represent the neural basis of a mechanism that creates a direct link between the sender of a message and its receiver. Thanks to this mechanism, actions done by other individuals become messages that are understood by an observer without any cognitive mediation.

On the basis of this property, Rizzolatti & Arbib (1998) proposed that the mirror-neuron system represents the neurophysiological mechanism from which language evolved. The theory of Rizzolatti & Arbib belongs to theories that postulate that speech evolved mostly from gestural communication (see Armstrong et al. 1995, Corballis 2002). Its novelty consists of the fact that it indicates a neurophysiological mechanism that creates a common (parity requirement), nonarbitrary, semantic link between communicating individuals.

The mirror-neuron system in monkeys is constituted of neurons coding object-directed actions. A first problem for the mirror-neuron theory of language evolution is to explain how this close, object-related system became an open system able to describe actions and objects without directly referring to them.

It is likely that the great leap from a closed system to a communicative mirror system depended upon the evolution of imitation (see Arbib 2002) and the related changes of the human mirror-neuron system: the capacity of mirror neurons to respond to pantomimes (Buccino et al. 2001, Grèzes et al. 2003) and to intransitive actions (Fadiga et al. 1995, Maeda et al. 2002) that was absent in monkeys.

The notion that communicative actions derived from object-directed actions is not new. Vygotski (1934), for example, explained that the evolution of pointing movements was due to attempts of children to grasp far objects. It is interesting to note that, although monkey mirror neurons do not discharge when the monkey observes an action that is not object directed, they do respond when an object is hidden, but the monkey knows that the action has a purpose (Kohler et al. 2002). This finding indicates that breaking spatial relations between effector and target does not impair the capacity of understanding the action meaning. The precondition for understanding pointing—the capacity to mentally represent the action goal—is already present in monkeys.

A link between object-directed and communicative action was also stressed by other authors (see McNeilage 1998, Van Hoof 1967; for discussion of this link from the mirror neurons perspective, see above).

Mirror Neurons and Speech Evolution

The mirror neuron communication system has a great asset: Its semantics is inherent to the gestures used to communicate. This is lacking in speech. In speech, or at least in modern speech, the meaning of the words and the phono-articulatory actions necessary to pronounce them are unrelated. This fact suggests that a necessary step for speech evolution was the transfer of gestural meaning, intrinsic to gesture itself, to abstract sound meaning. From this follows a clear neurophysiological prediction: Hand/arm and speech gestures must be strictly linked and must, at least in part, share a common neural substrate.

A number of studies prove that this is true. TMS experiments showed that the excitability of the hand motor cortex increases during both reading and spontaneous speech (Meister et al. 2003, Seyal et al. 1999, Tokimura et al. 1996). The effect is limited to the left hemisphere. Furthermore, no language-related effect is found in the leg motor area. Note that the increase of hand motor cortex excitability cannot be attributed to word articulation because, although word articulation recruits motor cortex bilaterally, the observed activation is strictly limited to the left hemisphere. The facilitation appears, therefore, to result from a coactivation of the dominant hand motor cortex with higher levels of language network (Meister et al. 2003).

Gentilucci et al. (2001) reached similar conclusions using a different approach. In a series of behavioral experiments, they presented participants with two 3-D objects, one large and one small. On the visible face of the objects there were either two crosses or a series of dots randomly scattered on the same area occupied by the crosses. Participants were required to grasp the objects and, in the condition in which the crosses appeared on the object, to open their mouth. The kinematics of hand, arm, and mouth movements was recorded. The results showed that lip aperture and the peak velocity of lip aperture increased when the movement was directed to the large object.

In another experiment of the same study Gentilucci et al. (2001) asked participants to pronounce a syllable (e.g., GU, GA) instead of simply opening their mouth. It was found that lip aperture was larger when the participants grasped a larger object. Furthermore, the maximal power of the voice spectrum recorded during syllable emission was also higher when the larger object was grasped.

Most interestingly, grasping movements influence syllable pronunciation not only when they are executed, but also when they are observed. In a recent study (Gentilucci 2003), normal volunteers were asked to pronounce the syllables BA or GA while observing another individual grasping objects of different size. Kinematics of lip aperture and amplitude spectrum of voice was influenced by the grasping movements of the other individual. Specifically, both lip aperture and voice peak amplitude were greater when the observed action was directed to larger objects. Control experiments ruled out that the effect was due to the velocity of the observed arm movement.

Taken together, these experiments show that hand gestures and mouth gestures are strictly linked in humans and that this link includes the oro-laryngeal movements used for speech production.

Auditory Modality and Mirror-Neuron Systems

If the meaning of manual gestures, understood through the mirror-neuron mechanism, indeed transferred, in evolution, from hand gestures to oro-laryngeal gestures, how did that transfer occur?

As described above, in monkeys there is a set of F5 mirror neurons that discharge in response to the sound of those actions that, when observed or executed by the monkey, trigger a given neuron (Kohler et al. 2002). The existence of these audio-visual mirror neurons indicates that auditory access to action representation is present also in monkeys.

However, the audio-visual neurons code only object-related actions. They are similar, in this respect, to the "classical" visual mirror neurons. But, as discussed above, object-related actions are not sufficient to create an efficient intentional communication system. Therefore, words should have derived mostly from association of sound with intransitive actions and pantomimes, rather than from object-directed actions.

An example taken from Paget (1930) may clarify the possible process at work. When we eat, we move our mouth, tongue, and lips in a specific manner. The observation of this combined series of motor actions constitutes the gesture whose meaning is transparent to everybody: "eat." If, while making this action, we blow air through the oro-laryngeal cavities, we produce a sound like "mnyam-mnyam," or "mnya-mnya," words whose meaning is almost universally recognized (Paget 1930). Thus through such an association mechanism, the meaning of an action, naturally understood, is transferred to sound.

It is plausible that, originally, the understanding of the words related to mouth actions occurred through activation of audio-visual mirror neurons related to ingestive behavior (see Ferrari et al. 2003). A fundamental step, however, toward speech acquisition was achieved when individuals, most likely thanks to improved imitation capacities (Donald 1991), became able to generate the sounds originally accompanied by a specific action without doing the action. This new capacity should have led to(and derived from)the acquisition of an auditory mirror system, developed on top of the original audio-visual one, but which progressively became independent of it.

More specifically, this scenario assumes that, in the case discussed above, the premotor cortex became progressively able to generate the sound "mnyam-mnyam" without the complex motor synergies necessary for producing ingestive action, and, in parallel, neurons developed able to both generate the sound and discharge (resonate) in response to that sound (echo-neurons). The incredibly confusing organization of Broca's area in humans, where phonology, semantics, hand actions, ingestive actions, and syntax are all intermixed in a rather restricted neural space (see Bookheimer 2002), is probably a consequence to this evolutive trend.

Is there any evidence that humans possess an echo-neuron system, i.e., a system that motorically "resonates" when the individual listens to verbal material? There is evidence that this is the case.

Fadiga et al. (2002) recorded MEPs from the tongue muscles in normal volunteers instructed to listen carefully to acoustically presented verbal and nonverbal stimuli. The stimuli were words, regular pseudowords, and bitonal sounds. In the middle of words and pseudowords either a double "f" or a double "r" were embedded. "F" is a labio-dental fricative consonant that, when pronounced, requires slight tongue mobilization, whereas "r" is linguo-palatal fricative consonant that, in contrast, requires a tongue movement to be pronounced. During the stimulus presentation the participants' left motor cortices were stimulated.

The results showed that listening to words and pseudowords containing the double "r" determines a significant increase of MEPs recorded from tongue muscles as compared to listening to words and pseudowords containing the double "f" and listening to bitonal sounds. Furthermore, the facilitation due to listening of the "r" consonant was stronger for words than for pseudowords.

Similar results were obtained by Watkins et al. (2003). By using TMS technique they recorded MEPs from a lip (*orbicularis oris*) and a hand muscle (first *interosseus*) in four conditions: listening to continuous prose, listening to nonverbal sounds, viewing speech-related lip movements, and viewing eye and brow movements. Compared to control conditions, listening to speech enhanced the MEPs recorded from the *orbicularis oris* muscle. This increase was seen only in response to stimulation of the left hemisphere. No changes of MEPs in any condition were observed following stimulation of the right hemisphere. Finally, the size of MEPs elicited in the first *interosseus* muscle did not differ in any condition.

Taken together these experiments show that an echo-neuron system exists in humans: when an individual listens to verbal stimuli, there is an activation of the speech-related motor centers.

There are two possible accounts of the functional role of the echo-neuron system. A possibility is that this system mediates only the imitation of verbal sounds. Another possibility is that the echo-neuron system mediates, in addition, speech perception, as proposed by Liberman and his colleagues (Liberman et al. 1967, Liberman & Mattingly 1985, Liberman & Wahlen 2000). There is no experimental evidence at present proving one or another of the two hypotheses. Yet, is hard to believe that the echo-system lost any relation with its original semantic function.

There is no space here to discuss the neural basis of action word semantics. However, if one accepts the evolutionary proposal we sketched above, there should be two roots to semantics. One, more ancient, is closely related to the action mirror-neuron system, and the other, more recent, is based on the echo-mirror-neuron system.

Evidence in favor of the existence of the ancient system in humans has been recently provided by EEG and fMRI studies. Pulvermueller (2001, 2002) compared EEG activations while subjects listened to face- and leg-related action verbs ("walking" versus "talking"). They found that words describing leg actions evoked stronger in-going current at dorsal sites, close to the cortical leg-area, whereas those of the "talking" type elicited the stronger currents at inferior sites, next to the motor representation of the face and mouth.

In an fMRI experiment, Tettamanti et al. (M. Tettamanti, G. Buccino, M.C. Saccuman, V. Gallese, M. Danna, P. Scifo, S.F. Cappa, G. Rizzolatti, D. Perani & F. Fazio, submitted manuscript) tested whether cortical areas active during action observation were also active during listening to action sentences. Sentences that describe actions performed with mouth, hand/arm, and leg were used. The presentation of abstract sentences of comparable syntactic structure was used as a control condition. The results showed activations in the precentral gyrus and in the posterior part of IFG. The activations in the precentral gyrus, and especially that during listening to hand-action sentences, basically corresponded to those found during the observation of the same actions. The activation of IFG was particularly strong during listening of mouth actions, but was also present during listening of actions done with other effectors. It is likely, therefore, that, in addition to mouth actions, in the inferior frontal gyrus there is also a more general representation of action verbs. Regardless of this last interpretation problem, these data provide clear evidence that listening to sentences describing actions engages visuo-motor circuits subserving action representation.

These data, of course, do not prove that the semantics is exclusively, or even mostly, due to the original sensorimotor systems. The devastating effect on speech of lesions destroying the perisylvian region testifies the importance in action understanding of the system based on direct transformation of sounds into speech motor gesture. Thus, the most parsimonious hypothesis appears to be that, during speech acquisition, a process occurs somehow similar to the one that, in evolution, gave meaning to sound. The meaning of words is based first on the old nonverbal semantic system. Subsequently, however, the words are understood even without a massive activation of the old semantic system. Experiments, such as selective inhibition through TMS or electrical stimulation of premotor and parietal areas, are needed to better understand the relative role of the two systems in speech perceptions.

ACKNOWLEDGMENTS

This study was supported by EU Contract QLG3-CT-2002-00746, Mirror, EU Contract IST-2000-28159, by the European Science Foundation, and by the Italian Ministero dell'Università e Ricerca, grants Cofin and Firb RBNEO1SZB4.

LITERATURE CITED

Altschuler EL, Vankov A, Hubbard EM, Roberts E, Ramachandran VS, Pineda JA. 2000. Mu wave blocking by observation of movement and its possible use as a tool to study theory of other minds. *Soc. Neurosci.* 68.1 (Abstr.)

Altschuler EL, Vankov A, Wang V, RamachandranVS,PinedaJA.1997.Personsee,person do: human cortical electrophysiological correlates of monkey see monkey do cell. *Soc. Neurosci.* 719.17 (Abstr.)

Amunts K, Schleicher A, Buergel U, Mohlberg H, Uylings HBM, Zilles K. 1999. Broca's region revisited: cytoarchitecture and inter-subject variability. *J. Comp. Neurol.* 412: 319–41

Arbib MA. 2002. Beyond the mirror system: imitation and evolution of language. In *Imitation in Animals and Artifacts,* ed. C Nehaniv, K Dautenhan, pp. 229–80. Cambridge MA: MIT Press

Armstrong AC, Stokoe WC, Wilcox SE. 1995. *Gesture and the Nature of Language.* Cambridge, UK: Cambridge Univ. Press

Baldissera F, CavallariP, CraigheroL, FadigaL. 2001. Modulation of spinal excitability during observation of hand actions in humans. *Eur. J. Neurosci.* 13:190–94

BinkofskiF, BuccinoG, PosseS, SeitzRJ, Rizzolatti G, Freund H. 1999. A fronto-parietal circuit for object manipulation in man: evidence from an fMRI-study. *Eur. J. Neurosci.* 11:3276–86

Bookheimer S. 2002. Functional MRI of language: new approaches to understanding the cortical organization of semantic processing. *Annu. Rev. Neurosci.* 25:151–88

Brass M, Bekkering H, Wohlschlager A, Prinz W. 2000. Compatibility between observed and executed finger movements: comparing symbolic, spatial, and imitative cues. *Brain Cogn.* 44:124–43

Buccino G, Binkofski F, Fink GR, Fadiga L, FogassiL,etal.2001. Action observation activates premotor and parietal are as in a somatotopic manner: an fMRI study. *Eur. J. Neurosci.* 13:400–4

BuccinoG, LuiF, CanessaN, PatteriI, Lagravinese G, et al. 2004a. Neural circuits involved in the recognition of actions performed by non-conspecifics: an fMRI study. *J. Cogn. Neurosci.* 16:1–14

Byrne RW.1995. *TheThinkingApe.EvolutionaryOriginsofIntelligence.* Oxford,UK:Oxford Univ. Press

Byrne RW. 2002. Seeing actions as hierarchically organized structures: great ape manual skills. See Meltzoff & Prinz 2002, pp. 122–40

Campbell AW. 1905. *Histological Studies on the Localization of Cerebral Function.* Cambridge, UK: Cambridge Univ. Press. 360 pp.

Cochin S, Barthelemy C, Lejeune B, Roux S, Martineau J. 1998. Perception of motion and qEEG activity in human adults. *Electroen-cephalogr. Clin. Neurophysiol.* 107:287–95

Cochin S, Barthelemy C, Roux S, Martineau J.1999. Observation and execution of movement: similarities demonstrated by quantified electroencephalograpy. *Eur. J. Neurosci.* 11:1839–42

Cohen-Seat G, Gastaut H, Faure J, Heuyer G. 1954. Etudes expérimentales de l'activité nerveuse pendant la projection cinémato-graphique. *Rev. Int. Filmologie* 5:7–64

Corballis MC. 2002. *From Hand to Mouth. The Origins of Language.* Princeton: Princeton Univ. Press. 257 pp.

Corbetta M. 1998. Frontoparietal cortical networks for directing attention and the eye to visual locations: identical, independent, or overlapping neural systems? *Proc. Natl. Acad. Sci. USA* 95:831–38

Craighero L, Bello A, Fadiga L, Rizzolatti G. 2002.Handactionpreparationinfluencesthe responses to hand pictures. *Neuropsychologia* 40:492–502

DecetyJ,ChaminadeT,GrezesJ,MeltzoffAN.2002. APET exploration of the neural mechanisms involved in reciprocal imitation. *Neu-roimage* 15:265–72

Di Pellegrino G, Fadiga L, Fogassi L, Gallese V, Rizzolatti G. 1992. Understanding motor events: a neurophysiological study. *Exp. Brain Res.* 91:176–80

Disbrow E, Roberts T, Krubitzer L. 2000. Somatotopic organization of cortical fields in the lateral sulcus of homo sapiens: evidence for SII and PV. *J. Comp. Neurol.* 418:1–21

Donald M. 1991. *Origin of the Modern Mind: Three Stages in the Evolution of Culture and Cognition.* Cambridge, MA: Harvard Univ. Press

Ehrsson HH, Fagergren A, Jonsson T, West-ling G, Johansson RS, Forssberg H. 2000. Cortical activity in precision-versus power-grip tasks: an fMRI study. *J. Neurophysiol.* 83:528–36

Fadiga L, Craighero L, Buccino G, Rizzolatti G. 2002. Speech listening specifically modulates the excitability of tongue muscles: a TMS study. *Eur. J. Neurosci.* 15:399–402

Fadiga L, Fogassi L, Pavesi G, Rizzolatti G. 1995. Motor facilitation during action observation: a magnetic stimulation study. *J. Neurophysiol.* 73:2608–11

Ferrari PF, Gallese V, Rizzolatti G, Fogassi L. 2003. Mirror neurons responding to the observation of ingestive and communicative mouth actions in the monkey ventral premotor cortex. *Eur. J. Neurosci.* 17:1703–14

Fogassi L, Gallese V, Fadiga L, Rizzolatti G. 1998. Neurons responding to the sight of goal directed hand/arm actions in the parietal area PF (7b) of the macaque monkey. *Soc. Neurosci.* 24:257.5 (Abstr.)

Galef BG. 1988. Imitation in animals: history, definition and interpretation of data from psychological laboratory. In *Comparative Social Learning,* ed. T Zental, BG Galef, pp. 3–28, Hillsdale, NJ: Erlbaum

Gallese V, Fadiga L, Fogassi L, Rizzolatti G. 1996.Action recognition in the premotor cortex. *Brain* 119:593–609

Gallese V, Fogassi L, Fadiga L, Rizzolatti G. 2002. Action representation and the inferior parietal lobule. In *Attention & Performance XIX. Common Mechanisms in Perception and Action,* ed. WPrinz, BHommel, pp. 247–66. Oxford, UK: Oxford Univ. Press

Gangitano M, Mottaghy FM, Pascual-LeoneA. 2001. Phase specific modulation of cortical motor output during movement observation. *NeuroReport* 12:1489–92

Gastaut HJ, Bert J. 1954. EEG changes during cinematographic presentation. *Electroencephalogr. Clin. Neurophysiol.* 6:433–44

Gentilucci M. 2003. Grasp observation influences speech production. *Eur. J. Neurosci.* 17:179–84

Gentilucci M, Benuzzi F, Gangitano M, Grimaldi S. 2001. Grasp with hand and mouth: a kinematic study on healthy subjects. *J. Neurophysiol.* 86:1685–99

Gerardin E, Sirigu A, Lehericy S, Poline JB, GaymardB, etal. 2000. Partially over lapping neural networks for real and imagined hand movements. *Cereb. Cortex* 10:1093–104

Grafton ST, Arbib MA, Fadiga L, Rizzolatti G. 1996. Localization of grasp representations in humans by PET: 2. Observation compared with imagination. *Exp. Brain Res.* 112:103–11

Grèzes J, Armony JL, Rowe J, Passingham RE. 2003. Activations related to "mirror" and "canonical" neurones in the human brain: an fMRI study. *Neuroimage* 18:928–37

Grèzes J, Costes N, Decety J. 1998. Top-down effect of strategy on the perception of human biological motion: a PET investigation. *Cogn. Neuropsychol.* 15:553–82

Grèzes J, Decety J. 2001. Functional anatomy of execution, mental simulation, observation, and verb generation of actions: a meta-analysis. *Hum. Brain Mapp.* 12:1–19

GrèzesJ, FonluptP, BertenthalB, Delon-Martin C, Segebarth C, Decety J. 2001. Does perception of biological motion rely on specific brain regions? *Neuroimage* 13:775–85

Greenfield PM. 1991. Language, tool and brain: the ontogeny and phylogeny of hierarchically organized sequential behavior. *Behav. Brain Sci.* 14:531–95

Hari R, Forss N, Avikainen S, Kirveskari S, Salenius S, Rizzolatti G. 1998. Activation of human primary motor cortex during action observation: a neuromagnetic study. *Proc. Natl. Acad. Sci. USA* 95:15061–65

Hari R, Salmelin R. 1997. Human cortical oscillations: a neuromagnetic view through the skull. *Trends Neurosci.* 20:44–49

Heiser M, Iacoboni M, Maeda F, Marcus J, Mazziotta JC. 2003. The essential role of Broca's area in imitation. *Eur. J. Neurosci.* 17:1123–28

HyvarinenJ. 1982. Posterior parietal lobe of the primate brain. *Physiol. Rev.* 62:1060–129

Iacoboni M, Koski LM, Brass M, Bekkering H, Woods RP, et al. 2001. Reafferent copies of imitated actions in the right superior temporal cortex. *Proc. Natl. Acad. Sci.USA* 98: 13995–99

Iacoboni M, Woods RP, Brass M, Bekkering H, Mazziotta JC, Rizzolatti G. 1999. Cortical mechanisms of human imitation. *Science* 286:2526–28

Jeannerod M. 1994. The representing brain. Neural correlates of motor intention and imagery. *Behav. Brain Sci.* 17:187–245

Jellema T, Baker CI, Wicker B, Perrett DI. 2000. Neural representation for the perception of the intentionality of actions. *Brain Cogn.* 442:280–302

Jellema T, Baker CI, Oram MW, Perrett DI. 2002. Cell populations in the banks of the superior temporal sulcus of the macaque monkey and imitation. See Meltzoff & Prinz 2002, pp. 267–90

Johnson Frey SH, Maloof FR, Newman-Norlund R, Farrer C, Inati S, Grafton ST. 2003. Actions or hand-objects interactions? Human inferior frontal cortex and action observation. *Neuron* 39:1053–58

Kalaska JF, Caminiti R, Georgopoulos AP. 1983. Cortical mechanisms related to the direction of two-dimensional arm movements: relations in parietal area 5 and comparison with motor cortex. *Exp. Brain Res.* 51:247–60

Kimmig H, Greenlee MW, Gondan M, Schira M, Kassubek J, Mergner T. 2001. Relationship between saccadic eye movements and cortical activity as measured by fMRI: quantitative and qualitative aspects. *Exp. Brain Res.* 141:184–94

Kohler E, Keysers C, Umiltà MA, Fogassi L, Gallese V, Rizzolatti G. 2002. Hearing sounds, understanding actions: action representation in mirror neurons. *Science* 297:846–48

Koski L, Iacoboni M, Dubeau MC, Woods RP, Mazziotta JC. 2003. Modulation of cortical activity during different imitative behaviors. *J. Neurophysiol.* 89:460–71

Koski L, Wohlschlager A, Bekkering H, Woods RP, Dubeau MC. 2002. Modulation of motor and premotor activity during imitation of target-directed actions. *Cereb. Cortex* 12: 847–55

Krams M, Rushworth MF, Deiber MP, Frackowiak RS, Passingham RE. 1998. The preparation, execution and suppression of copied movements in the human brain. *Exp. Brain Res.* 120:386–98

Lacquaniti F, Guigon E, Bianchi L, Ferraina S, Caminiti R. 1995. Representing spatial information for limb movement: role of area 5 in the monkey. *Cereb. Cortex* 5:391–409

Liberman AM, Cooper FS, Shankweiler DP, Studdert-Kennedy M. 1967. Perception of the speech code. *Psychol. Rev.* 74:431–61

Liberman AM, Mattingly IG. 1985. The motor theory of speech perception revised. *Cognition* 21:1–36

Liberman AM, Whalen DH. 2000. On the relation of speech to language. *Trends Cogn. Neurosci.* 4:187–96

MacNeilage PF. 1998. The frame/content theory of evolution of speech production. *Behav. Brain Sci.* 21:499–511

Maeda F, Kleiner-Fisman G, Pascual-Leone A. 2002. Motor facilitation while observing hand actions: specificity of the effect and role of observer's orientation. *J. Neurophys-iol.* 87:1329–35

Manthey S, Schubotz RI, von Cramon DY. 2003. Premotor cortex in observing erroneous action: an fMRI study. *Brain Res. Cogn. Brain Res.* 15:296–307

Meister IG, Boroojerdi B, Foltys H, Sparing R, Huber W, Topper R. 2003. Motor cortex hand area and speech: implications for the development of language. *Neuropsychologia* 41:401–6

Meltzoff AN, Prinz W. 2002. *The Imitative Mind. Development, Evolution and Brain Bases.* Cambridge, UK: Cambridge Univ. Press

Merleau-Ponty M. 1962. *Phenomenology of Perception.* Transl. C Smith. London: Rout-ledge (From French)

Mountcastle VB, Lynch JC, Georgopoulos A, Sakata H, Acuna C. 1975. Posterior parietal association cortex of the monkey: command functions for operations within extrapersonal space. *J. Neurophysiol.* 38:871–908

Nishitani N, Hari R. 2000. Temporal dynamics of cortical representation for action. *Proc. Natl. Acad. Sci. USA* 97:913–18

Nishitani N, Hari R. 2002. Viewing lip forms: cortical dynamics. *Neuron* 36:1211–20

Paget R. 1930. *Human Speech.* London: Kegan Paul, Trench

PatuzzoS, FiaschiA, ManganottiP. 2003. Modulation of motor cortex excitability in the left hemisphere during action observation: a single and paired-pulse transcranial magnetic stimulation study of self-and non-self action obervation. *Neuropsychologia* 41:1272–78

Paus T. 1996. Location and function of the human frontal eye-field: a selective review. *Neuropsychologia* 34:475–83

PeraniD, FazioF, BorgheseNA, TettamantiM, Ferrari S, et al. 2001. Different brain correlates for watching real and virtual hand actions. *Neuroimage* 14:749–58

Perrett DI, Harries MH, Bevan R, Thomas S, Benson PJ, et al. 1989. Frameworks of analysis for the neural representation of animate objectsandactions. *J.Exp.Biol.*146:87–113

Perrett DI, Mistlin AJ, Harries MH, Chitty AJ. 1990. Understanding the visual appearance and consequence of hand actions. In *Vision and Action: The Control of Grasping,* ed. MA Goodale, pp. 163–342. Norwood, NJ: Ablex

Petit L, Orssaud C, Tzourio N, Crivello F, Berthoz A, Mazoyer B. 1996. Functional anatomy of a prelearned sequence of horizontal saccades in humans. *J. Neurosci.* 16:3714–26

Petrides M, Pandya DN. 1984. Projections to the frontal cortex from the posterior parietal region in the rhesus monkey. *J. Comp. Neurol.* 228:105–16

Petrides M, Pandya DN. 1997. Comparative architectonic analysis of the human and the macaque frontal cortex. In *Handbook of Neuropsychology*, ed. F Boller, J Grafman, pp. 17–58. New York: Elsevier. Vol. IX

Prinz W. 2002. Experimental approaches to imitation. See Meltzoff & Prinz 2002, pp. 143–62

Pulvermueller F. 2001. Brain reflections of words and their meaning. *Trends Cogn. Sci.* 5:517–24

Pulvermueller F. 2002. *The Neuroscience of Language.* Cambridge, UK: Cambridge Univ. Press. 315 pp.

Rizzolatti G. 2004. The mirror-neuron system and imitation. In *Perspectives on Imitation: From Mirror Neurons to Memes,* ed. S Hurley, N Chater. Cambridge, MA: MIT Press. In press

RizzolattiG, ArbibMA. 1998. Language within our grasp. *Trends Neurosci.* 21:188–94

Rizzolatti G, Fadiga L, Fogassi L, Gallese V. 1996a. Premotor cortex and the recognition of motor actions. *Cogn. Brain Res.* 3:131–41

Rizzolatti G, FadigaL, MatelliM, Bettinardi V, PaulesuE, etal. 1996b. Localization of grasp representation in humans by PET: 1. Observation versus execution. *Exp. Brain Res.* 111:246–52

Rizzolatti G, Fogassi L, Gallese V. 2001. Neurophysiological mechanisms underlying the understanding and imitation of action. *Nat. Rev. Neurosci.* 2:661–70

RizzolattiG, FogassiL, GalleseV.2002.Motor and cognitive functions of the ventral premotor cortex. *Curr. Opin. Neurobiol.* 12:149–54

RizzolattiG, Luppino G. 2001. The cortical motor system. *Neuron* 31: 889–901

Rizzolatti G, Luppino G, Matelli M. 1998. The organization of the cortical motor system: new concepts. *Electroencephalogr. Clin. Neurophysiol.* 106:283–96

Rizzolatti G, Matelli M. 2003. Two different streams form the dorsal visual system. *Exp. Brain Res.* 153:146–57

SalmelinR, HariR. 1994. Spatiotemporal characteristics of sensorimotor neuromagnetic rhythms related to thumb movement. *Neuroscience* 60:537–50

SchubotzRI,vonCramonDY.2001.Functional organization of the lateral premotor cortex: fMRI reveals different regions activated by anticipation of object properties, location and speed. *Brain Res. Cogn. Brain Res.* 11:97–112

Schubotz RI, von Cramon DY. 2002a. A blueprint for target motion: fMRI reveals perceived sequential complexity to modulate premotor cortex. *Neuroimage* 16:920–35

Schubotz RI, von Cramon DY. 2002b. Predicting perceptual events activates corresponding motor schemes in lateral premotor cortex: an fMRI study. *Neuroimage* 15:787–96

SeyalM, MullB, BhullarN, AhmadT, GageB. 1999. Anticipation and execution of a simple reading task enhance corticospinal excitability. *Clin. Neurophysiol.* 110:424–29

Strafella AP, Paus T. 2000. Modulation of cortical excitability during action observation: a transcranial magnetic stimulation study. *NeuroReport* 11:2289–92

Tanné-Gariepy J, Rouiller EM, Boussaoud D. 2002. Parietal inputs to dorsal versus ventral premotor areas in the monkey: evidence for largely segregated visuomotor pathways. *Exp. Brain. Res.* 145:91–103

Thorndyke EL. 1898. Animal intelligence: an experimental study of the associative process in animals. *Psychol. Rev. Monogr.* 2:551–53

Tokimura H, Tokimura Y, Oliviero A, Asakura T, Rothwell JC. 1996. Speech-induced changes in corticospinal excitability. *Ann. Neurol.* 40:628–34

Tomaiuolo F, MacDonald JD, Caramanos Z, PosnerG, ChiavarasM, etal. 1999. Morphology, morphometry and probability mapping of the pars opercularis of the inferior frontal gyrus: an in vivo MRI analysis. *Eur. J. Neurosci.* 11:3033–46

Tomasello M, Call J. 1997. *Primate Cognition.* Oxford, UK: Oxford Univ. Press

Umiltà MA, Kohler E, Gallese V, Fogassi L, Fadiga L, et al. 2001. "I know what you are doing": a neurophysiological study. *Neuron* 32:91–101

Van Hoof JARAM. 1967. The facial displays of the catarrhine monkeys and apes. In *Primate Ethology,* ed. D Morris, pp. 7–68. London: Weidenfield & Nicolson

Visalberghi E, Fragaszy D. 2001. Do monkeys ape? Ten years after. In *Imitation in Animals and Artifacts,* ed. K Dautenhahn, C Nehaniv. Boston, MA: MIT Press

Von Bonin G, Bailey P. 1947. *The Neocortex of Macaca Mulatta.* Urbana: Univ. Ill. Press. 136 pp.

Von Economo C. 1929. *The Cytoarchitectonics of the Human Cerebral Cortex.* London: Oxford Univ. Press. 186 pp.

Vygotsky LS. 1934. *Thought and Language.* Cambridge, MA: MIT Press

Watkins KE, Strafella AP, Paus T. 2003. Seeing and hearing speech excites the motor system involved in speech production. *Neuropsychologia* 41:989–94

Whiten A, Ham R. 1992. On the nature and evolution of imitation in the animal kingdom: reappraisal of a century of research. In *Advances in the Study of Behavior,* ed. PBJ Slater, JS Rosenblatt, C Beer, M Milinski, pp. 239–83. San Diego: Academic

Wohlschlager A, Bekkering H. 2002. Is human imitation based on a mirror-neurone system? Some behavioural evidence. *Exp. Brain Res.* 143:335–41

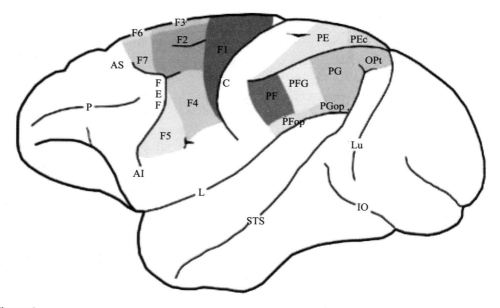

Figure 1

Lateral view of the monkey brain showing, in color, the motor areas of the frontal lobe and the areas of the posterior parietal cortex. For nomenclature and definition of frontal motor areas (F1–F7) and posterior parietal areas (PE, PEc, PF, PFG, PG, PF op, PG op, and Opt) see Rizzolatti et al. (1998). AI, inferior arcuate sulcus; AS, superior arcuate sulcus; C, central sulcus; L, lateral fissure; Lu, lunate sulcus; P, principal sulcus; POs, parieto-occipital sulcus; STS, superior temporal sulcus.

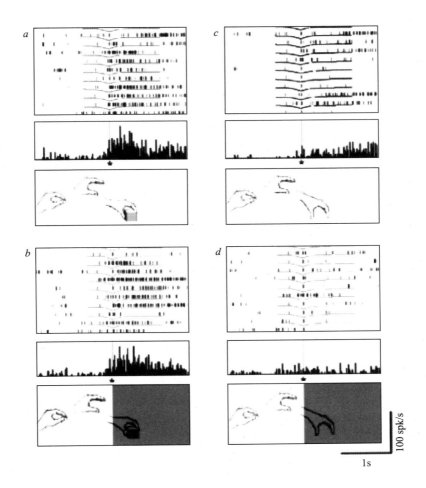

Figure 2

Mirror neuron responses to action observation in full vision (*A* and *C*) and in hidden condition (*B* and *D*). The lower part of each panel illustrates schematically the experimenter's action as observed from the monkey's vantage point. The asterisk indicates the location of a stationary marker attached to the frame. In hid-den conditions the experimenter's hand started to disappear from the monkey's vision when crossing this marker. In each panel above the illustration of the experimenter's hand, raster displays and histograms of ten consecutive trials recorded are shown. Above each raster, the colored line represents the kinematics of the experimenter's hand movements expressed as the distance between the hand of the experimenter and the stationary marker over time. Rasters and histograms are aligned with the moment when the experimenter's hand was closest to the marker. *Green vertical line:* movement onset; *red vertical line*: marker crossing; *blue vertical line:* contact with the object. Histograms bin width = 20 ms. The ordinate is in spike/s (From Umiltà et al. 2001).

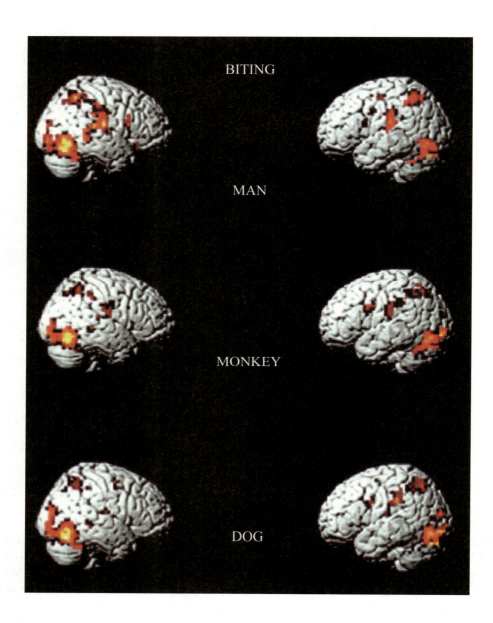

Figure 3
Cortical activations during the observation of biting made by a man, a monkey, and a dog. From Buccino et al. 2004.

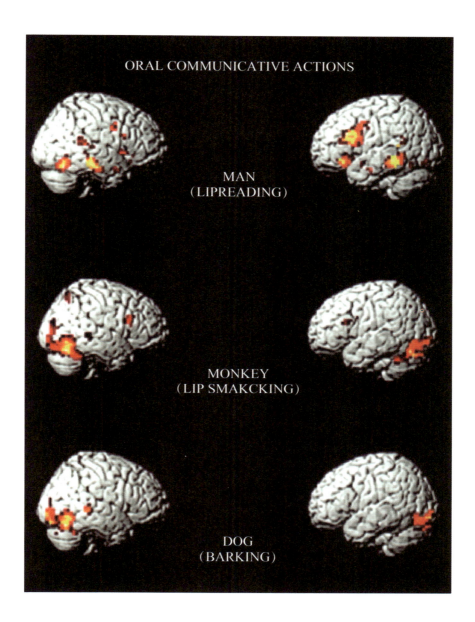

Figure 4

Cortical activations during the observation of communicative actions. For other explanations see text. From Buccino et al. 200

The Role of Organizers in Patterning the Nervous System

Clemens Kiecker and Andrew Lumsden

Medical Research Council (MRC) Center for Developmental Neurobiology, King's College, London SE1 1UL, United Kingdom; email: clemens.kiecker@kcl.ac.uk, andrew.lumsden@kcl.ac.uk

Key Words

morphogen signaling, gradients, Spemann, competence, neural tube, cell lineage restriction boundaries

Abstract

The foundation for the anatomical and functional complexity of the vertebrate central nervous system is laid during embryogenesis. After Spemann's organizer and its derivatives have endowed the neural plate with a coarse pattern along its anteroposterior and mediolateral axes, this basis is progressively refined by the activity of secondary organizers within the neuroepithelium that function by releasing diffusible signaling factors. Dorsoventral patterning is mediated by two organizer regions that extend along the dorsal and ventral midlines of the entire neuraxis, whereas anteroposterior patterning is controlled by several discrete organizers. Here we review how these secondary organizers are established and how they exert their signaling functions. Organizer signals come from a surprisingly limited set of signaling factor families, indicating that the competence of target cells to respond to those signals plays an important part in neural patterning .

INTRODUCTION

Although it remains astounding, even to the experienced neurobiologist, that a structure as complex as the human brain can arise from a single cell, work in different vertebrate model organisms has started to reveal a network of tissue and genetic interactions that

Dorsal blastopore lip: a group of dorsal mesodermal cells of the amphibian embryo where the involution of mesoderm and endoderm starts, marking the onset of gastrulation

engineer this extraordinary feat. During embryogenesis, the primordial neuroepithelium progressively subdivides into distinct regions in a patterning process governed by small groups of cells that regulate cell fate in surrounding tissues by releasing signaling factors. These local signaling centers are called organizers to reflect their ability to confer identity on neighboring tissues in a nonautonomous fashion.

SPEMANN'S ORGANIZER AND EARLY NEURAL PATTERNING

In 1935 Hans Spemann received the Nobel Prize in Medicine for his work with Hilde Mangold showing that transplantation of a small group of cells from the dorsal blastopore lip of a donor embryo to the ventral side of a host embryo is sufficient to induce a secondary body axis (reviewed in De Robertis & Kuroda 2004, Niehrs 2004, Stern 2001). Differently pigmented salamander embryos were used as donors and hosts, allowing for an easy distinction between cells of graft and host origin. Surprisingly, most tissues in the induced second axis were derived from the host, suggesting that the graft had induced surrounding tissue to form axial structures. Thus, Spemann named the dorsal blastopore lip the organizer, and tissues with comparable inductive activity have since been identified in all vertebrate model organisms and more recently also in some nonvertebrates (Darras et al. 2011, Meinhardt 2006, Nakamoto et al. 2011). Nowadays the term organizer is used more widely to describe groups of cells that can determine the fate of neighboring cell populations by emitting molecular signals.

AP: anteroposterior

DV: dorsoventral

Bone morphogenetic protein (BMP): subfamily of the transforming growth factor β superfamily of secreted signaling factors; initially identified by their promotion of bone and cartilage formation

Fibroblast growth factor (FGF): secreted signaling molecules that signal via tyrosine kinase receptors

Wnts: secreted lipid-modified glycoproteins that regulate multiple aspects of embryogenesis and adult homeostasis by activating several different signaling pathways

AME: axial mesendoderm

The ectopic twin induced in Spemann's experiment contained a complete CNS that was properly patterned along its anteroposterior (AP, head-to-tail) and dorsoventral (DV, back-to-belly) axes, indicating that the organizer harbors both neural-inducing and neural-patterning activities. More recently, a large number of factors that are expressed in Spemann's organizer have been identified, and several were found to be secreted inhibitors of bone morphogenetic proteins (BMPs). In combination with other findings in frog and fish embryos, this led to a model whereby Spemann's organizer induces the neural plate in the dorsal ectoderm by inhibiting BMPs, whereas the ventral ectoderm forms epidermis because it remains exposed to BMPs (De Robertis & Kuroda 2004, Muñoz-Sanjuán & Brivanlou 2002). Experiments in chick embryos have since added complexity to this default model for neural induction by implicating other signaling proteins such as fibroblast growth factors (FGFs) and Wnts as additional neural inducers (Stern 2006).

During gastrulation, the organizer region stretches out and gives rise to the axial mesendoderm (AME), which comes to underlie the midline of the neural plate along its AP axis. Otto Mangold found that different AP regions of the AME induced different parts of the embryonic axis when grafted into host embryos, leading to the idea of regionally specific inductions by the organizer (Niehrs 2004). This model was challenged in the 1950s when Nieuwkoop and others proposed that the CNS is patterned by a gradient of a transformer that travels within the plane of the neural plate and induces different neural fates in a dose-dependent manner such that forebrain, midbrain, hindbrain, and spinal cord form at increasing levels of this transformer (Stern 2001). FGFs (Mason 2007), retinoic acid (Maden 2007), and Wnts all posteriorize neuroectoderm dose-dependently, but Wnts appear to be the best candidates to fulfill this role in a manner consistent with Nieuwkoop's model (Kiecker & Niehrs 2001a). Spemann's organizer also secretes inhibitors of Wnts, in addition to BMP antagonists, and these factors remain expressed in the anterior AME but are absent from the posterior AME during gastrulation (Kiecker & Niehrs 2001b). Thus, the Spemann-Mangold model of regionally specific inductions and Nieuwkoop's gradientbased model turn out to be two sides of the same coin: The anterior AME induces the forebrain by acting as a sink for posteriorizing Wnts.

DORSOVENTRAL PATTERNING

The ectoderm that surrounds the neural plate expresses BMPs while Spemann's organizer and the extending AME express BMP inhibitors. Experiments in zebrafish embryos have suggested that this generates a gradient of BMP activity that defines mediolateral positions within the neural plate (Barth et al. 1999). Hence, a Cartesian coordinate system of two orthogonal gradients is established—a Wnt gradient along the AP axis and a BMP gradient along the mediolateral axis—and the AME defines the origin of this system by secreting BMP and Wnt antagonists (**Figure 1a**) (Meinhardt 2006, Niehrs 2010). It is clear that such global mechanisms can establish only a crude initial pattern, and we argue below that this pattern is increasingly refined through the establishment of local (or secondary) organizers in the neuroepithelium.

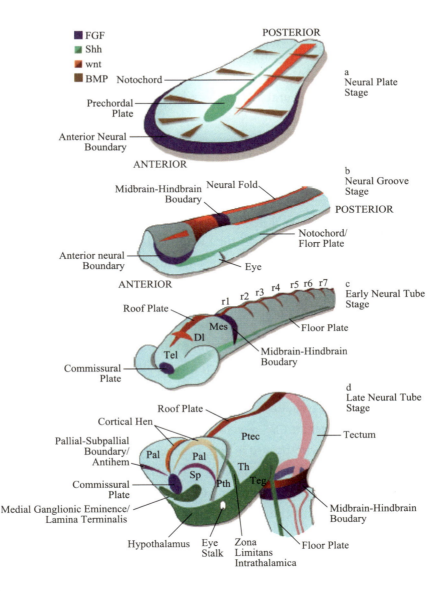

Figure 1

Main stages of neural development in a schematized amniote embryo. (*a*) Neural plate stage. Mediolateral gradients of BMP (*brown wedges*) and Shh (*green*) activity together with an anteroposterior gradient of Wnt activity (*red wedge*) establish a quasi-Cartesian coordinate system of positional information across the neural plate. (*b*) Neural groove stage. The interplay between Wnts and ANB-derived Wnt inhibitors patterns the area between the presumptive forebrain and midbrain (red wedge). (*c*) Early neural tube stage. (*d*) Late neural tube stage. Colors: FGF expression (*blue*), Shh expression (*green*), BMP expression (*brown*), Wnt expression (*orange/red*). Abbreviations: di, diencephalon; mes, mesencephalon; pal, pallium; ptec, pretectum; pth, prethalamus; r, rhombomere; sp, subpallium; teg, tegmentum; tel, telencephalon; th, thalamus.

Gastrulation is followed by neurulation during which the lateral folds of the neural plate roll up and fuse to form the neural tube (**Figure 1***b***,***c*). Thus, the initial mediolateral pattern is transposed into DV polarity: Cells that are medial in the plate end up ventral in the tube, whereas lateral neural plate cells end up dorsal.

The Notochord

During gastrulation and neurulation a large portion of the AME narrows to a thin rod, the notochord, which underlies the midline of the neural plate and later most of the neural tube (**Figure 1***a*–*c*). The prechordal plate, the anterior end of the AME that lies beneath the prospective anterior forebrain, is a bit wider than the notochord. The ventralmost cells of the neural tube that reside directly above the notochord form the floor plate on either side of which form motor neurons and various types of ventral interneurons.

Motor neurons: efferent neurons that control muscle activity

Interneurons: neurons that connect afferent and efferent neurons in multisynaptic pathways

Sbb: a secreted signaling factor; its active form is lipid modified and proteolytically processed

Although not a neural structure, the notochord was one of the first tissues shown to act as a local organizer of CNS development. Microsurgical experiments in chick embryos revealed that the notochord is both necessary and sufficient to induce ventral neural identity: Sections of the spinal cord from which the notochord had been removed developed without a floor plate or motor neurons; conversely, the transplantation of pieces of notochord beneath the lateral neural plate resulted in the induction of an ectopic floor plate and motor neurons above the graft (Placzek et al. 1990, van Straaten et al. 1985).

A breakthrough in understanding this action of the notochord came with the finding that *Sonic hedgehog* (*Shh*), a vertebrate ortholog of the *Drosophila* segment polarity gene *hedgehog*, is expressed in the notochord and a bit later also in the floor plate. *Shh* encodes a secreted signaling factor and is thus a prime candidate for the inductive signal released by the notochord. Overexpression of *Sbb* in mouse, zebrafish, and frog, or coculture of rat neuroectoderm with *Shh*-expressing cells result in ectopic floor plate, motor neuron, and ventral interneuron induction, indicating that Sbb mimics the effect of notochord grafts (Echelard et al. 1993, Krauss et al. 1993, Roelink et al. 1994). Conversely, mice carrying a mutation in the *Shh* gene fail to form a floor plate and lack multiple ventral neural cell types (Chiang et al. 1996). Taken together, these data strongly suggest that Shh mediates the organizer function of the notochord.

The Floor Plate

The floor plate is a strip of wedge-shaped glial cells along the ventral midline of the neural tube. Like the notochord, the floor plate expresses *Shh* and is therefore likely to contribute to the induction of ventral cell identities. Fate-mapping studies in the chick embryo in combination with detailed

Morphogen: factor that is released locally, forms a concentration gradient within a tissue, and induces different cell fates dose-dependently

examinations of cellular morphologies and marker gene expression have revealed that the floor plate consists of different cell populations along both its AP and mediolateral axes (Placzek & Briscoe 2005). For example, whereas the floor plate is devoid of neural progenitors in the spinal cord and hindbrain, dopaminergic neurons are generated in the floor plate of the midbrain (Ono et al. 2007, Puelles et al. 2004).

The origin of the floor plate remains controversial. The ablation and grafting experiments described above strongly suggest that it is induced by Shh signaling from the notochord; however, others have argued that the floor plate and notochord originate from a common precursor population in Spemann's organize and that floor plate cells are inserted into the midline of the neural plate as the AP axis of the embryo extends (Le Douarin & Halpern 2000). These views are likely not entirely incompatible: The common lineage of notochord and floor plate may endow them with shared properties, and both tissues are capable of inducing homeogenetic responses in neuroepithelium.

Ventral Neural Patterning by Shh: A Paradigm for Morphogen Signaling

The specification of multiple cell types in the ventral neural tube is arguably one of the most thoroughly studied examples of neural patterning. Considerable evidence gathered over the past 15 years indicates that Shh functions as a true morphogen in this process; i.e., it is released from a local source (notochord, floor plate) and forms a concentration gradient that specifies different cell fates in a dose-dependent fashion (Dessaud et al. 2008, Lupo et al. 2006). In mouse embryos that were genetically engineered to express fluorescently labeled Shh protein (Shh-GFP) from the *Shh* locus, a declining ventral-to-dorsal gradient of fluorescence is detectable within the ventral neural tube (Chamberlain et al. 2008).

The morphogen model is intuitively appealing because it explains a complex process pattern formation, drawing on a simple chemical activity—the diffusion of a single substance from a localized source. However, trying to understand the cellular mechanism of morphogen signaling raises a number of difficult issues. For example, how are different concentrations of a morphogen translated into distinct cell fates? In vertebrates, Shh activates the transcriptional activators Gli1 and Gli2 and antagonizes the repressor Gli3. Thus, the extracellular gradient of Shh is translated into opposing gradients of intracellular Gli1/2 and Gli3 activity along the DV axis of the neural tube (Fuccillo et al. 2006, Lei et al. 2004, Stamataki et al. 2005). These overlapping activities regulate two classes of transcriptional control genes: Class I genes such as *Pax*6, *Pax*7, and *Irx*3 are repressed, and class II genes, including *Foxa*2, *Nkx*2.2, *Olig*2, *Nkx*6.1, *Dbx*1, and *Dbx*2, are induced by Shh-Gli signaling. Different thresholds of Shh signaling are required for the repression or activation of individual class I and class II genes, resulting in a nested expression pattern of these genes along the DV axis of the spinal cord. Furthermore, several class I and class II genes cross-repress each other, resulting in a sharpening of the boundaries between their expression domains (Briscoe et al. 2000). Ultimately, the combinatorial expression of class I and class II genes at a specific DV location determines which type of neural progenitor will form.

Dessaud et al. (2008) recently suggested that the duration of exposure, in addition to the extracellular concentration of Shh, determines the fate of the receiving cell. This model is likely to reflect the gradual buildup of the Shh gradient in vivo better than would a static gradient model.

Ventral Patterning in the Hindbrain and Midbrain

Early in its development, the hindbrain becomes subdivided into a series of seven to eight segments called rhombomeres (r1–8) (reviewed in Kiecker & Lumsden 2005). Nevertheless, the topological organization of neurons along the DV axis of the hindbrain is similar to that of the spinal cord. The induction of hindbrain motor neurons (which contribute to the IVth–XIIth cranial nerves) also depends on Shh signaling from the notochord/floor plate and on the nested expression of various class I and class II genes (Osumi et al. 1997, Pattyn et al. 2003, Takahashi & Osumi 2002).

The ventral part of the midbrain (and of the posterior forebrain) gives rise to the tegmentum where neural progenitors are organized in a series of morphologically visible arcs that are characterized by periodic gene expression patterns (Sanders et al. 2002). Gain- and loss-of-function experiments in chick embryos have provided evidence that this pattern is also controlled by Shh signaling (Agarwala et al. 2001, Bayly et al. 2007).

Organizers of Ventral Forebrain Development

Like the notochord, the prechordal plate expresses *Shh*, and grafts of this AME tissue to the lateral forebrain result in ectopic expression of the ventral forebrain marker *Nkx*2.1 in chick embryos. However, this effect cannot be mimicked by implanting Shh-producing cells, suggesting that the inductive capacity of the prechordal plate reaches beyond mere secretion of Shh (Pera & Kessel 1997). A good candidate factor to mediate this difference is BMP7, which is expressed in the prechordal mesoderm but not in the posterior notochord and synergizes with Shh in inducing ventral forebrain identity (Dale et al. 1997). Furthermore, potent Wnt inhibitors are expressed in the prechordal plate, and these likely contribute to forebrain induction and ventralization (Kiecker & Niehrs 2001b).

The vertebrate forebrain becomes divided into the telencephalon anteriorly and the diencephalon posteriorly (**Figure 1c**). The dorsal part of the telencephalon, the pallium, gives rise to the hippocampus and the cerebral cortex (or functionally equivalent structures in nonmammalian species), whereas the ventral part, the subpallium, gives rise to the basal ganglia. The earliest distinction between pallial and subpallial identity is established at the neural plate stage and is mediated by Shh from the prechordal plate (Gunhaga et al. 2000). However, prechordal plate-derived Shh also induces a new domain of *Shh* expression in the most ventral part of the subpallium, the presumptive medial ganglionic eminence (MGE), in a manner that is very similar to the induction of the floor plate by Shh from the notochord. This telencephalic *Shh* domain is likely to act as a secondary organizer that refines the DV pattern of the telencephalon (**Figure 1d**) (Sousa & Fishell 2010).

The Roof Plate

Another specialized cell population, the roof plate, forms along the dorsal midline of the entire neural tube. Similar to the floor plate, the roof plate is induced by inductive signals from a nonneural tissue—in this case by BMPs and Wnts expressed in the ectoderm flanking the neural plate (Chizhikov & Millen 2004b). Neural tube closure brings the roof plate progenitors from either side of the neural plate together to form the dorsal midline of the neural tube. Identifying a specific genetic pathway for roof plate formation has been complicated by the fact that two other cell groups, the neural crest and some dorsal interneuron progenitors, also arise from the dorsal midline.

However, there is good evidence that the roof plate functions as an organizer of dorsal patterning in the spinal cord. Six populations of dorsal interneurons (dI1–6) are generated in the dorsal half of the spinal cord, and the coculture of roof plate and naive neural plate tissue leads to the induction of at least two of those (Liem et al. 1997). Genetic ablation of the roof plate in mice results in the lack of dI1–3 and expansion of dI4–6 interneurons, confirming that the roof plate is not only sufficient but also required for dorsal spinal cord patterning (K.J. Lee et al. 2000). *Lmx1a* mutant mice in which roof plate formation is disrupted show a similar, if somewhat milder phenotype (Millonig et al. 2000). Overexpression of *Lmx1a*—which encodes a LIM homeodomain transcription factor—in the chick spinal cord results in ectopic induction of dI1 at the expense of dI2–6 interneurons in the vicinity of the electroporated cells, suggesting that a diffusible signal is secondarily induced (Chizhikov& Millen 2004a).

Identification of the signal that mediates the organizer function of the roof plate has been more problematic than identifying that of the floor plate. The roof plate expresses several Wnts and a large number of BMPs, and the genetic inactivation of individual factors often results in entirely normal spinal cord development, probably owing to compensation by other family members. The BMP-type factor *GDF*7 is an exception because Gdf7$^{-/-}$ mutant mice lack dI1 interneurons (Lee et al. 1998). Zebrafish mutants with defects in the BMP pathway fail to form Rohon-Beard neurons, indicating that BMPs are required for the formation of at least some dorsal cell types (Nguyen et al. 2000). *Wnt1/Wnt3a* double mutant mice show a severe reduction of dI1–3 and a concomitant expansion of dI4 and dI5 interneurons and are therefore phenotypically similar to roof plate–ablated mice, although less severely so (Muroyama et al. 2002). However, the reduction of dI1–3 interneurons could be due to defective proliferation rather than patterning because Wnts act as mitogens in gain-of-function experiments in chick (Megason & McMahon 2002).

Boundary: an interface between two adjacent tissues that prevents intermingling between cells from either side of the boundary

Do roof plate signals act in a dose-dependent fashion, similar to Shh on the ventral side of the neural tube? Some evidence points toward graded effects by both BMPs and Wnts in the dorsal spinal cord (Liem et al. 1997, Megason & McMahon 2002, Timmer et al. 2002); however, the picture is far less conclusive than for Shh and it remains possible that qualitative as well as quantitative mechanisms are at work (i.e., individual BMPs or Wnts may specifically induce certain subpopulations of dorsal interneurons).

Dorsal Patterning in the Anterior Hindbrain

The dorsal part of the anterior hindbrain (r1) undergoes a series of complex morphological changes that result in the formation of the cerebellum. Granule cells, the most prevailing cell type in the cerebellum, are generated from the rhombic lip, a germinal zone at the interface between the roof plate and the r1 neuroepithelium (Hatten & Roussel 2011, Wingate 2001). The roof plate of r1, like that of the spinal cord, expresses several BMPs, and these are sufficient to initiate granule cell formation when added ectopically to r1 neuroepithelium (Alder et al. 1999). The locus coeruleus, the major nora-drenergic nucleus of the brain, is also induced in the dorsal half of r1 before its neurons migrate ventrally to reach their final destination in the lateral floor of the IVth ventricle. Application of BMP antagonists to the anterior hindbrain of chick embryos results in the disappearance of or a dorsal shift of locus coeruleus neurons, suggesting that the role of BMPs in the induction of these neurons may be dose dependent (Vogel-Höpker & Rohrer 2002).

The Cortical Hem

In the telencephalon, the roof plate sinks between the two cortical hemispheres and gives rise to the monolayered choroid plexus medially and to the cortical hem laterally. Immediately adjacent to the hem, which expresses BMPs and Wnts, the hippocampus is induced and the cerebral cortex forms next to that. Genetic ablation of the telencephalic roof plate in mice results in a severe undergrowth of the cortical primordium, suggesting a nonautonomous effect of the hem on cortical specification (Monuki et al. 2001).

Ectopic application of BMPs to the developing chick telencephalon leads to holoprosencephaly (a failure to separate the cerebral hemispheres), but this is likely to be a result of increased cell death rather than a change in patterning (Golden et al. 1999). A requirement for BMP signaling in telencephalic patterning has been tested by genetically disrupting a BMP receptor gene in the mouse forebrain. These mice fail to differentiate the choroid plexus, but all other telencephalic subdivisions develop normally, arguing against a role for BMPs as an organizer signal (Hé bert et al. 2002).

Wnt signaling specifies dorsal identity at the earliest stages of telencephalic development (Backman et al. 2005, Gunhaga et al. 2003), and the Wnt antagonists secreted by the prechordal plate likely help to set up early DV polarity by protecting the ventral forebrain from the dorsalizing activity of Wnts. At later stages, Wnt signaling is required for hippocampus differentiation (Galceran et al. 2000, S.M. Lee et al. 2000, Machon et al. 2003). The boundary between the pallium and the subpallium (PSB) begins to express Wnt inhibitors, raising the possibility that a gradient of Wnt activity is established across the pallium between the hem and the PSB (which has also been called the antihem) (Assimacopoulos et al. 2003, Frowein et al. 2002). However, there is little evidence for a later patterning function for Wnts beyond hippocampus induction (Chenn & Walsh 2002, Hirabayashi et al. 2004, Hirsch et al. 2007, Ivaniutsin et al. 2009, Machon et al. 2007,cerebellar structures in the posterior forebrain (Marín & Puelles 1994). Demonstrating a requirement for the MHB in the patterning of the surrounding tissues by microsurgery Muzio et al. 2005). The PSB also expresses several members of the epidermal growth factor family, transforming growth factor α and FGF7, but their roles in telencephalic development remain unknown (Assimacopoulos et al. 2003).

In summary, two major signaling centers organize DV patterning of the neural tube: the roof plate dorsally, which secretes BMPs and Wnts, and the floor plate ventrally, which secretes Shh. Both are induced by the same sets of molecular signals from the epidermis and the notochord, respectively. The notochord itself is an organizer that can mediate ventral neural patterning. In the telencephalon, an additional potential organizer, the PSB, is located midway along the DV axis.

ANTEROPOSTERIOR PATTERNING

In contrast with the DV axis, which is patterned by two signaling centers at opposite poles of the neural tube, the AP axis is patterned by several discrete local organizers.

The Midbrain-Hindbrain Boundary

The boundary between the midbrain and hindbrain (MHB) is characterized by a morphological constriction of the neural tube and is therefore also called the isthmus

MHB: midbrain-hindbrain boundary

(**Figure 1c,d**). The first experimental evidence that the MHB functions as an organizer came from microsurgical studies conducted in chick embryos: Transplantation of anterior hindbrain or posterior midbrain tissue into the posterior forebrain of a host embryo induced the formation of an ectopic isthmus and ectopic midbrain tissue around the graft (Bloch-Gallego et al. 1996, Gardner & Barald 1991, Martinez et al. 1991). Experimental rotation of the entire midbrain vesicle formed a double-posterior midbrain and induced ectopic midbrain and turned out to be less feasible because the isthmic organizer rapidly regenerates after surgical removal (Irving & Mason 1999).

Two signaling factors, Wnt1 and FGF8, are expressed on the anterior and posterior sides of the MHB, respectively. Both factors are required for midbrain-hindbrain development as *Wnt1* mutations in the mouse, and mutations in *Fgf8* in both mouse and zebrafish embryos result in defects in midbrain patterning and cerebellum formation (McMahon et al. 1992, Meyers et al. 1998, Picker et al. 1999, Reifers et al. 1998, Thomas et al. 1991). However, in gain-of-function experiments only FGF8 can mimic MHB organizer function (Crossley et al. 1996, Irving & Mason 2000, Lee et al. 1997, Martinez et al. 1999), whereas Wnt1 seems to promote cell growth and proliferation without affecting patterning (Panhuysen et al. 2004). Thus, FGF8 is the main organizer factor secreted from the MHB.

What determines the AP position of MHB formation? Many studies have uncovered a genetic module, including FGF8, Wnt1, and several transcription factors that stabilize MHB gene expression in a network of positive maintenance loops and mutually repressive interactions (Liu & Joyner 2001, Wurst & Bally-Cuif 2001). In particular, the homeodomain transcription factors Otx2 and *Gbx2* play a central role in MHB positioning. Otx2 is expressed in the forebrain and midbrain and Gbx2 in the anterior hindbrain, and the interface between the two expression domains presages the position of the MHB from neural plate stages onward. Various gene targeting experiments in the mouse have demonstrated that an experimental shift of this interface always results in a concomitant repositioning of the MHB (Acampora et al. 1997, Broccoli et al. 1999, Millet 1999, Wassarman et al. 1997). Knockdown and cell transplantation experiments in the fish have revealed that the *otx/gbx* interface is regulated by Wnt8 at the neural plate stage (Rhinn et al. 2005). Thus, the position of the MHB is directly defined by the early gradient of Wnt signaling that establishes the initial AP polarity of the neural plate. Similar to the mutual repression of class I and class II genes in the spinal cord, *Otx2* and *Gbx2* repress each other, thereby stabilizing the binary cell fate choice around the MHB (Liu & Joyner 2001, Wurst & Bally-Cuif 2001).

The Anterior Neural Boundary/Commissural Plate

Elegant cell ablation and transplantation experiments in zebrafish revealed an organizing function of the anterior border of the neural plate (ANB) (Houart et al. 1998). A Wnt inhibitor of the secreted Frizzled-related protein family is expressed in the ANB, and overexpression and depletion of this factor phenocopy the effects of ANB transplantation

ANB: anterior neural border
ZLI: zona limitans intrathalamica

and removal, respectively (Houart et al. 2002). *Wnt8B* is expressed in the presumptive midbrain and posterior forebrain at the stages when ANB signaling is required and is therefore likely to be the main antagonist of the ANB. Thus, after the global gradient of Wnt activity has established general AP polarity in the neural plate, Wnts regulate AP identity in a more localized fashion in the prospective forebrain-midbrain region (**Figure 1***b*).

In the mouse, a role for Wnt inhibition from the ANB has yet to be demonstrated; however, FGF8, which is also expressed there, can mimic the anteriorizing effects of ANB in explants in vitro (Shimamura & Rubenstein 1997). Studies in both mouse and zebrafish embryos have demonstrated a need for FGFs in forebrain patterning (Meyers et al. 1998, Walshe & Mason 2003).

After neural tube closure, the ANB becomes a patch of cells at the anterior end of the neural tube that will eventually form the commissural plate (CP), a scaffold for the formation of forebrain commissures at later stages. The CP continues to express FGF8, and impressive in utero electroporation experiments in the mouse have revealed that FGF8 promotes anterior at the expense of posterior cortical fates. An ectopic source of FGF8 at the posterior end of the cortex resulted in a partial mirrored duplication of anterior cortical areas (Fukuchi-Shimogori & Grove 2001). These experiments identified the CP as an organizer of cortical patterning via its secretion of FGF8. Toyoda et al. (2010) recently showed that FGF8 acts directly and at a long range during this process, i.e., as a true morphogen.

The Zona Limitans Intrathalamica

The zona limitans intrathalamica (ZLI) is a narrow stripe of *Shh*-expressing cells in the alar plate of the diencephalon, transecting the neuraxis between the presumptive prethalamus and the thalamus (Kitamura et al. 1997, Shimamura et al. 1995, Zeltser et al. 2001). Gain-and loss-of-function experiments in chick, zebrafish, and mouse embryos have revealed that the ZLI acts as an organizer of diencephalic development by secreting Shh (Kiecker & Lumsden 2004, Scholpp et al. 2006, Vieira et al. 2005, Vue et al. 2009). At least in the thalamus, the activity of Shh appears to be dose dependent, with higher levels of signaling inducing the gamma-aminobutyric acid (GABA)-ergic rostral thalamus and lower levels inducing the glutamatergic caudal thalamus (Hashimoto-Torii et al. 2003, Vue et al. 2009; but see Jeong et al. 2011).

Fgf8 is expressed in a small patch in the dorsal ZLI and contributes to regulating the fate decision between the rostral and caudal thalamus (Kataoka & Shimogori 2008). Furthermore, a plethora of *Wnts* shows sharp borders of expression at the ZLI, suggesting that Wnt signaling may also be involved in regulating the regionalization and/or proliferation of the diencephalic primordium (Bluske et al. 2009, Quinlan et al. 2009).

The apposition of any neural tissue anterior to the ZLI with any tissue between the ZLI and the MHB results in induction of *Shh*, indicating that planar interactions are sufficient to induce the ZLI organizer (Guinazu et al. 2007, Vieira et al. 2005). The ZLI forms at the interface between the expression domains of two classes of transcription factors: zinc finger proteins of the Fez family anteriorly and homeodomain proteins of the Irx family posteriorly (Hirata et al. 2006, Kobayashi et al. 2002, Rodríguez-Seguel et al. 2009, Scholpp et al. 2007). Expression of *Irx3* in chick is induced by Wnt signaling, suggesting that, as for *Gbx2* at the MHB, the early Wnt signal that posteriorizes the neural plate directly positions the ZLI (Braun et al. 2003).

Rhombomere Boundaries

Segmentation of the hindbrain into rhombomeres is controlled by graded retinoic acid signaling and by the reiterated and nested expression of tyrosine kinases and transcription factors, many of which are vertebrate orthologs of *Drosophila* gap and *Hox* genes (Kiecker & Lumsden 2005, Maden 2007). In zebrafish, the boundaries between rhombomeres express

> **Neurogenesis:** the process by which proliferating neural progenitors exit the cell cycle and differentiate into functional neurons

several Wnts (**Figure 1c**), and the knockdown of these factors results in disorganized neurogenesis adjacent to the boundaries, suggesting that they may function as organizers, although no patterning defects have been demonstrated within the rhombomeres of such embryos (Amoyel et al. 2005, Riley et al. 2004).

COMMON FEATURES OF NEUROEPITHELIAL ORGANIZERS

As discussed above, organizers influence cell fate in surrounding tissues by secreting diffusible signaling factors that often act in a morphogen-like fashion—that is, they induce different responses in receiving cells at different distances from the source. In addition to this defining feature of organizers, several other commonalities have been observed regarding the establishment, maintenance, and signaling properties of neuroepithelial organizers.

Organizers Form Along Cell Lineage Restriction Boundaries

One of the hallmarks of hindbrain segmentation is the formation of cell lineage-restricted boundaries that prevent cells from moving between adjacent rhombomeres (Fraser et al. 1990, Jimenez-Guri et al. 2010). This finding prompted a search for boundaries in other parts of the neural tube, which led to the discovery that such boundaries often coincide with organizers (Kiecker & Lumsden 2005): Cell lineage restriction at the MHB has been demonstrated by sophisticated time-lapse imaging in zebrafish embryos and genetic fate mapping in the mouse (Langenberg & Brand 2005, Zervas et al. 2004); the ZLI is flanked by boundaries on either side (Zeltser et al. 2001); and signaling functions have been reported for rhombomere boundaries (see above). Lineage restriction has not been tested at the ANB, but it seems unlikely that cells intermingle freely across the neuralepidermal border. All three DV organizers—floor plate, PSB, and roof plate—also show some degree of lineage restriction (Awatramani et al. 2003, Fishell et al. 1993, Fraser et al. 1990, Jimenez-Guri et al. 2010).

The molecular mechanisms underlying boundary formation in the neural tube are not well understood. However, specific signaling pathways have been implicated: Ephephrin signaling is essential for segmentation in the hindbrain (Cooke et al. 2005, Kemp et al. 2009, Xu et al. 1999), and the Notch pathway appears to be involved in boundary formation at the ZLI, at rhombomere boundaries, and at the MHB (Cheng et al. 2004, Tossell et al. 2011, Zeltser et al. 2001).

Cell lineage restriction at organizers probably serves a dual function. First, boundaries tend to minimize contact between flanking cell populations, which may help to keep organizers in a straight line, facilitating the generation of a consistent diffusion gradient. Second, cells on either side of the organizer are kept in separate immiscible pools, thereby stabilizing a pattern after it has been induced.

Positive Feedback Maintains Organizers

Another feature shared by several neuroepithelial organizers is that their maintenance depends on the signal they produce. Both the floor plate and the roof plate are induced by their own signals, BMP and Shh; the ZLI depends on ongoing Shh signaling (Kiecker & Lumsden 2004, Zeltser 2005); and MHB integrity depends on FGF signaling (Sunmonu et al. 2011, Trokovic et al. 2005). Alan Turing's classical model for pattern formation postulated a chemical network of local self-enhancement and long-range inhibition, and the autoinduction of neuroepithelial organizers fits the local self-enhancement component of this model rather well (Meinhardt 2009).

Intrinsic Factors Regulate Differential Responses to Organizer Signals

FGF signaling from the MHB establishes the tectum anteriorly and the cerebellum posteriorly, and Shh from the ZLI induces prethalamic gene expression anteriorly and patterns the thalamus posteriorly. How can one signal induce such asymmetric responses on either side of an organizer? Two orthologs of the *Drosophila* competence factor *iroquois*, *Irx2* and *Irx3*, are expressed posterior to the MHB and ZLI, respectively. Ectopic expression of these factors anterior to the organizer results in a conversion of tectum into cerebellum and of prethalamus into thalamus (Kiecker & Lumsden 2004, Matsumoto et al. 2004). These effects are dependent on the organizer signals FGF and Shh, suggesting that Irx2 and Irx3 are not patterning factors themselves but that they convey a prepattern that determines the competence of different subdivisions of the neural tube to respond to secreted signals.

FGF-soaked beads induce ectopic midbrain and hindbrain structures to form from posterior forebrain tissue, but in the anterior forebrain FGFs anteriorize the pallium. Similarly, Shh from the ventral midline induces the hypothalamus marker *Nkx2.1* anteriorly, whereas it induces *Nkx6.1* posteriorly. The limit between these two regions of differential competence to respond to FGFs coincides with the ZLI, and the homeobox genes *Irx3* and *Six3* were shown to mediate posterior versus anterior competence (Kobayashi et al. 2002).

Taken together, intrinsic factors establish a prepattern in the developing CNS that regulates the cellular response to organizer signals. These factors are often induced by the earliest signals that pattern the neural plate—for example, *Irx3* is induced and *Six3* is repressed by posteriorizing Wnt signaling (Braun et al. 2003)—thereby linking early and late stages of neural patterning. This does not mean that organizer signals are merely permissive triggers that determine the timing and extent of regional specialization, the identity of which is prepatterned; they are also responsible for evoking different responses within the same field (as exemplified by the induction of GABAergic versus glutamatergic neurons by different doses of Shh within the *Irx3*-positive thalamus).

Hes Genes Prevent Neurogenesis in Organizer Regions

Organizers typically coincide with boundaries that are characterized by slower proliferation and a delay or absence of neurogenesis (Guthrie et al. 1991, Lumsden & Keynes 1989). Transcription factors of the Hes family that mediate Notch signaling are required to inhibit neurogenesis at the MHB of zebrafish and frog embryos (Geling et al. 2004, Ninkovic et al. 2005, Takada et al. 2005). All neural progenitors express *Hes* genes, but they usually become downregulated when cells undergo neurogenesis. An analysis in mouse embryos has revealed that in boundary regions *Hes* genes remain expressed and that it is this strong persistent expression that sets boundaries apart and allows organizer regions to form (Baek et al. 2006).

NEUROEPITHELIAL ORGANIZERS ALSO REGULATE PROLIFERATION, NEUROGENESIS, AND AXON GUIDANCE

Many organizer signals also function as mitogens, suggesting that growth, in addition to patterning, is modulated by organizers. For example, *Shh* mutant mice show not only patterning defects, but also a structural lack of many ventral neural tissues (Chiang et al. 1996). Wnt1 promotes growth in the MHB region (Panhuysen et al. 2004), and Wnts from the roof plate and cortical hem are known to regulate proliferation of the spinal cord and pallium (Chenn & Walsh 2002, Ivaniutsin et al. 2009, Megason & McMahon 2002, Muzio et al. 2005). FGFs from the MHB promote growth of the midbrain and cerebellum (Partanen 2007), but they also serve as survival factors in the midbrain (Basson et al. 2008). Similarly, FGFs secreted from the CP prevent apoptosis and promote growth in the telencephalon (Paek et al. 2009, Thomson et al. 2009). Contrary to the proliferative effects of FGFs, Shh, and Wnts, the BMP pathway often induces apoptosis when ectopically activated (Anderson et al. 2002, Lim et al. 2005, Liu et al. 2004).

To complicate the picture even further, some organizer factors promote cell cycle exit and neurogenesis (Fischer et al. 2011, Hirabayashi et al. 2004, Machon et al. 2007, Munji et al. 2011, Xie et al. 2011). These seemingly contradictory effects of the same classes of signals may be explained by temporal changes in the competence of the target cells (Hirsch et al. 2007); however, in some cases different members of the same protein family exert opposing effects on the balance between proliferation and differentiation (Borello et al. 2008, David 2010). Thus, by releasing growth-promoting and growth-inhibiting cues from localized sources, organizers help to mold the increasingly complex shape of the neural tube and coordinate the temporal progression of neurogenesis in defined subdivisions of the neural tube (Scholpp et al. 2009).

Once their regional identity has been established, differentiated neurons need to wire up precisely to form functional networks. Organizers also play a role at this stage of CNS formation, for example, by expressing axon guidance factors such as the chemoattractant netrin, which is secreted by the floor plate to guide commissural axons (Dickson 2002, Tessier-Lavigne & Goodman 1996). More recently, many of the classical morphogens that are secreted by organizers have been found to double as axon guidance molecules at later stages (Charron & Tessier-Lavigne 2005, Osterfield et al. 2003, Sánchez-Camancho et al. 2005, Zou & Lyuksyutova 2007). Shh from the floor plate cooperates with netrin in attracting commissural axons, whereas BMPs from the roof plate repel them (Augsburger et al. 1999, Charron et al. 2003). After these axons have crossed the midline, Shh signaling repels them via Hedgehog-interacting protein (Bourikas et al. 2005). Wnts are expressed in an AP gradient in the floor plate and guide the same axons anteriorly after they have crossed the midline (Lyuksyutova et al. 2003), whereas corticospinal axons are directed posteriorly by a repulsive interaction between Wnts and the atypical Wnt receptor Ryk (Liu et al. 2005). Thus, organizers influence CNS formation not only at early patterning stages, but also at later stages when functional circuits are established.

Commissural axons: nerve fibers that cross the midline of the nervous system

NEUROEPITHELIAL ORGANIZERS IN EVOLUTION

Vertebrates possess the most complex of all brains; even their closest relatives, the tunicates and hemichordates, have relatively simple nervous systems (Meinertzhagen et al. 2004). Sets of transcription factors that mark AP and DV subdivisions of the neural tube are conserved far beyond the chordate phylum (Irimia et al. 2010, Lowe et al. 2003, Reichert 2005, Tomer et al. 2010, Urbach & Technau 2008), and several orthologs of AP marker genes are even found along the head-to-foot axis of the coelenterate *Hydra,* suggesting an ancient origin of the genetic modules that regulate neural patterning (Technau & Steele 2011). By contrast, local organizers appear to be far less conserved: For example, an equivalent of the MHB is present in the urochordate *Ciona* (Imai et al. 2009) but not in the cephalochordate *Amphioxus* (Holland 2009). Both *Ciona* and *Amphioxus* seem to lack an equivalent of the ZLI, whereas a comparable region has been identified in the hemichordate *Saccoglossus* (C. Lowe, personal communication). Some organizers are even missing in lower vertebrates; no *hedgehog* expression or MGE-like differentiation has been found in the telencephalon of the lamprey, suggesting that this ventroanterior organizer is a gnathostome invention (Sugahara et al. 2011).

These observations indicate that local organizers are more recent innovations than the basic AP/DV patterning network and that they show some evolutionary flexibility that may provide a driving force for morphological change. This idea is supported by the recent finding that differences in forebrain morphology among cichlid fishes from Lake Malawi are correlated with subtle changes in signal strength, timing of signal production, and the position of forebrain organizers (Sylvester et al. 2010). Similarly, loss of eyesight in a cave-dwelling morph of the tetra *Astyanax* was shown to be caused by changes in the forebrain expression of *fgf8* and *shh* (Pottin et al. 2011). Thus, although the basic subdivisions of the brain are likely to have developed a long time ago, organizers are a more recent acquisition that may have been imposed on the underlying pattern and allow evolutionary adaptation to ecological niches.

CONCLUSIONS

Almost 90 years have passed since Spemann discovered the amphibian gastrula organizer; however, the organizer concept is more topical than ever, in particular in the developing vertebrate CNS where multiple organizers regulate patterning, proliferation, neurogenesis, cell death, and axon pathfinding. Neural organizers are generated by inductive signaling events between neighboring tissues, and they often form along, or are stabilized by, cell lineage restriction boundaries. The pattern induced by an organizer results in the formation of different cell populations that can potentially form further organizers at their interfaces, thereby subdividing the neuroepithelium into increasingly more specialized regions. In many ways, neural development can be regarded as a self-organizing process: Once initial polarity has been established, all the interactions necessary to form a functional CNS occur within the neuroepithelium itself.

DISCLOSURE STATEMENT

The authors are not aware of any affiliations, memberships, funding, or financial holdings that might be perceived as affecting the objectivity of this review.

ACKNOWLEDGMENTS

We apologize to the many researchers whose work we could not cite due to space constraints.

LITERATURE CITED

Acampora D, Avantaggiato V, Tuorto F, Simeone A. 1997. Genetic control of brain morphogenesis through Otx gene dosage requirement. *Development* 124:3639–50

Agarwala S, Sanders TA, Ragsdale CW. 2001. Sonic hedgehog control of size and shape in midbrain pattern formation. *Science* 291:2147–50

Alder J, Lee KJ, Jessell TM, Hatten ME. 1999. Generation of cerebellar granule neurons in vivo by transplantation of BMP-treated neural progenitor cells. *Nat. Neurosci.* 2:535–40

Amoyel M, Cheng YC, Jiang YJ, Wilkinson DG. 2005. Wnt1 regulates neurogenesis and mediates lateral inhibition of boundary cell specification in the zebrafish hindbrain. *Development* 132:775–85

Anderson RM, Lawrence AR, Stottmann RW, Bachiller D, Klingensmith J. 2002. Chordin and noggin promote organizing centers of forebrain development in the mouse. *Development* 129:4975–87

Assimacopoulos S, Grove EA, Ragsdale CW. 2003. Identification of a Pax6-dependent epidermal growth factor family signaling source at the lateral edge of the embryonic cerebral cortex. *J. Neurosci.* 23:6399–403

Augsburger A, Schuchardt A, Hoskins S, Dodd J, Butler S. 1999. BMPs as mediators of roof plate repulsion of commissural neurons. *Neuron* 24:127–41

Awatramani R, Soriano P, Rodriguez C, Mai JJ, Dymecki SM. 2003. Cryptic boundaries in roof plate and choroid plexus identified by intersectional gene activation. *Nat. Genet.* 35:70–75

Backman M, Machon O, Mygland L, van den Bout CJ, Zhong W, et al. 2005. Effects of canonical Wnt signaling on dorso-ventral specification of the mouse telencephalon. *Dev. Biol.* 279:155–68

Baek JH, Hatakeyama J, Sakamoto S, Ohtsuka T, Kageyama R. 2006. Persistent and high levels of Hes1 expression regulate boundary formation in the developing central nervous system. *Development* 133:2467–76

Barth KA, Kishimoto Y, Rohr KB, Seydler C, Schulte-Merker S, Wilson SW. 1999. Bmp activity establishes a gradient of positional information throughout the entire neural plate. *Development* 126:4977–87

Basson MA, Echevarria D, Ahn CP, Sudarov A, Joyner AL, et al. 2008. Specific regions within the embryonic midbrain and cerebellum require different levels of FGF signaling during development. *Development* 135:889–98

Bayly RD, Ngo M, Aglyamova GV, Agarwala S. 2007. Regulation of ventral midbrain patterning by Hedgehog signaling. *Development* 134:2115–24

Bloch-Gallego E, Millet S, Alvarado-Mallart RM. 1996. Further observations on the susceptibility of diencephalic prosomeres to En-2 induction and on the resulting histogenetic capabilities. *Mech. Dev.* 58:51–63

Bluske KK, Kawakami Y, Koyano-Nakagawa N, Nakagawa Y. 2009. Differential activity of Wnt/β-catenin signaling in the embryonic mouse thalamus. *Dev. Dyn.* 238:3297–309

Borello U, Cobos I, Long JE, McWhirter JR, Murre C, Rubenstein JL. 2008. FGF15 promotes neurogenesis and opposes FGF8 function during neocortical development. *Neural Dev.* 3:17

Bourikas D, Pekarik V, Baeriswyl T, Grunditz A, Sadhu R, et al. 2005. Sonic hedgehog guides commissural axons along the longitudinal axis of the spinal cord. *Nat. Neurosci.* 8:297–304

Braun MM, Etheridge A, Bernard A, Robertson CP, Roelink H. 2003. Wnt signaling is required at distinct stages of development for the induction of the posterior forebrain. *Development.* 130:5579–87

Briscoe J, Pierani A, Jessell TM, Ericson J. 2000. A homeodomain protein code specifies progenitor cell identity and neuronal fate in the ventral neural tube. *Cell.* 101:435–45

Broccoli V, Boncinelli E, Wurst W. 1999. The caudal limit of *Otx2* expression positions the isthmic organizer. *Nature* 401:164–68

Chamberlain CE, Jeong J, Guo C, Allen BL, McMahon AP. 2008. Notochord-derived Shh concentrates in close association with the apically positioned basal body in neural target cells and forms a dynamic gradient during neural patterning. *Development* 135:1097–106

Charron F, Stein E, Jeong J, McMahon AP, Tessier-Lavigne M. 2003. The morphogen sonic hedgehog is an axonal chemoattractant that collaborates with netrin-1 in midline axon guidance. *Cell* 113:11–23

Charron F, Tessier-Lavigne M. 2005. Novel brain wiring functions for classical morphogens: a role as graded positional cues in axon guidance. *Development* 132:2251–62

Cheng YC, Amoyel M, Qiu X, Jiang YJ, Xu Q, Wilkinson DG. 2004. Notch activation regulates the segregation and differentiation of rhombomere boundary cells in the zebrafish hindbrain. *Dev. Cell* 6:539–50

Chenn A, Walsh CA. 2002. Regulation of cerebral cortical size by control of cell cycle exit in neural precursors. *Science* 297:365–69

Chiang C, Litingtung Y, Lee E, Young KE, Corden JL, et al. 1996. Cyclopia and defective axial patterning in mice lacking *Sonic hedgehog* gene function. *Nature* 383:407–13

Chizhikov VV, Millen KJ. 2004a. Control of roof plate formation by *Lmx1a* in the developing spinal cord. *Development* 131:2693–705

Chizhikov VV, Millen KJ. 2004b. Mechanisms of roof plate formation in the vertebrate CNS. *Nat. Rev. Neurosci.* 5:808–12

Cooke JE, Kemp HA, Moens CB. 2005. EphA4 is required for cell adhesion and rhombomere-boundary formation in the zebrafish. *Curr. Biol.* 15:536–42

Crossley PH, Martinez S, Martin GR. 1996. Midbrain development induced by FGF8 in the chick embryo. *Nature* 380:66–68

Dale JK, Vesque C, Lints TJ, Sampath TK, Furley A, et al. 1997. Cooperation of BMP7 and SHH in the induction of forebrain ventral midline cells by prechordal mesoderm. *Cell* 90:257–69

Darras S, Gerhart J, Terasaki M, Kirschner M, Lowe CJ. 2011. ß-catenin specifies the endomesoderm and defines the posterior organizer of the hemichordate *Saccoglossus kowalevskii*. *Development* 138:959–70

David MD. 2010. Wnt-3a and Wnt-3 differently stimulate proliferation and neurogenesis of spinal neural precursors and promote neurite outgrowth by canonical signaling. *J. Neurosci. Res.* 88:3011–23

De Robertis EM, Kuroda H. 2004. Dorsal-ventral patterning and neural induction in *Xenopus embryos*. *Annu. Rev. Cell Dev. Biol.* 20:285–308

Dessaud E, McMahon AP, Briscoe J. 2008. Pattern formation in the vertebrate neural tube: a sonic hedgehog morphogen-regulated transcriptional network. *Development* 135:2489–503

Dickson BJ. 2002. Molecular mechanisms of axon guidance. *Science* 298:1959–64

Echelard Y, Epstein DJ, St-Jacques B, Shen L, Mohler J, et al. 1993. Sonic hedgehog, a member of a family of putative signaling molecules, is implicated in the regulation of CNS polarity. *Cell* 75:1417–30

Fischer T, Faus-Kessler T, Welzl G, Simeone A, Wurst W, Prakash N. 2011. Fgf15-mediated control of neurogenic and proneural gene expression regulates dorsal midbrain neurogenesis. *Dev. Biol.* 350:496–510

Fishell G, Mason CA, Hatten ME. 1993. Dispersion of neural progenitors within the germinal zones of the forebrain. *Nature* 362:636–38

Fraser S, Keynes R, Lumsden A. 1990. Segmentation in the chick embryo hindbrain is defined by cell lineage restrictions. *Nature* 344:431–35

Frowein J, Campbell K, Götz M. 2002. Expression of *Ngn1*, *Ngn2*, *Cash1*, *Gsh2* and *Sfrp1* in the developing chick telencephalon. *Mech. Dev.* 110:249–52

Fuccillo M, Joyner AL, Fishell G. 2006. Morphogen to mitogen: the multiple roles of hedgehog signalling in vertebrate neural development. *Nat. Rev. Neurosci.* 7:772–83

Fukuchi-Shimogori T, Grove EA. 2001. Neocortex patterning by the secreted signaling molecule FGF8. *Science* 294:1071–74

Galceran J, Miyashita-Lin EM, Devaney E, Rubenstein JL, Grosschedl R. 2000. Hippocampus development and generation of dentate gyrus granule cells is regulated by LEF1. *Development* 127:469–82

Gardner CA, Barald KF. 1991. The cellular environment controls the expression of engrailed-like protein in the cranial neuroepithelium of quail-chick chimeric embryos. *Development* 113:1037–48

Geling A, Plessy C, Rastegar S, Strähle U, Bally-Cuif L. 2004. Her5 acts as a prepattern factor that blocks *neurogenin1* and *coe2* expression upstream of Notch to inhibit neurogenesis at the midbrain-hindbrain boundary. *Development* 131:1993–2006

Golden JA, Bracilovic A, McFadden KA, Beesley JS, Rubenstein JL, Grinspan JB. 1999. Ectopic bone morphogenetic proteins 5 and 4 in the chicken forebrain lead to cyclopia and holoprosencephaly. *Proc.*

Natl. Acad. Sci. USA 96:2439–44

Guinazu MF, Chambers D, Lumsden A, Kiecker C. 2007. Tissue interactions in the developing chick diencephalon. *Neural Dev.* 2:25

Gunhaga L, Jessell TM, Edlund T. 2000. Sonic hedgehog signaling at gastrula stages specifies ventral telencephalic cells in the chick embryo. *Development* 127:3283–93

Gunhaga L, Marklund M, Sjödal M, Hsieh JC, Jessell TM, Edlund T. 2003. Specification of dorsal telencephalic character by sequential Wnt and FGF signaling. *Nat. Neurosci.* 6:701–7

Guthrie S, Butcher M, Lumsden A. 1991. Patterns of cell division and interkinetic nuclear migration in the chick embryo hindbrain. *J. Neurobiol.* 22:742–54

Hashimoto-Torii K, Motoyama J, Hui CC, Kuroiwa A, Nakafuku M, Shimamura K. 2003. Differential activities of Sonic hedgehog mediated by Gli transcription factors define distinct neuronal subtypes in the dorsal thalamus. *Mech. Dev.* 120:1097–111

Hatten ME, Roussel MF. 2011. Development and cancer of the cerebellum. *Trends Neurosci.* 34:134–42

Hébert JM, Mishina Y, McConnell SK. 2002. BMP signaling is required locally to pattern the dorsal telencephalic midline. *Neuron* 35:1029–41

Hirabayashi Y, Itoh Y, Tabata H, Nakajima K, Akiyama T, et al. 2004. The Wnt/β-catenin pathway directs neuronal differentiation of cortical neural precursor cells. *Development* 131:2791–801

Hirata T, Nakazawa M, Muraoka O, Nakayama R, Suda Y, Hibi M. 2006. Zinc-finger genes Fez and Fez-like function in the establishment of diencephalon subdivisions. *Development* 133:3993–4004

Hirsch C, Campano LM, Wöhrle S, Hecht A. 2007. Canonical Wnt signaling transiently stimulates proliferation and enhances neurogenesis in neonatal neural progenitor cultures. *Exp. Cell Res.* 313:572–87

Holland LZ. 2009. Chordate roots of the vertebrate nervous system: expanding the molecular toolkit. *Nat. Rev. Neurosci.* 10:736–46

Houart C, Caneparo L, Heisenberg C, Barth K, Take-Uchi M, Wilson S. 2002. Establishment of the telencephalon during gastrulation by local antagonism of Wnt signaling. *Neuron* 35:255–65

Houart C, Westerfield M, Wilson SW. 1998. A small population of anterior cells patterns the forebrain during zebrafish gastrulation. *Nature* 391:788–92

Imai KS, Stolfi A, Levine M, Satou Y. 2009. Gene regulatory networks underlying the compartmentalization of the *Ciona* central nervous system. *Development* 136:285–93

Irimia M, Piñeiro C, Maeso I, Gómez-Skarmeta JL, Casares F, Garcia-Fernàndez J. 2010. Conserved developmental expression of *Fezf* in chordates and *Drosophila* and the origin of the zona limitans intrathalamica (ZLI) brain organizer. *Evodevo* 1:7

Irving C, Mason I. 1999. Regeneration of isthmic tissue is the result of a specific and direct interaction between rhombomere 1 and midbrain. *Development* 126:3981–89

Irving C, Mason I. 2000. Signalling by FGF8 from the isthmus patterns anterior hindbrain and establishes the anterior limit of *Hox* gene expression. *Development* 127:177–86

Ivaniutsin U, Chen Y, Mason JO, Price DJ, Pratt T. 2009. *Adenomatous polyposis coli* is required for early events in the normal growth and differentiation of the developing cerebral cortex. *Neural Dev.* 16:3

Jeong Y, Dolson DK, Waclaw RR, Matise MP, Sussel L, et al. 2011. Spatial and temporal requirements for sonic hedgehog in the regulation of thalamic interneuron identity. *Development* 138:531–41

Jimenez-Guri E, Udina F, Colas JF, Sharpe J, Padron-Barthe L, et al. 2010. Clonal analysis in mice underlines the importance of rhombomeric boundaries in cell movement restriction during hindbrain segmentation. *PLoS One* 5:e10112

Kataoka A, Shimogori T. 2008. Fgf8 controls regional identity in the developing thalamus. *Development* 135:2873–81

Kemp HA, Cooke JE, Moens CB. 2009. EphA4 and EfnB2a maintain rhombomere coherence by independently regulating intercalation of progenitor cells in the zebrafish neural keel. *Dev. Biol.* 327:313–26

Kiecker C, Lumsden A. 2004. Hedgehog signaling from the ZLI regulates diencephalic regional identity. *Nat. Neurosci.* 7:1242–49

Kiecker C, Lumsden A. 2005. Compartments and their boundaries in vertebrate brain development. *Nat. Rev.Neurosci.* 6:553–64

Kiecker C, Niehrs C. 2001a. A morphogen gradient of Wnt/β-catenin signalling regulates anteroposterior neural patterning in *Xenopus. Development* 128:4189–201

Kiecker C, Niehrs C. 2001b. The role of prechordal mesendoderm in neural patterning. *Curr. Opin. Neurobiol.* 11:27–33

Kitamura K, Miura H, Yanazawa M, Miyashita T, Kato K. 1997. Expression patterns of *Brx1* (Rieg gene), *Sonic hedgehog, Nkx2.2, Dlx1* and Arx during zona limitans intrathalamica and embryonic ventral lateral geniculate nuclear formation. *Mech. Dev.* 67:83–96

Kobayashi D, Kobayashi M, Matsumoto K, Ogura K, Nakafuku M, Shimamura K. 2002. Early subdivisions in the neural plate define distinct competence for inductive signals. *Development* 129:83–93

Krauss S, Concordet JP, Ingham PW. 1993. A functionally conserved homolog of the Drosophila segment polarity gene *hh* is expressed in tissues with polarizing activity in zebrafish embryos. *Cell* 75:1431–44

Langenberg T, Brand M. 2005. Lineage restriction maintains a stable organizer cell population at the zebrafish midbrain-hindbrain boundary. *Development* 132:3209–16

Le Douarin NM, Halpern ME. 2000. Discussion point. Origin and specification of the neural tube floor plate: insights from the chick and zebrafish. *Curr. Opin. Neurobiol.* 10:23–30

Lee KJ, Dietrich P, Jessell TM. 2000. Genetic ablation reveals that the roof plate is essential for dorsal interneuron specification. *Nature* 403:734–40

Lee KJ, Mendelsohn M, Jessell TM. 1998. Neuronal patterning by BMPs: a requirement for GDF7 in the generation of a discrete class of commissural interneurons in the mouse spinal cord. *Genes Dev.* 12:3394–407

Lee SM, Danielian PS, Fritzsch B, McMahon AP. 1997. Evidence that FGF8 signalling from the midbrain-hindbrain junction regulates growth and polarity in the developing midbrain. *Development* 124:959–69

Lee SM, Tole S, Grove EA, McMahon AP. 2000. A local Wnt-3a signal is required for development of the mammalian hippocampus. *Development* 127:457–67

Lei Q, Zelman AK, Kuang E, Li S, Matise MP. 2004. Transduction of graded Hedgehog signaling by a combination of Gli2 and Gli3 activator functions in the developing spinal cord. *Development* 131:3593–604

Liem KF Jr, Tremml G, Jessell TM. 1997. A role for the roof plate and its resident TGFβ-related proteins in neuronal patterning in the dorsal spinal cord. *Cell* 91:127–38

Lim Y, Cho G, Minarcik J, Golden J. 2005. Altered BMP signaling disrupts chick diencephalic development. *Mech. Dev.* 122:603–20

Liu A, Joyner AL. 2001. Early anterior/posterior patterning of the midbrain and cerebellum. *Annu. Rev.*

Neurosci. 24:869–96

Liu Y, Helms AW, Johnson JE. 2004. Distinct activities of Msx1 and Msx3 in dorsal neural tube development. *Development* 131:1017–28

Liu Y, Shi J, Lu CC, Wang ZB, Lyuksyutova AI, et al. 2005. Ryk-mediated Wnt repulsion regulates posterior- directed growth of corticospinal tract. *Nat. Neurosci.* 8:1151–59

Lowe CJ, Wu M, Salic A, Evans L, Lander E, et al. 2003. Anteroposterior patterning in hemichordates and the origins of the chordate nervous system. *Cell* 113:853–65

Lumsden A, Keynes R. 1989. Segmental patterns of neuronal development in the chick hindbrain. *Nature* 337:424–28

Lupo G, Harris WA, Lewis KE. 2006. Mechanisms of ventral patterning in the vertebrate nervous system. *Nat. Rev. Neurosci.* 7:103–14

Lyuksyutova AI, Lu CC, Milanesio N, King LA, Guo N, et al. 2003. Anterior-posterior guidance of commissural axons by Wnt-frizzled signaling. *Science* 302:1984–88

Machon O, Backman M, Machonova O, Kozmik Z, Vacik T, et al. 2007. A dynamic gradient of Wnt signaling controls initiation of neurogenesis in the mammalian cortex and cellular specification in the hippocampus. *Dev. Biol.* 311:223–37

Machon O, van den Bout CJ, Backman M, Kemler R, Krauss S. 2003. Role of β-catenin in the developing cortical and hippocampal neuroepithelium. *Neuroscience* 122:129–43

Maden M. 2007. Retinoic acid in the development, regeneration and maintenance of the nervous system. *Nat. Rev. Neurosci.* 8:755–65

Marín F, Puelles L. 1994. Patterning of the embryonic avian midbrain after experimental inversions: a polarizing activity from the isthmus. *Dev. Biol.* 163:19–37

Martinez S, Crossley PH, Cobos I, Rubenstein JL, Martin GR. 1999. FGF8 induces formation of an ectopic isthmic organizer and isthmocerebellar development via a repressive effect on Otx2 expression. *Development* 126:1189–200

Martinez S, Wassef M, Alvarado-Mallart RM. 1991. Induction of a mesencephalic phenotype in the 2-day-old chick prosencephalon is preceded by the early expression of the homeobox gene en. *Neuron* 6:971–81

Mason I. 2007. Initiation to end point: the multiple roles of fibroblast growth factors in neural development. *Nat. Rev. Neurosci.* 8:583–96

Matsumoto K, Nishihara S, Kamimura M, Shiraishi T, Otoguro T, et al. 2004. The prepattern transcription factor Irx2, a target of the FGF8/MAP kinase cascade, is involved in cerebellum formation. *Nat. Neurosci.* 7:605–12

McMahon AP, Joyner AL, Bradley A, McMahon JA. 1992. The midbrain-hindbrain phenotype of *Wnt-1⁻/Wnt-1⁻* mice results from stepwise deletion of engrailed-expressing cells by 9.5 days postcoitum. *Cell* 69:581–95

Megason SG, McMahon AP. 2002. A mitogen gradient of dorsal midline Wnts organizes growth in the CNS. *Development* 129:2087–98

Meinertzhagen IA, Lemaire P, Okamura Y. 2004. The neurobiology of the ascidian tadpole larva: recent developments in an ancient chordate. *Annu. Rev. Neurosci.* 27:453–85

Meinhardt H. 2006. Primary body axes of vertebrates: generation of a near-Cartesian coordinate system and the role of Spemann-type organizer. *Dev. Dyn.* 235:2907–19

Meinhardt H. 2009. Models for the generation and interpretation of gradients. *Cold Spring Harb. Perspect. Biol.* 1:a001362

Meyers EN, Lewandowski M, Martin GR. 1998. An *Fgf8* mutant allelic series generated by Cre- and Flp-mediated recombination. *Nat. Genet.* 18:136–41

Millet S. 1999. A role for *Gbx2* in repression of Otx2 and positioning the mid/hindbrain organizer. *Nature* 401:161–64

Millonig JH, Millen KJ, Hatten ME. 2000. The mouse Dreher gene *Lmx1a* controls formation of the roof plate in the vertebrate CNS. *Nature* 403:764–69

Monuki ES, Porter FD, Walsh CA. 2001. Patterning of the dorsal telencephalon and cerebral cortex by a roof plate-Lhx2 pathway. *Neuron* 32:591–604

Munji RN, Choe Y, Li G, Siegenthaler JA, Pleasure SJ. 2011. Wnt signaling regulates neuronal differentiation of cortical intermediate progenitors. *J. Neurosci.* 31:1676–87

Muñoz-Sanjuán I, Brivanlou AH. 2002. Neural induction, the default model and embryonic stem cells. *Nat. Rev. Neurosci.* 3:271–80

Muroyama Y, Fujihara M, Ikeya M, Kondoh H, Takada S. 2002. Wnt signaling plays an essential role in neuronal specification of the dorsal spinal cord. *Genes Dev.* 16:548–53

Muzio L, Soria JM, Pannese M, Piccolo S, Mallamaci A. 2005. A mutually stimulating loop involving emx2 and canonical wnt signalling specifically promotes expansion of occipital cortex and hippocampus. *Cereb. Cortex* 15:2021–28

Nakamoto A, Nagy LM, Shimizu T. 2011. Secondary embryonic axis formation by transplantation of D quadrant micromeres in an oligochaete annelid. *Development* 138:283–90

Nguyen VH, Trout J, Connors SA, Andermann P, Weinberg E, Mullins MC. 2000. Dorsal and intermediate neuronal cell types of the spinal cord are established by a BMP signaling pathway. *Development* 127:1209–20

Niehrs C. 2004. Regionally specific induction by the Spemann-Mangold organizer. *Nat. Rev. Genet.* 5:425–34

Niehrs C. 2010. On growth and form: a Cartesian coordinate system of Wnt and BMP signaling specifies bilaterian body axes. *Development* 137:845–57

Ninkovic J, Tallafuss A, Leucht C, Topczewski J, Tannhäuser B, et al. 2005. Inhibition of neurogenesis at the zebrafish midbrain-hindbrain boundary by the combined and dose-dependent activity of a new *hairy/E(spl)* gene pair. *Development* 132:75–88

Ono Y, Nakatani T, Sakamoto Y, Mizuhara E, Minaki Y, et al. 2007. Differences in neurogenic potential in floor plate cells along an anteroposterior location: midbrain dopaminergic neurons originate from mesencephalic floor plate cells. *Development* 134:3213–25

Osterfield M, Kirschner MW, Flanagan JG. 2003. Graded positional information: interpretation for both fate and guidance. *Cell* 113:425–28

Osumi N, Hirota A, Ohuchi H, Nakafuku M, Iimura T, et al. 1997. Pax-6 is involved in the specification of hindbrain motor neuron subtype. *Development* 124:2961–72

Paek H, Gutin G, Hébert JM. 2009. FGF signaling is strictly required to maintain early telencephalic precursor cell survival. *Development* 136:2457–65

Panhuysen M, Vogt Weisenhorn DM, Blanquet V, Brodski C, Heinzmann U, et al. 2004. Effects of Wnt1 signaling on proliferation in the developing mid-/hindbrain region. *Mol. Cell. Neurosci.* 26:101–11

Partanen J. 2007. FGF signalling pathways in development of the midbrain and anterior hindbrain. *J. Neurochem.* 101:1185–93

Pattyn A, Vallstedt A, Dias JM, Sander M, Ericson J. 2003. Complementary roles for Nkx6 and Nkx2 class proteins in the establishment of motoneuron identity in the hindbrain. *Development* 130:4149–59

Pera EM, Kessel M. 1997. Patterning of the chick forebrain anlage by the prechordal plate. *Development* 124:4153-62

Picker A, Brennan C, Reifers F, Clarke JD, Holder N, Brand M. 1999. Requirement for the zebrafish mid-hindbrain boundary in midbrain polarisation, mapping and confinement of the retinotectal projection. *Development* 126:2967–78

Placzek M, Briscoe J. 2005. The floor plate: multiple cells, multiple signals. *Nat. Rev. Neurosci.* 6:230–40

Placzek M, Tessier-Lavigne M, Yamada T, Jessell T, Dodd J. 1990. Mesodermal control of neural cell identity: floor plate induction by the notochord. *Science* 250:985–88

Pottin K, Hinaux H, Ré taux S. 2011. Restoring eye size in *Astyanax mexicanus* blind cavefish embryos through modulation of the Shh and Fgf 8 forebrain organising centres. *Development* 138:2467–76

Puelles E, Annino A, Tuorto F, Usiello A, Acampora D, et al. 2004. *Otx2* regulates the extent, identity and fate of neuronal progenitor domains in the ventral midbrain. *Development* 131:2037–48

Quinlan R, Graf M, Mason I, Lumsden A, Kiecker C. 2009. Complex and dynamic patterns of Wnt pathway gene expression in the developing chick forebrain. *Neural Dev.* 4:35

Reichert H. 2005. A tripartite organization of the urbilaterian brain: developmental genetic evidence from *Drosophila. Brain Res. Bull.* 66:491–94

Reifers F, Böhli H, Walsh EC, Crossley PH, Stainier DY, Brand M. 1998. *Fgf8* is mutated in zebrafish *acerebellar* (*ace*) mutants and is required for maintenance of midbrain-hindbrain boundary development and somitogenesis. *Development* 125:2381–95

Rhinn M, Lun K, Luz M, Werner M, Brand M. 2005. Positioning of the midbrain-hindbrain boundary organizer through global posteriorization of the neuroectoderm mediated by Wnt8 signaling. *Development* 132:1261–72

Riley BB, Chiang MY, Storch EM, Heck R, Buckles GR, Lekven AC. 2004. Rhombomere boundaries are Wnt signaling centers that regulate metameric patterning in the zebrafish hindbrain. *Dev. Dyn.* 231:278–91

Rodríguez-Seguel E, Alarcón P, Gómez-Skarmeta JL. 2009. *The Xenopus* Irx genes are essential for neural patterning and define the border between prethalamus and thalamus through mutual antagonism with the anterior repressors *Fezf* and *Arx. Dev. Biol.* 329:258–68

Roelink H, Augsburger A, Heemskerk J, Korzh V, Norlin S, et al. 1994. Floor plate and motor neuron induction by vhh-1, a vertebrate homolog of hedgehog expressed by the notochord. *Cell* 76:761–75

Sánchez-Camancho C, Rodríguez J, Ruiz JM, Trousse F, Bovolenta P. 2005. Morphogens as growth cone signalling molecules. *Brain Res. Brain Res. Rev.* 49:242–52

Sanders TA, Lumsden A, Ragsdale CW. 2002. Arcuate plan of chick midbrain development. *J. Neurosci.* 22:10742–50

Scholpp S, Delogu A, Gilthorpe J, Peukert D, Schindler S, Lumsden A. 2009. Her6 regulates the neurogenetic gradient and neuronal identity in the thalamus. *Proc. Natl. Acad. Sci. USA* 106:19895–900

Scholpp S, Foucher I, Staudt N, Peukert D, Lumsden A, Houart C. 2007. *Otx1l, Otx2* and Irx1b establish and position the ZLI in the diencephalon. *Development* 134:3167–76

Scholpp S, Wolf O, Brand M, Lumsden A. 2006. Hedgehog signalling from the zona limitans intrathalamica orchestrates patterning of the zebrafish diencephalon. *Development* 133:855–64

Shimamura K, Hartigan DJ, Martinez S, Puelles L, Rubenstein JL. 1995. Longitudinal organization of the anterior neural plate and neural tube. *Development* 121:3923–33

Shimamura K, Rubenstein JL. 1997. Inductive interactions direct early regionalization of the mouse forebrain. *Development* 124:2709–18

Sousa VH, Fishell G. 2010. Sonic hedgehog functions through dynamic changes in temporal competence in the developing forebrain. *Curr. Opin. Genet. Dev.* 20:391–99

Stamataki D, Ulloa F, Tsoni SV, Mynett A, Briscoe J. 2005. A gradient of Gli activity mediates graded Sonic hedgehog signaling in the neural tube. *Genes Dev.* 19:626–41

Stern CD. 2001. Initial patterning of the central nervous system: how many organizers? *Nat. Rev. Neurosci.* 2:92–98

Stern CD. 2006. Neural induction: 10 years on since the 'default model'. *Curr. Opin. Cell Biol.* 18:692–97

Sugahara F, Aota S, Kuraku S, Murakami Y, Takio-Ogawa Y, et al. 2011. Involvement of Hedgehog and FGF signalling in the lamprey telencephalon: evolution of regionalization and dorsoventral patterning of the vertebrate forebrain. *Development* 138:1217–26

Sunmonu NA, Li K, Guo Q, Li JY. 2011. *Gbx2* and *Fgf8* are sequentially required for formation of the midbrain-hindbrain compartment boundary. *Development* 138:725–34

Sylvester JB, Rich CA, Loh YH, van Staaden MJ, Fraser GJ, Streelman JT. 2010. Brain diversity evolves via differences in patterning. *Proc. Natl. Acad. Sci. USA* 107:9718–23

Takada H, Hattori D, Kitayama A, Ueno N, Taira M. 2005. Identification of target genes for the Xenopus Hes-related protein XHR1, a prepattern factor specifying the midbrain-hindbrain boundary. *Dev. Biol.* 283:253–67

Takahashi M, Osumi N. 2002. *Pax6* regulates specification of ventral neurone subtypes in the hindbrain by establishing progenitor domains. *Development* 129:1327–38

Technau U, Steele RE. 2011. Evolutionary crossroads in developmental biology: Cnidaria. *Development* 138:1447–58

Tessier-Lavigne M, Goodman CS. 1996. The molecular biology of axon guidance. *Science* 274:1123–33
Thomas KR, Musci TS, Neumann PE, Capecchi MR. 1991. Swaying is a mutant allele of the proto-oncogene *Wnt-1*. *Cell* 67:969–76

Thomson RE, Kind PC, Graham NA, Etherson ML, Kennedy J, et al. 2009. Fgf receptor 3 activation promotes selective growth and expansion of occipitotemporal cortex. *Neural Dev.* 4:4

Timmer JR, Wang C, Niswander L. 2002. BMP signaling patterns the dorsal and intermediate neural tube via regulation of homeobox and helix-loop-helix transcription factors. *Development* 129:2459–72

Tomer R, Denes AS, Tessmar-Raible K, Arendt D. 2010. Profiling by image registration reveals common origin of annelid mushroom bodies and vertebrate pallium. *Cell* 142:800–9

Tossell K, Kiecker C, Wizenmann A, Lang E, Irving C. 2011. Notch signalling stabilises boundary formation at the midbrain-hindbrain organiser. *Development* 138:3745–57

Toyoda R, Assimacopoulos S, Wilcoxon J, Taylor A, Feldman P, et al. 2010. FGF8 acts as a classic diffusible morphogen to pattern the neocortex. *Development* 137:3439–48

Trokovic R, Jukkola T, Saarimaki J, Peltopuro P, Naserke T, et al. 2005. Fgfr1-dependent boundary cells between developing mid- and hindbrain. *Dev. Biol.* 278:428–39

Urbach R, Technau GM. 2008. Dorsoventral patterning of the brain: a comparative approach. *Adv. Exp. Med.Biol.* 628:42–56

van Straaten HW, Hekking JW, Thors F, Wiertz-Hoessels EL, Drukker J. 1985. Induction of an additional floor plate in the neural tube. *Acta Morphol. Neerl. Scand.* 23:91–97

Vieira C, Garda AL, Shimamura K, Martinez S. 2005. Thalamic development induced by Shh in the chick embryo. *Dev. Biol.* 284:351–63

Vogel-Höpker A, Rohrer H. 2002. The specification of noradrenergic locus coeruleus (LC) neurones depends on bone morphogenetic proteins (BMPs). *Development* 129:983–91

Vue TY, Bluske K, Alishani A, Yang LL, Koyano-Nakagawa N, et al. 2009. Sonic hedgehog signaling controls thalamic progenitor identity and nuclei specification in mice. *J. Neurosci.* 29:4484–97

Walshe J, Mason I. 2003. Unique and combinatorial functions of Fgf3 and Fgf8 during zebrafish forebrain development. *Development* 130:4337–49

Wassarman KM, Lewandowski M, Campbell K, Joyner AL, Rubenstein JL, et al. 1997. Specification of the anterior hindbrain and establishment of a normal mid/hindbrain organizer is dependent on Gbx2 gene function. *Development* 124:2923–34

Wingate RJ. 2001. The rhombic lip and early cerebellar development. Curr. Opin. *Neurobiol.* 11:82–88

Wurst W, Bally-Cuif L. 2001. Neural plate patterning: upstream and downstream of the isthmic organizer. *Nat. Rev. Neurosci.* 2:99–108

Xie Z, Chen Y, Li Z, Bai G, Zhu Y, et al. 2011. Smad6 promotes neuronal differentiation in the intermediate zone of the dorsal neural tube by inhibition of the Wnt/β-catenin pathway. *Proc. Natl. Acad. Sci. USA* 108:12119–24

Xu Q, Mellitzer G, Robinson V, Wilkinson DG. 1999. In vivo cell sorting in complementary segmental domains mediated by Eph receptors and ephrins. *Nature* 399:267–71

Zeltser LM. 2005. Shh-dependent formation of the ZLI is opposed by signals from the dorsal diencephalon. *Development* 132:2023–33

Zeltser LM, Larsen CW, Lumsden A. 2001. A new developmental compartment in the forebrain regulated by *Lunatic fringe. Nat. Neurosci.* 4:683–84

Zervas M, Millet S, Ahn S, Joyner AL. 2004. Cell behaviors and genetic lineages of the mesencephalon and rhombomere 1. *Neuron* 43:345–57

Zou Y, Lyuksyutova AI. 2007. Morphogens as conserved axon guidance cues. *Curr. Opin. Neurobiol.* 17:22–28

Mechanisms and Functions of Theta Rhythms

Laura Lee Colgin

Center for Learning and Memory, The University of Texas, Austin, Texas 78712-0805; email: colgin@ mail.clm.utexas.edu

Key Words

oscillations, memory, hippocampus, place cells, phase precession, CA1

Abstract

The theta rhythm is one of the largest and most sinusoidal activity patterns in the brain. Here I survey progress in the field of theta rhythms research. I present arguments supporting the hypothesis that theta rhythms emerge owing to intrinsic cellular properties yet can be entrained by several theta oscillators throughout the brain. I review behavioral correlates of theta rhythms and consider how these correlates inform our understanding of theta rhythms' functions. I discuss recent work suggesting that one function of theta is to package related information within individual theta cycles for more efficient spatial memory processing. Studies examining the role of theta phase precession in spatial memory, particularly sequence retrieval, are also summarized. Additionally, I discuss how interregional coupling of theta rhythms facilitates communication across brain regions. Finally, I conclude by summarizing how theta rhythms may support cognitive operations in the brain, including learning.

INTRODUCTION

Theta rhythms:
~4–12 Hz, nearly sinusoidal patterns of electrical activity that are associated with active behaviors and REM sleep

Execution of complex cognitive functions by the brain requires coordination across many neurons in multiple brain areas. Brain rhythms (or oscillations) provide a mechanism for such coordination by linking the activity of related ensembles of neurons. One of the most intriguing brain rhythms is the theta rhythm. The ~4–12-Hz theta rhythms were first discovered in the rabbit by Jung & Kornmuller (1938). Although researchers did not understand what theta rhythms signified, they did notice the large amplitude and nearly sinusoidal regularity of these rhythms (**Figure 1**). Scientific interest in theta rhythms continued to grow during the following decades, and investigators found that theta rhythms occurred in other species as well, including cats, rats, and monkeys (Green & Arduini 1954, Grastyan et al. 1959, Vanderwolf 1969).

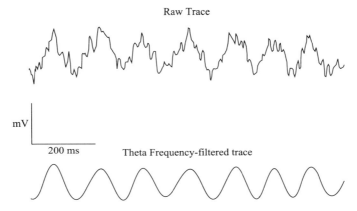

Figure 1

Theta recorded from hippocampal subfield CA1 of a freely exploring rat (L.L.Colgin, unpublished data). A raw trace (*top*) and a theta frequency (4–12 Hz)band-pass filtered trace (*bottom*) are shown.

In 1972, Landfield and colleagues reported that the degree to which rats remembered an aversive foot shock was correlated with the amount of theta recorded from screws implanted in the rats' skulls above the cortex (Landfield et al. 1972). The link between theta rhythms and memory was controversial at the time, but many studies have since supported the conclusion that theta rhythms are important for different types of learning and memory (Berry& Thompson 1978, Winson 1978, Macrides et al. 1982, Mitchell et al. 1982, Mizumori et al. 1990, M'Harzi & Jarrard 1992, Klimesch et al. 1996, Osipova et al. 2006, Robbe & Buzsáki 2009, Rutishauser et al. 2010, Liebe et al. 2012), as well as for synaptic plasticity (Larson et al. 1986, Staubli & Lynch 1987, Greenstein et al.1988, Pavlides et al. 1988, Orr et al. 2001, Hyman et al. 2003). However, it remains unclear why theta rhythms influence memory processing. Do theta rhythms promote memory, as they promote other cognitive states such as anxiety (Adhikari et al. 2010), by facilitating interregional interactions (Seidenbecher et al. 2003, Jones & Wilson 2005, Kay 2005, Benchenane et al. 2010, Hyman et al. 2010, Kim et al. 2011, Liebe et al. 2012)? Do theta rhythms affect memory directly by providing the correct timing required to induce changes in synaptic strength (Larson et al. 1986, Greenstein et al. 1988)? Do theta rhythms affect cognitive operations in general by providing a way to chunk information (Buzsáki 2006, Kepecs et al. 2006, Gupta et al. 2012)?

Here, I review the current status of research on theta rhythms. I begin by discussing mechanisms that generate theta rhythms. Understanding how theta rhythms are generated will provide clues regarding their function. I then review several proposed functions of theta rhythms. I discuss the effects of theta rhythms on the firing properties of hippocampal place cells, neurons that are activated in particular spatial locations (O'Keefe & Dostrovsky 1971). Functions of theta rhythms are also discussed with regard to cognition and behavior.

The question of whether theta rhythms exist in healthy humans was a subject of intense debate until fairly recently (Klimesch et al. 1994, Tesche & Karhu 2000). Consequently, research using animal models has established a longer-standing body of work on theta. This review focuses on studies of theta rhythms in lower mammals, primarily rats. I also place particular emphasis on theta rhythms in the hippocampus, a region that is critically involved in memory processing (Squire et al. 2004). Theta rhythms have been studied extensively in the hippocampus, and the hippocampal electroencephalogram displays prominent theta rhythms during active behaviors (Figure 1).

MECHANISMS OF THETA RHYTHMS

Place cells: principal neurons of the hippocampus that fire selectively in specific spatial locations

Hippocampus:
a region of the medial temporal lobe that is essential for spatial and episodic memory

Medial septum (MS):
a subcortical region providing cholinergic and GABAergic inputs to cortical structures including the hippocampus and entorhinal cortex

HCN channels:
Hyperpolarization-activated and cyclic nucleotide-gated nonselective cation channels

EC: entorhinal cortex

The Role of the Medial Septum in Theta Generation

Until recently, researchers generally agreed that the medial septum (MS) generates theta rhythms (see Vertes & Kocsis 1997 for a review) because lesioning or inactivating the MS disrupts theta (Green & Arduini 1954, Petsche et al. 1962, Mitchell et al. 1982, Mizumori et al. 1990). MS pacemaker cells are believed to be GABAergic inhibitory interneurons (Toth et al. 1997), which express hyperpolarization-activated and cyclic nucleotide-gated nonselective cation channels (HCN channels) (Varga et al. 2008). These HCN-expressing interneurons fire rhythmically at theta frequencies and are phase-locked to theta rhythms in the hippocampus (Hangya et al. 2009). Cholinergic neurons of the MS, on the other hand, do not fire rhythmically at theta frequencies (Simon et al. 2006) and are thus unlikely to act as theta pacemakers. Cholinergic neurons may instead modulate the excitability of other neurons in a way that promotes their theta rhythmic firing. Backprojections from the hippocampus to the MS may be important to keep the two regions coupled (Toth et al. 1993). The importance of other subcortical regions for theta generation has been addressed in another review (Vertes et al. 2004) and is not discussed here.

Recent work has brought into question the belief that the MS is responsible for theta generation. Goutagny and associates (2009) found that theta rhythms emerge in vitro in an intact hippocampus preparation lacking any connections with the MS (**Figure 2**). The in vitro theta rhythms appeared spontaneously (i.e., without application of any drugs). Also, theta activity persisted in hippocampal subfield CA1 after neighboring subfield CA3 was removed, indicating that an excitatory recurrent collateral network was not a required component of the theta-generating machinery. Local inactivation of an intermediate portion of the longitudinal axis did not eliminate theta from either the septal or the temporal pole. Instead, theta rhythms persisted in both septal and temporal regions but were no longer coherent. Theta in the septal hippocampus was faster than theta in the temporal hippocampus after the inactivation procedure. Whole-cell recordings of CA1 pyramidal cells revealed rhythmic inhibitory postsynaptic potentials, whereas rhythmic excitatory postsynaptic potentials were recorded in interneurons. These results suggest that theta rhythms are produced by local interactions between hippocampal interneurons and pyramidal cells.

Although the hippocampus may possess the machinery necessary to produce theta intrinsically in vitro, much evidence indicates that the MS is involved in theta generation in behaving animals. Lesioning or inactivating the

MEC: medial entorhinal cortex

MS disrupts theta in structures that receive MS projections, including the entorhinal cortex (EC) and the hippocampus (Green & Arduini 1954, Petsche et al. 1962, Mitchell et al. 1982, Mizumori et al. 1990, Brandon et al. 2011, Koenig et al. 2011). The MS transitions to the theta state ~500 ms before theta appears in the hippocampus (Bland et al. 1999), supporting the idea that septohippocampal projections initiate theta rhythms. Spikes of MS pacemaker interneurons are maximally phase-locked to hippocampal theta occurring ~80 ms later (Hangya et al. 2009), supporting the idea that MS interneurons drive theta in the hippocampus (Toth et al. 1997), albeit with some delay. The substantial delay may reflect the time required to recruit a significant proportion of cells into a synchronized theta network (Hangya et al. 2009).

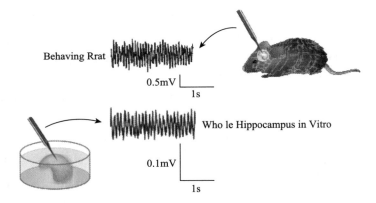

Figure 2

Theta rhythms recorded by Goutagny et al. (2009) in CA1 of a whole hippocampus in vitro preparation are highly similar to theta rhythms recorded from CA1 in vivo. Modified with permission from Colgin & Moser(2009).

A Multitude of Theta Oscillators

Other mechanisms in addition to MS inputs appear to be involved in hippocampal theta rhythm generation in vivo. Current source density analyses of hippocampal local field potentials (LFPs) from freely exploring rats indicate that multiple current dipoles coexist during theta (Kamondi et al. 1998; see figure 4 in Buzsáki 2002). Presumably active current sinks are seen in stratum lacunosum-moleculare and stratum radiatum and are thought to reflect excitatory inputs from the EC and CA3, respectively. At the same time, a putative current source reflecting inhibitory inputs can be seen in stratum pyramidale. These inhibitory inputs are likely hippocampal interneurons firing rhythmically at theta frequencies during periods when they are released from theta inhibition imposed by MS interneurons (Toth et al. 1997). Following surgical removal of the EC, the stratum lacunosum-moleculare dipole disappears. Theta rhythms that remain after EC lesioning are suppressed by the cholinergic antagonist atropine, in contrast with atropine-resistant theta rhythms that normally occur during exploration (Kramis et al. 1975). Atropine-sensitive theta rhythms are normally observed during behavioral immobility and during anesthesia induced by urethane (Kramis et al. 1975), a drug that suppresses input from the EC (Ylinen et al. 1995). These findings suggest that the EC provides sufficient excitatory drive to entrain theta rhythms during active behaviors but that another source of excitatory drive is required in the absence of entorhinal inputs. This excitatory drive may come from MS cholinergic inputs, which produce slow depolarizations in

LFP: local field potential

hippocampal neurons (Madison et al. 1987).

A problem with this interpretation is that theta rhythms in the whole hippocampus in vitro are not blocked by atropine, yet the isolated hippocampus receives no input from the EC (Goutagny et al. 2009). Thus, it is unclear from where the excitatory drive originates in the isolated hippocampal preparation. One possibility is that neurons are more excitable in this preparation than in other in vitro preparations. The potassium concentration used in the artificial cerebrospinal fluid (aCSF) in the Goutagny et al. (2009) study (3.5–5 mM) was higher than the potassium concentration(2.5 mM) used in typical hippocampal slice studies (e.g., Frerking et al. 2001, Brager & Johnston 2007). Studies of spontaneous sharp-wave ripples in hippocampal slices used similarly high potassium concentrations in the aCSF (4.25–4.75 mM; Kubota et al. 2003, Colgin et al. 2004, Ellender et al. 2010). It is not obvious, however, why a relatively high concentration of potassium in the aCSF would lead to theta rhythms in an isolated whole hippocampal preparation and to sharp-wave ripples in hippocampal slices. One difference between preparations is a more intact CA3 recurrent collateral system in the isolated whole hippocampus. However, CA3 was essential for sharp waves in slices (Colgin et al. 2004) but not for theta rhythms in the isolated whole hippocampus (Goutagny et al. 2009), indicating that differences in spontaneous in vitro activity are not due to preservation of CA3 recurrent collaterals. One likely explanation is that a higher number of connections between interneurons and pyramidal neurons are preserved in the whole hippocampus preparation. A high degree of pyramidal cell-interneuron interconnectivity may be essential for theta generation but not for sharp-wave ripples. Of particular interest in this regard are horizontal interneurons (Maccaferri 2005, Goutagny et al. 2009). Horizontal interneurons are activated almost exclusively by CA1 axon collaterals (Blasco-Ibáñez & Freund 1995). Horizontal interneurons fire spontaneously, without requiring fast excitatory input (Maccaferri & McBain 1996), and thus would be likely to fire in the intact hippocampal preparation. Moreover, horizontal interneurons show resonance at theta frequencies (Pike et al. 2000). Thus, horizontal interneurons may be essential for generating theta rhythms intrinsically in the hippocampus.

A common feature of horizontal interneurons in the hippocampus and pacemaker interneurons in the MS is the expression of HCN channels (Maccaferri & McBain 1996, Varga et al. 2008). HCN channels are nonselective cation channels that are activated by hyperpolarization (Robinson & Siegelbaum 2003). Activation of HCN channels leads to slow depolarizing currents (Ih currents) that can drive the membrane back to threshold and trigger action potentials. In neurons that express HCN channels, a repeating sequence of events can emerge that consists of an action potential, afterhyperpolarization, depolarization via HCN channels, and another action potential that starts the cycle again. In this way, HCN channels may facilitate rhythmic firing in neurons. Many neurons that exhibit theta rhythmic firing express HCN channels (Maccaferri & McBain 1996, Dickson et al. 2000, Hu et al. 2002, Varga et al. 2008). Theta frequency membrane-potential oscillations are disrupted by pharmacological blockade (Dickson et al. 2000) or genetic deletion (Giocomo & Hasselmo 2009) of HCN channels. Moreover, the time course of activation and deactivation of Ih currents may determine the particular theta frequency of membrane-potential oscillations in medial entorhinal cortex (MEC) neurons; that is, faster time constants correspond to higher-frequency theta rhythms (Giocomo et al. 2007, Giocomo & Hasselmo 2008). Together, these findings suggest that HCN channels may be a key cellular mechanism that contributes to theta rhythm generation.

Effects of HCN channels provide one potential explanation of the seemingly contradictory findings that theta rhythms depend on MS inputs yet can be generated intrinsically in the hippocampus. Perhaps any network of neurons that expresses HCN channels is primed to participate in theta rhythms, provided that the neurons are sufficiently depolarized to initiate the cycle described above. Depolarization could come from a variety of sources, depending on experimental conditions. Rhythmically firing cells would recruit other cells, and theta rhythms would spread across the network. In this scenario, local neuronal ensembles would actually be separate oscillators; the range of synchronization would be determined by the extent of pacemaker projections. Although the simplicity of this explanation is attractive, theta rhythm generation appears to be more complicated, considering that theta rhythms in vivo are not decreased in HCN1 knockout mice (Nolan et al. 2004, Giocomo et al. 2011). Nevertheless, it seems likely that some combination of intrinsic conductances and network mechanisms allows theta oscillatory activity to originate from a variety of different sources.

Support for the idea of multiple theta oscillators comes from recent studies of theta oscillations in the hippocampus. In freely behaving rats, theta rhythms are not synchronized across the hippocampus. Instead, systematic phase shifts are observed across the septotemporal axis (Lubenov & Siapas 2009, Patel et al. 2012). That is, hippocampal theta rhythms are traveling waves that consistently propagate toward the temporal pole of the hippocampus. The mechanisms of traveling wave generation remain unknown, but investigators have proposed several possibilities (Lubenov & Siapas 2009, Patel et al. 2012). One possibility is that theta waves emerge first in the septal hippocampus because MS inputs reach the septal hippocampus first. This proposed mechanism is problematic for several reasons (Patel et al. 2012). A more likely mechanism is that traveling theta waves reflect interactions within a network of weakly coupled oscillators (Lubenov & Siapas 2009, Patel et al. 2012). Separate oscillators may relate to differences in intrinsic conductances between septal (also called dorsal) and temporal (also called ventral) hippocampi. Maurer et al. (2005) found that intrinsic theta oscillations have a lower frequency in the intermediate hippocampus than they do in the septal hippocampus in freely behaving rats. This finding is consistent with septal-to-temporal propagation of traveling waves (but see Marcelin et al. 2012 for contradictory in vitro results). Frequencies of intrinsic theta oscillations are lower in ventral MEC than in dorsal MEC (Giocomo et al. 2007, Giocomo & Hasselmo 2008), leading Lubenov & Siapas (2009) to predict that MEC theta is a traveling wave also.

Recordings of theta rhythms in the isolated whole hippocampus also provide support for the weakly coupled oscillators mechanism (Goutagny et al. 2009). Coherence between septal and temporal theta rhythms in CA1 decreased significantly when an intermediate CA1 location was inactivated. After the inactivation, septal and temporal theta rhythms became uncoupled, and the frequency of temporal theta rhythms decreased significantly. Prior to the inactivation, septal theta rhythms led temporal theta rhythms by ~50 ms. This finding suggests that the septal hippocampus entrains the temporal hippocampus during theta rhythms, an idea that is consistent with septotemporal propagation of traveling theta waves. These results suggest that theta oscillators are coupled by local interactions between pyramidal cells and interneurons, not by MS projections. However, different coupling mechanisms likely exist in behaving animals, including entrainment of hippocampal theta by MS and EC oscillators. Coupling of EC and hippocampal theta oscillators may involve excitatory inputs from the EC to the hippocampus (Buzsáki 2002) or long-range inhibitory connections between the EC and the hippocampus (Melzer et al. 2012).

Despite decades of research dedicated to uncovering the mechanisms of theta rhythm generation, new and surprising findings regarding theta mechanisms continue to be discovered. Achieving a deeper understanding of the mechanisms of theta rhythms is an essential step toward fully understanding the functions of this complex rhythm. Additionally, a thorough understanding of theta mechanisms will help investigators develop methods to block theta rhythm generation selectively, with minimal side effects (e.g., loss of neuromodulatory inputs, changes in mean firing rates). Such a selective blockade would allow researchers to assess how functions are altered when theta rhythms are not present and to reveal causal links between theta and its functions.

FUNCTIONS OF THETA RHYTHMS

Behavioral Correlates of Theta Rhythms

Since the early days of theta rhythms research, scientists have looked for clues regarding theta functions by determining the behavioral correlates of theta rhythms (Buzsáki 2005). Theta rhythms arise during movement (Vanderwolf 1969) and exhibit higher amplitudes during active movement than during passive movement (Terrazas et al. 2005). Theta rhythms are also present during behaviors associated with intake of stimuli. The frequency of theta matches the frequency of whisking (Berg & Kleinfeld 2003) and sniffing (Macrides et al. 1982) in rats, as well as saccadic eye movements in humans (Otero-Millan et al. 2008). Theta rhythms selectively correlate with active, and not passive, sampling of stimuli. In rats sampling behaviorally relevant stimuli or stimuli associated with reward, theta rhythms were temporally correlated with rhythmic sniffing (Macrides et al. 1982) and whisking (Ganguly & Kleinfeld 2004). However, no relationship was observed between theta and whisking when rats were aimlessly whisking in air (Berg et al. 2006).

Curiously, theta rhythms also occur during rapid eye movement (REM) sleep (Vanderwolf 1969, Winson 1974), but the role of theta in REM sleep is less clear. One possibility is that theta rhythms play a role in memory consolidation during REM sleep (Louie & Wilson 2001, Montgomery et al. 2008), but this idea is controversial (Vertes 2004). This review does not cover functions of theta during REM sleep and instead focuses on functions of theta during waking states.

Information Packaging by Theta Rhythms

The hippocampus is at the end of an information-processing stream that begins when information is taken in from the environment. The hippocampus receives convergent information from multiple areas and thus requires a mechanism to coordinate related activity. The relationship between theta rhythms and rhythmic behaviors involved in stimuli intake suggests that each theta cycle contains a discrete sample of related sensory information (Kepecs et al. 2006). Theta rhythms may also link this information from different sensory modalities with information related to motivational or emotional states. Each theta cycle may thus represent a discrete processing entity that contains information about the current conditions at any given moment. This idea has recently received empirical support from two studies.

In the first study, Jezek and associates (2011) used an innovative behavioral task to investigate how CA3 neuronal ensembles respond to abrupt changes in environmental stimuli. In this task, rats were trained in two boxes that were distinguished by different sets of light cues. Initially, the two boxes were placed in different locations, connected by a corridor through which the rats could freely pass. This setup ensured that distinct ensembles of place cells were active in the two boxes, indicating that separate representations for the two environments had formed (Colgin et al. 2010). At a later stage of training, investigators presented an identical box that contained both sets of light cues in a common central location, midway between the two original locations. The light cues could be switched on and off to present one or the other environment, and separate representations for each environment were maintained. By switching the light cues, the authors could instantaneously "teleport" the rat from one environment to the other. After the switch, the network usually transitioned to the spatial representation that matched the new set of cues. In some cases, though, the network would flicker back and forth for a few seconds between the two representations, the one that matched the current set of cues and the one that matched the previous set of cues. During these episodes of instability, the different representations were separated by the period of a theta cycle, indicating that theta rhythms modulated the switching between distinct representations. In ~99% of cases, each theta cycle contained information related to only one of the memory representations, which suggests that theta rhythms link representations of sets of stimuli related to one environment and segregate them from representations of the other environment. Thus, these results support the hypothesis that theta cycles package related information.

Another study provided support for the hypothesis that theta cycles chunk information during spatial memory processing. In this study, Gupta and colleagues (2012) recorded CA1 place cells in rats running on a T-maze with two choice points. The maze contained several landmarks including two reward locations on the return arm. The authors analyzed theta cycles during which three or more place cells were active to find evidence of place cell sequences (i.e., sequences of spikes with a significant spatial and temporal structure related to the structure of behaviors on the maze). Place cell sequences within a theta cycle were found to represent particular paths on the maze. Longer theta cycles were associated with longer path lengths, suggesting that the theta cycle was the processing unit used to organize representations of discrete paths. Each theta cycle represented a path that began just behind the animal's current location and ended just in front of the animal. Theta sequences represented more space ahead of an animal leaving a landmark and more space behind an animal approaching a landmark, meaning that segments of the maze between landmarks were overrepresented. Thus, individual theta cycles contained representations of task-relevant maze segments, which suggests that a path containing different yet related spatial locations may be more efficiently encoded within a theta cycle as a coherent spatial concept (e.g., path between two feeders).

Gupta and colleagues (2012) also showed that some theta cycles heavily represented upcoming locations, whereas other theta cycles were biased toward locations in the recent past. The authors hypothesized that the theta sequences coding upcoming locations reflect a look-ahead mode, which signifies anticipation of an upcoming location. Sequences coding locations in the recent past may reflect a look-behind mode, which encodes recent experiences. The sequences studied by Gupta et al. (2012) may provide insights about place cell sequences during a phenomenon known as theta phase precession, as explained in the next section.

Theta phase precession:
a phenomenon in which spikes from an individual cell occur on progressively earlier theta phases across successive theta cycles

Theta Phase Precession

In theta phase precession, spikes from a place cell initially occur at late theta phases as the rat enters the cell's place field and then occur at progressively earlier phases on subsequent theta cycles (O'Keefe & Recce 1993, Skaggs et al. 1996) (**Figure 3***a*). Because each theta cycle is associated with spikes from multiple place cells with sequentially occurring place fields, individual theta cycles contain compressed representations of space (Skaggs et al. 1996; "theta sequences" in Foster & Wilson 2007). The compressed representation consists of a code for current location (represented by the most active place cells) preceded by representations of recently visited locations on earlier theta phases and succeeded by representations of upcoming locations on later theta phases (Dragoi & Buzsáki 2006). An alternative viewpoint is that spikes on early theta phases represent the current location, whereas spikes on later theta phases represent upcoming locations (Lisman & Redish 2009). The findings by Gupta et al. (2012), showing that some theta cycles contain sequences that largely represent space behind the rat, whereas sequences on other theta cycles represent more space ahead of the rat, seem to support the former interpretation of theta phase precession (for additional discussion of the predictive component of phase precession, see Lisman & Redish 2009). Here, the discussion of theta phase precession focuses on CA1 because the majority of phase precession studies have been conducted in CA1. However, it is important to note that theta phase precession has been reported in other regions as well, including the dentate gyrus (Yamaguchi et al. 2002, Mizuseki et al. 2009), CA3 (Mizuseki et al. 2009, 2012), the subiculum (Kim et al. 2012), the EC (Hafting et al. 2008, Mizuseki et al. 2009), and the ventral striatum (van der Meer & Redish 2011).

Many different models of theta phase precession propose various mechanisms for the phenomenon (O'Keefe & Recce 1993, Jensen & Lisman 1996, Tsodyks et al. 1996,

Wallenstein & Hasselmo 1997, Kamondi et al. 1998, Magee 2001, Harris et al. 2002, Mehta et al. 2002, Maurer & McNaughton 2007, Navratilova et al. 2012). Thorough comparison and contrast of the different models are beyond the scope of this review. I refer the reader to several other articles that discuss strengths and weaknesses of the various models (Maurer & McNaughton 2007, Lisman & Redish 2009, Burgess & O'Keefe 2011, Navratilova et al. 2012). Here, I instead discuss how recent whole-cell recordings of place cells in mice running in a virtual reality environment revealed mechanisms of phase precession that must be accounted for in future models (Harvey et al. 2009). First, depolarization of the place cell membrane potential during the first half of the place field ramped up asymmetrically (**Figure 3***b*), similar to asymmetric excitatory input proposed in the phase precession model of Mehta and colleagues (2002). Next, the power of membrane-potential theta oscillations was approximately two times higher within the cell's place field than that outside it. Last, spikes were phase-locked to intracellularly recorded theta and showed phase precession relative to LFP theta (**Figure 3***c–e*), consistent with predictions of "dual oscillator" models (Burgess & O'Keefe 2011).

Although this intracellular report provides extremely useful information regarding the mechanisms of theta phase precession, some questions remain. The intrinsic membrane-potential theta oscillations are thought to reflect rhythmic basket cell inhibition at the soma (Kamondi et al. 1998). Yet, Harvey and associates (2009) report that the frequency and amplitude of intracellular theta oscillations increase in the cell's place field, and it is not obvious how such increases would involve basket cell inhibition. Membrane-potential theta oscillations likely reflect basket cell inhibition in anesthetized animals (Kamondi et al. 1998) or at times when the animal is not in a cell's place field. When an animal enters a cell's place field, however, strong depolarization in the apical dendrites likely induces higher-frequency theta oscillations that result from intrinsic dendritic properties (Kamondi et al. 1998, Burgess & O'Keefe 2011). These faster dendritic oscillations may then entrain theta recorded at the soma, which would explain why spikes within the place field were so consistently phase-locked to the somatic oscillation (Harvey et al. 2009), even though inputs carrying spatial information arrive at the dendrites. Further support for this idea comes from recordings of theta under anesthesia (Kamondi et al. 1998), a condition in which excitatory dendritic inputs are suppressed (Ylinen et al. 1995). Under these conditions, intracellular and LFP theta oscillations share the same frequency (see figure 2*b* in Kamondi et al. 1998).

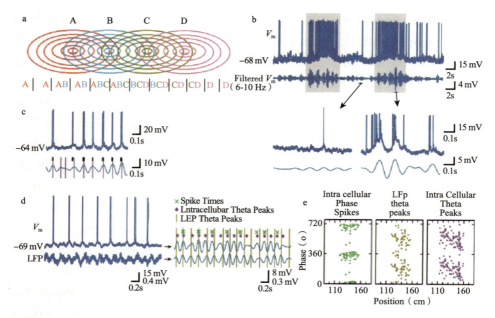

Figure 3

(*a*) Schematic of theta phase precession. Four partially overlapping place fields are shown for an animal running to the right. Repeated activation of place cell sequences on theta cycles (*vertical lines*) is depicted below. Reproduced with permission from Skaggs et al. (1996).(*b*) Example of unfiltered and theta-filtered (6–10 Hz) whole cell recordings from a mouse exploring a virtual environment during runs through the cell's place field (*gray*). Magnified recordings for the periods indicated with horizontal bars are shown below. (*c*) Example of raw (*top*) and theta-filtered membrane-potential recordings (*bottom*) illustrate phase-locking of spikes to intracellular theta. Black vertical lines indicate spike times. Peaks of filtered theta are indicated by purple vertical lines. (*d*) Raw (*left*) and theta-filtered (*right*) traces are shown for simultaneous whole cell (*top*) and local field potential (LFP) recordings (*bottom*). Note how spikes are locked to the peak of intracellular theta oscillations and show phase precession with respect to LFP theta. The top and bottom calibration bar labels indicate the scale for top and bottom traces, respectively. (*e*) Spikes' intracellular theta phases at positions along the virtual environment are plotted (*left*); note how theta phase-locking, but not theta phase precession, is seen. Spikes' phases with respect to LFP theta at positions in the virtual environment are plotted (*middle*); theta phase precession is apparent. The LFP theta phases associated with intracellularly detected theta peaks are plotted on the y axis, and corresponding positions are plotted on the x axis (*right*); note that intracellular theta shows phase precession with respect to LFP theta. Panels *b, c, d,* and e are modified with permission from Harvey et al. (2009).

The question remains about how these proposed mechanisms of phase precession relate to the proposed sequence retrieval function of phase precession. One possibility is that gamma oscillations are involved in both theta phase precession and sequence retrieval. Gamma oscillations co-occur with theta oscillations (Buzsáki et al. 1983, Bragin et al. 1995) and may play a role in theta phase precession (Jensen & Lisman 1996, Dragoi & Buzsáki 2006). Spiking of some CA1 place cells is modulated by gamma oscillations during theta phase precession (Senior et al. 2008). Additionally, recent work indicates that theta-modulated gamma is the optimal pattern for dendritic signal propagation (Vaidya & Johnston 2012), raising the possibility that theta-modulated gamma inputs arriving in the dendrites (Bragin et al. 1995, Colgin et al. 2009) enhance membrane-potential oscillations during theta phase precession. Gamma oscillations have also been linked to sequence retrieval. Strong gamma rhythms occur when place cells represent upcoming sequences of locations as rats pause at decision points on a maze (Johnson & Redish 2007), and slow gamma rhythms promote sequence reactivation during nontheta states (Carr et al. 2012).

If gamma-mediated sequence retrieval occurs during theta phase precession, phase-phase coupling (Jensen & Colgin 2007) is expected between theta and gamma oscillations because spikes would be phase-locked to gamma while still maintaining phase precession relationships with theta. In support of this idea, Belluscio and colleagues (2012) recently reported phase-phase coupling between theta and gamma in the hippocampus. When they subdivided gamma into different frequency bands, the slow gamma band displayed the strongest phase-phase coupling (see figure *7a* in Belluscio et al. 2012). Sequence reactivation in the absence of theta involved slow gamma oscillations but no other gamma frequencies (Carr et al. 2012). Slow gamma rhythms couple CA3 and CA1 (Colgin et al. 2009), and CA3 is thought to be critical for retrieving representations of upcoming locations (Jensen & Lisman 1996, Dragoi & Buzsáki 2006). Thus, slow gamma rhythms may transmit retrieved sequences of upcoming locations from CA3 to CA1 during theta phase precession; in this scheme, representations of consecutive locations would be activated on successive gamma cycles (Jensen & Lisman 1996, Dragoi & Buzsáki 2006). Inputs from the EC provide information about current location (Hafting et al. 2005) and likely serve as a cue for CA3 recall of upcoming locations (Jensen & Lisman 1996, Dragoi & Buzsáki 2006). Layer III EC inputs to CA1 may contribute to phase precession in the second half of the place field because CA3 cells do not receive layer III input and do not show phase precession across the full extent of the theta cycle (Mizuseki et al. 2012).

Work described above demonstrates how intrinsic cellular properties may combine with synaptic interactions to facilitate hippocampal functions such as sequence retrieval. Thus far, most results discussed here involve effects within an individual region, mainly CA1. The next section discusses how theta rhythms affect interactions across different brain regions.

Theta Coupling between Regions

Theta oscillations are seen in structures involved in initial stages of sensory processing as well as in regions further down the processing stream (Jung & Kornmuller 1938, Vanderwolf 1969, Jones & Wilson 2005, Kay 2005). Much evidence suggests that theta rhythms are involved in facilitating the transfer of information from one brain region to another during sensory information processing. Theta-enhanced transmission across brain regions may be important for several different functions. Here, I focus on the role of interregional theta coupling in memory operations involving the hippocampus.

Consistent with the hippocampus's key role in spatial memory processing, several studies have demonstrated links between interregional theta coupling and performance on a variety of spatial memory tasks. In a spatial working-memory task, medial prefrontal

mPFC: medial prefrontal cortex

cortex (mPFC) neurons were more strongly phase-locked to CA1 theta rhythms during correct-choice trials than during error trials (Jones & Wilson 2005). Additionally, Jones & Wilson (2005) observed theta coherence between CA1 and mPFC during choice trials but not during forced-turn trials. Similarly, in a delayed nonmatch to position task, Hyman and colleagues (2010) found that 94% of theta-modulated mPFC neurons were significantly more phase-locked to hippocampal theta during correct trials than during error trials. Moreover, theta coherence between CA1 and mPFC increased significantly after rule acquisition at the choice point in a Y-maze with periodically switching reward contingency rules (Benchenane et al. 2010). In another study involving the hippocampus and the mPFC, the proportion of mPFC neurons that were phase-locked to hippocampal theta oscillations increased after successful learning of an object-place association (Kim et al. 2011).

The mPFC is not the only region that exhibits theta coupling with the hippocampus during spatial memory processing. Theta coherence between the striatum and the hippocampus increased during the period between the tone and the selected turn in a tone-cued T-maze task, but only in rats that successfully learned the task (DeCoteau et al. 2007). In this same task, coupling between striatal theta phase and hippocampal gamma amplitude was observed during the tone onset period (Tort et al. 2008). The authors suggested that striatal theta phase-hippocampal gamma amplitude coupling may signify times during decision making when the striatum accesses spatial information from the hippocampus.

The studies described thus far in this section involve spatial memory, but interregional theta coupling appears to be involved in other types of memory processing as well. The results of one study indicate that theta coordination of the hippocampus and the amygdala is important for fear memory retrieval. Theta coupling between the lateral amygdala and CA1 was significantly increased when conditioned fear stimuli were presented, and significant correlations between CA1 and lateral amygdala theta were seen in animals that displayed behavioral signs of fear (i.e., freezing; Seidenbecher et al. 2003). Interregional theta coupling also affects memory processing of nonaversive stimuli. Kay (2005) found positive correlations between performance on a two-odor discrimination task and theta coherence of the olfactory bulb and the hippocampus, suggesting that theta coupling enhances communication between the olfactory bulb and the hippocampus during olfactory memory processing. In a trace conditioning task using a visual conditioned stimulus, theta synchronization across different mPFC sites increased after investigators presented the conditioned stimulus only after learning had occurred (Paz et al. 2008). Because CA1 and the subiculum project to the mPFC (Swanson 1981, Jay & Witter 1991), the authors hypothesized that enhanced theta coupling with hippocampal inputs was responsible for the increased synchronization across mPFC.

The above-described studies show theta coordination during memory processing involving the hippocampus. However, theta coupling across regions appears to be a more general mechanism that is useful for other cognitive operations and physiological states, including visual short-term memory (Liebe et al. 2012) and anxiety (Adhikari et al. 2010). Reductions in theta coupling across regions may also relate to behavioral deficits in diseases such as schizophrenia (Dickerson et al. 2010, Sigurdsson et al. 2010). Remarkably, a new study using a rat model of schizophrenia found that interventions that restore normal performance on a cognitive control task also reestablish healthy theta synchrony between left and right hippocampi (Lee et al. 2012).

How does theta coupling enhance communication across brain areas? Theta synchronization of neurons in a given brain area likely leads to a more effective activation of downstream targets. Interregional theta coupling likely also ensures that downstream neurons will be excitable when inputs arrive. Theta's relatively slow time scale permits long synaptic delays and thus can feasibly sustain coupling across distributed brain regions. These points, taken together with the results reviewed above, support the conclusion that theta rhythms promote coordination across distributed brain areas during different types of information processing.

CONCLUDING REMARKS

Studies in recent years have provided breakthroughs in our understanding of the mechanisms and functions of theta rhythms. Theta-modulated neurons likely express intrinsic properties that prime them to produce theta oscillations in response to a variety of extrinsic inputs. Projections from interneurons in the MS are thought to pace theta in most neurons in the hippocampus at a given time. However, when an animal is in a cell's place field, excitatory CA3 and entorhinal inputs in the dendrites may produce stronger and faster theta oscillations that entrain the cell's firing. This flexibility in theta entrainment likely allows inputs to select the appropriate cell ensembles during particular cognitive tasks. Theta synchronization of related cell ensembles may then promote effective activation of target structures and thereby facilitate cognitive operations such as learning.

Although much progress has been made, many questions remain regarding the mechanisms and functions of theta rhythms. How important are intrinsic neuronal mechanisms for theta generation? It would be interesting to determine how theta rhythms in vivo are affected by manipulations that selectively and reversibly inhibit conductances that are tuned to theta. Regarding theta functions, major questions persist about how results from rodents pertain to other species. Theta rhythms are continuous in rodents during active behaviors (Buzsáki et al. 1983, Buzsáki 2002) but occur only in short bouts in humans (Kahana et al. 1999), bats (Yartsev et al. 2011), and monkeys (Killian et al. 2012). It seems unlikely that complex functions such as learning could be achieved by short theta bouts in bats and primates and yet require continuous theta in rats. One possibility is that the conditions used in typical primate experiments are not optimal for engaging theta machinery. Rodents in experimental settings often engage in behaviors that are similar to their natural behaviors (e.g., foraging for food). Human studies of theta often involve virtual reality environments. These uncommon behaviors may not readily activate circuits involved in theta rhythm generation. Moreover, foraging tasks require rodents to use a combination of sensory cues (e.g., olfactory, tactile, visual) and thus would be expected to engage a mechanism that packages related information from different sensory modalities. In contrast, investigations of theta in primates typically employ tasks requiring only one sensory modality, vision. Perhaps theta-generating machinery is less easily triggered in such tasks. Wireless monitoring devices implanted in patients undergoing deep brain stimulation may provide a way for researchers to measure theta rhythms during ordinary human behaviors. However, a recent study has shown that the activity of hippocampal place cells in bats is not modulated by theta rhythms during flying, bats' natural exploratory behavior (Yartsev & Ulanovsky 2013). Moreover, bat MEC neurons, unlike rat MEC neurons, do not exhibit membrane-potential resonance at theta frequencies (Heys et al. 2013). The puzzles that remain regarding theta functions across species underscore the importance of ongoing theta research in humans and in animal models.

DISCLOSURE STATEMENT

The author is not aware of any affiliations, memberships, funding, or financial holdings that might be perceived as affecting the objectivity of this review.

ACKNOWLEDGMENTS

I thank the Klingenstein Fund and the Sloan Foundation for their support.

LITERATURE CITED

Adhikari A, Topiwala MA, Gordon JA. 2010. Synchronized activity between the ventral hippocampus and the medial prefrontal cortex during anxiety. *Neuron* 65:257–69

Belluscio MA, Mizuseki K, Schmidt R, Kempter R, Buzsáki G. 2012. Cross-frequency phase-phase coupling between theta and gamma oscillations in the hippocampus. *J. Neurosci.* 32:423–35

Benchenane K, Peyrache A, Khamassi M, Tierney PL, Gioanni Y, et al. 2010. Coherent theta oscillations and reorganization of spike timing in the hippocampal-prefrontal network upon learning. *Neuron* 66:921–36

Berg RW, Kleinfeld D. 2003. Rhythmic whisking by rat: retraction as well as protraction of the vibrissae is under active muscular control. *J. Neurophysiol.* 89:104–17

Berg RW, Whitmer D, Kleinfeld D. 2006. Exploratory whisking by rat is not phase locked to the hippocampal theta rhythm. *J. Neurosci.* 26:6518–22

Berry SD, Thompson RF. 1978. Prediction of learning rate from the hippocampal electroencephalogram. *Science* 200: 1298–300

Bland BH, Oddie SD, Colom LV. 1999. Mechanisms of neural synchrony in the septo-hippocampal pathways underlying hippocampal theta generation. *J. Neurosci.* 19:3223–37

Blasco-Ibá ñez JM, Freund TF. 1995. Synaptic input of horizontal interneurons in stratum oriens of the hippocampal CA1 subfield: structural basis of feed-back activation. *Eur. J. Neurosci.* 7:2170–80

Brager DH, Johnston D. 2007. Plasticity of intrinsic excitability during long-term depression is mediated through mGluR-dependent changes in I(h) in hippocampal CA1 pyramidal neurons. *J. Neurosci.* 27:13926–37

Bragin A, Jando G, Nadasdy Z, Hetke J, Wise K, Buzsáki G. 1995. Gamma (40–100 Hz) oscillation in the hippocampus of the behaving rat. *J. Neurosci.* 15:47–60

Brandon MP, Bogaard AR, Libby CP, Connerney MA, Gupta K, Hasselmo ME. 2011. Reduction of theta rhythm dissociates grid cell spatial periodicity from directional tuning. *Science* 332:595–99

Burgess N, O'Keefe J. 2011. Models of place and grid cell firing and theta rhythmicity. *Curr. Opin. Neurobiol.* 21:734–44

Buzsáki G. 2002. Theta oscillations in the hippocampus. *Neuron* 33:325–40

Buzsáki G. 2005. Theta rhythm of navigation: link between path integration and landmark navigation, episodic and semantic memory. *Hippocampus* 15:827–40

Buzsáki G. 2006. *Rhythms of the Brain.* New York: Oxford Univ. Press. 448 pp.

Buzsáki G, Leung LW, Vanderwolf CH. 1983. Cellular bases of hippocampal EEG in the behaving rat. *Brain Res.* 287:139–71

Carr MF, Karlsson MP, Frank LM. 2012. Transient slow gamma synchrony underlies hippocampal memory replay. *Neuron* 75:700–13

Colgin LL, Denninger T, Fyhn M, Hafting T, Bonnevie T, et al. 2009. Frequency of gamma oscillations routes flow of information in the hippocampus. *Nature* 462:353–57

Colgin LL, Kubota D, Jia Y, Rex CS, Lynch G. 2004. Long-term potentiation is impaired in rat hippocampal slices that produce spontaneous sharp waves. *J. Physiol.* 558:953–61

Colgin LL, Leutgeb S, Jezek K, Leutgeb JK, Moser EI, et al. 2010. Attractor-map versus autoassociation based attractor dynamics in the hippocampal network. *J. Neurophysiol.* 104:35–50

Colgin LL, Moser EI. 2009. Hippocampal theta rhythms follow the beat of their own drum. *Nat. Neurosci.* 12:1483–84

DeCoteau WE, Thorn C, Gibson DJ, Courtemanche R, Mitra P, et al. 2007. Learning-related coordination of striatal and hippocampal theta rhythms during acquisition of a procedural maze task. *Proc. Natl. Acad. Sci. USA* 104:5644–49

Dickerson DD, Wolff AR, Bilkey DK. 2010. Abnormal long-range neural synchrony in a maternal immune activation animal model of schizophrenia. *J. Neurosci.* 30:12424–31

Dickson CT, Magistretti J, Shalinsky MH, Fransén E, Hasselmo ME, Alonso A. 2000. Properties and role of I(h) in the pacing of subthreshold oscillations in entorhinal cortex layer II neurons. *J. Neurophysiol.* 83:2562–79

Dragoi G, Buzsáki G. 2006. Temporal encoding of place sequences by hippocampal cell assemblies. *Neuron* 50:145–57

Ellender TJ, Nissen W, Colgin LL, Mann EO, Paulsen O. 2010. Priming of hippocampal population bursts by individual perisomatic-targeting interneurons. *J. Neurosci.* 30:5979–91

Foster DJ, Wilson MA. 2007. Hippocampal theta sequences. *Hippocampus* 17:1093–99

Frerking M, Schmitz D, Zhou Q, Johansen J, Nicoll RA. 2001. Kainate receptors depress excitatory synaptic transmission at CA3→CA1 synapses in the hippocampus via a direct presynaptic action. *J. Neurosci.* 21:2958–66

Ganguly K, Kleinfeld D. 2004. Goal-directed whisking increases phase-locking between vibrissa movement and electrical activity in primary sensory cortex in rat. *Proc. Natl. Acad. Sci. USA* 101:12348–53

Giocomo LM, Hasselmo ME. 2008. Time constants of h current in layer II stellate cells differ along the dorsal to ventral axis of medial entorhinal cortex. *J. Neurosci.* 28:9414–25

Giocomo LM, Hasselmo ME. 2009. Knock-out of HCN1 subunit flattens dorsal-ventral frequency gradient of medial entorhinal neurons in adult mice. *J. Neurosci.* 29:7625–30

Giocomo LM, Hussaini SA, Zheng F, Kandel ER, Moser MB, Moser EI. 2011. Grid cells use HCN1 channels for spatial scaling. *Cell* 147:1159–70

Giocomo LM, Zilli EA, Fransé n E, Hasselmo ME. 2007. Temporal frequency of subthreshold oscillations scales with entorhinal grid cell field spacing. *Science* 315:1719–22

Goutagny R, Jackson J, Williams S. 2009. Self-generated theta oscillations in the hippocampus. *Nat. Neurosci.* 12:1491–93

Grastyan E, Lissak K, Madarasz I, Donhoffer H. 1959. Hippocampal electrical activity during the development of conditioned reflexes. *Electroencephalogr. Clin. Neurophysiol.* 11:409–30

Green JD, Arduini AA. 1954. Hippocampal electrical activity in arousal. *J. Neurophysiol.* 17:533–57

Greenstein YJ, Pavlides C, Winson J. 1988. Long-term potentiation in the dentate gyrus is preferentially induced at theta rhythm periodicity. *Brain Res.* 438:331–34

Gupta AS, van der Meer MA, Touretzky DS, Redish AD. 2012. Segmentation of spatial experience by hippocampal theta sequences. *Nat. Neurosci.* 15:1032–39

Hafting T, Fyhn M, Bonnevie T, Moser MB, Moser EI. 2008. Hippocampus-independent phase precession in entorhinal grid cells. *Nature* 453:1248–52

Hafting T, Fyhn M, Molden S, Moser MB, Moser EI. 2005. Microstructure of a spatial map in the entorhinal cortex. *Nature* 436:801–6

Hangya B, Borhegyi Z, Szilagyi N, Freund TF, Varga V. 2009. GABAergic neurons of the medial septum lead the hippocampal network during theta activity. *J. Neurosci.* 29:8094–102

Harris KD, Henze DA, Hirase H, Leinekugel X, Dragoi G, et al. 2002. Spike train dynamics predicts theta-related phase precession in hippocampal pyramidal cells. *Nature* 417:738–41

Harvey CD, Collman F, Dombeck DA, Tank DW. 2009. Intracellular dynamics of hippocampal place cells during virtual navigation. *Nature* 461:941–46

Heys JG, MacLeod KM, Moss CF, Hasselmo ME. 2013. Bat and rat neurons differ in theta-frequency resonance despite similar coding of space. *Science* 340:363–67

Hu H, Vervaeke K, Storm JF. 2002. Two forms of electrical resonance at theta frequencies, generated by M-current, h-current and persistent Na+ current in rat hippocampal pyramidal cells. *J. Physiol.* 545:783–805

Hyman JM, Wyble BP, Goyal V, Rossi CA, Hasselmo ME. 2003. Stimulation in hippocampal region CA1 in behaving rats yields long-term potentiation when delivered to the peak of theta and long-term depression when delivered to the trough. *J. Neurosci.* 23:11725–31

Hyman JM, Zilli EA, Paley AM, Hasselmo ME. 2010. Working memory performance correlates with prefrontal-hippocampal theta interactions but not with prefrontal neuron firing rates. *Front. Integr. Neurosci.* 4:2

Jay TM, Witter MP. 1991. Distribution of hippocampal CA1 and subicular efferents in the prefrontal cortex of the rat studied by means of anterograde transport of Phaseolus vulgaris-leucoagglutinin. *J. Comp. Neurol.* 313:574–86

Jensen O, Colgin LL. 2007. Cross-frequency coupling between neuronal oscillations. *Trends Cogn. Sci.* 11:267–69

Jensen O, Lisman JE. 1996. Hippocampal CA3 region predicts memory sequences: accounting for the phase precession of place cells. *Learn. Mem.* 3:279–87

Jezek K, Henriksen EJ, Treves A, Moser EI, Moser MB. 2011. Theta-paced flickering between place-cell maps in the hippocampus. *Nature* 478:246–49

Johnson A, Redish AD. 2007. Neural ensembles in CA3 transiently encode paths forward of the animal at a decision point. *J. Neurosci.* 27:12176–89

Jones MW, Wilson MA. 2005. Theta rhythms coordinate hippocampal-prefrontal interactions in a spatial memory task. *PLoS Biol.* 3(12):e402

Jung R, Kornmuller AE. 1938. Eine Methodik der ableitung lokalisierter Potentialschwankungen aus subcorticalen Hirngebieten. *Arch. Psychiat. Nervenkrankh.* 109:1–30

Kahana MJ, Sekuler R, Caplan JB, Kirschen M, Madsen JR. 1999. Human theta oscillations exhibit task dependence during virtual maze navigation. *Nature* 399:781–84

Kamondi A, Acsády L, Wang XJ, Buzsáki G. 1998. Theta oscillations in somata and dendrites of hippocampal pyramidal cells in vivo: activity-dependent phase-precession of action potentials. *Hippocampus* 8:244–61

Kay LM. 2005. Theta oscillations and sensorimotor performance. *Proc. Natl. Acad. Sci. USA* 102:3863–68

Kepecs A, Uchida N, Mainen ZF. 2006. The sniff as a unit of olfactory processing. *Chem. Senses* 31:167–79

Killian NJ, Jutras MJ, Buffalo EA. 2012. A map of visual space in the primate entorhinal cortex. *Nature* 491:761–64

Kim J, Delcasso S, Lee I. 2011. Neural correlates of object-in-place learning in hippocampus and prefrontal cortex. *J. Neurosci.* 31:16991–7006

Kim SM, Ganguli S, Frank LM. 2012. Spatial information outflow from the hippocampal circuit: distributed spatial coding and phase precession in the subiculum. *J. Neurosci.* 32:11539–58

Klimesch W, Doppelmayr M, Russegger H, Pachinger T. 1996. Theta band power in the human scalp EEG and the encoding of new information. *NeuroReport* 7:1235–40

Klimesch W, Schimke H, Schwaiger J. 1994. Episodic and semantic memory: an analysis in the EEG theta and alpha band. *Electroencephalogr. Clin. Neurophysiol.* 91:428–41

Koenig J, Linder AN, Leutgeb JK, Leutgeb S. 2011. The spatial periodicity of grid cells is not sustained during reduced theta oscillations. *Science* 332:592–95

Kramis R, Vanderwolf CH, Bland BH. 1975. Two types of hippocampal rhythmical slow activity in both the rabbit and the rat: relations to behavior and effects of atropine, diethyl ether, urethane, and pentobarbital. *Exp. Neurol.* 49:58–85

Kubota D, Colgin LL, Casale M, Brucher FA, Lynch G. 2003. Endogenous waves in hippocampal slices. *J. Neurophysiol.* 89:81–89

Landfield PW, McGaugh JL, Tusa RJ. 1972. Theta rhythm: a temporal correlate of memory storage processes in the rat. *Science* 175:87–89

Larson J, Wong D, Lynch G. 1986. Patterned stimulation at the theta frequency is optimal for the induction of hippocampal long-term potentiation. *Brain Res.* 368:347–50

Lee H, Dvorak D, Kao HY, Duffy AM, Scharfman HE, Fenton AA. 2012. Early cognitive experience prevents adult deficits in a neurodevelopmental schizophrenia model. *Neuron* 75:714–24

Liebe S, Hoerzer GM, Logothetis NK, Rainer G. 2012. Theta coupling between V4 and prefrontal cortex predicts visual short-term memory performance. *Nat. Neurosci.* 15:456–62

Lisman J, Redish AD. 2009. Prediction, sequences and the hippocampus. *Philos. Trans. R. Soc. Lond. B* 364: 1193–201

Louie K, Wilson MA. 2001. Temporally structured replay of awake hippocampal ensemble activity during rapid eye movement sleep. *Neuron* 29:145–56

Lubenov EV, Siapas AG. 2009. Hippocampal theta oscillations are travelling waves. *Nature* 459:534–39

Maccaferri G. 2005. Stratum oriens horizontal interneurone diversity and hippocampal network dynamics . *J. Physiol.* 562:73–80

Maccaferri G, McBain CJ. 1996. The hyperpolarization-activated current (Ih) and its contribution to pacemaker activity in rat CA1 hippocampal stratum oriens-alveus interneurones. *J. Physiol.* 497(Pt.

1):119–30

Macrides F, Eichenbaum HB, Forbes WB. 1982. Temporal relationship between sniffing and the limbic theta rhythm during odor discrimination reversal learning. *J. Neurosci.* 2:1705–17

Madison DV, Lancaster B, Nicoll RA. 1987. Voltage clamp analysis of cholinergic action in the hippocampus. *J. Neurosci.* 7:733–41

Magee JC. 2001. Dendritic mechanisms of phase precession in hippocampal CA1 pyramidal neurons. *J. Neurophysiol.* 86:528–32

Marcelin B, Lugo JN, Brewster AL, Liu Z, Lewis AS, et al. 2012. Differential dorsoventral distributions of Kv4.2 and HCN proteins confer distinct integrative properties to hippocampal CA1 pyramidal cell distal dendrites. *J. Biol. Chem.* 287:17656–61

Maurer AP, McNaughton BL. 2007. Network and intrinsic cellular mechanisms underlying theta phase precession of hippocampal neurons. *Trends Neurosci.* 30:325–33

Maurer AP, Vanrhoads SR, Sutherland GR, Lipa P, McNaughton BL. 2005. Self-motion and the origin of differential spatial scaling along the septo-temporal axis of the hippocampus. *Hippocampus* 15:841–52

Mehta MR, Lee AK, Wilson MA. 2002. Role of experience and oscillations in transforming a rate code into a temporal code. *Nature* 417:741–46

Melzer S, Michael M, Caputi A, Eliava M, Fuchs EC, et al. 2012. Long-range-projecting GABAergic neurons modulate inhibition in hippocampus and entorhinal cortex. *Science* 335:1506–10

M'Harzi M, Jarrard LE. 1992. Effects of medial and lateral septal lesions on acquisition of a place and cue radial maze task. *Behav. Brain Res.* 49:159–65

Mitchell SJ, Rawlins JN, Steward O, Olton DS. 1982. Medial septal area lesions disrupt theta rhythm and cholinergic staining in medial entorhinal cortex and produce impaired radial arm maze behavior in rats. *J. Neurosci.* 2:292–302

Mizumori SJ, Perez GM, Alvarado MC, Barnes CA, McNaughton BL. 1990. Reversible inactivation of the medial septum differentially affects two forms of learning in rats. *Brain Res.* 528:12–20

Mizuseki K, Royer S, Diba K, Buzsáki G. 2012. Activity dynamics and behavioral correlates of CA3 and CA1 hippocampal pyramidal neurons. *Hippocampus* 22:1659–80

Mizuseki K, Sirota A, Pastalkova E, Buzsáki G. 2009. Theta oscillations provide temporal windows for local circuit computation in the entorhinal-hippocampal loop. *Neuron* 64:267–80

Montgomery SM, Sirota A, Buzsáki G. 2008. Theta and gamma coordination of hippocampal networks during waking and rapid eye movement sleep. *J. Neurosci.* 28:6731–41

Navratilova Z, Giocomo LM, Fellous JM, Hasselmo ME, McNaughton BL. 2012. Phase precession and variable spatial scaling in a periodic attractor map model of medial entorhinal grid cells with realistic after-spike dynamics. *Hippocampus* 22:772–89

Nolan MF, Malleret G, Dudman JT, Buhl DL, Santoro B, et al. 2004. A behavioral role for dendritic integration: HCN1 channels constrain spatial memory and plasticity at inputs to distal dendrites of CA1 pyramidal neurons. *Cell* 119:719–32

O'Keefe J, Dostrovsky J. 1971. The hippocampus as a spatial map. Preliminary evidence from unit activity in the freely-moving rat. *Brain Res.* 34:171–75

O'Keefe J, Recce ML. 1993. Phase relationship between hippocampal place units and the EEG theta rhythm. *Hippocampus* 3:317–30

Orr G, Rao G, Houston FP, McNaughton BL, Barnes CA. 2001. Hippocampal synaptic plasticity is modulated by theta rhythm in the fascia dentata of adult and aged freely behaving rats. *Hippocampus* 11:647–54

Osipova D, Takashima A, Oostenveld R, Fernández G, Maris E, Jensen O. 2006. Theta and gamma oscillations predict encoding and retrieval of declarative memory. *J. Neurosci.* 26:7523–31

Otero-Millan J, Troncoso XG, Macknik SL, Serrano-Pedraza I, Martinez-Conde S. 2008. Saccades and microsaccades during visual fixation, exploration, and search: foundations for a common saccadic generator. *J. Vis.* 8:1–18

Patel J, Fujisawa S, Berenyi A, Royer S, Buzsáki G. 2012. Traveling theta waves along the entire septotemporal axis of the hippocampus. *Neuron* 75:410–17

Pavlides C, Greenstein YJ, Grudman M, Winson J. 1988. Long-term potentiation in the dentate gyrus is induced preferentially on the positive phase of theta-rhythm. *Brain Res.* 439:383–87

Paz R, Bauer EP, Pare D. 2008. Theta synchronizes the activity of medial prefrontal neurons during learning. *Learn. Mem.* 15:524–31

Petsche H, Stumpf C, Gogolak G. 1962. The significance of the rabbit's septum as a relay station between the midbrain and the hippocampus. I. The control of hippocampus arousal activity by the septum cells. *Electroencephalogr. Clin. Neurophysiol.* 14:202–11

Pike FG, Goddard RS, Suckling JM, Ganter P, Kasthuri N, Paulsen O. 2000. Distinct frequency preferences of different types of rat hippocampal neurones in response to oscillatory input currents. *J. Physiol.* 529 (Pt. 1):205–13

Robbe D, Buzsáki G. 2009. Alteration of theta timescale dynamics of hippocampal place cells by a cannabinoid is associated with memory impairment. *J. Neurosci.* 29:12597–605

Robinson RB, Siegelbaum SA. 2003. Hyperpolarization-activated cation currents: from molecules to physiological function. *Annu. Rev. Physiol.* 65:453–80

Rutishauser U, Ross IB, Mamelak AN, Schuman EM. 2010. Human memory strength is predicted by theta-frequency phase-locking of single neurons. *Nature* 464:903–7

Seidenbecher T, Laxmi TR, Stork O, Pape HC. 2003. Amygdalar and hippocampal theta rhythm synchronization during fear memory retrieval. *Science* 301:846–50

Senior TJ, Huxter JR, Allen K, O'Neill J, Csicsvari J. 2008. Gamma oscillatory firing reveals distinct populations of pyramidal cells in the CA1 region of the hippocampus. *J. Neurosci.* 28:2274–86

Sigurdsson T, Stark KL, Karayiorgou M, Gogos JA, Gordon JA. 2010. Impaired hippocampal-prefrontal synchrony in a genetic mouse model of schizophrenia. *Nature* 464:763–67

Simon AP, Poindessous-Jazat F, Dutar P, Epelbaum J, Bassant MH. 2006. Firing properties of anatomically identified neurons in the medial septum of anesthetized and unanesthetized restrained rats. *J. Neurosci.* 26:9038–46

Skaggs WE, McNaughton BL, Wilson MA, Barnes CA. 1996. Theta phase precession in hippocampal neuronal populations and the compression of temporal sequences. *Hippocampus* 6:149–72

Squire LR, Stark CE, Clark RE. 2004. The medial temporal lobe. *Annu. Rev. Neurosci.* 27:279–306

Staubli U, Lynch G. 1987. Stable hippocampal long-term potentiation elicited by 'theta' pattern stimulation. *Brain Res.* 435:227–34

Swanson LW. 1981. A direct projection from Ammon's horn to prefrontal cortex in the rat. *Brain Res.* 217:150–54

Terrazas A, Krause M, Lipa P, Gothard KM, Barnes CA, McNaughton BL. 2005. Self-motion and the hippocampal spatial metric. *J. Neurosci.* 25:8085–96

Tesche CD, Karhu J. 2000. Theta oscillations index human hippocampal activation during a working memory task. *Proc. Natl. Acad. Sci. USA* 97:919–24

Tort AB, Kramer MA, Thorn C, Gibson DJ, Kubota Y, et al. 2008. Dynamic cross-frequency couplings of local field potential oscillations in rat striatum and hippocampus during performance of a T-maze task. *Proc. Natl. Acad. Sci. USA* 105:20517–22

Toth K, Borhegyi Z, Freund TF. 1993. Postsynaptic targets of GABAergic hippocampal neurons in the medial septum-diagonal band of Broca complex. *J. Neurosci.* 13:3712–24

Toth K, Freund TF, Miles R. 1997. Disinhibition of rat hippocampal pyramidal cells by GABAergic afferents from the septum. *J. Physiol.* 500(Pt. 2):463–74

Tsodyks MV, Skaggs WE, Sejnowski TJ, McNaughton BL. 1996. Population dynamics and theta rhythm phase precession of hippocampal place cell firing: a spiking neuron model. *Hippocampus* 6:271–80

Vaidya SP, Johnston D. 2012. HCN channels contribute to the spatial synchrony of theta frequency synaptic inputs in CA1 pyramidal neurons. *Soc. Neurosci. Annu. Meet. Abstr.* 435.15 (Abstr.)

van der Meer MA, Redish AD. 2011. Theta phase precession in rat ventral striatum links place and reward information. *J. Neurosci.* 31:2843–54

Vanderwolf CH. 1969. Hippocampal electrical activity and voluntary movement in the rat. *Electroencephalogr. Clin. Neurophysiol.* 26:407–18

Varga V, Hangya B, Kranitz K, Ludanyi A, Zemankovics R, et al. 2008. The presence of pacemaker HCN channels identifies theta rhythmic GABAergic neurons in the medial septum. *J. Physiol.* 586:3893–915

Vertes RP. 2004. Memory consolidation in sleep: dream or reality. *Neuron* 44:135–48

Vertes RP, Hoover WB, Viana Di Prisco G. 2004. Theta rhythm of the hippocampus: subcortical control and functional significance. *Behav. Cogn. Neurosci. Rev.* 3:173–200

Vertes RP, Kocsis B. 1997. Brainstem-diencephalo-septohippocampal systems controlling the theta rhythm of the hippocampus. *Neuroscience* 81:893–926

Wallenstein GV, Hasselmo ME. 1997. GABAergic modulation of hippocampal population activity: sequence learning, place field development, and the phase precession effect. *J. Neurophysiol.* 78:393–408

Winson J. 1974. Patterns of hippocampal theta rhythm in the freely moving rat. *Electroencephalogr. Clin. Neurophysiol.* 36:291–301

Winson J. 1978. Loss of hippocampal theta rhythm results in spatial memory deficit in the rat. *Science* 201:160–63

Yamaguchi Y, Aota Y, McNaughton BL, Lipa P. 2002. Bimodality of theta phase precession in hippocampal place cells in freely running rats. *J. Neurophysiol.* 87:2629–42

Yartsev MM, Ulanovsky N. 2013. Representation of three-dimensional space in the hippocampus of flying bats. *Science* 340:367–72

Yartsev MM, Witter MP, Ulanovsky N. 2011. Grid cells without theta oscillations in the entorhinal cortex of bats. *Nature* 479:103–7

Ylinen A, Soltesz I, Bragin A, Penttonen M, Sik A, Buzsáki G. 1995. Intracellular correlates of hippocampal theta rhythm in identified pyramidal cells, granule cells, and basket cells. *Hippocampus* 5:78–90

Neural Basis of the Perception and Estimation of Time

Hugo Merchant,[1] *Deborah L. Harrington,*[2,3] *and Warren H. Meck*[4]

[1] Instituto de Neurobiología, UNAM, Campus Juriquilla, México; email: hugomerchant@unam.mx
[2] VA San Diego Healthcare System, San Diego, California 92161; email: dharrington@ucsd.edu
[3] Department of Radiology, University of California, San Diego, La Jolla, California 92093
[4] Department of Psychology and Neuroscience, Duke University, Durham, North Carolina 27701; email: meck@psych.duke.edu

Key Words

temporal processing, interval tuning, medial premotor cortex, cortico-thalamic-basal ganglia circuit, striatal beat-frequency model

Abstract

Understanding how sensory and motor processes are temporally integrated to control behavior in the hundredths of milliseconds-to-minutes range is a fascinating problem given that the basic electrophysiological properties of neurons operate on a millisecond timescale. Single-unit recording studies in monkeys have identified localized timing circuits, whereas neuropsychological studies of humans who have damage to the basal ganglia have indicated that core structures, such as the cortico-thalamic-basal ganglia circuit, play an important role in timing and time perception. Taken together, these data suggest that a core timing mechanism interacts with context-dependent areas. This idea of a temporal hub with a distributed network is used to investigate the abstract properties of interval tuning as well as temporal illusions and intersensory timing. We conclude by proposing that the interconnections built into this core timing mechanism are designed to provide a form of degeneracy as protection against injury, disease, or age-related decline.

INTRODUCTION

Timing is everything. The flow of information through time structures how information is perceived, experienced, and remembered. Throughout normal development we gradually acquire a sense of duration and rhythm that is basic to many facets of behavior such as speaking, driving a car, dancing to or playing music, and performing physical activities (Allman et al. 2012, Meck 2003). Yet there is no specific biological system that senses time as there are for sight, hearing, and taste. As such, there has been an explosion of research into the neural underpinnings of timing. Initially, a driving force behind many studies was scalar timing theory, which defined sources and forms of timing variability that were derived from clock, memory, and decision processes (Gibbon et al. 1984). However, as an understanding of the neurobiological bases of timing developed, so did a neurophysiological model of timing (Matell & Meck 2004), which captured the intrinsic interactive nature of interval-timing circuits as well as Weber's law and the scalar property of interval timing (Brannon et al. 2008, Cheng & Meck 2007, Gu et al. 2013, Meck & Malapani 2004).

This article reviews recent progress toward elucidating the neural mechanisms of interval timing, wherein durations in the range of hundreds of milliseconds to multi-seconds are perceived, estimated, or reproduced. We begin by discussing advances in psychophysics and in cell-ensemble recording that address the fundamental debate about whether explicit timing is governed by a common system or by distributed context-dependent networks. Next, we review current research into the functional significance of neuroanatomical systems that govern interval timing. We then consider emerging investigations into brain mechanisms underlying distortions in temporal resolution, which emanate from intersensory timing or factors that produce illusions of time. This body of research indicates that interval timing emerges from interactions of a core timing center with distributed brain regions. To this end, the striatal-beat frequency model of interval timing (Matell & Meck 2004) is discussed, which captures the psychophysiological and neuroanatomical properties of timing networks.

EVIDENCE FAVORING A COMMON TIMING MECHANISM

The psychophysics of the perception and estimation of time started in the late nineteenth century (Fraisse 1984) and evolved as many timing tasks and species were studied to test the boundaries of categorical scaling (Penney et al. 2008) and the existence of one or multiple clocks (Buhusi & Meck 2009b). One central finding was that the variability of interval timing is proportional to the duration of the interval used. This scalar property implies that the standard deviation of the quantified intervals increases proportionally with the average of the intervals and follows Weber's law. Weber's law is given as $SD(T) = kT$, where k is a constant corresponding to the Weber fraction. The coefficients of variation (σ/μ) or the Weber fractions show similar values in the range of hundreds of milliseconds in a variety of tasks, sensory modalities, and species, suggesting a common timing mechanism for this, and perhaps other, time scales (Gibbon et al. 1997). The conceptual framework behind this hypothesis proposes that the temporal resources of a common internal clock are shared in a variety of timing tasks, thereby producing similar temporal variance (**Figure 1*a***).

Another major finding was that the overall variability in a timing task can be dissociated into time-dependent (e.g., clock) and time-independent (e.g., motor) sources (Repp 2005). Different quantitative and paradigmatic strategies were used to distinguish components of performance variability. The slope method, for instance, uses a generalized form of Weber's law, wherein investigators compute a linear regression between the variability and the squared interval duration. The resulting slope corresponds to the time-dependent component, whereas the intercept corresponds to the time-independent processes. Slopes of interval discrimination and synchronization-continuation tapping tasks are similar for a range of intervals from 325 to 550 ms (Ivry & Hazeltine 1995), supporting the view of a common clock in a variety of contexts (**Figure 2*a***). Moreover, temporal variability among an individual's performance correlates between different explicit timing tasks, including self-paced tapping tasks using the finger, foot, and heel (Keele et al. 1985), tapping and phasic figure drawing (Zelaznik et al. 2002), and duration-discrimination and tapping tasks (Keele et al. 1985). This correlation implies that participants who are good timers in one behavioral context are also good timers in another, again in support of a common timing mechanism.

The study of perceptual learning and generalization to other behaviors and modalities has provided important insights into the neural underpinnings of temporal processing (**Figure 1*b***). For example, intensive duration-discrimination learning can generalize across untrained auditory frequencies (Karmarkar & Buonomano 2003), sensory modalities, and stimulus locations (Nagarajan et al. 1998), and even from sensory to motor-timing tasks (Meegan et al. 2000). However, these studies found no reliable generalization toward untrained durations, which concurs with a study of duration-production learning that reported a smooth decrease in generalization as the untrained interval deviated from the trained duration and suggested the existence of neural circuits that are tuned to specific durations (Bartolo & Merchant 2009). Overall, these findings support the notion of a common or a partially overlapping distributed timing mechanism, but they also introduce the concept of duration-specific circuits. Of course, these two features are not mutually exclusive when a large neural network is considered

(Karmarkar & Buonomano 2007, Matell & Meck 2004).

EVIDENCE FAVORING A UBIQUITOUS TIMING ABILITY OF CORTICAL NETWORKS

Other studies support the hypothesis that timing is an inherent computational ability of every cortical circuit and that it can be performed locally (**Figure 2b**). This notion, based on network simulations, implies that cortical networks can tell time during perception

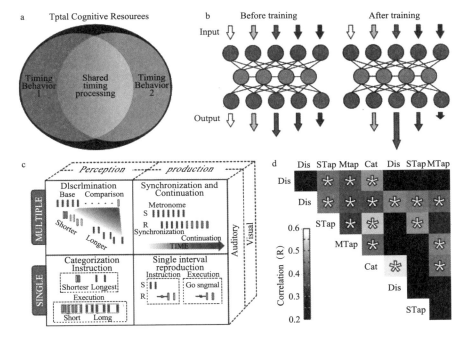

Figure 1

Psychophysical tools to study the neural underpinnings of interval timing using a black-box approach. (*a*) The hypothesis in some experimental psychology studies is that because some cognitive resources are shared across timing tasks there should be a common timing mechanism. (*b*) The characterization of the generalization properties of specific overlearned timing tasks across modalities, stimulation sites and properties, interval durations, and timing contexts; also has been used to test the existence of a common or multiple interval clocks. (*c*) Timing tasks. Four timing tasks, performed with auditory or visual interval markers, were used to evaluate the influence of three factors on timing performance: visual versus auditory modality, single versus multiple intervals, and perception versus production of the intervals. (*d*) Correlation matrix showing the Pearson R value in a grayscale matrix (*inset, bottom left*) for all possible pairwise task comparisons. Asterisks indicate significant correlations (*P* < 0.05) between specific pairs of tasks. Open and closed fonts correspond to tasks with auditory and visual markers, respectively. Abbreviations: Dis, discrimination; Cat, categorization; STap, single interval reproduction; MTap, synchronization and continuation task. Adapted from Merchant et al. 2008c.

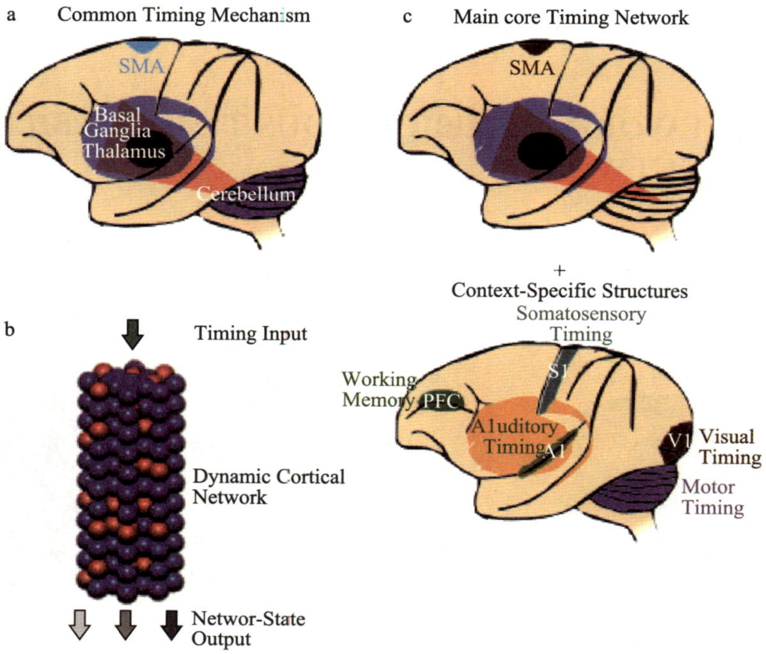

Figure 2

Three possible timing mechanisms. (*a*) Common timing mechanism. Psychophysical and lesion studies have suggested the existence of a dedicated timing mechanism that depends on one neural structure such as the cerebellum or the basal ganglia and that is engaged in temporal processing in a wide range of timing behaviors. However, fMRI studies have suggested that this general timing mechanism is distributed and depends on the activation of a large network including the supplementary motor area (SMA), the parietal and prefrontal cortices, as well as the basal ganglia and the cerebellum. (*b*) Ubiquitous timing. Modeling and cell-culture recordings have suggested that timing is an intrinsic property of cortical network dynamics and therefore that there is no dedicated neural circuit for temporal integration. (*c*) Partially shared timing mechanism. This model proposes that temporal estimation depends on the interaction of multiple areas, including regions that are consistently involved in temporal processing across timing context and that conform the main core timing network and areas that are activated in a context-dependent fashion. The main core timing network consists of the SMA and the basal ganglia. The interaction between the two sets of structures gives the specific temporal performance in a task. The red triangles correspond to the dopaminergic system innervating the basal ganglia.

tasks as a result of time-dependent changes in synaptic and cellular properties, which influence the population response to sensory events in a history-dependent manner(Karmakar & Buonomano2007). In contrast, motor timing is thought to depend on the activity of cortical recurrent networks with strong internal connections capable of self-sustained activity (Buonomano & Laje 2010). Chronic stimulation in cortical slices produces changes in the temporal structure of the cell activity that reflect the durations used during training (Johnson et al. 2010). Hence, this in vitro experimental evidence suggests that recurrent cortical circuits have inherent timing ability in the hundreds of milliseconds.

Psychophysical experiments also suggest that sensory timing is local. For example, the apparent duration of a visual stimulus can be modified in a local region of the visual field by adapting to oscillatory motion or flicker, which suggests a spatially localized temporal mechanism for time perception of visual events (Burr et al. 2007, Johnston et al. 2006). Furthermore, learning to discriminate temporal modulation rates is accompanied not only by a specific learning transfer to duration discrimination, but also by an increase in the amplitude of the early auditory evoked responses to trained stimuli (van Wassenhove & Nagarajan 2007). These studies emphasize the concept of timing as a local, context-dependent process.

AN INTERMEDIATE HYPOTHESIS: A MAIN CORE TIMING MECHANISM INTERACTS WITH CONTEXT-DEPENDENT AREAS

Other research suggests that a hybrid model may better account for temporal performance variability in different contexts. Merchant and colleagues (2008c) conducted a multidimensional analysis of performance variability on four timing tasks that differed in sensorimotor processing, the number of durations, and the modality of the stimuli that defined the intervals (**Figure 1c**). Though

> **CTBG:**
> cortico-thalamic-basal ganglia
> **MPC:** medial premotor areas
> **SMA:** supplementary motor area

variability increased linearly as a function of duration in all tasks, compliance with the scalar property was accompanied by a strong effect of the nontemporal variables on temporal accuracy (Merchant et al. 2008b,c). Intersubject analyses comparing performance variability between pairs of tasks revealed a complex set of correlations between many, but not all, tasks, irrespective of stimulus modality (**Figure 1d**). These results can be interpreted neither as evidence for multiple context-dependent timing mechanisms, nor as evidence for a common timing mechanism. Rather, in accordance with neuroimaging research described below, the findings suggest a partially distributed timing mechanism, integrated by core structures such as the cortico-thalamic-basal ganglia (CTBG) circuit and areas that are selectively engaged by different behavioral contexts (Buhusi & Meck 2005, Coull et al. 2011) (**Figure 2c**). Task-dependent areas may interact with the core timing system to produce the characteristic pattern of performance variability in a specific timing paradigm and the pattern of intertask correlations depicted in **Figure 1d**.

NEUROPHYSIOLOGICAL BASIS OF TIME ESTIMATION: RAMPING ACTIVITY IN THE CORE TIMING CIRCUIT

Cell activity changes associated with temporal processing in behaving monkeys are found in the cerebellum (Perrett 1998), basal ganglia (Jin et al. 2009), thalamus (Tanaka 2007), posterior parietal cortex (Leon & Shadlen 2003), prefrontal cortex (Brody et al. 2003; Genovesio et al. 2006, 2009; Oshio et al. 2008), dorsal premotor cortex (Lucchetti & Bon 2001), motor cortex (Lebedev et al. 2008), and medial premotor areas (MPC), namely the supplementary (SMA) and presupplementary motor areas (preSMA) (Mita et al. 2009). These areas form different circuits that are linked to sensorimotor processing via the skeletomotor or oculomotor effector systems. Most studies report climbing activity during a variety of timing tasks. Therefore, the increase or decrease in instantaneous activity with the passage of time is a property present in many cortical and subcortical areas that may be involved in different aspects of temporal processing in the hundreds of milliseconds range.

Recently, the activity of MPC cells was recorded during a synchronization-continuation tapping task (SCT) that includes basic sensorimotor components of rhythmic behaviors (Zarco et al. 2009) (**Figure 1c**). Different types of neurons exhibit ramping activity before or after the button press in the SCT (Merchant et al. 2011). A large group of cells shows ramps before movement onset that are similar across produced durations and the sequential structure of the task and are considered motor ramps (Perez et al. 2013) (**Figure 3a**). Another cell population exhibits an increase in ramp duration but a decrease in slope as a function of the monkey's produced duration, reaching a particular discharge magnitude at a specific time before the button press. These cells are called relative-timing cells because their ramping profile appears to signal how much time is left to trigger the button press (**Figure 3b**). Another group of cells shows a consistent increase followed by a decrease in their instantaneous discharge rate when the neural activity was aligned to the previous button press. In these absolute-timing cells, the duration of the up-down profile of activation increases as a function of the produced interval (**Figure 3c**), whereas in the time-accumulator cells we see an additional increase in the magnitude of the ramps' peak (**Figure 3d**). Therefore, these cells could be representing the passage of time since the previous movement, using two different encoding strategies: one functioning as an accumulator of elapsed time where the peak magnitude is directly associated with the time passed, and another where the duration of the activation period is encoding the length of the time passed since the previous movement. The noisy character of ramping activity implies that whatever is reading this temporal information downstream cannot be relying on single cells to quantify the passage of time or to produce accurately timed movements. Therefore, a population code is suggested for encoding elapsed time, by which the reading network adds the magnitudes of a population of individual ramps over time, resulting in a ramp population function (Merchant et al. 2011) **(Figure 3)**.

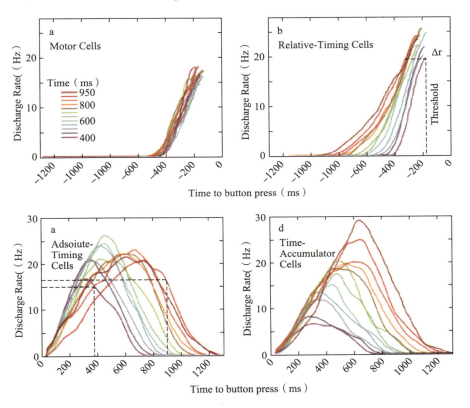

Figure 3

Ramp population functions for motor (*a*) and relative-timing (*b*) cells aligned to the next button press. Ramp population functions for absolute-timing (*c*) and time-accumulator (*d*) cells aligned to the previous button press. The color code in the inset of panel *a* corresponds to the duration of the produced intervals during the synchronization-continuation tapping task (SCT). The ramp population functions are equal to the total magnitude of individual ramps over time. Adapted from Merchant et al. 2011.

The rhythmic structure of the SCT may impose the need to predict when to trigger the next tap to generate an interval, but also to quantify the time passed from the previous movement to have a cohesive mechanism to generate repetitive tapping behavior. Indeed, the cells encoding elapsed (absolute-timing) and remaining time (relative-timing) interacted during the repetitive phases of the SCT, supporting the proposed rhythmical timing mechanism (Merchant et al. 2011).

The ubiquitous increases or decreases in cell discharge rate as a function of time across different timing tasks and areas of the core CTBG timing circuit suggest that ramping activity is a fundamental element of the timing mechanism. Key characteristics of ramping activity are its instantaneous nature and the fact that it normally peaks at the time of an anticipated motor response. Although the absolute-timing and the time-accumulator cells (**Figure 3c,d**) are encoding elapsed time since the previous motor event, ramping cells are engrained in the temporal construction of motor intentions and actions (Merchant et al. 2004). This integration is crucial because interval-timing tasks require a response to express a perceptual decision or produce a timed response. Therefore, the ramping activity may be part of the temporal apparatus that gaits the motor responses. However, more abstract tasks such as interval tuning may represent the more cognitive aspects of temporal processing that can be dissociated from the motor response and the corresponding ramping activity.

INTERVAL TUNING: AN ABSTRACT SIGNAL OF TEMPORAL COGNITION

Investigators have observed a graded modulation in the cell discharge rate as a function of event duration during the SCT in MPC. **Figure 4a,c** shows the activation profile of a cell in the preSMA of a monkey performing this task. The neuron shows a larger discharge rate for the longest durations, with a preferred interval around 900 ms (**Figure 4e**). In fact, a large population of MPC cells is tuned to different signal durations during the SCT, with a distribution of preferred durations that covers all intervals in the hundreds of milliseconds. These observations suggest that the MPC contains a representation of event duration, where different populations of duration-tuned cells are activated depending on the duration of the produced interval (Merchant et al. 2012b). Most of these cells also showed selectivity to the sequential organization of the task, a property that has been described in sequential motor tasks in MPC (Tanji 2001). The cell in **Figure 4a,c** also shows an increase in activity during the last produced interval of the task's continuation phase. Again, at the cell-population level, all the possible preferred ordinal sequences were covered. These findings support the proposal that MPC can multiplex event duration with the number of elements in a sequence during rhythmic tapping (Merchant et al. 2013).

Researchers have evaluated the existence of a common timing mechanism or a set of context-dependent neural clocks on interval-selective cells, comparing their tuning properties during the execution of two tasks: the SCT and a single duration reproduction task (SIRT) (**Figure 1c**) (Merchant et al. 2013). A large group of neurons showed similar preferred durations in both timing contexts. **Figure 4a–d** shows that the cell that is tuned to long durations during the SCT is similarly tuned during the SIRT. Furthermore, the tuning curves for the cells in the SCT and the SIRT are similar for auditory and visual markers (**Figure 4e**). These findings confirm that MPC is part of a core timing circuit and suggest that interval tuning can be used to represent the duration of intervals in different timing tasks. Additional experiments are needed to determine whether tuning to event duration is an emergent property of MPC cells that depends on the local integration of graded inputs or occurs throughout the CTBG circuit (Matell et al. 2003, 2011). In summary, cell tuning is an encoding mechanism used by the cerebral cortex to represent different sensory, motor, and cognitive features, including event duration (Merchant et al. 2012). This signal must be integrated as a population code, in which the cells can vote in favor of their preferred duration to establish the interval for rhythmic tapping (Merchant et al. 2013).

NEUROANATOMICAL AND NEUROCHEMICAL SYSTEMS FOR INTERVAL TIMING

Turning to neuroanatomical and neurochemical levels of analysis, investigations in humans and other animals emphasize the centrality of the striatum and dopamine (DA) neurotransmission in explicit timing (Agostino et al. 2011; Balci et al. 2013; Cheng et al. 2007; Coull et al. 2011, 2012; Gu et al. 2011; Hinton & Meck 1997, 2004; Höhn et al. 2011; Jones & Jahanshahi 2011; Lake & Meck 2012; Meck 1996, 2006a,b; Pleil et al. 2011). These findings are compatible with reports of timing dysfunction in disorders of the basal ganglia, including Parkinson's disease (PD) (Gu et al. 2013; Harrington et al. 1998b, 2011b; Jones et al. 2008; Koch et al. 2004, 2008; Meck & Benson 2002; Smith et al. 2007) and prodromal

DA: dopamine
PD: Parkinson's disease

Huntington's disease (HD) (Paulsen et al. 2004, Rowe et al. 2010, Zimbelman et al. 2007). However, investigators have observed exceptions notably in PD, which has been studied the most extensively. For example, normal performance was reported on a test of motor timing (Spencer & Ivry 2005) and on several different test of time perception (Wearden et al. 2008). The reasons for this variation are unclear but may be due to the insensitivity of tasks when temporal discriminations are easy and/or when feedback is frequently provided (Wearden et al. 2008). Another important consideration is that timing deficits in PD correlate with disease severity (Artieda et al. 1992). Numerous studies have tested

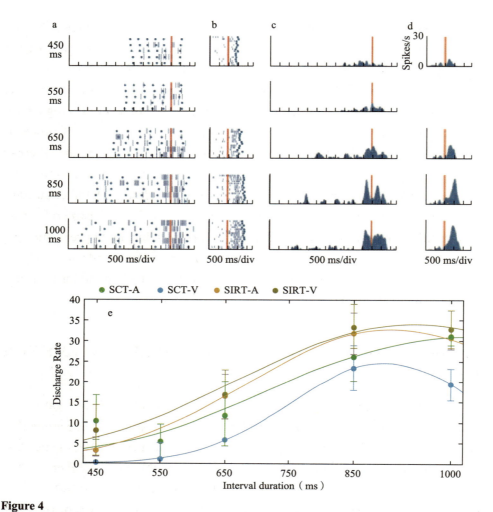

Figure 4

Interval tuning across tasks and sensory modalities. Responses of an interval-tuned cell with a long preferred interval across different temporal contexts. (*a*) Raster histograms (*blue*) aligned (*red line*) to the third tap of the continuation phase during the synchronization continuation tapping task (SCT) in the visual condition. (*b*) Raster histograms (*blue*) aligned (*red line*) to the first tapping movement during the single duration reproduction task (SIRT) in the auditory condition. (*c, d*) Average spike-density functions of the responses shown in panels *a* and *b*, respectively. (*e*) Tuning functions for the same cell, where the mean (± SEM) of the discharge rate is plotted as a function of the target interval duration. The continuous lines correspond to the significant Gaussian fittings: SCT, visual SCT, auditory SIRT, visual SIRT.

early-stage PD patients, who, despite considerable DA cell loss, may have the capacity to compensate for cognitive difficulties. The cerebellum may be one compensatory route (Kotz & Schwartze 2011, Yu et al. 2007), possibly because it predicts and finely tunes behavioral states on the basis of an efferent copy of sensory and motor information. Cortical systems may support compensatory processing in PD, as well. These issues of degeneracy aside (see sidebar, Degeneracy and the Neural Representation of Time), the finding that subgroups of PD patients do and do not exhibit timing disturbances (Merchant et al. 2008a) resonates with the considerable heterogeneity in the disease's clinical symptoms, their day-to-day fluctuations, and individual differences in response to DA therapy.

DEGENERACY AND THE NEURAL REPRESENTATION OF TIME

Lewis & Meck (2012) have recently proposed a form of degeneracy in timing systems that parallels the degeneracy observed in many other aspects of brain function, as well as in the genetic code and immune systems (Mason 2010, Price & Friston 2002). Degeneracy is a type of functional compensation that is distinct from redundancy in that structurally different mechanisms are involved in degeneracy, whereas multiple copies of identical mechanisms are the basis for redundancy. The fundamental importance of timing to all perception and action necessitates degeneracy and may explain the ease with which timing can be performed by a range of different neural architectures (Wiener et al. 2011b). Degeneracy also allows for predictions about timing hierarchies and the sources and forms of variability in timing as a function of manipulations such as rTMS and reversible pharmacological treatments (Jahanshahi et al. 2010a,b; Teki et al. 2012). The applicability of this line of thought to time perception should be obvious: It is very difficult to abolish time perception completely, especially as a result of focal, unilateral brain lesions where, in addition to degeneracy, redundancy from the opposite hemisphere likely contributes to recovery (Aparicio et al. 2005; Coslett et al. 2010; Meck et al. 1984, 2013; Yin & Troger 2011).

Neurodegenerative disorders eventually alter cortical functioning, which may be another source of timing disturbances in PD and prodromal HD. Indeed, damage to the right hemisphere of the prefrontal (dorsolateral prefrontal and premotor) and the inferior parietal cortices disrupts time perception (Harrington et al. 1998b). Moreover, in patients with right but not left hemisphere damage, elevated temporal discrimination thresholds correlated with a weakened ability to reorient attention but did not correlate with deficits in pitch discrimination. This finding suggests that frontoparietal systems may govern attention and working memory, which interact with time keeping processes(Lustig et al. 2005). Unfortunately, systematic investigations into cortical regions that are essential for timing in humans have been hampered by difficulties in obtaining sufficient samples of patients with focal damage, particularly to some brain regions now thought to be critical.

The advent of functional magnetic resonance imaging (fMRI) has been a welcome development to the study of timing. To date, most of this research has been conducted in healthy adults and has focused on regional activation patterns that are associated with various timing tasks. Research shows fairly good consensus that, apart from the basal ganglia, the SMA, an element of the CTBG (**Figure 2c**), is

fMRI: functional magnetic resonance imaging

routinely engaged during interval timing (Coull et al. 2011; Harrington et al. 2004, 2010; Meck et al. 2008; Wiener et al. 2010c). Carefully conducted meta-analyses further suggest that the rostral SMA, namely preSMA, may be more engaged by perceptually based timing within the suprasecond range, whereas caudal SMA may be more engaged by sensorimotor-based timing within the subsecond range (Schwartze et al. 2012). However, mounting evidence indicates that timing is governed by more distributed neural networks that are recruited depending on the behavioral context and stage of learning (Allman et al. 2012, MacDonald et al. 2012). Consequently, the major challenge has been to unravel their functional significance.

FUNCTIONAL SIGNIFICANCE OF BRAIN CIRCUITS THAT GOVERN INTERVAL TIMING

Temporal processing unavoidably engages a host of cognitive and sensorimotor processes that activate multiple brain regions. For this reason, it has been difficult in functional imaging to distinguish core-timing systems from those that support other interacting processes. One approach to the dilemma is to exploit the temporal resolution of event-related fMRI by linking the acquisition of brain images across time to different components of a task that are assumed to engage certain processes. The ordinal-comparison procedure commonly used to study time perception involves two steps: encoding an anchor or standard duration (encoding phase), followed by encoding a comparison interval and judging whether it is longer or shorter than the standard (decision phase) as illustrated in **Figure 5a** (Gu & Meck 2011, Harrington et al. 2004). A hypothetical hemodynamic response function associated with each of the phases is illustrated in **Figure 5b**. One assumption is that activation in core-timing systems should be seen while encoding the standard and the comparison intervals. Activation in systems that support executive or decisional processes may be more apparent when comparing the two intervals during the decision phase when a judgment is made. To control for the engagement of cognitive and low-level sensorimotor processes, the time course of activation is compared with control tasks that contain the same stimuli but require different decisions. If auditory signals are timed, tasks that control for processing involved in perceptual decision making (pitch perception) or low-level sensorimotor processes (sensorimotor control tasks) may be used (**Figure 5a**). An early fMRI study sought to distinguish timing-related brain activation using these methods (Rao et al. 2001). In the time-perception task, a 1,200-ms standard interval was compared with four shorter and longer intervals that were ±5% increments of the standard. For pitch perception,

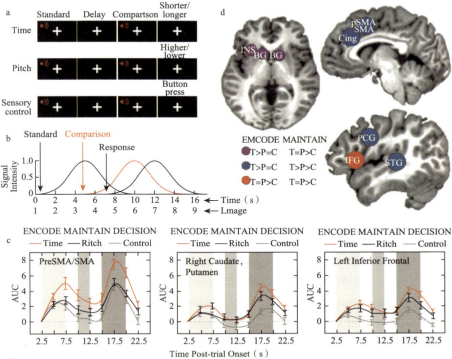

Figure 5

Functional significance of brain circuits that govern interval timing. (*a*) Illustration of the trial events in time-perception and control tasks. In the time-perception task, a standard and a comparison interval are successively presented and separated by a delay, which varies across studies. In the example, intervals are designated by tones. To control for brain activation related to cognitive and low-level sensorimotor processing, activation is compared with tasks composed of the same trial events but with different processing requirements. In the pitch perception task, the frequency of the two tones is discriminated. In the sensory control task, a button press is made following the presentation of the tones. A fixation cross remains on the screen during imaging. (*b*) The three hypothetical time-course functions illustrate the expected hemodynamic response associated with encoding the standard interval (*black*), encoding the comparison interval (*red*), and making a response (*gray*). Arrows leading from each event designate their onset. The hemodynamic response peaks 4 to 6s after event onset. An image of the entire brain is acquired every 2s.(c)Time course of activation (area under the curve; AUC) for the time-perception (*red*), pitch perception (*black*), and control (*gray*) tasks in representative regions (adapted from Harrington et al. 2010). Gray boxes on the x axis denote the epochs used to calculate AUC for a trial's encode, maintain, and decision phases. (*d*) Illustration of regional activation patterns for a trial's encode and maintain phases in the time (T), pitch (P), and control (C) tasks (Harrington et al. 2010). Activation in purple regions was related to interval encoding (T > P = C for encode; T = P > C for maintain). Activation in blue regions was related to accumulation and maintenance of sensory time codes (T > P = C for encode and maintain). Activation in the red region was related to inhibitory control (T = P > C for encode and maintain). For all regions except the IFG, activation was greater for the time than for the pitch and control tasks in the decision phase. Abbreviations: BG, basal ganglia; cing, anterior cingulate; IFG, inferior frontal gyrus; INS, anterior insula; PCG, precentral gyrus; preSMA, presupplementary motor area; SMA, supplementary motor area; STG, superior temporal gyrus.

a 700-Hz standard tone was compared with 4 higher or lower comparison pitches. The results showed that caudate and putamen activation evolved early in association with standard-interval encoding and was sustained during the trial's decision phase. Activation of the right parietal cortex also began at the onset of the standard interval, possibly owing to its role in attention. In contrast, cerebellar activation unfolded later, just before and during movement execution, suggesting that it did not govern interval encoding. Subsequent studies using this general approach also linked striatal activation to interval timing (Coull et al. 2008, 2011; Harrington et al. 2004).

At the same time, independent empirical work (Lewis et al. 2004) and theoretical developments (Hazy et al. 2006) implicate the striatum in working memory. Owing to the short delays between the encoding and the decision phases of the time-perception tasks in earlier studies (Coull et al. 2008, Harrington et al. 2004, Rao et al. 2001), brain systems involved in the maintenance of temporal information could not be ascertained. To address this issue, Harrington et al. (2010) inserted 10-s and 12.5-s delays between the standard and comparison interval, thereby permitting a better separation of activation associated with a trial's encoding, maintenance, and decision periods, as illustrated in **Figure 5c**. Pairwise subtractions of brain activation during time discrimination (T), pitch discrimination (P), and the sensorimotor control (C) tasks were conducted for each period. Task difficulty did not differ between the time and the pitch tasks. The fMRI results showed that the striatum's pattern of timing-related activation during the different epochs could be distinguished from that found for most other brain regions. Striatal (bilateral caudate and putamen) activation was greater for the timing task than for the two control tasks (T > P = C) during encoding but did not differ between the time and pitch tasks (T = P > C) during maintenance (**Figure 5c,d**). Thus, timing-related activity was specific to interval encoding, consistent with the positive correlation of striatal activation and timing proficiency noted during this same period (Coull et al. 2008, Harrington et al. 2004). This result also concurs with the view that passive maintenance of working memory is controlled via recurrent excitation of the prefrontal cortex and sensory neurons, and active maintenance over longer periods involves recurrent thalamocortical activity (Hazy et al. 2006). The only other region exhibiting the same activation pattern as did the striatum was the anterior insula. This region is situated to integrate processing from disparate domains (e.g., interoception, emotion, and cognition), including time (Kosillo & Smith 2010; Wittmann et al. 2010a,b), via its dense interconnections with most association centers and the basal ganglia. Moreover, anterior insula connectivity with frontal cognitive-control centers suggests that it also assists in the perceptual analysis of sensory information (Eckert et al. 2009). In contrast, preSMA/

SMA, precentral, and superior temporal activation was greatest for the timing task (T > P = C) during both encoding and maintenance periods (**Figure 5c,d**). This finding suggests that the preSMA and SMA modulate nontemporal aspects of processing, consistent with reports that timing-related activation is lost in the face of more difficult control tasks (Livesey et al. 2007). One prospect is that the SMA and association areas support accumulation and maintenance of sensory information, which is more demanding for interval timing. Another important finding was that activation of the inferior frontal gyrus did not differ between the time and pitch tasks during all three periods of the trial, although it was greater than in the control condition (T = P > C) (**Figure 5c,d**). The inferior frontal gyrus is an inhibitory control center with fiber tracts to preSMA. This pathway may be a route to control the accumulation of sensory information into preSMA. Last, during the trial's decision phase, activation in almost all regions including the striatum was greater during the timing task (T > P > C), perhaps signifying the more significant integration demands of duration than pitch processing. In addition, timing-related activation of the cerebellum and a classic frontoparietal executive network did not emerge until the decision phase (T > P > C). The latter finding supports the positive correlation of frontoparietal region activity with time-discrimination difficulty (Harrington et al. 2004).

Taken together, interval timing was governed by distributed brain networks that flexibly altered activation, depending on task demands. Although some have argued that much brain activation during temporal processing relates to cognitive or sensory processes, timing emerges from the communication among brain regions rather than from processing in a single region. The use of control tasks can identify regional activity that is more dominant during timing; however, sensory and cognitive centers that are vital for interval timing can be overlooked. For example, using the subtractive method has led some to conclude that the inferior parietal cortex is not a key element of interval encoding (Coull et al. 2008, Harrington et al. 2010, Livesey et al. 2007). Yet applying repetitive transcranial magnetic stimulation (rTMS) to the supramarginal gyrus of the right hemisphere dilated perceived duration owing to its effect on interval encoding (Wiener et al. 2010a, 2012). This result indicates that the right supramarginal gyrus is a critical element of the neural circuitry that encodes time, which corroborates the detrimental effect of right parietal damage on time perception (Harrington et al. 1998b). Furthermore, as we begin to explore how the brain construes time under circumstances that alter the resolution of perceived duration, interactions among larger-scale brain networks will likely be considerably important.

ILLUSIONS OF TIME

An area of research that has potential to enrich our understanding of timing networks concerns illusions of time, which are important because they reveal how the brain normally organizes and interprets information depending on internal states, past experiences, or properties of stimuli. Psychophysical studies have long reported that the experience of time is not isomorphic to physical time, but rather depends on many factors. For example, emotionally aversive events are perceived as lasting longer than their physical duration (Cheng et al. 2008a, Droit-Volet & Meck 2007). Larger magnitude, more complex, or intense stimuli also expand perceived duration, whereas repeated, high-probability, and non-salient stimuli compress time (Eagleman 2008, van Wassenhove et al. 2008). The mechanisms of temporal illusions continue to be debated, but the consensus indicates that attention and arousal are key factors. In pacemaker-counter models, attention and arousal are thought to speed up or slow down timing by closing or opening a switch that allows pulses generated from a "clock" to be accumulated and counted (Buhusi & Meck 2009a, Ulrich et al. 2006). For instance, heightened levels of physiological arousal induced by psychologically negative sounds expand subjective duration (Mella et al. 2011). Reducing the level of attention devoted to timing compresses perceived duration and attenuates activation in the CTBG timing-related circuit, but also in the frontal, temporal, and parietal cortices (Coull et al. 2004, 2011).

Despite this impressive body of work, little is understood about brain mechanisms that bring about temporal illusions. Emerging neuroimaging research suggests that the mechanisms may be partially context dependent. One fMRI study investigated the neural mechanisms of time dilation produced by a looming visual stimulus (Wittmann et al. 2010b), which captures attention possibly because it signals a potential intrinsic threat to organisms. Participants viewed a series of five discs; all but the fourth disc, which was a looming or receding disc, were static, and participants judged whether the fourth stimulus was longer or shorter than the others. Relative to the receding control condition, activation was greater in rostral-medial frontal areas and medial-posterior cortices, including the posterior cingulate. These areas are tightly interconnected with the limbic system, which governs a variety of functions including emotion and motivation processing. The results were attributed to the arousing effect of looming signals, which have an inherently emotional component.

Another study investigated mechanisms of time dilation produced by emotionally aversive stimuli (Dirnberger et al. 2012). Participants judged which of two pictures was displayed for a longer or shorter amount of time. Perceived duration was dilated when one of the pictures was aversive (aversive-neutral) relative to a control condition, in which both pictures were neutral. On a subsequent recognition test, overestimation of time was associated with better memory of a picture, indicating that time dilation enhanced memory encoding. A region-of-interest analysis revealed that activation of rostral-medial frontal areas (superior frontal, preSMA/SMA) and lateral inferior frontal cortex was greater for aversive-neutral than control pairs. Brain activation was further distinguished by the accuracy of time discriminations, whereby right amygdala, anterior insula, and putamen activation was greater on trials in which time was overestimated than on correct trials. Thus, when time was dilated, the limbic system (amygdala) and tightly interconnected regions, notably the anterior insula and medial frontal cortex, were more engaged. Taken together, both studies suggest that time dilation via stimuli that have an emotionally threatening connotation is partly brought about by heightened activity in elements of the limbic system and interconnecting medial cortical areas.

Temporal distortions also emerge in contexts that have no emotional overtone, such as the illusion that auditory signals are perceived as lasting longer than visual signals of the same physical duration when they are compared. A recent fMRI study sought to investigate the neural basis of the illusion (Harrington et al. 2011a), which is of interest because it may elucidate how synchrony is maintained across the senses. The audiovisual effect on perceived duration has been attributed to a pacemaker-accumulator system that runs faster for auditory than for visual signals, possibly owing to an attentional switch that permits a faster accumulation of pulses (Lustig & Meck 2011). In this study, participants judged whether an auditory (A) or visual (V) comparison interval was longer or shorter than a standard interval, which was either of the same or a different modality (**Figure 6a**). Time was dilated relative to all other conditions when the duration of an A comparison interval was judged relative to a V standard (V-A), and time was compressed when the duration of a V comparison interval was judged relative to an A standard (A-V) (**Figure 6b**). Regional analyses of brain activation showed that audiovisual distortions were governed by frontal cognitive-control centers (preSMA, middle/inferior frontal), where activation was greater when time was compressed, and higher association centers (superior temporal cortex, posterior insula, middle occipital cortex), where the level of activation was driven by the modality of the comparison interval (**Figure 6c**). Although this study identified regional activation differences between time dilation and compression, timing emerges from communication among brain networks, to which conventional regional analyses of activation are insensitive. As such, the effective connectivity of these regions was examined to determine if the strength of their interactions with other brain regions differed between the time dilation and compression conditions. Effective connectivity was not found for frontal cortical areas, possibly because, as supramodal control centers, they flexibly direct attention and executive resources, though more so when time discriminations are demanding (A-V). Rather, connectivity of bilateral association areas with frontoparietal and temporal cortices and the striatum was typically stronger when perceived duration was dilated than when compressed (**Figure 6d**). This result may be due to the salience of auditory signals in the context of timing (Repp & Penel 2002). This attention-based explanation would cause "clock pulses" to accumulate faster (pacemaker/accumulator models) or perhaps increase cortico-cortical oscillatory frequencies (Allman & Meck 2012), thereby producing an overestimation of time for auditory signals. Altogether, the finding reveals a basic principal of functional organization that produces distortions in the experience of time.

Notably, regional activation of the striatum did not differ between the time dilation and compression conditions (Harrington et al. 2011a). Although this result suggests that the integration of cortical oscillatory states by the striatum may not be faster for auditory than for visual signals, it leaves open the question of whether the strength of striatal connectivity is modulated by time dilation and compression (Matell & Meck 2004, N'Diaye et al. 2004). This area of inquiry is important for future research because measures of brain connectivity can be more sensitive to effects of psychological variables than are conventional regional analyses of activation. Indeed, it appears that the striatum may influence time dilation in emotionally aversive contexts (Dirnberger et al. 2012).

INTERSENSORY TIMING

The synthesis of temporal information across the senses is essential for perception and cognition, yet little is understood about how the brain maintains temporal synchrony among modalities. The striatum may be central to governing intersensory timing because it is involved in multisensory integration (Nagy et al. 2006) and is thought to integrate cortical oscillatory activity that comprises the clock signal (Coull et al. 2011). These proposals agree with a report that the bilateral striatum and also the thalamus and SMA exhibit greater activation during unimodal as compared with cross-modal timing (Harrington et al. 2011a) (**Figure 6e**). The result was not compatible with an attention-switching model of striatal function (van Schouwenburg et al. 2010), which would predict greater activation in the cross-modal than in the unimodal condition. The activation pattern may develop if intersensory integration of time codes is less stable or noisier

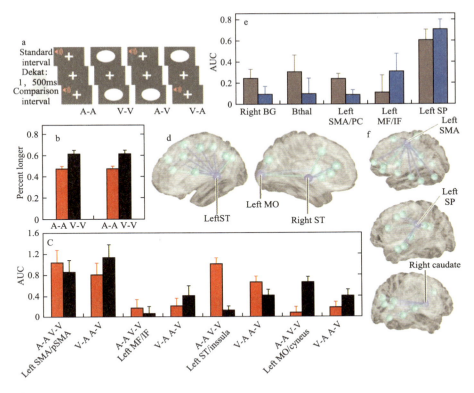

Figure 6

Brain networks that govern distortions in time and intersensory timing. The content contained in this figure is adapted from Harrington et al. (2011a). (*a*) Trial events for four conditions of a time-perception task. Pairs of auditory (A) and visual (V) stimuli were successively presented. The standard interval and comparison interval were of the same modality in the unimodal conditions (A-A,V-V) and were different in the cross-modal conditions (A-V, V-A). The four standard durations (1467, 1540, 1620, and 1710 ms) were each paired with three shorter and three longer comparison durations that were ±7% increments of each standard interval. A fixation cross remained on the screen during imaging. (*b*) Mean percent longer responses. Relative to all other conditions, time was significantly dilated for the V-A condition and compressed for the A-V condition. (*c, d*) Results from analysis of time dilation and compression effects on brain activity. (*c*) Mean area under the curve (AUC) in representative regions showing an interaction of comparison interval modality × timing condition (unimodal versus cross-modal). Horizontal bars denote significant differences between conditions. (*d*) Connectivity map of the left and right superior temporal cortex and left middle occipital cortex with representative regions. Effective connectivity of the bilateral superior temporal cortex and insula and the left middle occipital cortex was typically stronger for the time dilation than for the compression condition. (*e, f*) Results from analyses of brain systems that govern intersensory timing. (*e*) Mean AUC in representative regions showing differences in activation between unimodal and cross-modal timing. (*f*) Connectivity maps of left SMA, left superior parietal cortex, and right caudate with representative regions. Effective connectivity for all regions was stronger for cross-modal than for unimodal timing. Abbreviations: BG, basal ganglia (caudate, putamen); IF, inferior frontal; MF, middle frontal; MO, middle occipital cortex; PC, precentral cortex; preSMA, presupplementary motor area; SMA, supplementary motor area; SP, superior parietal; ST, superior temporal; Thal, thalamus.

relative to intrasensory timing. By comparison, activation of a classic frontoparietal working-memory network was greater for cross-modal than for unimodal timing, likely because of greater attention and executive demands of intersensory timing. The effective connectivity of these regions was then examined to determine if the strength of their interactions with other brain regions differed between the timing conditions. All regions showed stronger connectivity for cross-modal than for unimodal timing, perhaps owing to the more effortful demands of multimodal temporal integration. **Figure 6*f*** illustrates that caudate, SMA, and superior parietal connectivity was stronger with frontal cognitive-control centers, association centers, visual areas, and a memory system (precuneus, posterior cingulate, parahippocampus). Thus, the synthesis of audiovisual time codes in core timing and attention networks involves interactions with extensive brain networks.

STRIATAL BEAT-FREQUENCY MODEL OF INTERVAL TIMING

The inherent interactive nature of timing networks revealed by the above studies is embodied by the striatal beat-frequency (SBF) model of interval timing (Allman & Meck 2012, Matell & Meck 2004, Oprisan & Buhusi 2011, van Rijn et al. 2011). The SBF model uses medium spiny neurons located in the dorsal striatum, which is typically thought to be involved in decision making and executive function. Each spiny neuron receives ~30,000 inputs from cortical neurons, and it is this level of convergence (many to one) that allows the medium spiny neurons to serve as coincidence detectors of activity patterns engaged in by the cortical neurons. One easily detectable activity pattern is the oscillatory firing pattern of cortical neurons that is typically synchronized to the onset of relevant stimuli by DA release from the substantia nigra pars compacta (Jahanshahi et al. 2006). Given this initial synchronization (i.e., temporary alignment of the downbeat of neural firing), the subsequent evolution of neural firing will reflect the inherent rhythmical structure of each neuron's tendency to fire as well as random drift or desynchronization of firing among individual neurons. Despite the variability in this pattern of neural firing, which grows as a function of the time since the initial synchronization, the medium spiny neurons can still detect different patterns of neural firing on the basis of the high degree of redundancy in the system due to the convergence of 30,000 inputs. This coincident detection involves the ability of medium spiny neurons to sense temporal patterns of

SBF:
striatal beat-frequency (model)

simultaneous activity across their spatially arranged receptive fields. Individual synapses within these receptive fields are trained to detect and respond to specific patterns of oscillatory input on the basis of previous experience and the influence of long-term potentiation and depression—two well-known neurobiological mechanisms for the encoding of event durations (Matell & Meck 2000, 2004).

Each time period from milliseconds to seconds or minutes to hours will be reflected by different patterns of neural activity that can be repeatedly reproduced as long as the initial synchronization retains its efficacy and the pattern is not reset or interrupted by subsequent stimulus onsets affecting that particular set of detectors. Multiple durations can be timed simultaneously by assuming multiple detectorsor timers (i.e., spiny neurons) within the striatum that display chronotopy, a preference for particular ranges of durations. The readout of such a timing system is provided by frontal cortex monitoring of the firing activity of this chronotopically arranged time line— thereby completing the CTBG circuit. In this respect, the SBF model is an important advancement because of its veracity at both behavioral and physiological levels. Previous timing models either provided a good description of timing behaviors, but contained components that were inconsistent with the properties of the brain structures involved, or were neurobiologically feasible, but made inaccurate behavioral predictions. The computational version of the SBF model as described by Matell & Meck (2004) is constructed such that its mechanisms are consistent with neural regions thought to be involved in timing (e.g., frontal cortex and striatum), and its output is consistent with physiological recordings and behavioral results from interval-timing experiments (Matell et al. 2003, 2011; Meck et al. 2012). Indeed, striatal neurons fire in a peak-shaped manner centered on the target duration, following the predictions of the SBF model. Critically, the SBF model reproduces the scalar property, the hallmark of interval timing (Gibbon et al. 1984, 1997; Van Rijn et al. 2013). Preservation of this scalar property is critical for experimental manipulations thought to influence the timing system per se (i.e., deviations are considered diagnostic of effects on other systems that influence behavior). In the SBF model, the scalar property occurs because of variability in the firing patterns of striatal neurons and because cortical activity is assumed to be oscillatory, such that firing patterns at the harmonics (i.e., 1/2, 2/3, etc.) are similar but not identical to those occurring at the target duration. Teki et al. (2012) recently proposed a unified model of interval timing based on coordinated activity in the core striatal and ancillary olivocerebellar networks that takes advantage of the interconnections between these networks and the cerebral cortex. Timing in this model posits a type of degeneracy (Lewis & Meck 2012) that involves the initiation and maintenance of timing by beat-based striatal activation that is adjusted by olivocerebellar mechanisms that can substitute for the striatal timing system as a function of neural deactivation if needed (Allman & Meck 2012, Jahanshahi et al. 2010a) and/or by genetic modifications of neurotransmitter/receptor function (Liao & Cheng 2005, Sysoeva et al. 2010, Wiener et al. 2011a) and aging (Balci et al. 2009; Cheng et al. 2008b, 2011a).

CONCLUSIONS

Recent research concurs with modern neurophysiological models whereby the capacity to perceive and estimate time is thought to emerge from interactions of a core CTBG timing circuit with brain regions that provide signals needed to time events (Allman & Meck 2012, Lustig & Meck 2005). Important advances from cell recordings further indicate that elements of the CTBG not only display chronotopy, but also represent organizational features of the context that permits more abstract timing behaviors (Merchant et al. 2011). At the neuroanatomical level, basal ganglia and SMA functioning were dissociated by differential activity that was respectively linked to fluctuations in the task's interval timing and working-memory demands (Harrington et al. 2010). Thus, elements of the CTBG timing circuit display different context-dependent activation dynamics that warrant further inquiry. The functional significance of networks engaged by timing will also be advanced by studies of how the brain construes time in situations that influence the resolution of perceived duration. Emerging research indicates that temporal distortions produced by emotionally charged events or stimuli that capture attention are partly brought about by activity in emotion or association networks (Dirnberger et al. 2012, Cheng et al. 2011b, Harrington et al. 2011a, Wittmann et al. 2010b). Although the role of the striatum remains debated in these studies, temporal distortions that emanate from intersensory timing are driven by the CTBG circuit. Likewise, degeneracy in timing revealed by disease (Jahanshahi et al. 2010a) and individual differences in the expression of neurotransmitter function (Sysoeva et al. 2010, Wiener et al. 2011a) also hold promise for uncovering neurophysiological mechanisms of timing networks. More generally, it is important for future research to study brain connectivity, which more fully characterizes the communication of timing circuits with other brain networks (Cheng et al. 2011b; Harrington et al. 2011a,b; MacDonald & Meck 2004).

DISCLOSURE STATEMENT

The authors are not aware of any affiliations, memberships, funding, or financial

holdings that might be perceived as affecting the objectivity of this review.

LITERATURE CITED

Agostino PV, Golombek DA, Meck WH. 2011. Unwinding the molecular basis of interval and circadian timing. *Front. Integr. Neurosci.* 5:64

Allman MJ, Meck WH. 2012. Pathophysiological distortions in time perception and timed performance. *Brain* 135:656–77

Allman MJ, Pelphrey KA, Meck WH. 2012. Developmental neuroscience of time and number: implications for autism and other neurodevelopmental disabilities. *Front. Integr. Neurosci.* 6:7

Aparicio P, Diedrichsen J, Ivry RB. 2005. Effects of focal basal ganglia lesions on timing and force control. *Brain Cogn.* 58:62–74

Artieda J, Pastor MA, Lacruz F, Obeso JA. 1992. Temporal discrimination is abnormal in Parkinson's disease. *Brain* 115:199–210

Balci F, Meck WH, Moore H, Brunner D. 2009. Timing deficits in aging and neuropathology. In *Animal Models of Human Cognitive Aging,* ed. JL Bizon, A Woods, pp. 161–201. Totowa, NJ: Humana

Balci F, Wiener M, C¸ avdarogˇ lu B, Coslett HB. 2013. Epistasis effects of dopamine genes on interval timing and reward magnitude in humans. *Neuropsychologia* 51:293–308

Bartolo R, Merchant H. 2009. Learning and generalization of time production in humans: rules of transfer across modalities and interval durations. *Exp. Brain Res.* 197:91–100

Brannon EM, Libertus ME, Meck WH, Woldorff MG. 2008. Electrophysiological measures of time processing in infant and adult brains: Weber's law holds. *J. Cogn. Neurosci.* 20:193–203

Brody CD, Hernández A, Zainos A, Romo R. 2003. Timing and neural encoding of somatosensory parametric working memory in macaque prefrontal cortex. *Cereb. Cortex* 13:1196–207

Buhusi CV, Meck WH. 2005. What makes us tick? Functional and neural mechanisms of interval timing. *Nat. Rev. Neurosci.* 6:755–65

Buhusi CV, Meck WH. 2009a. Relative time sharing: new findings and an extension of the resource allocation model of temporal processing. *Philos. Trans. R. Soc. Lond. B* 364:1875–85

Buhusi CV, Meck WH. 2009b. Relativity theory and time perception: single or multiple clocks? *PLoS ONE* 4(7):e6268

Buonomano DV, Laje R. 2010. Population clocks: motor timing with neural dynamics. *Trends Cogn. Sci.* 14:520–27

Burr D, Tozzi A, Morrone M. 2007. Neural mechanisms for timing visual events are spatially selective in real-world coordinates. *Nat. Neurosci.* 10:423–25

Cheng RK, Ali YM, Meck WH. 2007. Ketamine "unlocks" the reduced clock-speed effect of cocaine following extended training: evidence for dopamine-glutamate interactions in timing and time perception. *Neurobiol. Learn. Mem.* 88:149–59

Cheng RK, Dyke AG, McConnell MW, Meck WH. 2011a. Categorical scaling of duration as a function of temporal context in aged rats. *Brain Res.* 1381:175–86

Cheng RK, Jesuthasan S, Penney TB. 2011b. Time for zebrafish. *Front. Integr. Neurosci.* 5:40

Cheng RK, MacDonald CJ, Williams CL, Meck WH. 2008a. Prenatal choline supplementation alters the timing, emotion, and memory performance (TEMP) of adult male and female rats as indexed by differential reinforcement of low-rate schedule behavior. *Learn. Mem.* 15:153–62

Cheng RK, Meck WH. 2007. Prenatal choline supplementation increases sensitivity to time by reducing non-scalar sources of variance in adult temporal processing. *Brain Res.* 1186:242–54

Cheng RK, Scott AC, Penney TB, Williams CL, Meck WH. 2008b. Prenatal choline availability differentially modulates timing of auditory and visual stimuli in aged rats. *Brain Res.* 1237:167–75

Coslett HB, Wiener M, Chatterjee A. 2010. Dissociable neural systems for timing: evidence from subjects with basal ganglia lesions. *PLoS ONE* 5(4):e10324

Coull JT, Cheng RK, Meck WH. 2011. Neuroanatomical and neurochemical substrates of timing. *Neuropsychopharmacology* 36:3–25

Coull JT, Hwang HJ, Leyton M, Dagher A. 2012. Dopamine precursor depletion impairs timing in healthy volunteers by attenuating activity in putamen and SMA. *J. Neurosci.* 32:16704–15

Coull JT, Nazarian B, Vidal F. 2008. Timing, storage, and comparison of stimulus duration engage discrete anatomical components of a perceptual timing network. *J. Cogn. Neurosci.* 20:2185–97

Coull JT, Vidal F, Nazarian B, Macar F. 2004. Functional anatomy of the attentional modulation of time estimation. *Science* 303:1506–8

Dirnberger G, Hesselmann G, Roiser JP, Preminger S, Jahanshahi M, Paz R. 2012. Give it time: neural evidence for distorted time perception and enhanced memory encoding in emotional situations. *NeuroImage* 63:591–99

Droit-Volet S, Meck WH. 2007. How emotions colour our perception of time. Trends Cogn. Sci. 11:504–13

Eagleman DM. 2008. Human time perception and its illusions. *Curr. Opin. Neurobiol.* 18:131–36

Eckert MA, Menon V, Walczak A, Ahlstrom J, Denslow S, et al. 2009. At the heart of the ventral attention system: the right anterior insula. *Hum. Brain Mapp.* 30:2530–41

Fraisse P. 1984. Perception and estimation of time. *Annu. Rev. Psychol.* 35:1–36

Genovesio A, Tsujimoto S, Wise SP. 2006. Neuronal activity related to elapsed time in prefrontal cortex. *J. Neurophysiol.* 95:3281–85

Genovesio A, Tsujimoto S, Wise SP. 2009. Feature- and order-based timing representations in the frontal cortex. *Neuron* 63:254–66

Gibbon J, Church RM, Meck WH. 1984. Scalar timing in memory. *Ann. NY Acad. Sci.* 423:52–77

Gibbon J, Malapani C, Dale CL, Gallistel CR. 1997. Toward a neurobiology of temporal cognition: advances and challenges. *Curr. Opin. Neurobiol.* 7:170–84

Gu BM, Cheng RK, Yin B, Meck WH. 2011. Quinpirole-induced sensitization to noisy/sparse periodic input: temporal synchronization as a component of obsessive-compulsive disorder. *Neuroscience* 179:143–50

Gu BM, Jurkowski AJ, Lake JI, Malapani C, Meck WH. 2013. Bayesian models of interval timing and distortions in temporal memory as a function of Parkinson's disease and dopamine-related error processing. In *Time Distortions in Mind: Temporal Processing in Clinical Populations,* ed. A Vatakis, MJ Allman. Boston, MA: Brill. In press

Gu BM, Meck WH. 2011. New perspectives on Vierordt's law: memory-mixing in ordinal temporal comparison tasks. *Lect. Notes Comp. Sci.* 6789:67–78

Harrington DL, Boyd LA, Mayer AR, Sheltraw DM, Lee RR, et al. 2004. Neural representation of interval encoding and decision making. *Cogn. Brain Res.* 21:193–205

Harrington DL, Castillo GN, Fong CH, Reed JD. 2011a. Neural underpinnings of distortions in the experience of time across senses. *Front. Integr. Neurosci.* 5:32

Harrington DL, Castillo GN, Greenberg PA, Song DD, Lessig S, et al. 2011b. Neurobehavioral mechanisms of temporal processing deficits in Parkinson's disease. *PLoS ONE* 6:e17461

Harrington DL, Haaland KY, Hermanowicz N. 1998a. Temporal processing in the basal ganglia. *Neuropsychology* 12:3–12

Harrington DL, Haaland KY, Knight RT. 1998b. Cortical networks underlying mechanisms of time perception. *J. Neurosci.* 18:1085–95

Harrington DL, Zimbelman JL, Hinton SC, Rao SM. 2010. Neural modulation of temporal encoding, maintenance, and decision processes. *Cereb. Cortex* 20:1274–85

Hazy TE, Frank MJ, O'Reilly RC. 2006. Banishing the homunculus: making working memory work. *Neuroscience* 139:105–18

Hinton SC, Meck WH. 1997. How time flies: function and neural mechanisms of interval timing. *Adv. Psychol.* 120:409–57

Hinton SC, Meck WH. 2004. Frontal-striatal circuitry activated by human peak-interval timing in the supraseconds range. *Cogn. Brain Res.* 21:171–82

Höhn S, Dallé rac G, Faure A, Urbach Y, Nguyen HP, et al. 2011. Behavioral and in vivo electrophysiological evidence for presymptomatic alteration of prefronto-striatal processing in the transgenic rat model for Huntington disease. *J. Neurosci.* 31:8986–97

Ivry RB, Hazeltine RE. 1995. Perception and production of temporal intervals across a range of durations: evidence of a common timing mechanism. *J. Exp. Psychol. Hum. Percept. Perform.* 21:3–18

Jahanshahi M, Jones CR, Dirnberger G, Frith CD. 2006. The substantia nigra pars compacta and temporal processing. *J. Neurosci.* 26:12266–73

Jahanshahi M, Jones CR, Zijlmans J, Katsenschlager R, Lee L, et al. 2010a. Dopaminergic modulation of striato-frontal connectivity during motor timing in Parkinson's disease. *Brain* 133:727–45

Jahanshahi M, Wilkinson L, Gahir H, Dharminda A, Lagnado DA. 2010b. Medication impairs probabilistic classification learning in Parkinson's disease. *Neuropsychologia* 48:1096–103

Jin DZ, Fujii N, Graybiel AM. 2009. Neural representation of time in cortico-basal ganglia circuits. *Proc. Natl. Acad. Sci. USA* 106:19156–61

Johnson HA, Goel A, Buonomano DV. 2010. Neural dynamics of in vitro cortical networks reflects experienced temporal patterns. Nat. Neurosci. 13:917–19

Johnston A, Arnold DH, Nishida S. 2006. Spatially localized distortions of event time. *Curr. Biol.* 16:472–79

Jones CR, Malone TJ, Dirnberger G, Edwards M, Jahanshahi M. 2008. Basal ganglia, dopamine and temporal processing: performance on three timing tasks on and off medication in Parkinson's disease. *Brain Cogn.* 68:30–41

Jones CRG, Jahanshahi M. 2011. Dopamine modulates striato-frontal functioning during temporal processing. *Front. Integr. Neurosci.* 5:70

Karmarkar UR, Buonomano DV. 2003. Temporal specificity of perceptual learning in an auditory discrimination task. *Learn. Mem.* 10:141–47

Karmarkar UR, Buonomano DV. 2007. Timing in the absence of clocks: encoding time in neural network states. *Neuron* 53:427–38

Keele S, Nicoletti R, Ivry R, Pokorny R. 1985. Do perception and motor production share common timing mechanisms? A correlational analysis. *Acta Psychol.* 60:173–91

Koch G, Brusa L, Caltagirone C, Oliveri M, Peppe A, et al. 2004. Subthalamic deep brain stimulation improves time perception in Parkinson's disease. *NeuroReport* 15:1071–73

Koch G, Costa A, Brusa L, Peppe A, Gatto I, et al. 2008. Impaired reproduction of second but not millisecond time intervals in Parkinson's disease. *Neuropsychologia* 46:1305–13

Kosillo P, Smith AT. 2010. The role of the human anterior insular cortex in time processing. *Brain Struct. Funct.* 214:623–28

Kotz SA, Schwartze M. 2011. Differential input of the supplementary motor area to a dedicated temporal processing network: functional and clinical implications. *Front. Integr. Neurosci.* 5:86

Lake JI, Meck WH. 2012. Differential effects of amphetamine and haloperidol on temporal reproduction: dopaminergic regulation of attention and clock speed. *Neuropsychologia* 51:284–92

Lebedev MA, O'Doherty JE, Nicolelis MA. 2008. Decoding of temporal intervals from cortical ensemble activity. *J. Neurophysiol.* 99:166–86

Leon MI, Shadlen MN. 2003. Representation of time by neurons in the posterior parietal cortex of the macaque. *Neuron* 38:317–27

Lewis PA, Meck WH. 2012. Time and the sleeping brain. *Psychologist* 25:594–97

Lewis SJ, Dove A, Robbins TW, Barker RA, Owen AM. 2004. Striatal contributions to working memory: a functional magnetic resonance imaging study in humans. *Eur. J. Neurosci.* 19:755–60

Liao RM, Cheng RK. 2005. Acute effects of d-amphetamine on the differential reinforcement of low-rate (DRL) schedule behavior in the rat: comparison with selective dopamine receptor antagonists. *Chin. J. Physiol.* 48:41–50

Livesey AC, Wall MB, Smith AT. 2007. Time perception: manipulation of task difficulty dissociates clock functions from other cognitive demands. *Neuropsychologia* 45:321–31

Lucchetti C, Bon L. 2001. Time-modulated neuronal activity in the premotor cortex of macaque monkeys. *Exp. Brain Res.* 141:254–60

Lustig C, Matell MS, Meck WH. 2005. Not "just" a coincidence: frontal-striatal synchronization in working memory and interval timing. *Memory* 13:441–48

Lustig C, Meck WH. 2005. Chronic treatment with haloperidol induces working memory deficits in feedback effects of interval timing. *Brain Cogn.* 58:9–16

Lustig C, Meck WH. 2011. Modality differences in timing and temporal memory throughout the lifespan. *Brain Cogn.* 77:298–303

MacDonald CJ, Cheng RK, Meck WH. 2012. Acquisition of "Start" and "Stop" response thresholds in peak-interval timing is differentially sensitive to protein synthesis inhibition in the dorsal and ventral striatum. *Front. Integr. Neurosci.* 6:10

MacDonald CJ, Meck WH. 2004. Systems-level integration of interval timing and reaction time. *Neurosci. Biobehav. Rev.* 28:747–69

Mason PH. 2010. Degeneracy at multiple levels of complexity. *Biol. Theory* 5:277–88

Matell MS, Meck WH. 2000. Neuropsychological mechanisms of interval timing behaviour. *BioEssays* 22:94–103

Matell MS, Meck WH. 2004. Cortico-striatal circuits and interval timing: coincidence detection of oscillatory processes. *Cogn. Brain Res.* 21:139–70

Matell MS, Meck WH, Nicolelis MAL. 2003. Interval timing and the encoding of signal duration by ensembles of cortical and striatal neurons. *Behav. Neurosci.* 117:760–73

Matell MS, Shea-Brown E, Gooch C, Wilson AG, Rinzel J. 2011. A heterogeneous population code for elapsed time in rat medial agranular cortex. *Behav. Neurosci.* 125:54–73

Meck WH. 1996. Neuropharmacology of timing and time perception. *Cogn. Brain Res.* 3:227–42

Meck WH. 2003. *Functional and Neural Mechanisms of Interval Timing.* Boca Raton, FL: CRC Press

Meck WH. 2006a. Frontal cortex lesions eliminate the clock speed effect of dopaminergic drugs on interval timing. *Brain Res.* 1108:157–67

Meck WH. 2006b. Neuroanatomical localization of an internal clock: a functional link between mesolimbic, nigrostriatal, and mesocortical dopaminergic systems. *Brain Res.* 1109:93–107

Meck WH, Benson AM. 2002. Dissecting the brain's internal clock: how frontal-striatal circuitry keeps time and shifts attention. *Brain Cogn.* 48:195–211

Meck WH, Cheng RK, MacDonald CJ, Gainetdinov RR, Caron MG, C̦evik MÖ. 2012. Gene-dose dependent effects of methamphetamine on interval timing in dopamine-transporter knockout mice. *Neuropharmacology* 62:1221–29

Meck WH, Church RM, Matell MS. 2013. Hippocampus, time, and memory—a retrospective analysis. *Behav. Neurosci.* In press

Meck WH, Church RM, Olton DS. 1984. Hippocampus, time, and memory. *Behav. Neurosci.* 98:3–22

Meck WH, Malapani C. 2004. Neuroimaging of interval timing. *Cogn. Brain Res.* 21:133–37

Meck WH, Penney TB, Pouthas V. 2008. Cortico-striatal representation of time in animals and humans. *Curr. Opin. Neurobiol.* 18:145–52

Meegan DV, Aslin RN, Jacobs RA. 2000. Motor timing learned without motor training. *Nat. Neurosci.* 3:860–62

Mella N, Conty L, Pouthas V. 2011. The role of physiological arousal in time perception: psychophysiological evidence from an emotion regulation paradigm. *Brain Cogn.* 75:182–87

Merchant H, Battaglia-Mayer A, Georgopoulos AP. 2004. Neural responses during interception of real and apparent circularly moving stimuli in motor cortex and area 7a. *Cereb. Cortex* 14:314–31

Merchant H, de Lafuente V, Peña-Ortega F, Larriva-Sahd J. 2012. Functional impact of interneuronal inhibition in the cerebral cortex of behaving animals. *Prog. Neurobiol.* 99:163–78

Merchant H, Luciana M, Hooper C, Majestic S, Tuite P. 2008a. Interval timing and Parkinson's disease: heterogeneity in temporal performance. *Exp. Brain Res.* 184:233–48

Merchant H, Pérez O, Zarco W, Gámez J. 2013. Interval tuning in the primate medial premotor cortex as a general timing mechanism. *J. Neurosci.* 33:9082–96

Merchant H, Zarco W, Bartolo R, Prado L. 2008b. The context of temporal processing is represented in the multidemensional relationships between timing tasks. *PLoS One* 3(9):e3169

Merchant H, Zarco W, Perez O, Prado L, Bartolo R. 2011. Measuring time with multiple neural chronometers during a synchronization-continuation task. *Proc. Natl. Acad. Sci. USA* 108:19784–89

Merchant H, Zarco W, Prado L. 2008c. Do we have a common mechanism for measuring time in the hundreds of millisecond range? Evidence from multiple timing tasks. *J. Neurophysiol.* 99:939–49

Mita A, Mushiake H, Shima K, Matsuzaka Y, Tanji J. 2009. Interval time coding by neurons in the presupplementary and supplementary motor areas. *Nat. Neurosci.* 12:502–7

Nagarajan SS, Blake DT, Wright BA, Byl N, Merzenich MM. 1998. Practice related improvements in somatosensory interval discrimination are temporally specific but generalize across skin location, hemisphere, and modality. *J. Neurosci.* 18:1559–70

Nagy A, Eordegh G, Paroczy Z, Markus Z, Benedek G. 2006. Multisensory integration in the basal ganglia. *Eur. J. Neurosci.* 24:917–24

N'Diaye K, Ragot R, Garnero L, Pouthas V. 2004. What is common to brain activity evokes by the perception of visual and auditory filled durations? A study with MEG and EEG co-recordings. *Cogn. Brain Res.* 21:250–68

Oprisan SA, Buhusi CV. 2011. Modeling pharmacological clock and memory patterns of interval timing in a striatal beat-frequency model with realistic, noisy neurons. *Front. Integr. Neurosci.* 5:52

Oshio K, Chiba A, Inase M. 2008. Temporal filtering by prefrontal neurons in duration discrimination. *Eur. J. Neurosci.* 28:2333–43

Paulsen JS, Zimbelman JL, Hinton SC, Langbehn DR, Leveroni CL, et al. 2004. fMRI biomarker of early neuronal dysfunction in presymptomatic Huntington's disease. *Am. J. Neuroradiol.* 25:1715–21

Penney TB, Gibbon J, Meck WH. 2008. Categorical scaling of duration bisection in pigeons (*Columba livia*), mice (*Mus musculus*), and humans (*Homo sapiens*). *Psychol. Sci.* 19:1103–9

Perez O, Kass R, Merchant H. 2013. Trial time warping to discriminate stimulus-related from movement-related neural activity. *J. Neurosci. Methods* 212:203–10

Perrett SP. 1998. Temporal discrimination in the cerebellar cortex during conditioned eyelid responses. *Exp. Brain Res.* 121:115–24

Pleil KE, Cordes S, Meck WH, Williams CL. 2011. Rapid and acute effects of estrogen on time perception in male and female rats. *Front. Integ. Neurosci.* 5:63

Price CJ, Friston KJ. 2002. Degeneracy and cognitive anatomy. *Trends Cogn. Sci.* 6:416–21

Rao SM, Mayer AR, Harrington DL. 2001. The evolution of brain activation during temporal processing. *Nat. Neurosci.* 4:317–23

Repp BH. 2005. Sensorimotor synchronization: a review of the tapping literature. *Psychon. Bull. Rev.* 12:969–92

Repp BH, Penel A. 2002. Auditory dominance in temporal processing: new evidence from synchronization with simultaneous visual and auditory sequences. *J. Exp. Psychol. Hum. Percept. Perform.* 28:1085–99

Rowe KC, Paulsen JS, Langbehn DR, Duff K, Beglinger LJ, et al. 2010. Self-paced timing detects and tracks change in prodromal Huntington disease. *Neuropsychology* 24:435–42

Schwartze M, Rothermich K, Kotz SA. 2012. Functional dissociation of pre-SMA and SMA-proper in temporal processing. *NeuroImage* 60:290–98

Smith JG, Harper DN, Gittings D, Abernethy D. 2007. The effect of Parkinson's disease on time estimation as a function of stimulus duration range and modality. *Brain Cogn.* 64:130–43

Spencer RM, Ivry RB. 2005. Comparison of patients with Parkinson's disease or cerebellar lesions in the production of periodic movements involving event-based or emergent timing. *Brain Cogn.* 58:84–93

Sysoeva OV, Tonevitsky AG, Wackermann J. 2010. Genetic determinants of time perception mediated by the serotonergic system. *PLoS ONE* 5(9):e12650

Tanaka M. 2007. Cognitive signals in the primate motor thalamus predict saccade timing. *J. Neurosci.* 27:12109–18

Tanji J. 2001. Sequential organization of multiple movements: involvement of cortical motor areas. *Annu.*

Rev. Neurosci. 24:631–51

Teki S, Grube M, Griffiths TD. 2012. A unified model of time perception accounts for duration-based and beat-based timing mechanisms. *Front. Integr. Neurosci.* 5:90

Ulrich R, Nitschke J, Rammsayer T. 2006. Crossmodal temporal discrimination: assessing the predictions of a general pacemaker-counter model. *Percept. Psychophys.* 68:1140–52

van Rijn H, Gu BM, Meck WH. 2013. Dedicated clock/timing-circuit theories of interval timing. In *Neurobiology of Interval Timing,* ed. H Merchant, V de Lafuente. New York: Springer-Verlag. In press

van Rijn H, Kononowicz TW, Meck WH, Ng KK, Penney TB. 2011. Contingent negative variation and its relation to time estimation: a theoretical evaluation. *Front. Integr. Neurosci.* 5:91

van Schouwenburg MR, den Ouden HE, Cools R. 2010. The human basal ganglia modulate frontal-posterior connectivity during attention shifting. *J. Neurosci.* 30:9910–18

van Wassenhove V, Buonomano DV, Shimojo S, Shams L. 2008. Distortions of subjective time perception within and across senses. *PLoS ONE* 3:e1437

van Wassenhove V, Nagarajan SS. 2007. Auditory cortical plasticity in learning to discriminate modulation rate. *J. Neurosci.* 27:2663–72

Wearden JH, Smith-Spark JH, Cousins R, Edelstyn NM, Cody FW, O'Boyle DJ. 2008. Stimulus timing by people with Parkinson's disease. *Brain Cogn.* 67:264–79

Wiener M, Hamilton R, Turkeltaub P, Matell MS, Coslett HB. 2010a. Fast forward: Supramarginal gyrus stimulation alters time measurement. *J. Cogn. Neurosci.* 22:23–31

Wiener M, Kliot D, Turkeltaub PE, Hamilton RH, Wolk DA, Coslett HB. 2012. Parietal influence on temporal encoding indexed by simultaneous transcranial magnetic stimulation and electroencephalography. *J. Neurosci.* 32:12258–67

Wiener M, Lohoff FW, Coslett HB. 2011a. Double dissociation of dopamine genes and timing in humans. *J. Cogn. Neurosci.* 23:2811–21

Wiener M, Matell MS, Coslett HB. 2011b. Multiple mechanisms for temporal processing. *Front. Integr. Neurosci.* 5:31

Wiener M, Turkeltaub P, Coslett HB. 2010c. The image of time: a voxel-wise meta-analysis. *NeuroImage* 49:1728–40

Wittmann M, Simmons AN, Aron JL, Paulus MP. 2010a. Accumulation of neural activity in the posterior insula encodes the passage of time. *Neuropsychologia* 48:3110–20

Wittmann M, van Wassenhove V, Craig AD, Paulus MP. 2010b. The neural substrates of subjective time dilation. *Front. Hum. Neurosci.* 4:2

Yin B, Troger AB. 2011. Exploring the 4th dimension: hippocampus, time, and memory revisited. *Front. Integr. Neurosci.* 5:36

Yu H, Sternad D, Corcos DM, Vaillancourt DE. 2007. Role of hyperactive cerebellum and motor cortex in Parkinson's disease. *NeuroImage* 35:222–33

Zarco W, Merchant H, Prado L, Mendez JC. 2009. Subsecond timing in primates: comparison of interval production between human subjects and Rhesus monkeys. *J. Neurophysiol.* 102:3191–202

Zelaznik HN, Spencer RMC, Ivry RB. 2002. Dissociation of explicit and implicit timing in repetitive tapping and drawing movements. *J. Exp. Psychol.: Hum. Percept. Perform.* 28:575–88

Zimbelman JL, Paulsen JS, Mikos A, Reynolds NC, Hoffmann RG, Rao SM. 2007. fMRI detection of early neural dysfunction in preclinical Huntington's disease. *J. Int. Neuropsychol. Soc.* 13:758–69

RELATED RESOURCES

Buonomano DV. 2007. The biology of time across different scales. *Nat. Chem. Biol.* 3:594–97

Gorea A. 2011. Ticks per thought or thoughts per tick? A selective review of time perception with hints on future research. *J. Physiol.* 105:153–63

Grondin S. 2010. Timing and time perception: a review of recent behavioral and neuroscience findings and theoretical directions. *Atten. Percept. Psychophys.* 72:561–82

Macar F, Vidal F. 2009. Timing processes: an outline of behavioural and neural indices not systematically considered in timing models. *Can. J. Exp. Psychol.* 63: 227–39

Mauk MD, Buonomano DV. 2004. The neural basis of temporal processing. *Annu. Rev. Neurosci.* 27:304–40

Wittmann M. 2013. The inner sense of time: how the brain creates a representation of duration. *Nat. Rev. Neurosci.* 14:217–23.

Long-range Neural Synchrony in Behavior

Alexander Z. Harris[1,2] and Joshua A. Gordon[1,2]

Alexander Z. Harris: ah2835@columbia.edu; Joshua A. Gordon: jg343@columbia.edu
[1]Department of Psychiatry, Columbia University, New York, NY 10032
[2]Division of Integrative Neuroscience, New York State Psychiatric Institute, New York, NY 10032

Key Words

Oscillations; Coherence; Hippocampus; Prefrontal Cortex; Gamma; Theta

Abstract

Long-range synchrony between distant brain regions accompanies multiple forms of behavior. This review compares and contrasts the methods by which long-range synchrony is evaluated in both humans and model animals. Three examples of behaviorally-relevant long-range synchrony are discussed in detail: gamma-frequency synchrony during visual perception; hippocampal-prefrontal synchrony during working memory; and prefrontal-amygdala synchrony during anxiety. Implications for circuit mechanism, translation, and clinical relevance are discussed.

INTRODUCTION

The rich tapestry formed by the trillions of connections between far-flung brain regions demonstrates the complexity of the brain. New imaging techniques artfully display these connections (Figure 1A), which facilitate cooperation amongst distributed elements of neural systems. Yet the connectivity of the brain does not statically derive from these anatomical pathways; it dynamically fluctuates with mood and cognitive states, influenced by stimuli and influencing behavior. Detecting and quantifying connectivity provides a key to understanding this dynamism.

Studying long-range neural synchrony has proven invaluable for this purpose. This research assays the degree to which brain regions are functionally connected by measuring the degree to which their neural activity patterns are synchronized. Neural synchrony can be quantified using a wide range of tools, including magnetoencephalography (MEG), electroencephalography (EEG) and functional neuroimaging (Figure 1B), as well as direct neurophysiological recordings (Figure 1C).

Numerous studies have established compelling if correlative links between synchrony and behavioral states, identifying several common themes. The brain shows high synchrony even at rest; such baseline or "resting state" synchrony tends to generally follow from anatomical connectivity. During tasks, synchrony typically changes within the activated regions, often in specific task phases or with specific perceptual or cognitive demands. Disease states may have altered synchrony, often correlated with associated alterations in behavior. These themes suggest that long-range synchrony, while supported by anatomical connectivity, changes on behavioral timescales. This review focuses on three sets of studies that illustrate these themes. In citing key examples of the relationship between long-range synchrony and behavior, this review builds on the growing momentum in the literature. Both animal and human studies identify functional connectivity correlates of behavior. Increasingly, researchers have applied such methods to large clinical samples and sophisticated animal models to examine the role of dysconnectivity in neuropsychiatric disease. Moreover, the advent of technologies to manipulate specific circuits and cell types allows direct testing of causal hypotheses generated from these data. These simultaneous developments promise to move the field beyond beautiful pictures and elegant correlations, towards establishing the causal relationship between long-range neural synchrony and behavior.

METHODS: MEASURING LONG-RANGE SYNCHRONY

The advent of technologies that measure neural activity over time make examining long-range synchrony possible. These technologies include functional neuroimaging (primarily functional magnetic resonance imaging [fMRI]), magnetoencephalography (MEG), and electroencephalography (EEG) in humans. In model animals, studies of long-range synchrony mostly use intracranial electrophysiological techniques that permit simultaneous, fast measurements of neural activity in multiple brain regions during behavior. Rare human intracranial recordings of neural activity supplement the data acquired through non-invasive methods. The temporal and anatomical precision of these technologies differ considerably, both because of the origins of the biological signals they rely on, and the techniques used to capture these signals.

A few pertinent details of non-invasive approaches to measuring neural activity in humans will aid the discussion of neural synchrony (for full review see (Bandettini 2009, Ioannides 2007, Pan et al 2011, Sakkalis 2011)). fMRI relies on anatomically localized changes in blood flow induced by neural activity. Increases in metabolic demand (driven by neural activity) boost blood flow, which in turn raises the blood oxygen level. The resulting signal, blood-oxygen-level dependent (BOLD) contrast, changes slowly relative to neural activity. By contrast, EEG, MEG and intracranial recordings measure electrical activity directly, or via the magnetic fields that activity induces. The extracranial signals picked up by EEG and MEG result from the coordinated (and often rhythmic) activity of large numbers of neurons; only the current generated by the activation of a substantial number of neurons can induce currents or magnetic fields large enough to be detected outside the skull. Electrical measures have sub-millisecond time resolutions, compared to the seconds-long time course of the BOLD signal. However, locating the origin of these electrical signals depends on modeling how intracranial sources give rise to extracranial signals, a challenging endeavor (Pascual-Marqui et al 2002).

Intracranial recordings can measure neural activity simultaneously from multiple brain regions with a high degree of anatomical and temporal precision. These electrodes yield two types of neural activity: spikes and local field potentials (LFPs). Spikes represent the extracellular manifestation of action potentials, while slow, large voltage fluctuations caused by synchronous synaptic activity of many neurons produce LFPs. Depth electrodes can capture both spikes and LFPs in targeted structures, while electrocorticography (ECOG) captures LFPs from the surface of the brain using flexible, multi-channel arrays.

The principle behind long-range synchrony is to measure activity simultaneously from multiple sites, and ask if activity at these distributed sites tends to change in a coordinated manner. For fMRI, one tracks activity in individual voxels or regions of interest (ROIs) and measures correlations in these time series. Correlations can be positive (meaning activity in the two areas tends to go up and down together) or negative (meaning activity in one area correlates with less activity in the second area, and vice-versa). Given the slow timescale of the BOLD signal, fluctuations in fMRI activity occur slowly (0.01–0.1 Hz).

LFPs and EEGs recorded from the behaving brain consist of oscillatory activity in characteristic frequency ranges, such as delta (0.5–4 Hz), theta (4–8 Hz), alpha (8–12 Hz), beta (12–30 Hz) and gamma (30–120 Hz) (Figure 2A). We call LFPs recorded simultaneously from two regions *synchronous* if their peaks and valleys align (which we will call *phase coherence*) or if their amplitudes correlate (*power correlation*). The mathematical term, *coherence*-typically calculated as a function of frequency-encompasses both of these aspects of LFP-LFP synchrony. The coherence spectrum between, for example, the hippocampus and prefrontal cortex, demonstrates peaks in coherence at delta, theta and gamma frequencies ((Adhikari et al 2010, Jones & Wilson 2005a, O'Neill et al 2013, Sigurdsson et al 2010); Figure 2B), indicating that these two regions strongly synchronize in these frequency ranges.

Spikes can be used to measure synchrony in two ways. *Phase-locking* refers to the temporal relationship between spikes and LFP oscillations, quantified by the degree to which the spikes align with particular phases of the oscillation ((Jones & Wilson 2005a, O'Keefe & Recce 1993); Figure 2C). Alternatively, *cross-correlations* of spikes recorded simultaneously in both regions measure the degree to which neurons in the two regions synchronize their action potentials (Brown et al 2004); Figure 2D.

Simultaneous fMRI and electrophysiological synchrony studies have related these two measures. Task-evoked fMRI synchrony seems to arise from common fluctuations in gamma oscillations (Goense & Logothetis 2008, Nir et al 2007, Shmuel & Leopold 2008). But this description can be misleading. Electrophysiological measures rely on *fast* synchrony–precise, rapid alignment of neuronal activity across brain regions on the order of milliseconds. By contrast, synchrony measured with BOLD seems to respond to *correlations* in power: BOLD signals reflect *slow* fluctuations in the *strength* of these faster oscillations. Theoretically, for example, the strength of gamma oscillations might rise and fall together in two brain regions, while the oscillations themselves need not by synchronized. Such methodological differences may be crucial to understanding how circuits perform computations, and how such computations go awry in disease. Given the emphasis in the animal literature on long-range coherence in behavior and the utility of fMRI for assaying synchrony in disease states, clarifying the relationship between slow and fast synchrony has important translational implications.

SYNCHRONY IN THE VISUAL SYSTEM

Binding by Synchrony: Evidence from Animal Models

The framework for studying long-range neural synchrony emerged from visual system research. Over 25 years ago, Singer and colleagues recorded synchronous neural activity in neurons with non-overlapping receptive fields located in cortical columns up to 7 mm apart in the cat primary visual cortex (Gray et al 1989). Visual stimuli moving in the same direction across distant receptive fields induced weak synchrony, while the same stimuli moving in opposite directions failed to do so. A single, long stimulus that simultaneously activated both receptive fields resulted in robust synchrony.

These physiological findings evoke a behavioral phenomenon: that of binding the disparate features of visual stimuli into a unified perception of an object. The possibility that gamma synchrony could solve the "binding problem" is grounded in previous theoretical models (Grossberg 1976, Malsburg 1981, Milner 1974). Subsequent work found stimulus-induced synchrony in the middle temporal area of awake, behaving monkeys (Kreiter & Singer 1996), and inter-regional synchrony between activity in primary visual cortex and a visual association area in the cat (Engel et al 1991a, Engel et al 1991b). Together, these studies suggest that synchronous neural firing emerges under the Gestalt psychophysical principles that define the perception of a single object, such as spatial continuity and coherent motion (Wagemans et al 2012).

If synchrony underlies the perception of bound objects, rather than mere receptive field stimulation, one would expect synchrony only from stimuli that are perceived as bound. In strabismic cats, each eye forms a separate representation of a stimulus; the discrepancy, known as "binocular rivalry", is resolved by suppressing the image from the non-dominant eye. Fries and colleagues (Fries et al 1997) predicted that synchrony should only exist for objects perceived through the dominant eye. Indeed, neural activity shows enhanced gamma-range synchrony only in the receptive fields of the dominant eye (Fries et al 1997).

If synchrony underlies the perception of bound objects, attending to an object should enhance perceptual binding and thus neural synchrony. Indeed, when monkeys attend to one visual stimulus while ignoring a distractor, gamma power in the spike-triggered average of LFPs recorded in area V4 increases only for attended stimuli (Fries et al 2001). The strength of this increase in local synchrony correlates with the monkeys' performance on the task (Womelsdorf et al 2006). Moreover, the LFP activity recorded in V4 is coherent in the gamma range with LFPs recorded in V1, again only for attended stimuli (Bosman et al 2012). Granger causality analysis, a method to assess the directionality of information flow between brain regions, suggested that V1 activity drove V4 activity. Collectively, these data suggest that neural synchrony between brain regions reflects behaviorally relevant circuit communication, although they do not distinguish neural synchrony driven by task demands from synchrony specifically devoted to binding visual objects.

Synchrony in Human Studies of Perception

Studies conducted in animals share the limitation that the subjects cannot directly report their perceptual experience. However, long-range synchrony studies in humans rely on techniques with less temporal and spatial resolution than those used in animals. As noted above, structured EEG signals result from large groups of simultaneously active neurons; non-synchronous firing would result in little net activity. Thus, the power of oscillatory signals serves a proxy for *local* neural synchrony. Gamma power increases in the central lead with the perception of a coherent shape, even if that shape is generated with illusory contours (Tallon-Baudry et al 1996). Similarly, beta and gamma power in both occipital and frontal leads increases as subjects find a hidden image (Tallon-Baudry et al 1997). Synchronous activity across EEG leads provides additional evidence of long-range synchrony. For instance, both gamma power within, and phase synchrony between, frontal and posterior electrodes increases during facial perception (Rodriguez et al 1999). Similarly, attending to a unilaterally presented visual stimulus induces widespread increases in gamma phase synchronization of EEG electrodes located over the posterior visual cortical areas contralateral to the stimulus. These data imply that perception of a coherent image results in widespread increases in gamma power and synchrony.

Not all studies find widespread gamma synchrony with visual perception. Von Stein and Sarnthein (von Stein & Sarnthein 2000) observed increases in EEG gamma and alpha power that remained localized to the occipital leads. Long-range coherence in the theta and low beta ranges only developed with tasks designed to involve multiple brain regions, such as visuo-spatial working memory or cross-sensory object recognition (von Stein & Sarnthein 2000). One of the few human studies to use intracortical electrodes to measure stimulus-evoked synchrony in the visual cortex found *decreases* in local gamma power followed by increases in beta synchrony during a visual working memory task (Tallon-Baudry et al 2001). Thus, while most studies agree that inter-area synchrony marks long-range circuit communication for visual tasks, the precise frequency range of this synchrony may vary.

As discussed earlier, animal studies show gamma range synchrony as binocular rivalry resolves. To study binocular rivalry in humans, Tononi and colleagues "frequency tagged" visual stimuli monocularly presented to each eye by flickering sets of images at slightly different frequencies (Tononi et al 1998). Doing so generates a sharp increase in MEG signal recorded at the presented frequencies. By measuring the topographic distribution and coherence of these tagged signals, two different groups found evidence for widespread increased synchrony as binocular rivalry resolved (Cosmelli et al 2004, Srinivasan et al 1999). Similarly, transient gamma and long lasting theta phase synchrony between frontal and posterior leads emerges prior to the resolution of binocular rivalry (Doesburg et al 2005). Broadly, these experiments support the conclusion that long-range synchrony develops as subjects perceive bound objects.

This observation of transient gamma followed by long-lasting theta synchrony implies that gamma synchrony might signal shifts in perception, while lower frequencies may maintain a given percept. Supporting this hypothesis, parietal and frontal EEG electrodes develop transient gamma phase synchrony, followed by sustained occipital lead alpha power increases just as subjects report a perceptual switch of a Necker cube—an image that spontaneously shifts three-dimensional perspective (Nakatani & van Leeuwen 2006). Similar transient increases in gamma power in both frontal and occipital EEG leads also occurred as subjects viewed another image with changing perspectives (Keil et al 1999).

Perhaps because of the transient nature of neural activity and resultant changes in power and synchrony, few studies of visually-induced synchrony have linked electrical activity with fMRI. Although evoked fMRI BOLD signal correlates with LFP activity, especially in the gamma range (Goense & Logothetis 2008, Ossandon et al 2011, Scheeringa et al 2011), fMRI connectivity correlates best with fluctuations in the power of low frequency (0.1–4 Hz) electrical activity (He et al 2008, Nir et al 2008, Scholvinck et al 2010). One study that did simultaneously record MEG and fMRI compared brain activity as subjects either fixated on a target of an otherwise blank screen or watched a short segment of a popular movie (Betti et al 2013). MEG synchrony was measured by calculating power in different frequency ranges and examining how this power fluctuates on slower timescales (<0.3 Hz). Broadly, the authors found that watching the movie caused a decrease in connectivity in the resting state network when assayed both with fMRI and with slow fluctuations in MEG alpha/beta power. Conversely, there were parallel increases in specific node-to-node MEG power correlations in the theta, beta and gamma regions, which often, but not always, corresponded to fMRI connectivity. This study serves as a proof of principle that one can tailor methods to find agreement across MRI and MEG synchrony measures.

Does Synchrony Amount to Binding?

The idea that synchronous activity encodes binding remains controversial, with critics raising objections on both theoretical and experimental grounds. In an influential paper, Shadlen and Movshon argue that the primary visual cortex lacks the requisite information to determine if elements belong to a single object and therefore synchronous firing in V1 cannot represent binding (Shadlen & Movshon 1999). They pointedly note that the perception of an object often requires binding receptive fields with opposite motion, such as the two ends of a spinning propeller (Merker 2013), yet such stimuli reportedly decrease synchrony. Moreover, they argue that in the active brain, synchronous spike activity often occurs by chance, rendering a system that uses synchrony to code information implausible. Instead, they suggest that observed synchrony reflects shared connectivity (Shadlen & Movshon 1999).

Several studies do not support the binding-by-synchrony hypothesis (Dong et al 2008, Lamme & Spekreijse 1998, Palanca & DeAngelis 2005, Roelfsema et al 2004, Thiele & Stoner 2003). For instance, Palanca and DeAngelis (2005) recorded multiunit activity in the medial temporal area as monkeys watched a single object moving versus unconnected objects with similar receptive field properties. Although they found a modest increase in coherence in the single object condition, binding-associated synchrony accounted for only 0.1% of the variance in a generalized linear model, while basic visual stimulus properties, such as overlapping receptive fields and preferred directions, accounted for up to 56%. Moreover, in a clever experiment designed to address whether binding *per se* was associated with enhanced neural synchrony, the authors showed monkeys single and unconnected objects, allowing the animals to form perceptual binding before the objects disappeared behind a mask with apertures that made equivalent features visible. This experiment varies only whether the monkeys formed perceptual binding on the object, keeping the visual features constant. Under these conditions, bound single objects are not associated with enhanced synchrony (Palanca & DeAngelis 2005).

Moving Beyond Binding: Function and Mechanism

The data presented so far suggest that while synchrony may not fully encode binding (see (Uhlhaas et al 2009)), visual tasks are indeed associated with local and long-range synchrony. How can we interpret the observed synchrony? A conservative explanation is that LFP oscillations simply represent the "hum" of "wheels turning" during local neural activity (Merker 2013). Synchrony reflects the simultaneous participation, and likely communication, between distant regions. Synchrony is nonetheless still important; to turn Merker's analogy around: just as one could infer a vehicle's direction and speed by analyzing changes in wheel hum, one could determine temporal and spatial characteristics of functional neural circuit connectivity by studying long range synchrony.

Yet the ubiquitous nature of neural oscillations as well as the propensity of particular neural subtypes to resonate at specific frequencies (reviewed in (Wang 2010)), suggests that oscillatory synchrony may not only reflect neural communication, but also facilitate it. Fries (Fries 2005) proposed that long-range coherence of oscillations ensures that a given region provides input in a temporal window when the downstream target is appropriately receptive. Along similar lines, Lisman and Jensen (Lisman & Jensen 2013) suggested that fast oscillations provide a temporal window within which the most excited cells fire synchronously, punctuated by pauses; this scheme avoids mixing messages from multiple inputs. Nesting fast (i.e., gamma) oscillations within slower (i.e., theta) oscillations provides a framework for encoding and faithfully transmitting sequenced information.

LONG-RANGE SYNCHRONY AND WORKING MEMORY

Hippocampal-prefrontal Synchrony in Rodents

Another canonical example of long-range synchrony occurs between the hippocampus and prefrontal cortex during spatial working memory in the rodent. Spatial working memory involves the short-term storage of spatial information in order to solve a task. In the rodent, spatial working memory requires both the hippocampus and medial prefrontal cortex (mPFC) (Aggleton et al 1986, Izaki et al 2001, Lee & Kesner 2003, Yoon et al 2008). Disconnection experiments show spatial working memory requires cooperation between the two structures. The hippocampus and mPFC have predominantly unilateral connections (Thierry et al 2000, Verwer et al 1997). Bilaterally disrupting either the hippocampus or the mPFC significant impairs performance on spatial working memory, while unilateral disruption on the same side has no effect, suggesting that the surviving pair of structures suffices to support spatial working memory. Disrupting the hippocampus on one side, and the mPFC on the other, impairs task performance, suggesting that connectivity between the two structures is required (Floresco et al 1997, Wang & Cai 2006).

Inspired by these findings, Matt Wilson and colleagues set out to measure connectivity between the hippocampus and mPFC during working memory. They demonstrated that mPFC neurons synchronize with theta-frequency oscillations in the mPFC (Jones & Wilson 2005b, Siapas et al 2005), and that the strength of this synchrony increases during the choice phase of a task, in which rats must execute a rule-based decision using their memory of the previous reward location (Jones & Wilson 2005a). Several other groups have confirmed this increase in synchrony, adding refinements that speak to the role it might play in the behavior (Gordon 2011, Hyman et al 2005, Hyman et al 2010). Theta-frequency synchrony increases gradually throughout the choice phase, peaking as the animal makes its decision (Benchenane et al 2010); gamma synchrony between the entorhinal cortex and hippocampus peaks at the same time point (Yamamoto et al 2014). Greater synchrony in both frequency ranges appears during correct trials compared to error trials (Jones & Wilson 2005c, Korotkova et al 2010, Yamamoto et al 2014). And deficits in synchrony accompany deficits in working memory performance in various models (Belforte et al 2010, Korotkova et al 2010, Sigurdsson et al 2010, Yamamoto et al 2014).

These experiments provide evidence for synchrony of these two regions. But what does this synchrony actually mean? The prevailing thought is that it reflects information flow. Consistent with this notion, synchrony in the hippocampal-prefrontal system exists not only in working memory, but also in other behavioral states, including sleep (Siapas & Wilson 1998), when consolidation of long-term memory is thought to take place. The evidence from these studies suggests that synchrony reflects effective connectivity, and perhaps effective information transfer between the two structures. The specific frequency (theta, ripple, etc.) involved may simply reflect the dominant mode of activity within the hippocampus at the time. Since theta-frequency oscillations dominate the hippocampus during navigation, spatial working memory during navigation results in theta-frequency synchrony.

Evidence that synchrony reflects information transfer from the hippocampus to the prefrontal cortex comes from attempts to determine the directionality of these interactions by examining the temporal relationship between activity in the two regions. Activity in the hippocampus tends to lead activity in the prefrontal cortex (Jones & Wilson 2005a, Siapas et al 2005, Sigurdsson et al 2010). Moreover, hippocampal lesions disrupt some forms of task-relevant coding in the mPFC (Burton et al 2009). Among other task-relevant information, mPFC neurons encode information about spatial location (Jung et al 1998); our lab, among several others, is currently investigating whether such place information requires hippocampal input.

Inferences from Human Electrophysiological Studies

Given the plethora of studies linking hippocampal-prefrontal synchrony to working memory in rodents, multiple studies have examined this system in humans. The results have been mixed. Typically, working memory studies in people utilize either visuospatial or linguistic tasks. The former relies primarily on visual decoding of object location on a 2-dimensional screen. The latter relies on remembering short strings of digits, letters or words. Neither of these tasks utilize the kind of place-based coding seen in the rodent hippocampus. Perhaps accordingly, they do not typically implicate the hippocampal-prefrontal circuit.

Working memory tasks in humans do induce changes in oscillatory strength and synchrony in EEGs. The strength of theta-, alpha-, and gamma-frequency activity increases with working memory demand across multiple leads (Raghavachari et al

2001, Roux & Uhlhaas 2014, Watrous et al 2014). Coherence between pairs of leads also increases, though the details vary from study to study. Payne and Kounios (Payne & Kounios 2009), using a visual letter recognition task, found that theta coherence between frontal and parietal leads increased with memory load. Using a similar visuospatial task, Sauseng et al. (Sauseng et al 2005) also saw increases in theta coherence between frontal and parietal sites, along with decreases in alpha coherence frontally. By contrast, alpha coherence increases in a working memory task requiring interpretation of semantic meaning (Haarmann & Cameron 2005).

The interpretation of these working memory-related coherence changes between distant EEG leads is unclear. Since the resistance of the skull creates significant problems with volume conduction, strong oscillations emerging from a single site could be recorded from multiple distant leads, causing an artifactual increase in coherence. Ideally, convincing evidence of active synchronization would show pockets of highly coherent leads in two distinct regions with asynchronous activity in between. Unfortunately, where studies do examine the distribution of coherence, they typically show large, contiguous swaths of cortex oscillating synchronously (c.f., Fig. 2 in (Sauseng et al 2005)). Indeed, in one careful examination of this issue (Raghavachari et al 2006), intracranial recordings from cortical surface arrays found that coherence decreases monotonically as a function of distance between electrode pairs, suggesting that volume conduction accounts for any synchrony. These caveats make EEG studies difficult to interpret.

MEG offers distinct advantages over EEG, as magnetic fields more easily cross the skull, reducing issues of volume conduction and permitting better source localization. Findings from MEG studies have confirmed and extended EEG findings, recapitulating increases in oscillatory power at various frequencies in frontal areas (Jensen et al 2002, Jensen & Tesche 2002, Kaiser et al 2003, Roux & Uhlhaas 2014), as well as other regions (Bonnefond & Jensen 2012, Haegens et al 2010, Roux et al 2012). Using source reconstruction techniques, Guitart-Masip et al.(Guitart-Masip et al 2013) mapped the increase in theta power to discrete sources in the anterior hippocampus and anterior cingulate cortex; they then showed that the hippocampal theta source synchronizes with several prefrontal regions, particularly during a reversal-learning task that requires maintenance of shape-location associations. Few studies of synchrony in other frequency bands exist, though one network-based analysis suggested that lower (<25 Hz) frequency bands comprise the main effects of working memory (Palva et al 2010).

Intracranial Recordings and Imaging

Intracranial recordings offer perhaps the most definitive method for characterizing long-range synchrony between defined brain regions, and have begun to corroborate the extracranial studies. In a series of recordings from patients with epilepsy, Axmacher, Fell and colleagues reported increased synchrony between LFPs in the hippocampus and other temporal lobe structures, including the rhinal cortices and the inferior temporal cortex (Axmacher et al 2008, Fell et al 2008). In a few subjects also implanted with ECOG arrays over the prefrontal cortex, they also found evidence of gamma-frequency synchrony between the hippocampus and prefrontal sites (Axmacher et al 2008). By contrast, Rissman et al. (Rissman et al 2008) found that as working memory load increases in a face recognition working memory task, so does connectivity between the hippocampus and the fusiform face area. These findings suggest that the hippocampus synchronizes with specific cortical targets of relevance to the particular task used.

Similar to electrophysiological studies, imaging studies fail to consistently identify synchrony between the hippocampus and prefrontal cortex. Some fMRI studies have demonstrated increases in hippocampal-prefrontal synchrony with working memory demand, for example, during face or letter recognition tasks (Finn et al 2010, Rissman et al 2008). Others have reported decreases in synchrony between the prefrontal cortex and hippocampus, during a face recognition task with fMRI (Axmacher et al 2008) or a numerical recognition task with PET (Meyer-Lindenberg et al 2001, Meyer-Lindenberg et al 2005).

One possibility, given the findings above, is that hippocampal-prefrontal synchrony is specifically involved in spatial forms of working memory. To test this, Meyer-Lindenberg and colleagues recently devised a virtual reality version of the classic rodent spatial working memory test (Bähner et al 2015). Using this task, they found increased synchrony between the hippocampus and dorsolateral prefrontal cortex, as well as an extended prefrontal-parietal network.

Clinical Relevance: Deficits in Synchrony in Schizophrenia

The rough correspondence between findings in humans and rodents raises the possibility of translational relevance. Working memory disruptions accompany multiple neuropsychiatric illnesses, including schizophrenia (Krieger et al 2005, Piskulic et al 2007). It is possible that deficits in long-range synchrony may underlie such behavioral deficits; indeed, studies have suggested alterations in hippocampal-prefrontal synchrony (Ford et al 2002, Lawrie et al 2002, Meyer-Lindenberg et al 2001, Meyer-Lindenberg et al 2005) as well as deficits in global connectivity (Argyelan et al 2014, Bassett et al 2012, Venkataraman et al 2012) in schizophrenia patients. Understanding the neurobiological mechanisms underlying these deficits may reveal novel therapeutic targets.

Deficits in hippocampal-prefrontal synchrony have been identified in both genetic and environmental animal models of schizophrenia predisposition. Mice modeling a microdeletion on chromosome 22 that increases the risk of schizophrenia about 30-fold (Karayiorgou et al 1995) have working memory deficits (Stark et al 2008), as do patients with the microdeletion (Lajiness-O'Neill et al 2005, Lewandowski et al 2007, Sobin et al 2005, van Amelsvoort et al 2004). Hippocampal-prefrontal synchrony is reduced in these mice (Figure 3A, (Sigurdsson et al 2010)), and the reduction correlates with deficits in working memory.

Infection during gestation is a significant environmental risk factor for schizophrenia, increasing the risk for contracting the disorder by 2–3 fold (Canetta & Brown 2012). The risk seems similar regardless of the infectious agent, suggesting that activation of the maternal immune system, rather than infection itself, is deleterious. Accordingly, offpring of female rats exposed to immune activation during pregnancy develop a panoply of behavioral abnormalities reminiscent of schizophrenia, including working memory deficits (Canetta & Brown 2012). These offspring also have deficits in hippocampal-prefrontal synchrony remarkably similar to those seen in the genetic model (Figure 3B; (Dickerson et al 2010)). Long range synchrony thus appears to be a shared pathophysiological consequence of at least two risk models, suggesting that synchrony could be an intermediate phenotype of relevance to schizophrenia in general.

FEAR AND ANXIETY

Fear Conditioning as a Means to Probe Anxiety-related Circuitry

Pavlovian fear conditioning, a well-characterized model applicable across many species, provides a rich avenue of insight into the neural mechanisms of fear (reviewed in (Maren 2001)). In this paradigm, a neutral cue such as a tone (the conditioned stimulus, or CS) is paired with an aversive experience, such as a mild shock (the unconditioned stimulus, or UCS) that elicits behavioral manifestations of fear, such as freezing or escape. Over time, the subject associates the CS with the aversive stimulus and exhibits anxious behavior when presented with the CS, even in the absence of the UCS. Extinction, a second form of learning pertinent to anxiety disorders, occurs after fear conditioning, when the CS is repeatedly experienced without a paired UCS. The subject learns that the CS no longer predicts the UCS and stops exhibiting anxious behaviors.

Long-range neural synchrony has been observed in the circuits involved in fear conditioning in both humans and model animals. Fear conditioning induces synchronous

EEG activity in healthy human subjects (Keil et al 2001, Keil et al 2007, Miltner et al 1999, Mueller et al 2014). For example, coherence increases between EEG leads overlying the visual and somatosensory cortices during the establishment of light (CS)–shock (UCS) associations (Miltner et al 1999). Moreover, this synchrony disappears with extinction; it seems specific to the expression of fear.

The limited spatial resolution of EEG precludes probing subcortical structures with known involvement in fear circuitry, such as the amygdala (Mueller et al 2014). However, fMRI can reveal the connectivity of the amygdala with cortical structures during fear conditioning. One recent study examined large-scale network connectivity during predictable (CS paired with UCS) and unpredictable (UCS alone) threat (Wheelock et al 2014); unpredictable threat produced more intense anxiety. The authors analyzed the network activity of 15 brain regions activated by both kinds of threats. For predictable threats, the dorso*lateral* prefrontal cortex formed an outward hub, communicating with the greatest number of structures, including the insular cortex, which served as a secondary hub. By contrast, for unpredictable threat, the dorso*medial* prefrontal cortex (dmPFC) formed the primary hub, while the amygdala comprised a secondary hub (Wheelock et al 2014). These results suggest that the dorsolateral prefrontal cortex and insular cortex provide emotional regulation during predictable stress. For unpredictable stress, however, the amygdala and dmPFC orchestrate a reactive stress response.

While the slow time course of fMRI signals prevents a real-time parsing of amygdala-prefrontal interactions, recent work in model animals provides insight into their dynamics. Paz and colleagues (Klavir et al 2013) recorded neural activity in the amygdala and dorsal anterior cingulate cortex in monkeys that had been trained that a particular CS predicted the UCS, while a different CS predicted its absence. They then switched the contingencies and observed the firing responses of neurons in both regions to the surprising mismatches of CS and unexpected outcome. Some neurons in the amygdala fired to any type of mismatch between CS and outcome (representing "unsigned prediction errors", which merely indicate surprise). Others fired only for specific types of mismatch (representing "signed prediction errors," whether positive or negative). Amygdala neurons that represent unsigned prediction errors fired slightly before neurons in the anterior cingulate, while amygdala neurons that represent signed prediction errors fired slightly after anterior cingulate neurons (Klavir et al 2013). These data suggest that the amygdala provides information to the dorsal anterior cingulate cortex about surprising threat contingencies while the cingulate cortex instructs the amygdala about valance (positive or negative value). Interestingly, anterior cingulate-to-amygdala directionality is associated with the resistance to extinction (Livneh & Paz

2012), raising the intriguing possibility that pathologically persistent fear represents a failure of the anterior cingulate to properly connect with the amygdala.

Recent work from our laboratory also investigated synchronous neural activity between the mPFC and the amygdala during discrimination between safe and threatening cures (Likhtik et al 2014, Stujenske et al 2014). Mice were trained with a CS+ paired with a mild shock, and an explicitly unpaired CS-. Mice that distinguished the CS+ and CS- had elevated theta (4–5 Hz) coherence between the mPFC and the amygdala. The directionality of this synchrony is modulated dynamically. During the CS+, the two structures equally influence each other; during the CS-, a predominant mPFC-to-amygdala directionality emerges (Likhtik et al 2014). This mPFC lead inversely correlates with freezing, suggesting that it represents safety. During the CS-, mPFC input synchronizes a subset of amygdala neurons, generating a local gamma-frequency oscillation that is coupled to theta in the mPFC (Stujenske et al 2014). Extinction is also associated with an mPFC lead (Lesting et al 2013, Narayanan et al 2011, Stujenske et al 2014). Collectively, these data suggest that synchrony in the mPFC-to-BLA circuit reflects a dynamic rivalry of signals conveying fear and safety.

Synchrony also arises with fear memory consolidation. Studies by Pape and colleagues showed that fear conditioned stimuli induce synchrony between low theta-frequency (4–5 Hz) oscillations in LFPs recorded from the prefrontal cortex, hippocampus, and amygdala (Lesting et al 2013, Narayanan et al 2007a, Seidenbecher et al 2003). Elevated hippocampal-amygdala synchrony appears 24 hours after training, not earlier or later (Narayanan et al 2007a), matching the timecourse of memory consolidation. Synchrony also increases 24 hours after the memory reactivation (Narayanan et al 2007b), implicating hippocampal-amygdala circuits in reconsolidation, the fascinating phenomenon which renders memories labile (Nader et al 2000). Consistent with a role for long-range synchrony in memory consolidation, Paré and colleagues (Popa et al 2010) found that synchrony across the hippocampal-prefrontal-amygdala circuit emerges during REM sleep after training. The extent and directionality of synchrony within the circuit predicts the strength of the resultant fear memory.

Innate and Stress-induced Anxiety

Animals and people both have innate defensive reactions to stimuli that do not require learning. Such stimuli include predator smells or dark, confined spaces (for people; rodents find bright, open spaces aversive). While both forms of anxiety appear to rely on the amygdala and prefrontal cortex (Deacon et al 2002, Gonzalez et al 2000, Lacroix et al 2000, Shah et al 2004, Shah & Treit 2003), studies in rodents suggest that innate forms of anxiety also require an extended network of additional brain regions, including the hippocampus and bed nucleus of the stria terminals (Bannerman et al 2003, Deacon et al 2002, File & Gonzalez 1996, Kim et al 2013, Kjelstrup et al 2002). Consistent with these findings, we have found increased LFP coherence and spike-LFP phase locking between the mPFC and the hippocampus in mice exploring anxiety-provoking environments, such as the open field and the elevated plus maze, which both rely on open spaces to induce anxiety (Adhikari et al 2010). Similar increases in synchrony are also seen between the mPFC and the amygdala (Likhtik et al 2014, Stujenske et al 2014), where, just as in learned fear, an mPFC-to-amygdala directionality predominates during relative safety.

In humans, studies of anxiety find differences in baseline connectivity between individuals with high and low anxiety. High anxiety individuals show positive amygdala-ventromedial prefrontal cortex fMRI correlations and negative amygdala-dorsomedial cortex correlations, while low anxiety individuals show roughly the opposite results (Kim et al 2011). Amygdala-anterior insular cortex synchrony correlates with state (moment-to-moment) anxiety, while increased anatomical connectivity, measured with diffusion tensor imaging (DTI), correlates with trait (lifetime tendency) anxiety (Baur et al 2013). Similarly, adolescents with high cortisol reactivity have higher resting state fMRI connectivity between the salience network (including the anterior cingulate cortex and the insula) and the subgenual cingulate cortex (Thomason et al 2011). These results suggest that individuals with anxiety have both structural and functional differences in connectivity involving the insular cortex, prefrontal cortex and amygdala, even in the absence of a specific anxiogenic task.

These studies lead to the conclusion that in highly anxious individuals, "resting state" functional connectivity patterns may reflect free-floating anxiety, rather than a neutral state (Baur et al 2013, Kim et al 2011). The finding that functional connectivity (measured by synchrony) correlates with anxiety state while anatomical connectivity (as measured by DTI) correlates with anxiety trait emphasizes the distinction between dynamics and structure. In anxiety disorders, stress might evoke anxiety states by altering dynamic connectivity, overlaid on a baseline of disturbed structural connectivity in susceptible individuals.

Indeed, stress evokes changes in synchrony within an extended network that includes the amygdala, insula and prefrontal cortex. In healthy volunteers, watching a stressful video increases cortisol, noradrenergic activity, and fMRI connectivity in a "salience network" that includes the insula, cingulate, amygdala, midbrain and thalamus. The strength of network connectivity correlates with cortisol and negative affect (Hermans et al 2011). Blocking noradrenergic activity, but not cortisol, reduces this salience network (Hermans et al 2011).

These data imply that the salience network reflects attention to stressful stimuli, rather than the stress response itself.

Further insight into the relationship between networks mediating attentiveness and aversiveness comes from studies of network activity during anxious anticipation. One study, utilizing factor analysis of fMRI data, suggested three successive phases that unfold over time: first, the salience network increases in activity and connectivity; next the anterior insula and bed nucleus synchronize; finally, nucleus accumbens activity (typically associated with reward-related activity) decreases (McMenamin et al 2014). In this study, amygdala activity did not increase, but its connectivity with other brain regions widened (McMenamin et al 2014). A similar study did find increases in amygdala activity, as well as increased synchrony between the amygdala and parts of dorsal prefrontal cortex; the strength of this connectivity correlated both with reaction time speed for identifying emotional stimuli as well trait anxiety scores (Robinson et al 2012).

Functional Connectivity in Generalized Anxiety Disorder

Data from both rodents and humans suggest that long-range synchronous activity conveys neural communication in circuits active during anxiety. The networks identified in both learned fear and innate anxiety appear similar. However, differing anxiety levels within healthy subjects can manifest as different functional connectivity patterns. These findings raise the possibility that anxiety disorders result from disruption in long-range synchrony within these circuits.

Several studies have compared the functional connectivity patterns of healthy controls and patients with generalized anxiety disorder (GAD). The findings are in broad agreement, though again, the specifics vary. In a combined fMRI and DTI study, healthy controls had the typical negative correlation between activity in the dorsal prefrontal cortex and amygdala (Tromp et al 2012), suggesting that one of these regions tends to suppress activity in the other. In patients with GAD, this negative correlation was weaker (Tromp et al 2012). A similar negative correlation between activity in the amygdala and ventrolateral prefrontal cortex was induced in healthy adolescents exposed to emotional pictures; this correlation was also weaker in adolescents with GAD (Monk et al 2008). These deficits in connectivity appear to correspond with decreased cognitive control over emotional responses (Etkin et al 2010).

These studies of circuit dysfunction in GAD also tie in to an emerging literature examining the molecular and circuit basis of anxiety in genetic mouse models. Mice lacking either of two components of the serotonin system, the serotonin 1A receptor or the serotonin transporter, have phenotypes of increased innate anxiety. Recordings from the mPFC of serotonin 1A receptor knockout mouse suggest a failure of the mPFC to form representations of anxiety-provoking environments (Adhikari et al 2011), while multi-site recordings from serotonin transporter knockout mice show altered amygdala-mPFC synchrony (Narayanan et al 2011). Further studies investigating the mechanisms by which long-range synchrony is altered in these or other animal models may well provide greater insight into the neurobiology of GAD, and identify novel approaches toward improved treatment of anxiety disorders.

OPEN QUESTIONS AND FUTURE DIRECTIONS

The three examples discussed above demonstrate a substantial amount of concordance across paradigms and species regarding the relationship between synchrony and behavior. The take home message is that long-range synchrony–between visual cortical areas and parietal areas during perception; between the hippocampus and prefrontal cortex during working memory; and between the amygdala and prefrontal cortex during anxiety–correlates with behavior. What this means, in terms of mechanism, causality, translatability, and clinical relevance, remains to be determined.

First, we need a better mechanistic understanding of how cross-regional synchrony emerges. A better circuit level understanding of long-range synchrony would help guide experiments aimed at testing whether such synchrony is necessary for behavior. After identifying specific circuit elements—cell types, synapses, pathways, etc. —one could disrupt and augment synchrony with the basic neuroscientists' arsenal of tools. For example, optogenetic inhibition of a specific connection could disrupt synchrony between directly connected brain regions. Conversely, stimulating the same connection (at specific frequencies) could enhance synchrony. The resulting effects of these manipulations on behavior could then be assayed. Already, this approach has begun to clarify the circuit mechanisms underlying gamma synchrony and its relevance to behavior (Cardin et al 2009, Colgin et al 2009, Lasztoczi & Klausberger 2014, Sohal et al 2009, Yamamoto et al 2014). A similar approach for lower frequency oscillations would help resolve the meaning of lower frequency power changes, such as alpha and theta (reviewed in (Lisman & Jensen 2013, Merker 2013, Palva & Palva 2007)).

While understanding mechanisms will require animal models, differences in how synchrony is measured raise issues of how to properly translate findings between animals and humans. Of particular concern, findings obtained through fMRI and those obtained from electrophysiological methods have vastly different time courses. fMRI measures activity changes that occur over a few seconds; electrophysiological synchrony is measured in milliseconds. How do they agree at all? As noted above, activity within an area measured by fMRI seems to correlate with the *strength* of gamma oscillations (Goense & Logothetis 2008, Nir et al 2007, Shmuel & Leopold 2008). Thus, correlations in fMRI-measured activity may represent common fluctuations in gamma strength. This explanation provides a framework with which to interpret some of the findings in the literature. For example, while we and others have found increases in theta-frequency synchrony between the hippocampus and prefrontal cortex during the retrieval phase of spatial working memory tasks in rodents (Benchenane et al 2010, Hyman et al 2005, Hyman et al 2010, Jones & Wilson 2005b, Sigurdsson et al 2010), the Meyer-Lindenberg group found increases in synchrony during the encoding portion of their task (Bähner et al 2015). The discrepancy may reflect species or task differences. However, it is also possible that the fMRI captures gamma, rather than theta synchrony. Consistent with this notion, we recently found enhanced gamma-frequency synchrony between the hippocampus and mPFC during the encoding phase of a spatial working memory task in mice (Spellman et al, submitted).

These details have direct clinical relevance, since deficits in synchrony at different frequencies may have far different circuit-level mechanisms, and developing new treatments will require first understanding which specific circuit elements to target. The findings presented here make it clear that long-range synchrony accompanies behavior. The next steps, establishing mechanism, evaluating causality, translating findings across species and determining clinical relevance, are already underway, and promise to have considerable impact on our understanding of how the networked brain functions, and how that function goes awry in disease.

ACKNOWLEDGMENTS

Supported by NIMH R01 MH081968; R01 MH096274; P50 MH096891; and T32 MH015144; and the Hope for Depression Research Foundation.

BIBLIOGRAPHY

Adhikari A, Topiwala MA, Gordon JA. Synchronized activity between the ventral hippocampus and the medial prefrontal cortex during anxiety. Neuron. 2010; 65:257–69. [PubMed: 20152131]
Adhikari A, Topiwala MA, Gordon JA. Single Units in the Medial Prefrontal Cortex with Anxiety-Related Firing Patterns Are Preferentially Influenced by Ventral Hippocampal Activity. Neuron. 2011;

71:898–910. [PubMed: 21903082]

Aggleton JP, Hunt PR, Rawlins JN. The effects of hippocampal lesions upon spatial and non-spatial tests of working memory. Behav Brain Res. 1986; 19:133–46. [PubMed: 3964405]

Argyelan M, Ikuta T, DeRosse P, Braga RJ, Burdick KE, et al. Resting-state fMRI connectivity impairment in schizophrenia and bipolar disorder. Schizophr Bull. 2014; 40:100–10. [PubMed: 23851068]

Axmacher N, Schmitz DP, Wagner T, Elger CE, Fell J. Interactions between medial temporal lobe, prefrontal cortex, and inferior temporal regions during visual working memory: a combined intracranial EEG and functional magnetic resonance imaging study. J Neurosci. 2008; 28:7304–12. [PubMed: 18632934]

Bähner F, Plichta MM, Demanuele C, Schweiger J, Gerchen MF, et al. Hippocampal-dorsolateral prefrontal coupling as a species-conserved cognitive mechanism: A human translational imaging study. Neuropsychopharmacology. 2015 in press.

Bandettini PA. What's new in neuroimaging methods? Ann N Y Acad Sci. 2009; 1156:260–93. [PubMed: 19338512]

Bannerman DM, Grubb M, Deacon RM, Yee BK, Feldon J, Rawlins JN. Ventral hippocampal lesions affect anxiety but not spatial learning. Behav Brain Res. 2003; 139:197–213. [PubMed: 12642189]

Bassett DS, Nelson BG, Mueller BA, Camchong J, Lim KO. Altered resting state complexity in schizophrenia. Neuroimage. 2012; 59:2196–207. [PubMed: 22008374]

Baur V, Hanggi J, Langer N, Jancke L. Resting-state functional and structural connectivity within an insula-amygdala route specifically index state and trait anxiety. Biol Psychiatry. 2013; 73:85–92. [PubMed: 22770651]

Belforte JE, Zsiros V, Sklar ER, Jiang Z, Yu G, et al. Postnatal ablation of NMDA receptors in corticolimbic interneurons leads to schizophrenia-related phenotypes. Nat Neurosci. 2010

Benchenane K, Peyrache A, Khamassi M, Tierney PL, Gioanni Y, et al. Coherent Theta Oscillations and Reorganization of Spike Timing in the Hippocampal-Prefrontal Network upon Learning. Neuron. 2010; 66:921–36. [PubMed: 20620877]

Betti V, Della Penna S, de Pasquale F, Mantini D, Marzetti L, et al. Natural scenes viewing alters the dynamics of functional connectivity in the human brain. Neuron. 2013; 79:782–97. [PubMed: 23891400]

Bonnefond M, Jensen O. Alpha oscillations serve to protect working memory maintenance against anticipated distracters. Curr Biol. 2012; 22:1969–74. [PubMed: 23041197]

Bosman CA, Schoffelen JM, Brunet N, Oostenveld R, Bastos AM, et al. Attentional stimulus selection through selective synchronization between monkey visual areas. Neuron. 2012; 75:875–88. [PubMed: 22958827]

Brown EN, Kass RE, Mitra PP. Multiple neural spike train data analysis: state-of-the-art and future challenges. Nat Neurosci. 2004; 7:456–61. [PubMed: 15114358]

Burton BG, Hok V, Save E, Poucet B. Lesion of the ventral and intermediate hippocampus abolishes anticipatory activity in the medial prefrontal cortex of the rat. Behav Brain Res. 2009; 199:222–34. [PubMed: 19103227]

Canetta SE, Brown AS. Prenatal Infection, Maternal Immune Activation, and Risk for Schizophrenia. Transl Neurosci. 2012; 3:320–27. [PubMed: 23956839]

Cardin JA, Carlen M, Meletis K, Knoblich U, Zhang F, et al. Driving fast-spiking cells induces gamma rhythm and controls sensory responses. Nature. 2009; 459:663–7. [PubMed: 19396156]

Colgin LL, Denninger T, Fyhn M, Hafting T, Bonnevie T, et al. Frequency of gamma oscillations routes flow of information in the hippocampus. Nature. 2009; 462:353–7. [PubMed: 19924214]

Cosmelli D, David O, Lachaux JP, Martinerie J, Garnero L, et al. Waves of consciousness: ongoing cortical patterns during binocular rivalry. Neuroimage. 2004; 23:128–40. [PubMed: 15325359]

Deacon RM, Bannerman DM, Rawlins JN. Anxiolytic effects of cytotoxic hippocampal lesions in rats. Behav Neurosci. 2002; 116:494–7. [PubMed: 12049331]

Dickerson DD, Wolff AR, Bilkey DK. Abnormal long-range neural synchrony in a maternal immune activation animal model of schizophrenia. J Neurosci. 2010; 30:12424–31. [PubMed: 20844137]

Doesburg SM, Kitajo K, Ward LM. Increased gamma-band synchrony precedes switching of conscious perceptual objects in binocular rivalry. Neuroreport. 2005; 16:1139–42. [PubMed: 16012336]

Dong Y, Mihalas S, Qiu F, von der Heydt R, Niebur E. Synchrony and the binding problem in macaque visual cortex. J Vis. 2008; 8:30, 1–16. [PubMed: 19146262]

Engel AK, Konig P, Kreiter AK, Singer W. Interhemispheric synchronization of oscillatory neuronal responses in cat visual cortex. Science. 1991a; 252:1177–9. [PubMed: 2031188]

Engel AK, Kreiter AK, Konig P, Singer W. Synchronization of oscillatory neuronal responses between striate and extrastriate visual cortical areas of the cat. Proc Natl Acad Sci U S A. 1991b; 88:6048–52. [PubMed: 2068083]

Etkin A, Prater KE, Hoeft F, Menon V, Schatzberg AF. Failure of anterior cingulate activation and connectivity with the amygdala during implicit regulation of emotional processing in generalized anxiety disorder. Am J Psychiatry. 2010; 167:545–54. [PubMed: 20123913]

Fell J, Ludowig E, Rosburg T, Axmacher N, Elger CE. Phase-locking within human mediotemporal lobe predicts memory formation. Neuroimage. 2008; 43:410–9. [PubMed: 18703147]

File SE, Gonzalez LE. Anxiolytic effects in the plus-maze of 5-HT1A-receptor ligands in dorsal raphe and ventral hippocampus. Pharmacol Biochem Behav. 1996; 54:123–8. [PubMed: 8728549]

Finn AS, Sheridan MA, Kam CL, Hinshaw S, D'Esposito M. Longitudinal evidence for functional specialization of the neural circuit supporting working memory in the human brain. J Neurosci. 2010; 30:11062–7. [PubMed: 20720113]

Floresco SB, Seamans JK, Phillips AG. Selective roles for hippocampal, prefrontal cortical, and ventral striatal circuits in radial-arm maze tasks with or without a delay. J Neurosci. 1997; 17:1880–90. [PubMed: 9030646]

Ford JM, Mathalon DH, Whitfield S, Faustman WO, Roth WT. Reduced communication between frontal and temporal lobes during talking in schizophrenia. Biol Psychiatry. 2002; 51:485–92. [PubMed: 11922884]

Fox MD, Snyder AZ, Vincent JL, Corbetta M, Van Essen DC, Raichle ME. The human brain is intrinsically organized into dynamic, anticorrelated functional networks. Proc Natl Acad Sci U S A. 2005; 102:9673–8. [PubMed: 15976020]

Fries P. A mechanism for cognitive dynamics: neuronal communication through neuronal coherence. Trends Cogn Sci. 2005; 9:474–80. [PubMed: 16150631]

Fries P, Reynolds JH, Rorie AE, Desimone R. Modulation of oscillatory neuronal synchronization by selective visual attention. Science. 2001; 291:1560–3. [PubMed: 11222864]

Fries P, Roelfsema PR, Engel AK, Konig P, Singer W. Synchronization of oscillatory responses in visual cortex correlates with perception in interocular rivalry. Proc Natl Acad Sci U S A. 1997; 94:12699–704. [PubMed: 9356513]

Goense JB, Logothetis NK. Neurophysiology of the BOLD fMRI signal in awake monkeys. Curr Biol.2008; 18:631–40. [PubMed: 18439825]

Gonzalez LE, Rujano M, Tucci S, Paredes D, Silva E, et al. Medial prefrontal transection enhances social interaction. I: behavioral studies. Brain Res. 2000; 887:7–15. [PubMed: 11134584]

Gordon JA. Oscillations and hippocampal-prefrontal synchrony. Curr Opin Neurobiol. 2011; 21:486–91. [PubMed: 21470846]

Gray CM, Konig P, Engel AK, Singer W. Oscillatory responses in cat visual cortex exhibit inter-columnar synchronization which reflects global stimulus properties. Nature. 1989; 338:334–7. [PubMed: 2922061]

Grossberg S. Adaptive pattern classification and universal recoding: II. Feedback, expectation, olfaction, illusions. Biol Cybern. 1976; 23:187–202. [PubMed: 963125]

Guitart-Masip M, Barnes GR, Horner A, Bauer M, Dolan RJ, Duzel E. Synchronization of medial temporal lobe and prefrontal rhythms in human decision making. J Neurosci. 2013; 33:442–51. [PubMed: 23303925]

Haarmann HJ, Cameron KA. Active maintenance of sentence meaning in working memory: evidence from EEG coherences. Int J Psychophysiol. 2005; 57:115–28. [PubMed: 15939498]

Haegens S, Osipova D, Oostenveld R, Jensen O. Somatosensory working memory performance in humans depends on both engagement and disengagement of regions in a distributed network. Hum Brain Mapp. 2010; 31:26–35. [PubMed: 19569072]

He BJ, Snyder AZ, Zempel JM, Smyth MD, Raichle ME. Electrophysiological correlates of the brain's intrinsic large-scale functional architecture. Proc Natl Acad Sci U S A. 2008; 105:16039–44. [PubMed: 18843113]

Hermans EJ, van Marle HJ, Ossewaarde L, Henckens MJ, Qin S, et al. Stress-related noradrenergic activity prompts large-scale neural network reconfiguration. Science. 2011; 334:1151–3. [PubMed: 22116887]

Hyman JM, Zilli EA, Paley AM, Hasselmo ME. Medial prefrontal cortex cells show dynamic modulation with the hippocampal theta rhythm dependent on behavior. Hippocampus. 2005; 15:739–49. [PubMed: 16015622]

Hyman JM, Zilli EA, Paley AM, Hasselmo ME. Working memory performance correlates with prefrontal-hippocampal theta interactions but not with prefrontal neuron firing rates. Frontiers in Integrative Neuroscience. 2010; 4

Ioannides AA. Dynamic functional connectivity. Curr Opin Neurobiol. 2007; 17:161–70. [PubMed: 17379500]

Izaki Y, Maruki K, Hori K, Nomura M. Effects of rat medial prefrontal cortex temporal inactivation on a delayed alternation task. Neurosci Lett. 2001; 315:129–32. [PubMed: 11716980]

Jensen O, Gelfand J, Kounios J, Lisman JE. Oscillations in the alpha band (9–12 Hz) increase with memory load during retention in a short-term memory task. Cereb Cortex. 2002; 12:877–82. [PubMed: 12122036]

Jensen O, Tesche CD. Frontal theta activity in humans increases with memory load in a working memory task. Eur J Neurosci. 2002; 15:1395–9. [PubMed: 11994134]

Jones M, Wilson M. Theta rhythms coordinate hippocampal-prefrontal interactions in a spatial working memory task. PLoS Biol. 2005a; 2:e402. [PubMed: 16279838]

Jones MW, Wilson MA. Phase precession of medial prefrontal cortical activity relative to the hippocampal theta rhythm. Hippocampus. 2005b; 15:867–73. [PubMed: 16149084]

Jones MW, Wilson MA. Theta rhythms coordinate hippocampal-prefrontal interactions in a spatial memory task. PLoS Biol. 2005c; 3:e402. [PubMed: 16279838]

Jung MW, Qin Y, McNaughton BL, Barnes CA. Firing characteristics of deep layer neurons in prefrontal cortex in rats performing spatial working memory tasks. Cereb Cortex. 1998; 8:437–50. [PubMed: 9722087]

Kaiser J, Ripper B, Birbaumer N, Lutzenberger W. Dynamics of gamma-band activity in human magnetoencephalogram during auditory pattern working memory. Neuroimage. 2003; 20:816–27. [PubMed: 14568454]

Karayiorgou M, Morris MA, Morrow B, Shprintzen RJ, Goldberg R, et al. Schizophrenia susceptibility associated with interstitial deletions of chromosome 22q11. Proc Natl Acad Sci U S A. 1995; 92:7612–6. [PubMed: 7644464]

Keil A, Muller MM, Gruber T, Wienbruch C, Stolarova M, Elbert T. Effects of emotional arousal in the cerebral hemispheres: a study of oscillatory brain activity and event-related potentials. Clin Neurophysiol. 2001; 112:2057–68. [PubMed: 11682344]

Keil A, Muller MM, Ray WJ, Gruber T, Elbert T. Human gamma band activity and perception of a gestalt. J Neurosci. 1999; 19:7152–61. [PubMed: 10436068]

Keil A, Stolarova M, Moratti S, Ray WJ. Adaptation in human visual cortex as a mechanism for rapid discrimination of aversive stimuli. Neuroimage. 2007; 36:472–9. [PubMed: 17451974]

Kim MJ, Gee DG, Loucks RA, Davis FC, Whalen PJ. Anxiety dissociates dorsal and ventral medial prefrontal cortex functional connectivity with the amygdala at rest. Cereb Cortex. 2011; 21:1667–73. [PubMed: 21127016]

Kim SY, Adhikari A, Lee SY, Marshel JH, Kim CK, et al. Diverging neural pathways assemble a behavioural state from separable features in anxiety. Nature. 2013; 496:219–23. [PubMed: 23515158]

Kjelstrup KG, Tuvnes FA, Steffenach HA, Murison R, Moser EI, Moser MB. Reduced fear expression after lesions of the ventral hippocampus. Proc Natl Acad Sci U S A. 2002; 99:10825–30. [PubMed: 12149439]

Klavir O, Genud-Gabai R, Paz R. Functional connectivity between amygdala and cingulate cortex for adaptive aversive learning. Neuron. 2013; 80:1290–300. [PubMed: 24314732]

Korotkova T, Fuchs EC, Ponomarenko A, von Engelhardt J, Monyer H. NMDA receptor ablation on parvalbumin-positive interneurons impairs hippocampal synchrony, spatial representations, and working memory. Neuron. 2010; 68:557–69. [PubMed: 21040854]

Kreiter AK, Singer W. Stimulus-dependent synchronization of neuronal responses in the visual cortex of the awake macaque monkey. J Neurosci. 1996; 16:2381–96. [PubMed: 8601818]

Krieger S, Lis S, Janik H, Cetin T, Gallhofer B, Meyer-Lindenberg A. Executive function and cognitive subprocesses in first-episode, drug-naive schizophrenia: an analysis of N-back performance. Am J Psychiatry. 2005; 162:1206–8. [PubMed: 15930072]

Lacroix L, Spinelli S, Heidbreder CA, Feldon J. Differential role of the medial and lateral prefrontal cortices in fear and anxiety. Behav Neurosci. 2000; 114:1119–30. [PubMed: 11142644]

Lajiness-O'Neill RR, Beaulieu I, Titus JB, Asamoah A, Bigler ED, et al. Memory and learning in children with 22q11.2 deletion syndrome: evidence for ventral and dorsal stream disruption? Child Neuropsychol. 2005; 11:55–71. [PubMed: 15823983]

Lamme VA, Spekreijse H. Neuronal synchrony does not represent texture segregation. Nature. 1998; 396:362–6. [PubMed: 9845071]

Lasztoczi B, Klausberger T. Layer-specific GABAergic control of distinct gamma oscillations in the CA1 hippocampus. Neuron. 2014; 81:1126–39. [PubMed: 24607232]

Lawrie SM, Buechel C, Whalley HC, Frith CD, Friston KJ, Johnstone EC. Reduced frontotemporal functional connectivity in schizophrenia associated with auditory hallucinations. Biol Psychiatry. 2002; 51:1008–11. [PubMed: 12062886]

Lee I, Kesner RP. Time-dependent relationship between the dorsal hippocampus and the prefrontal cortex in spatial memory. J Neurosci. 2003; 23:1517–23. [PubMed: 12598640]

Lesting J, Daldrup T, Narayanan V, Himpe C, Seidenbecher T, Pape HC. Directional theta coherence in prefrontal cortical to amygdalo-hippocampal pathways signals fear extinction. PLoS One. 2013; 8:e77707. [PubMed: 24204927]

Lewandowski KE, Shashi V, Berry PM, Kwapil TR. Schizophrenic-like neurocognitive deficits in children and adolescents with 22q11 deletion syndrome. Am J Med Genet B Neuropsychiatr Genet. 2007; 144:27–36. [PubMed: 17034021]

Likhtik E, Stujenske JM, Topiwala MA, Harris AZ, Gordon JA. Prefrontal entrainment of amygdala activity signals safety in learned fear and innate anxiety. Nat Neurosci. 2014; 17:106–13. [PubMed: 24241397]

Lisman JE, Jensen O. The theta-gamma neural code. Neuron. 2013; 77:1002–16. [PubMed: 23522038]

Livneh U, Paz R. Amygdala-prefrontal synchronization underlies resistance to extinction of aversive memories. Neuron. 2012; 75:133–42. [PubMed: 22794267]

Malsburg, C. The correlation theory of brain function. Göttingen, West Germany: Max Planck Institute for Biophysical Chemistry; 1981.

Maren S. Neurobiology of Pavlovian fear conditioning. Annu Rev Neurosci. 2001; 24:897–931. [PubMed: 11520922]

McMenamin BW, Langeslag SJ, Sirbu M, Padmala S, Pessoa L. Network organization unfolds over time during periods of anxious anticipation. J Neurosci. 2014; 34:11261–73. [PubMed: 25143607]

Merker B. Cortical gamma oscillations: the functional key is activation, not cognition. Neurosci Biobehav Rev. 2013; 37:401–17. [PubMed: 23333264]

Meyer-Lindenberg A, Poline JB, Kohn PD, Holt JL, Egan MF, et al. Evidence for abnormal cortical functional connectivity during working memory in schizophrenia. Am J Psychiatry. 2001; 158:1809–17. [PubMed: 11691686]

Meyer-Lindenberg AS, Olsen RK, Kohn PD, Brown T, Egan MF, et al. Regionally specific disturbance of dorsolateral prefrontal-hippocampal functional connectivity in schizophrenia. Arch Gen Psychiatry. 2005; 62:379–86. [PubMed: 15809405]

Milner PM. A model for visual shape recognition. Psychol Rev. 1974; 81:521–35. [PubMed: 4445414]

Miltner WH, Braun C, Arnold M, Witte H, Taub E. Coherence of gamma-band EEG activity as a basis for associative learning. Nature. 1999; 397:434–6. [PubMed: 9989409]

Monk CS, Telzer EH, Mogg K, Bradley BP, Mai X, et al. Amygdala and ventrolateral prefrontal cortex activation to masked angry faces in children and adolescents with generalized anxiety disorder. Arch Gen Psychiatry. 2008; 65:568–76. [PubMed: 18458208]

Mueller EM, Panitz C, Hermann C, Pizzagalli DA. Prefrontal oscillations during recall of conditioned and extinguished fear in humans. J Neurosci. 2014; 34:7059–66. [PubMed: 24849342]

Nader K, Schafe GE, Le Doux JE. Fear memories require protein synthesis in the amygdala for reconsolidation after retrieval. Nature. 2000; 406:722–6. [PubMed: 10963596]

Nakatani H, van Leeuwen C. Transient synchrony of distant brain areas and perceptual switching in ambiguous figures. Biol Cybern. 2006; 94:445–57. [PubMed: 16532332]

Narayanan RT, Seidenbecher T, Kluge C, Bergado J, Stork O, Pape HC. Dissociated theta phase synchronization in amygdalo-hippocampal circuits during various stages of fear memory. Eur J Neurosci. 2007a; 25:1823–31. [PubMed: 17408428]

Narayanan RT, Seidenbecher T, Sangha S, Stork O, Pape HC. Theta resynchronization during reconsolidation of remote contextual fear memory. Neuroreport. 2007b; 18:1107–11. [PubMed: 17589308]

Narayanan V, Heiming RS, Jansen F, Lesting J, Sachser N, et al. Social defeat: impact on fear extinction and amygdala-prefrontal cortical theta synchrony in 5-HTT deficient mice. PLoS One. 2011; 6:e22600. [PubMed: 21818344]

Nir Y, Fisch L, Mukamel R, Gelbard-Sagiv H, Arieli A, et al. Coupling between neuronal firing rate, gamma LFP, and BOLD fMRI is related to interneuronal correlations. Curr Biol. 2007; 17:1275–85. [PubMed: 17686438]

Nir Y, Mukamel R, Dinstein I, Privman E, Harel M, et al. Interhemispheric correlations of slow spontaneous neuronal fluctuations revealed in human sensory cortex. Nat Neurosci. 2008; 11:1100–8. [PubMed: 19160509]

O'Keefe J, Recce M. Phase relationship between hippocampal place units and the EEG theta rhythm. Hippocampus. 1993; 3:317–30. [PubMed: 8353611]

O'Neill PK, Gordon JA, Sigurdsson T. Theta oscillations in the medial prefrontal cortex are modulated by spatial working memory and synchronize with the hippocampus through its ventral subregion. J Neurosci. 2013; 33:14211–24. [PubMed: 23986255]

Ossandon T, Jerbi K, Vidal JR, Bayle DJ, Henaff MA, et al. Transient suppression of broadband gamma power in the default-mode network is correlated with task complexity and subject performance. J Neurosci. 2011; 31:14521–30. [PubMed: 21994368]

Palanca BJ, DeAngelis GC. Does neuronal synchrony underlie visual feature grouping? Neuron. 2005; 46:333–46. [PubMed: 15848810]

Palva S, Monto S, Palva JM. Graph properties of synchronized cortical networks during visual working memory maintenance. Neuroimage. 2010; 49:3257–68. [PubMed: 19932756]

Palva S, Palva JM. New vistas for alpha-frequency band oscillations. Trends Neurosci. 2007; 30:150–8. [PubMed: 17307258]

Pan H, Epstein J, Silbersweig DA, Stern E. New and emerging imaging techniques for mapping brain circuitry. Brain Res Rev. 2011; 67:226–51. [PubMed: 21354205]

Pascual-Marqui RD, Esslen M, Kochi K, Lehmann D. Functional imaging with low-resolution brain electromagnetic tomography (LORETA): a review. Methods Find Exp Clin Pharmacol. 2002; 24(Suppl

C):91–5. [PubMed: 12575492]

Payne L, Kounios J. Coherent oscillatory networks supporting short-term memory retention. Brain Res. 2009; 1247:126–32. [PubMed: 18976639]

Piskulic D, Olver JS, Norman TR, Maruff P. Behavioural studies of spatial working memory dysfunction in schizophrenia: a quantitative literature review. Psychiatry Res. 2007; 150:111–21. [PubMed: 17292970]

Popa D, Duvarci S, Popescu AT, Lena C, Pare D. Coherent amygdalocortical theta promotes fear memory consolidation during paradoxical sleep. Proc Natl Acad Sci U S A. 2010; 107:6516–9. [PubMed: 20332204]

Raghavachari S, Kahana MJ, Rizzuto DS, Caplan JB, Kirschen MP, et al. Gating of human theta oscillations by a working memory task. J Neurosci. 2001; 21:3175–83. [PubMed: 11312302]

Raghavachari S, Lisman JE, Tully M, Madsen JR, Bromfield EB, Kahana MJ. Theta oscillations in human cortex during a working-memory task: evidence for local generators. J Neurophysiol. 2006; 95:1630–8. [PubMed: 16207788]

Rissman J, Gazzaley A, D'Esposito M. Dynamic adjustments in prefrontal, hippocampal, and inferior temporal interactions with increasing visual working memory load. Cereb Cortex. 2008; 18:1618–29. [PubMed: 17999985]

Robinson OJ, Charney DR, Overstreet C, Vytal K, Grillon C. The adaptive threat bias in anxiety: amygdala-dorsomedial prefrontal cortex coupling and aversive amplification. Neuroimage. 2012; 60:523–9. [PubMed: 22178453]

Rodriguez E, George N, Lachaux JP, Martinerie J, Renault B, Varela FJ. Perception's shadow: long-distance synchronization of human brain activity. Nature. 1999; 397:430–3. [PubMed: 9989408]

Roelfsema PR, Lamme VA, Spekreijse H. Synchrony and covariation of firing rates in the primary visual cortex during contour grouping. Nat Neurosci. 2004; 7:982–91. [PubMed: 15322549]

Roux F, Uhlhaas PJ. Working memory and neural oscillations: alpha-gamma versus theta-gamma codes for distinct WM information? Trends Cogn Sci. 2014; 18:16–25. [PubMed: 24268290]

Roux F, Wibral M, Mohr HM, Singer W, Uhlhaas PJ. Gamma-band activity in human prefrontal cortex codes for the number of relevant items maintained in working memory. J Neurosci. 2012; 32:12411–20. [PubMed: 22956832]

Sakkalis V. Review of advanced techniques for the estimation of brain connectivity measured with EEG/MEG. Comput Biol Med. 2011; 41:1110–7. [PubMed: 21794851]

Sauseng P, Klimesch W, Schabus M, Doppelmayr M. Fronto-parietal EEG coherence in theta and upper alpha reflect central executive functions of working memory. Int J Psychophysiol. 2005; 57:97–103. [PubMed: 15967528]

Scheeringa R, Fries P, Petersson KM, Oostenveld R, Grothe I, et al. Neuronal dynamics underlying high- and low-frequency EEG oscillations contribute independently to the human BOLD signal. Neuron. 2011; 69:572–83. [PubMed: 21315266]

Scholvinck ML, Maier A, Ye FQ, Duyn JH, Leopold DA. Neural basis of global resting-state fMRI activity. Proc Natl Acad Sci U S A. 2010; 107:10238–43. [PubMed: 20439733]

Seidenbecher T, Laxmi TR, Stork O, Pape HC. Amygdalar and hippocampal theta rhythm synchronization during fear memory retrieval. Science. 2003; 301:846–50. [PubMed: 12907806]

Setsompop K, Kimmlingen R, Eberlein E, Witzel T, Cohen-Adad J, et al. Pushing the limits of in vivo diffusion MRI for the Human Connectome Project. Neuroimage. 2013; 80:220–33. [PubMed: 23707579]

Shadlen MN, Movshon JA. Synchrony unbound: a critical evaluation of the temporal binding hypothesis. Neuron. 1999; 24:67–77. 111–25. [PubMed: 10677027]

Shah AA, Sjovold T, Treit D. Inactivation of the medial prefrontal cortex with the GABAA receptor agonist muscimol increases open-arm activity in the elevated plus-maze and attenuates shock-probe burying in rats. Brain Res. 2004; 1028:112–5. [PubMed: 15518648]

Shah AA, Treit D. Excitotoxic lesions of the medial prefrontal cortex attenuate fear responses in the elevated-plus maze, social interaction and shock probe burying tests. Brain Res. 2003; 969:183–94. [PubMed: 12676379]

Shmuel A, Leopold DA. Neuronal correlates of spontaneous fluctuations in fMRI signals in monkey visual cortex: Implications for functional connectivity at rest. Hum Brain Mapp. 2008; 29:751–61. [PubMed: 18465799]

Siapas AG, Lubenov EV, Wilson MA. Prefrontal phase locking to hippocampal theta oscillations. Neuron. 2005; 46:141–51. [PubMed: 15820700]

Siapas AG, Wilson MA. Coordinated interactions between hippocampal ripples and cortical spindles during slow-wave sleep. Neuron. 1998; 21:1123–8. [PubMed: 9856467]

Sigurdsson T, Stark KL, Karayiorgou M, Gogos JA, Gordon JA. Impaired hippocampal-prefrontal synchrony in a genetic mouse model of schizophrenia. Nature. 2010; 464:763–67. [PubMed: 20360742]

Sobin C, Kiley-Brabeck K, Daniels S, Khuri J, Taylor L, et al. Neuropsychological characteristics of children with the 22q11 Deletion Syndrome: a descriptive analysis. Child Neuropsychol. 2005; 11:39–53. [PubMed: 15823982]

Sohal VS, Zhang F, Yizhar O, Deisseroth K. Parvalbumin neurons and gamma rhythms enhance cortical circuit performance. Nature. 2009; 459:698–702. [PubMed: 19396159]

Srinivasan R, Russell DP, Edelman GM, Tononi G. Increased synchronization of neuromagnetic responses during conscious perception. J Neurosci. 1999; 19:5435–48. [PubMed: 10377353]

Stark K, Xu B, Bagchi A, Lai W, Liu H, et al. Altered brain microRNA biogenesis contributes to phenotypic deficits in a 22q11-deletion mouse model. Nat Genet. 2008; 40:751–60. [PubMed: 18469815]

Stujenske JM, Likhtik E, Topiwala MA, Gordon JA. Fear and safety engage competing patterns of theta-gamma coupling in the basolateral amygdala. Neuron. 2014; 83:919–33. [PubMed: 25144877]

Tallon-Baudry C, Bertrand O, Delpuech C, Permier J. Oscillatory gamma-band (30–70 Hz) activity induced by a visual search task in humans. J Neurosci. 1997; 17:722–34. [PubMed: 8987794]

Tallon-Baudry C, Bertrand O, Delpuech C, Pernier J. Stimulus specificity of phase-locked and non-phase-locked 40 Hz visual responses in human. J Neurosci. 1996; 16:4240–9. [PubMed:8753885]

Tallon-Baudry C, Bertrand O, Fischer C. Oscillatory synchrony between human extrastriate areas during visual short-term memory maintenance. J Neurosci. 2001; 21:RC177. [PubMed: 11588207]

Thiele A, Stoner G. Neuronal synchrony does not correlate with motion coherence in cortical area MT. Nature. 2003; 421:366–70. [PubMed: 12540900]

Thierry AM, Gioanni Y, Degenetais E, Glowinski J. Hippocampo-prefrontal cortex pathway: anatomical and electrophysiological characteristics. Hippocampus. 2000; 10:411–9. [PubMed: 10985280]

Thomason ME, Hamilton JP, Gotlib IH. Stress-induced activation of the HPA axis predicts connectivity between subgenual cingulate and salience network during rest in adolescents. J Child Psychol Psychiatry. 2011; 52:1026–34. [PubMed: 21644985]

Tononi G, Srinivasan R, Russell DP, Edelman GM. Investigating neural correlates of conscious perception by frequency-tagged neuromagnetic responses. Proc Natl Acad Sci U S A. 1998; 95:3198–203. [PubMed: 9501240]

Tromp DP, Grupe DW, Oathes DJ, McFarlin DR, Hernandez PJ, et al. Reduced structural connectivity of a major frontolimbic pathway in generalized anxiety disorder. Arch Gen Psychiatry. 2012; 69:925–34. [PubMed: 22945621]

Uhlhaas PJ, Pipa G, Lima B, Melloni L, Neuenschwander S, et al. Neural synchrony in cortical networks: history, concept and current status. Front Integr Neurosci. 2009; 3:17. [PubMed: 19668703]

van Amelsvoort T, Henry J, Morris R, Owen M, Linszen D, et al. Cognitive deficits associated with schizophrenia in velo-cardio-facial syndrome. Schizophr Res. 2004; 70:223–32. [PubMed: 15329299]

Venkataraman A, Whitford TJ, Westin CF, Golland P, Kubicki M. Whole brain resting state functional connectivity abnormalities in schizophrenia. Schizophr Res. 2012; 139:7–12. [PubMed: 22633528]

Verwer RW, Meijer RJ, Van Uum HF, Witter MP. Collateral projections from the rat hippocampal formation to the lateral and medial prefrontal cortex. Hippocampus. 1997; 7:397–402. [PubMed: 9287079]

von Stein A, Sarnthein J. Different frequencies for different scales of cortical integration: from local gamma to long range alpha/theta synchronization. Int J Psychophysiol. 2000; 38:301–13. [PubMed: 11102669]

Wagemans J, Elder JH, Kubovy M, Palmer SE, Peterson MA, et al. A century of Gestalt psychology in visual perception: I. Perceptual grouping and figure-ground organization. Psychol Bull. 2012; 138:1172–217. [PubMed: 22845751]

Wang GW, Cai JX. Disconnection of the hippocampal-prefrontal cortical circuits impairs spatial working memory performance in rats. Behav Brain Res. 2006; 175:329–36. [PubMed: 17045348]

Wang XJ. Neurophysiological and computational principles of cortical rhythms in cognition. Physiol Rev. 2010; 90:1195–268. [PubMed: 20664082]

Watrous AJ, Fell J, Ekstrom AD, Axmacher N. More than spikes: common oscillatory mechanisms for content specific neural representations during perception and memory. Curr Opin Neurobiol. 2014; 31C:33–39. [PubMed: 25129044]

Wheelock MD, Sreenivasan KR, Wood KH, Ver Hoef LW, Deshpande G, Knight DC. Threat-related learning relies on distinct dorsal prefrontal cortex network connectivity. Neuroimage. 2014; 102(Pt 2):904–12. [PubMed: 25111474]

Womelsdorf T, Fries P, Mitra PP, Desimone R. Gamma-band synchronization in visual cortex predicts speed of change detection. Nature. 2006; 439:733–6. [PubMed: 16372022]

Yamamoto J, Suh J, Takeuchi D, Tonegawa S. Successful execution of working memory linked to synchronized high-frequency gamma oscillations. Cell. 2014; 157:845–57. [PubMed: 24768692]

Yoon T, Okada J, Jung MW, Kim JJ. Prefrontal cortex and hippocampus subserve different components of working memory in rats. Learn Mem. 2008; 15:97–105. [PubMed: 18285468]

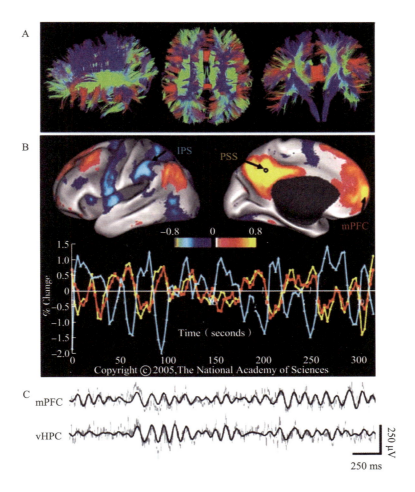

Figure 1

Structural and dynamic connectivity in the brain. **A**. Diffusion tensor imaging tractography in the human brain. Reproduce d with permission from (Setsompop et al 2013). **B**. Top, Resting state fMRI image illustrating the correlations observed for a single resting subject between a seed region in the posterior cingulate/precuneus (PCC) and all other voxels in the brain. Warm colors represent positive correlations while cool colors reflect negative correlations. Bottom, An example time course of the PCC (yellow) signal, along with a positively correlated region, the medial prefrontal cortex (mPFC, orange), and a negatively correlated region, the intraparietal sulcus (IPS, blue). Reproduced with permission from (Fox et al 2005). **C**. Simultaneously recorded local field potentials from depth electrode in the mPFC (top) and ventral hippocampus (vHPC; bottom) in a mouse during active exploration. Raw traces are plotted in gray and theta filtered traces are overlaid in black (adapted with permission from (Adhikari et al 2010)).

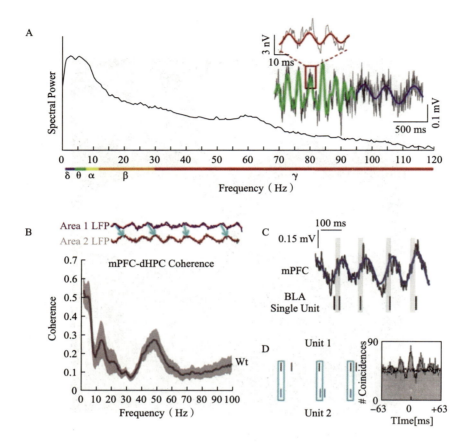

Figure 2

Oscillations and synchrony in local field potentials. **A**. The power spectrum of a local field potential (LFP) recorded from the nucleus accumbens of an actively exploring mouse. The frequency range conventions are color coded below the x-axis (blue: delta, green: theta, yellow: alpha, orange: beta, red: gamma). Right, The raw local field potential is plotted in gray and a band-filtered trace is overlaid to highlight segments with prominent theta (green), delta (blue) and gamma (red) oscillations. **B**. Top, Two cartoon LFP traces displaying a consistent phase relationship. Bottom, The mPFC and dHPC show peaks in coherence in the delta, theta and gamma frequency range (adapted with permission from Sigurdsson et al 2010). **C**. Raw (gray) and theta-filtered (blue) mouse mPFC LFP traces, along with simultaneously recorded basolateral amygdala (BLA) single-unit activity illustrating phase locking. Gray bars are aligned on zero phase (reproduced with permission from Stujenske et al 2014). **D**. Left, schematic of synchronously firing spike trains. Right, cross-correlations of two neurons recorded in middle temporal area of a monkey watching a visual stimulus. The black line outlining the cross-correlogram represents the fitted function used to quantify correlation strength, and the thin black line corresponds to that coherence expected by chance (adapted with permission from (Kreiter & Singer 1996)).

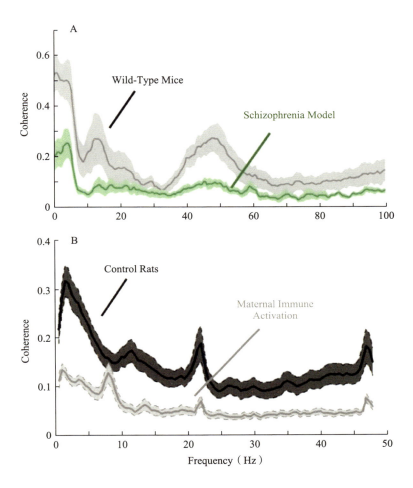

Figure 3

Synchrony deficits in rodent models of schizophrenia predisposition. **A**. Coherence between LFPs recorded from the hippocampus and medial prefrontal cortex of a mouse model of a microdeletion that raises the risk of schizophrenia by 30-fold (green) and wild-type littermates (gray). Shaded areas are +/− s.e.m. Adapted from Sigurdsson et al., 2010. **B**. Coherence between hippocampal and medial prefrontal LFPs in a rat model of prenatal infection, a risk factor that raises the risk of schizophrenia by 2–3 fold (gray) and wild-type controls (black). Conventions as in A. Adapted from Dickerson et al., 2010, with permission.

3-D Maps and Compasses in the Brain

Arseny Finkelstein, Liora Las, and Nachum Ulanovsky

Department of Neurobiology, Weizmann Institute of Science, Rehovot 76100, Israel; email: nachum. ulanovsky@weizmann.ac.il

Key Words

spatial cognition, 3-D topology, place cells, grid cells, head-direction cells, rodents, primates, bats

Abstract

The world has a complex, three-dimensional(3-D) spatial structure, but until recently the neural representation of space was studied primarily in planar horizontal environments. Here we review the emerging literature on allocentric spatial representations in 3-D and discuss the relations between 3-D spatial perception and the underlying neural codes. We suggest that the statistics of movements through space determine the topology and the dimensionality of the neural representation, across species and different behavioral modes. We argue that hippocampal place-cell maps are metric in all three dimensions, and might be composed of 2-D and 3-D fragments that are stitched together into a global 3-D metric representation via the 3-D head-direction cells. Finally, we propose that the hippocampal formation might implement a neural analogue of a Kalman filter, a standard engineering algorithm used for 3-D navigation.

INTRODUCTION

Most animals, including insects, fish, birds, and mammals, navigate actively through space to find food, mating partners, and shelter. Decades of research on the neurobiology of navigation focused on two-dimensional (2-D) navigation on flat surfaces, which laid the foundations for understanding the neural basis of 2-D spatial representations (Hafting et al. 2005, O'Keefe & Dostrovsky 1971, O'Keefe & Nadel 1978, Taube et al. 1990a, Tolman 1948). However, we live in a 3-D world, and many animals need to orient and move through 3-D space—but until recently, very little was known about how 3-D space is encoded in the brain. The extra dimension may complicate the required brain computations and poses technical challenges to the experimenters who wish to study 3-D navigation. Here we review the emerging data on the neural basis of 3-D navigation in mammals, primarily from rodents and bats, and propose a synthesis of findings across species. Notably, although this review focuses on neural representations of allocentric (absolute) navigable 3-D space, similar questions were also studied for egocentric (body-referenced) 3-D space (see the sidebar, 3-D Maps of Egocentric Space).

The review consists of four parts. First, we describe the neural basis of maps and compasses for allocentric space during different navigation modes in 3-D, focusing on 3-D place cells and head-direction cells. Second, we discuss whether 3-D maps and compasses are local or global and offer predictions for 3-D grid cells. Third, we discuss the possible principles guiding the development of neural maps and compasses for 3-D space. Specifically, we propose that the neural representations of 3-D maps and 3-D compasses are flexible and are set largely by the ongoing behavior of the animal and by the behavioral repertoire experienced during ontogeny—and not only by phylogeny. Fourth,

Place cells: hippocampal neurons that discharge when the animal passes through a specific location in the environment, called the place field

Head-direction cells: neurons found in multiple brain regions that discharge when the animal's head points toward a specific absolute (allocentric) direction, providing a compass signal

Grid cells: neurons that discharge when the animal passes through the vertices of a periodic hexagonal lattice that tiles the environment

we describe how 3-D maps and compasses might interact, and we draw functional parallels to engineering approaches to 3-D navigation.

3-D MAPS OF EGOCENTRIC SPACE

In addition to the emerging study of the neural representation of 3-D allocentric space, which is the focus of this review, a large body of research points to the existence of a metric 3-D representation of egocentric, body-centered space—primarily in the parietal cortex (Buneo et al. 2002, Whitlock et al. 2012). This body-centered 3-D reference frame supports eye movements and arm reaching in primates and may underlie our ability to catch a flying ball in 3-D (Blohm et al. 2009, Breveglieri et al. 2012, Rosenberg et al. 2013). In fact, humans are ecologically quite 3-D animals when it comes to stereoscopic vision and our ability to perform complex 3-D movements in egocentric space (Cohen & Rosenbaum 2004, Klatzky & Giudice 2013), although we note the imperfections of visual depth perception (Foley 1980). The existence of a 3-D metric representation of egocentric space in monkey parietal cortex likely results from the high ethological importance of precise calculations of 3-D spatial movements for primates (Bueti & Walsh 2009). How egocentric and allocentric 3-D representations are related remains an important open question (Chen et al. 2013, Nitz 2012, Wang 2012). The 3-D head-direction system in the dorsal presubiculum (Finkelstein et al. 2015) may be important for linking these two major 3-D reference frames. An important future challenge will be to understand how the allocentric 3-D map and 3-D compass signals in the hippocampal formation are integrated with the 3-D egocentric system in the parietal cortex to enable 3-D hand and eye movements and whole-body locomotion.

3-D NAVIGATION MODES AND THEIR NEURAL CORRELATES

Navigation Modes in 3-D Space

All animals live in a 3-D world. However, 3-D navigation is not a unitary behavior but consists of several categories that may give rise to different neural representations. We propose to distinguish between three different behavioral modes of 3-D navigation: planar, multilayered, and volumetric navigation (**Figure 1a–c**).

1.Planar navigation implies movement along a 2-D surface embedded in 3-D space (**Figure 1a**) (O'Keefe & Nadel 1978, Tolman 1948). Planar navigation is 2-D in nature—whether it is performed on a horizontal, tilted, or vertical surface—and therefore may not require any explicit information about the dimension that is orthogonal to the locomotion plane.

2.Multilayered navigation requires movement across several interconnected planes (**Figure 1b**) (Thibault et al. 2012). Although navigation along each branch or layer of a multilayered space can be described as being 1-D or 2-D, navigating a multilayered environment requires an explicit 3-D representation—including the ability to discriminate between planes (e.g., floors), understand their 3-D spatial relations, and make novel 3-D shortcuts (Montello & Pick 1993).

3.Volumetric navigation is not limited by locomotion surfaces and involves an unconstrained movement through 3-D space (**Figure 1c**) (Grobéty & Schenk 1992, Jovalekic et al. 2011). Nevertheless, it can still be restricted by the 3-D boundaries of the environment, as well as by the animal's ability to jump, glide, fly, or swim in 3-D. Similar to multilayered navigation, successful volumetric navigation requires an explicit 3-D representation of space.

Although some species exhibit only one of these navigation modes (e.g., horses navigate always in a planar fashion, and dolphins always navigate volumetrically), most mammals can flexibly switch between navigation modes according to the spatial layout of the environment and the behavioral needs of the animal. For example, rats can switch from planar navigation during open-field foraging to multilayered navigation in burrows or dense vegetation; and other rodents, such as squirrels, also navigate volumetrically (**Figure1*c***). Monkeys are typically engaged in all three modes of 3-D navigation, moving in a planar fashion on the ground and in multilayered and volumetric modes in complex environments (e.g., the tree canopy). Humans exhibit terrestrial planar navigation, as well as multilayered navigation in modern buildings, and can navigate volumetrically as they swim, dive, or pilot an aircraft (**Figure 1*a–c***). Finally, bats— mammals capable of both crawling and flying—have mastered all three navigation modes, as their habitat spans from rock surfaces to complex caves, foliage, and large 3-D open spaces; this makes bats an ideal animal model to study the neural basis of 3-D navigation (Geva-Sagiv et al. 2015).

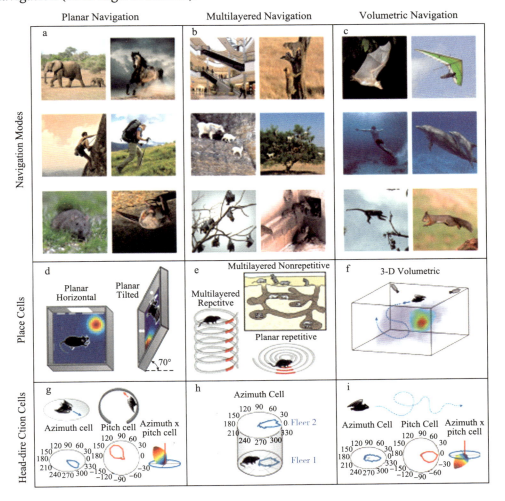

Figure 1

3-D navigation modes and their neural correlates. (*a–c*) Navigation modes in 3-D space across mammalian species: (*a*)planar,(*b*) multilayered, and (*c*) volumetric navigation. (*d–f*) 3-D tuning of place cells during the different 3-D navigation modes. (*d*) In planar navigation, place cells in a horizontal arena (*left*; Whitlock et al. 2008) and on a tilted 2-D surface (*right*; Ulanovsky & Moss 2011) exhibit circular place fields within the 2-D plane. (*e*) During multilayered navigation, place cells were tested in repetitive multilayered environments (*left schematic*) (Hayman et al. 2011), yielding repetitive place fields (*red*, spikes); this could stem from the repetitive nature of the environment because repetitive place fields are also observed in planar repetitive environments (*bottom right*) (Cowen & Nitz 2014). Place cells have yet to be recorded in multilayered nonrepetitive environments, such as a rat's burrow system (top right). (f) During volumetric navigation, place cells were recorded in a bat flying through 3-D volumetric space (5.8 × 4.6 × 2.7-m room); these neurons exhibited spherical (isotropic) tuning to all three dimensions, as shown in this example (Yartsev & Ulanovsky 2013). (*g–i*) 3-D tuning of head-direction cells during the different 3-D navigation modes. (*g*) Head-direction cells tuned to azimuth, pitch, or azimuth × pitch were recorded in bats during planar navigation on a horizontal-arena floor or on a vertical ring (Finkelstein et al. 2015). (*h*) Azimuthal head-direction cells exhibit the same directional preference during multilayered navigation in rats tested on two different floors (Stackman et al. 2000). (*i*) In bats flying through 3-D volumetric space, head-direction cells exhibited similar types of tuning to those found in planar navigation (shown in panel *g*; numbers indicate degrees); namely, neurons were tuned to azimuth, pitch, or azimuth × pitch (Finkelstein et al. 2015). 3-D head-direction cells thus provide a global 3-D compass signal that could underlie 3-D navigation in bats. Such 3-D head-direction tuning was found so far only in the dorsal presubiculum of bats (Finkelstein et al. 2015), but 3-D tuning has never been tested in this brain region in other species; it therefore remains to be seen whether 3-D head-direction cells will be found in the presubiculum of mice, which are well adapted to 3-D multilayered navigation, and monkeys, which can also navigate in 3-D volumetrically. Panel d (*left*) adapted with permission from Whitlock et al. (2008); copyright (2008) National Academy of Sciences U.S.A. Panel *f* adapted from Yartsev & Ulanovsky (2013), adapted with permission from AAAS. Panel *h* based on data from Stackman et al. (2000)

Neural Correlates of 3-D Navigation

Decades of neurophysiological studies in rodents revealed that the hippocampal formation is crucial for navigation and contains neurons whose functional properties can support map-and-compass navigation (Moser et al. 2008, O'Keefe & Nadel 1978). These neurons include place cells (O'Keefe & Dostrovsky 1971), head-direction cells (Taube et al. 1990a), grid cells (Hafting et al. 2005), and border/boundary cells (Lever et al. 2009, Solstad et al. 2008).

Place cells discharge selectively when the animal passes through a certain spatial location

Border/boundary cells: neurons that signal environmental geometry by discharging whenever the animal is close to a salient border of the environment

Dorsal presubiculum (postsubiculum): a cortical area where head-direction cells were first discovered in rats; it provides the major head-direction input to the medial entorhinal cortex

on a 2-D horizontal surface and were suggested to be the neural substrate of a "cognitive map" of the environment (O'Keefe & Nadel 1978, Tolman 1948). Similarly, during planar navigation on tilted or vertical surfaces, place cells in rodents, monkeys, and bats exhibit circular place fields at a specific vertical × horizontal position—suggesting sensitivity of place cells to the vertical dimension (see **Figure 1*d***) (Jeffery et al. 2005, Knierim & McNaughton 2001, Knierim et al. 2000, Ludvig et al. 2004, Ulanovsky & Moss 2007). In contrast, during multilayered navigation in rats moving along a vertical helix, Hayman et al. (2011) suggested that place cells were tuned in the horizontal dimension but were relatively insensitive to the vertical dimension, with place fields of a given neuron appearing stacked above each other (illustrated schematically in **Figure 1*e***, *left*). Notably, in this experiment, the different layers (loops) of the helical apparatus were exact repetitions of each other (**Figure 1*e***, *left:* multilayered repetitive). Such a geometric repetition is known in 2-D horizontal environments to result in repeating place fields in equivalent segments (**Figure 1*e***, *bottom:* planar repetitive) (Cowen & Nitz 2014, Derdikman et al. 2009, Singer et al. 2010, Spiers et al. 2015)—and therefore the repetitiveness of place fields on the vertical helix might have resulted from the repetitive nature of the setup, rather than reflecting an underlying lack of altitude encoding in rats. It remains to be seen how place cells encode a naturalistic, multilayered environment not composed of repeating elements (**Figure 1*e***, *top*: multilayered nonrepetitive). Finally, our recent recordings in the hippocampal formation of freely flying bats demonstrated the existence of 3-D place cells under fully volumetric 3-D conditions (Yartsev & Ulanovsky 2013). These neurons discharged when the animal passed through a 3-D spherical place field (**Figure 1*f***, *red portion of space*)—thus providing an explicit 3-D neural representation, including the encoding of altitude, that could support 3-D volumetric navigation.

Head-direction cells (Ranck 1984, Taube et al. 1990a) are neurons that discharge selectively when the animal's head points at a specific azimuthal direction on a 2-D planar surface, and researchers have suggested these cells are the neural analogue of a compass (Taube 2007). Studies in subcortical regions of the rat brain showed that azimuthal head-direction cells are insensitive to head pitch; moreover, no pitch cells that span the behaviorally relevant pitch range were found in rodents (Stackman & Taube 1998). However, our recent study of a cortical area—the dorsal presubiculum of bats—revealed azimuth cells (**Figure 1*g***, *left*) that resembled the head-direction cells found in rodents (Taube et al. 1990a), but we also found neurons tuned to head pitch (**Figure 1*g***, *middle*) (Finkelstein et al. 2015). Additionally, bat dorsal presubiculum contained 3-D head-direction cells that responded to a particular combination of azimuth × pitch, thus representing the direction of the head vector in 3-D space (**Figure 1*g***, *right*). Notably, all the 3-D studies in rats were conducted in subcortical nuclei, whereas our study in bats was the first to examine 3-D coding in the presubiculum of any species. Because rodents move their head extensively in 3-D (Wallace et al. 2013), it would be interesting to examine 3-D representations in the presubiculum of rodents, where 3-D tuning has never been studied, as well as in the presubiculum of primates.

During multilayered navigation in an apparatus comprised of two visually distinct floors, head-direction cells of rats fired in the same azimuth direction on both floors (**Figure 1h**) (Stackman et al. 2000). This result is consistent with the notion that the head-direction signal is global and is not compartmentalized (as shown also in 2-D; see Whitlock & Derdikman 2012, Yoder et al. 2011)—a point to which we return below.

Head-direction tuning during 3-D volumetric navigation has been studied to date only in flying bats (Finkelstein et al. 2015). This study found head-direction cells that were tuned to azimuth, pitch, or conjunctively to azimuth × pitch (**Figure 1i**)—similar to the tuning found during planar navigation (**Figure 1g**). Thus, the head-direction signal in dorsal presubiculum of bats is invariably 3-D during both planar and volumetric navigation, suggesting the head-direction signal is global and does not anchor to a specific locomotion plane. In contrast, we argue below that the dimensionality of spatial representation by place cells would switch flexibly from 2-D to 3-D, according to the navigation mode. Furthermore, because many mammalian species navigate in 3-D, we predict that similar representations will be found in other mammals, including mice and primates—both of which orient very well in 3-D.

A MOSAIC OF 2-D AND 3-D MAPS STITCHED BY A GLOBAL 3-D COMPASS

An important question is, how are the map and compass signals related to each other? Do they integrate into a global 3-D representation of the environment, and if so, how? For instance, does the brain use a global 3-D map and compass or a mosaic of local representations for different locomotion planes? In the next two sections, we discuss the intriguing possibility that separate sets of maps exist for 2-D and 3-D space, and these can be combined into a coherent spatial representation using a global 3-D compass.

Multiple Maps for 3D Space

As discussed above, navigation is not a unitary behavior: Animals can switch abruptly between 2-D and 3-D navigation modes (**Figure 2a**). This suggests three possible neuronal representations. The first possibility is that a 3-D volumetric neural map is present at all times, during both 3-D and 2-D behavioral modes. According to this option, all place cells have 3-D tuning, including silent cells—those neurons that are inactive in a given 2-D environment (Alme et al. 2014)—which in fact might be 3-D place cells anchored to a global map whose 3-D place field happens to lie outside the 2-D locomotion surface. A second possibility is that rapid switches between behavioral modes imply rapid switches between the dimensionality of the neural representations. According to this possibility, place cells will have a 2-D tuning during planar navigation and a 3-D tuning during 3-D volumetric or multilayered navigation (Ulanovsky & Finkelstein 2013). A third possibility is that of attentional switches: For example, even during navigation on a 2-D surface, we may expect switches between a 2-D tuning when the animal attends to local surface cues (e.g., local odors and textures on the floor) and a 3-D tuning when the animal attends to global cues (e.g., distal visual landmarks that are embedded in 3-D space). This latter notion is consistent with the known flexibility of the hippocampal spatial code, manifested in switches of the placecell map between local and global reference frames (Knierim & Hamilton 2011) (see the sidebar, Reference Frames).

Notably, switches between reference frames might complicate the interpretation of studies on 3-D place cells during 2-D navigation. Specifically, two different studies in rats attempted to measure 3-D tuning of place cells by moving the plane of locomotion in 3-D space. In the first study, a horizontal arena was rotated in pitch (tilted upward by 45°), and many place cells in hippocampal area CA1 changed their firing pattern at a certain altitude (Knierim & McNaughton 2001)—indicating either 3-D tuning or a change in the anchoring reference frame, leading to remapping. In the second study, the arena was shifted upward by translation, which did not alter the firing pattern of place cells (Knierim & Rao 2003). These different results may stem from variations in how the locomotion plane was moved in 3-D space and by the identity of the sensory cues that determined the place-field tuning. For instance, if place cells are anchored to visual cues on the room's ceiling, a large part of these cues may become occluded once the arena is tilted, which will likely lead to remapping, as was observed in the first study. In contrast, shifting the arena upward (without rotating it) should not alter substantially the visual cues available above the animal, and hence the place-cell tuning would remain stable, as was shown in the second study. These differences illustrate the inherent difficulties posed by studying animals moving on 2-D planar surfaces (horizontal, tilted, or vertical) to elucidate the underlying 3-D representation.

The problem of a global map versus many different local maps is also relevant for tuning of grid cells in 3-D. Although grid-cell activity during 3-D volumetric navigation has not been tested experimentally yet, several models predicted that 3-D grid cells would exhibit regular 3-D lattice patterns (Horiuchi & Moss 2015, Mathis et al. 2015); for further details on possible optimal 3-D lattices, see Jeffery et al. (2015). If 3-D grid cells in the brain indeed exhibit a regular 3-D lattice, an interesting question would be, how will the 2-D borders of the environment affect this putative 3-D lattice? One possibility is that 2-D grid patterns observed while animals locomote on the floor or walls of the environment are simply a cross-section through the full 3-D lattice (**Figure 2***b*, *left*).

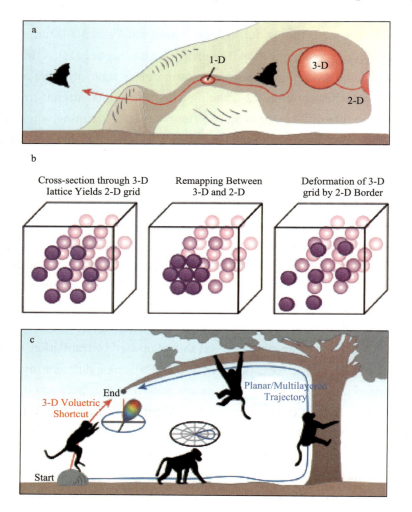

Figure 2

A mosaic of multiple 2-D and 3-D maps anchored by a global 3-D compass. (*a*) Animals can switch between different navigation modes and different dimensions (1-D, 2-D, and 3-D navigation), and the neural representations may switch accordingly. I llustrated here are hypothetical examples of a place cell with a 1-D place field during 1-D movement through a narrow tunnel, a 2-D place field on the surfaces of the environment (e.g., the wall of a cave), and a 3-D volumetric place field during volumetric 3-D flight. (*b*) Different hypotheses on the possible relations between the 2-D grids of grid cells (e.g., on the cave wall) and 3-D grids (e.g., during flight). (*Left*) Grid fields in 3-D are assumed to be arranged according to a 3-D lattice, and the 2-D grid on a plane is suggested to be a direct cross-section of that plane through the full 3-D volumetric grid. (*Middle*) 3-D grid cells might undergo remapping when the animal switches from 3-D to 2-D locomotion, such that the 2-D grid pattern is unrelated to the 3-D grid. (*Right*) The 3-D grid might be deformed by 2-D environmental borders, and thus its fields would appear in 2-D as a deformed 2-D grid. (*c*)Azimuthal head-direction cells referenced to the locomotion plane may be sufficient for orienting the animal during planar or even during multilayered navigation (*blue arrow*), but they would not allow volumetric 3-D shortcuts (*red arrow*). Instead, we propose that 3-D shortcuts are computed based on 3-D head-direction cells tuned to both azimuth and pitch, which provide a global 3-D compass that encodes the direction of the heading vector in 3-D (see also Finkelstein et al. 2015).

REFERENCE FRAMES

A reference frame defines the coordinate system in which measurements are conducted. An allocentric reference frame describes the orientation of the animal in external (world) coordinates (e.g., "north"), whereas an egocentric reference frame defines the orientation of one body part relative to another (e.g., right turn of the head with respect to the body). Furthermore, for an allocentric reference frame, the external coordinate system can be anchored to local cues (e.g., tactile cues on the floor of the behavioral apparatus) or global cues (e.g., cave walls for a bat navigating indoors, or distal landmarks such as mountains during navigation outdoors).

This notion is consistent with studies suggesting that grid-cell firing patterns on 1-D linear tracks are cuts through a perfect 2-D grid lattice (Domnisoru et al. 2013, Yoon et al. 2016)—suggesting that perhaps grid-cell representations might be of a higher dimensionality than the dimensionality of the experimental apparatus. However, a recent study in rats running on horizontal versus sloped surfaces did not find the changes in grid pattern that would be expected from planar 2-D cuts through a global 3-D volumetric map, arguing against this possibility (Hayman et al. 2015). A second possibility is that there will be remapping (phase-shift/rescaling/rotation) between 3-D and 2-D because of the changes in behavior under different locomotion modes, so that the grid patterns in 3-D and 2-D will be uncorrelated (**Figure 2*b***, *middle*). A third possibility is that the 2-D walls will deform the 3-D grid (**Figure 2*b***, *right*), akin to the border-induced shearing deformation of 2-D grids reported recently in rats (Stensola et al. 2015). It remains to be seen which of these possibilities (or perhaps a combination) holds at the interface between 3-D volumes and their 2-D borders. Importantly, here we assumed a highly regular lattice structure for 3-D grid cells—but another theoretical study (Stella & Treves 2015) predicted that in fact, 3-D grid cells would exhibit spherical firing fields, but these fields would be arranged much less regularly than expected from optimal-packing lattices. An important future challenge will therefore be to record from grid cells in 3-D.

Assuming a separate set of maps for 2-D and 3-D, how are these multiple maps stitched together? This also relates to the question of how multiple 2-D maps for different compartments are stitched in 2-D. A recent study in rats running between two interconnected 2-D compartments showed that each compartment was encoded initially by a distinct local grid map, but as the animals acquired experience in the environment, these two grid maps merged gradually into one global grid representation that was coherent across both compartments (Carpenter et al. 2015). Likewise, with experience, the 2-D map fragments (e.g., fragments representing open spaces, passageways, or different locomotion planes such as floors, walls, and ceilings) may merge into a global 3-D representation of the navigable environment. Such fusion is expected to occur as a result of the continuity of movements through 2-D and 3-D space and may be aided by a 3-D neural compass, as discussed in the next section.

In summary, we suggest that space is represented by a mosaic of multiple neural maps. This mosaic is composed of distinct 2-D planar maps and 3-D volumetric maps. Switching between such 2-D planar maps and 3-D volumetric maps may stem from toggling in reference frames and, in turn, could support rapid switching between different navigation modes. Finally, with experience, the different 2-D and 3-D maps may merge into one global representation.

A Global 3-D Compass

We propose that the correct alignment and stitching of different map fragments into a coherent representation might depend on the existence of a global 3-D compass system, required for preserving the overall sense of direction across environments. Such a global 3-D compass signal might be implemented neurally by the 3-D head-direction system in the dorsal presubiculum (Finkelstein et al. 2015). The 3-D head-direction cells that we found in bats (tuned to azimuth × pitch) were tuned to 3-D regardless of the navigation mode—during both planar and volumetric navigation. Moreover, pitch cells were found to maintain their preferred pitch angle during navigation in a complex environment that consisted of a horizontal arena and a vertical ring positioned inside the arena, which enabled full pitch maneuvers (**Figure 1g**, *middle*) (Finkelstein et al. 2015), suggesting a global allocentric 3-D head-direction signal. Similarly, in rats, the azimuthal head-direction signal was maintained across multiple compartments, rooms, and floors—as long as the animal could move actively between the different parts of the environment (**Figure 1h**) (Stackman et al. 2000, Yoder et al. 2011). Furthermore, in rats running in a hairpin maze, the head-direction signal was maintained in all the subsections of the maze, despite fragmentation of the place-cell map and grid map (Whitlock & Derdikman 2012). Finally, rat head-direction cells were shown to be controlled by global distal cues and ignored the local cues unless distal cues became unavailable (Zugaro et al. 2001). Taken together, the studies from bats and rats suggest that the head-direction signal is global and is maintained across different compartments of the environment.

Is the head-direction signal global under all circumstances? Although during active navigation the head-direction signal seems to be global, under passive-translocation conditions the head-direction cells might stop using the global (room) reference frame and become anchored to the local locomotion plane, thus undergoing remapping (Taube et al. 2013). This remapping was shown for azimuthal head-direction cells; it would be interesting to examine whether the azimuth and pitch components of the 3-D compass might remap differently. Specifically, the preferred direction of azimuth cells is set primarily by visual landmarks (Zugaro et al. 2001), but pitch cells might in principle compute the vertical tilt of the head from gravity alone (Laurens et al. 2013). Therefore, whereas azimuth cells may undergo remapping (Taube et al. 1990b), pitch cells might always exhibit the same preferred direction, set by gravity—and therefore, the pitch signal might always be global, even when the azimuth signal is not. Notably, during natural navigation outdoors, the azimuthal component of the compass can be set reliably by global sensory cues, such as the direction of the sun, stars, wind, or distal mountains (Childs & Buchler 1981, Wallraff 2005). Together with gravity, these cues could create a global 3-D reference frame that may always be available for anchoring a global 3-D compass for outdoor navigation.

But how does the 3-D neural compass integrate horizontal and vertical information? For rats performing planar navigation on vertical surfaces, the preferred azimuthal head direction on the vertical wall follows the preferred direction on the horizontal surface after a 90° pitch rotation (Calton & Taube 2005, Stackman et al. 2000); this has led to the suggestion that in rats, the vertical wall might be encoded as an unfolded extension of the horizontal floor, as if the vertical walls were felled down (Taube et al. 2013). In light of these findings, Jeffery and colleagues (2013) proposed that the brain contains not a 3-D volumetric map but a mosaic of connected planar maps whose relative direction is updated via the azimuthal head-direction signal. Consequently, to move from point A to a higher or lower point B, a surface-dwelling animal would always have to follow a route along interconnected locomotion planes. However, this does not account for the ability of many animals to perform 3-D volumetric shortcuts, such as monkeys leaping in arboreal environments (**Figure 2c**, 3-D volumetric shortcut) (Channon et al. 2010) or humans' ability to point toward the 3-D direction of a hidden target (Montello & Pick 1993, Wilson et al. 2004). Such 3-D shortcut behaviors require 3-D metric knowledge and a 3-D compass. This is consistent with our recent findings in bats, which demonstrated that head-direction cells in the presubiculum encode both azimuth and pitch directions, providing a direction in 3-D space (Finkelstein et al. 2015). Thus, the 3-D head-direction signal in the presubiculum can serve as an omnipresent 3-D compass that stays invariant between the different navigation modes. Taken together, such a global 3-D compass, anchored to a global 3-D reference frame, could be used to interconnect between the different locomotion planes during planar navigation or to calculate 3-D shortcuts during volumetric navigation (**Figure 2c**, 3-D volumetric shortcut)—facilitating the binding of different 2-D and 3-D map fragments into a coherent spatial representation.

PROPERTIES OF 3-D MAPS AND COMPASSES MAY DEPEND ON THE STATISTICS OF BEHAVIOR

Topology of 3-D Spatial Representations

During ontogeny, the movements of animals typically become increasingly more complex and 3-D. Therefore, an important question is, Do the properties of 3-D maps and compasses depend on ontogeny and on adult behavior, or are they hard-wired phylogenetic determinants? The head-direction signal appears in rat pups even before eye opening, suggesting that its formation is independent of visual cues (Bjerknes et al. 2015, Tan et al. 2015). In contrast, place cells develop later in ontogeny, followed by the appearance of adult-like grid cells (Langston et al. 2010, Wills et al. 2010). These findings may imply that the head-direction signal is hard-wired, whereas place and grid representations are experience-dependent. Alternatively, these data are also consistent with the idea that experience during ontogeny influences all these spatial codes. According to this notion, the head-direction signal develops early because pups move their head early in ontogeny (even in utero), whereas the formation of place and grid representations is delayed because pups start to move their center of mass voluntarily only later in life.

Do the statistics of spatial behaviors during ontogeny also shape the detailed properties of the neural representation? First, some theoretical models suggest that the metric of grid cells should reflect the metric of the environment—for example, grid cell development in non-Euclidian hyperbolic spaces was predicted to result in hyperbolic grids (Urdapilleta et al. 2015), and experiments are under way to test grid cells in rats raised in spherical environments (Kruge et al. 2013). Second, both theoretical and experimental studies have suggested that the temporal structure of place-cell population dynamics may be shaped by the statistics of the animal's movement through space and would therefore reflect the topology of the environment (Chen et al. 2012, Curto & Itskov 2008, Dabaghian et al. 2014, Poucet 1993, Stella et al. 2013)—again supporting the importance of movement statistics and movement topology for understanding the properties of spatial mapping.

The 3-D head-direction system in bats (**Figure1g,i**) (Finkelstein et al. 2015) provides an interesting case study for how movement statistics influence neural representations. During both flight and crawling, bats maneuver across 360° of azimuth (**Figure 3a**, *top*) and 360° of pitch, with pitch angles spanning both upright and inverted positions (**Figure 3a**, *bottom*). Importantly, the simplest way to describe such continuous angular rotations of 360° azimuth × 360° pitch is by using a toroidal coordinate system (**Figure3b,c**) (see the sidebar, Toroidal Coordinates for 3-D Head Direction). In such coordinates, every combination of azimuth and pitch (each spanning a cyclical range of 360°) is described as a point on the toroidal manifold. Thus, angular movement in azimuth and pitch follows a continuous trajectory along the toroidal manifold—in sharp contrast to spherical coordinates, in which any rotation in pitch beyond ±90° (beyond the poles of the sphere) will cause an abrupt 180° switch in the azimuth direction. If continuity of movement determines the continuity of the neural representation, one would predict that head-direction cells in bats would represent the 3-D head direction in toroidal, continuous coordinates—as indeed we found in the bat dorsal presubiculum (Finkelstein et al. 2015).

The continuity of movement through space can be translated into a continuous neural representation by means of spike-timing-dependent plasticity (STDP) (Dan & Poo 2004). Specifically, a continuous spatial movement will lead to short temporal delays between the activity of similarly tuned head-direction cells, which would selectively potentiate the synapses between similarly tuned neurons—and this in turn would lead to continuity of the neural representation. Such an STDP-based model can explain mechanistically how a continuous toroidal representation (**Figure 3b**) emerges in an animal that maneuvers in a continuous angular manner (**Figure 3a**). Therefore, we predict that head-direction cells with similar preferred directions (in azimuth × pitch) would become more strongly interconnected during ontogeny—akin to the experience-dependent increase in connectivity between co-tuned neurons in visual cortex (Ko et al. 2013). Conversely, if the animal has not experienced a specific orientation during the development of the network, the appropriate neural representation might not form. Indeed, in rats (which do not typically locomote upside-down), the majority of azimuthal head-direction cells shut down when the rat assumed an inverted orientation (Calton & Taube 2005), suggesting a hemitorus model that does not represent the inverted pose (Stackman et al. 2000). Interestingly, some cells did not abolish their firing in inverted rats but became moderately tuned to the opposite direction as compared to the upright orientation (a 180° azimuth shift; Taube et al. 2004), consistent with a full toroidal topology. This suggests that perhaps with enough experience, a toroidal representation would be formed in rats.

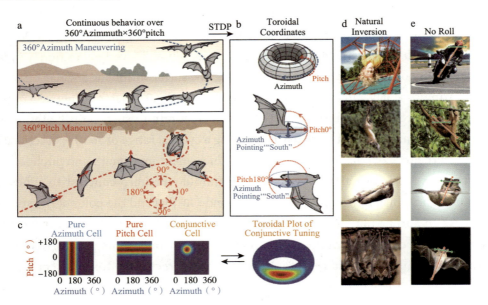

Figure 3

Topology of 3-D natural movements translates to topology of 3-D neural representations. (*a*) During 3-D volumetric navigation, bats maneuver across all possible combinations of 360° azimuth (top) and 360° pitch (*bottom*), creating a continuous behavioral coverage of 360° azimuth×360° pitch. (*b*) The neural representation of head direction (*top*) in toroidal coordinates covers 360° azimuth × 360° pitch cyclically and can result from spike-timing-dependent plasticity (STDP) mechanisms combined with the continuous behavioral coverage of 360° azimuth × 360° pitch. Toroidal azimuth (*middle*) is defined as the azimuthal direction of the interaural axis (blue arrow, pointing "south"), whereas pitch is defined as the angle of the naso-occipital axis above or below the horizontal plane (red arrow). In toroidal coordinates, rotations in pitch (*bottom*) do not change the azimuth (note the naso-occipital axis, in blue, is still pointing "south"). The decoupling of the two axes of azimuth and pitch allows a continuous representation of azimuth × pitch, without any discontinuities (singularities) at the poles. (*c*) A neuronal representation of head direction in toroidal coordinates spans 360° azimuth × 360° pitch and can be represented on unfolded plots of the torus (see three leftmost plots). Shown schematically are three head-direction cells: a pure azimuth cell, a pure pitch cell, and a conjunctive cell tuned to a particular 3-D combination of azimuth × pitch. The conjunctive azimuth ×pitch cell is also plotted on the toroidal manifold (right). (*d*) Many animals maneuver upside-down, exhibiting natural inversion. (*e*) Many animals, including humans, actively stabilize their head to avoid head-roll and maintain horizontal head (*green lines*) even during extreme body postures (*red lines*).

TOROIDAL COORDINATES FOR 3-D HEAD DIRECTION
A toroidal coordinate system utilizes two independent axes for azimuth and pitch, each spanning a range of 360°—in contrast to the spherical coordinate system, in which azimuth spans 360° but pitch spans only 180° (see Finkelstein et al. 2015). Toroidal azimuth is defined by the horizontal direction of the interaural axis (i.e., the axis going through the ears), and the toroidal pitch is defined as the vertical angle of the naso-occipital axis (i.e., going from tail to nose). Importantly, pitching of the naso-occipital axis does not change the azimuthal angle of the interaural axis, and vice versa—making these axes independent. Note that the toroidal coordinates use only two of the Euler angles—yaw (azimuth) and pitch—while ignoring the third angle, roll; this is justified in the context of navigation, because for a navigating animal, azimuth and pitch define the vector of heading-direction in 3-D space, with respect to external room coordinates, whereas roll is merely a rotation around this vector and hence is less important for navigation.

Although behavioral experience can shape neural representations, the converse may also be true: The neural representation can shape and constrain behavior. For example, if the animal developed a toroidal representation during ontogeny, it will likely be able to maneuver behaviorally over 360° of pitch and navigate while inverted. However, if an animal did not experience a certain behavioral state, this may prevent the formation of the appropriate neural representation, which might later restrict the animal from navigating under similar conditions. This was observed, for example, in rats raised in laboratory cages; these animals were severely impaired during inverted navigation (Valerio et al. 2010). In contrast, many species—including monkeys and mice (which climb upside-down naturally)—can locomote in the inverted pose while maneuvering extensively in pitch (**Figure 3d**). Because such azimuth × pitch movements are continuous in space, we predict that mice and primates might also develop a toroidal representation of head azimuth × pitch, which would support 3-D navigation. Taken together, toroidal encoding of the 3-D compass might exist in the brain of multiple species; its existence in bats demonstrates empirically how topology of movement could translate into topology of the neural representation.

Commutativity of the 3-D Neural Compass

Coding of 3-D angles poses a potentially difficult problem. A 3-D orientation of a rigid body in space can be described by the three Euler angles, corresponding to azimuth (yaw), pitch, and roll rotation angles. This 3-D rotations group is mathematically noncommutative, meaning that changing the order of rotations (e.g., whether you first rotate in azimuth, then pitch, then roll, or vice versa) will lead to a different final orientation. Such strong dependence on movement history would make any directional-trajectory computations very difficult. One possible solution to this noncommutativity problem is to avoid rotations in roll and to use the toroidal representation, which does not incorporate roll, because toroidal azimuth × pitch coordinates are commutative (Finkelstein et al. 2015). Consistent with this idea, we found that only a small fraction of neurons in the bat presubiculum were modulated by roll. Furthermore, behaviorally, many terrestrial and flying animals (across all phyla, including insects, birds, and many mammals) stabilize their head actively to avoid roll (**Figure 3e**) (Dunbar et al. 2008, Iriarte-Díaz & Swartz 2008, Kress & Egelhaaf 2012, Pozzo et al. 1990, Viollet & Zeil 2013). Thus, we propose that the need for a commutative neural representation of head direction explains why most animals typically exhibit very small roll angles (**Figure 3e**). This proposal complements more classical explanations of roll avoidance, such as sensory stabilization (Laurens et al. 2013). Taken together, the need for commutativity might explain why most animals avoid roll and why azimuth × pitch directions are represented in the brain in toroidal coordinates.

Isotropy and Scale of 3-D Spatial Representations

In the section above, we suggested that movement statistics shape the topology of the neural representation. Another aspect of movement statistics is that animal navigational behavior may be nonisotropic (not identical in all directions), which could potentially create a 3-D spatial representation with different resolution in different directions. In addition to nonisotropic behaviors, sensory inputs are also nonisotropic—as exemplified by vertical gravity information in vestibular processing and by the limited depth accuracy provided by stereoscopic vision along the line of sight. This problem might be alleviated during navigation, when additional depth cues are present (e.g., motion parallax). This highlights the importance of studying the possible factors governing the resolution and isotropy of allocentric spatial codes.

Many animals process the horizontal and vertical dimensions differently during 3-D navigation (Davis et al. 2014, Grobéty & Schenk 1992, Jovalekic et al. 2011, Wilson et al. 2015). For instance, rats were shown to explore a 3-D lattice maze using a layer-like strategy, performing horizontal movements first and then vertical movements—creating a preference for horizontal routes (Jovalekic et al. 2011); however, another study in a similar task showed that rats exhibited a higher spatial accuracy in the vertical dimension (Grobéty & Schenk 1992). Behavioral studies in humans showed that in pointing experiments on imagined 3-D directions in familiar multilayered environments, the mental representation is distorted in both the horizontal and the vertical plane, resulting in imagined buildings being taller and narrower (Brandt et al.

2015). Importantly, the way we explore and learn 3-D routes affects our performance: Horizontal exploration results in better horizontal accuracy, whereas vertical exploration yields better vertical accuracy (Thibault et al. 2012), highlighting the fact that movement statistics affect the nature of spatial representations.

So is the neural map isotropic? When rats were trained to move on a vertical wall with protruding pegs, hippocampal place fields were found to be elongated in the vertical dimension, suggesting that under these conditions, place fields are tuned nonisotropically and provide less information about the animal's altitude (Hayman et al. 2011). However, in this experimental setup, the rats explored the vertical wall using mostly horizontal movements and jumped vertically between rows of pegs only occasionally (Hayman et al. 2011)—a nonisotropic sampling of the horizontal and vertical dimensions. In contrast, during navigation on tilted surfaces, when the movement was likely more isotropic for both the vertical and horizontal components, place-field tuning appeared to be isotropic (circular) in both rats and bats (Jeffery et al. 2005, Knierim & McNaughton 2001, Ulanovsky & Moss 2007). Furthermore, 3-D place fields in bats were isotropic (spherical) during 3-D volumetric navigation (**Figure 1*f***), exhibiting similar tuning width in all dimensions (**Figure 4*a***) (Yartsev & Ulanovsky 2013). Likewise, 3-D head-direction representation in bats is also isotropic, with similar tuning width for azimuth cells and pitch cells (**Figure 4*b***) (Finkelstein et al. 2015). Notably, the observation that under certain conditions the spatial resolution of place cells can be nonisotropic (Hayman et al. 2011, Jeffery et al. 2013) does not necessarily mean it is nonmetric. This is analogous to the global positioning system (GPS) that provides a metric 3-D position in all dimensions (x, y, z coordinates in GPS are all given literally in meters), although this metric information is more accurate in longitude and latitude (x and y) than in altitude (z) (Kaplan & Hegarty 2005). Therefore, we propose that the accumulated evidence on 3-D representations across species suggests the following: First, nonisotropic movements lead to nonisotropic neural representations, whereas isotropic movements result in isotropic representations (Ulanovsky 2011). Second, the 3-D spatial map is likely represented metrically in all three dimensions.

Another key factor that can influence the tuning width of 3-D place cells is the physical scale (size) of the environment. This point was reviewed in detail by Geva-Sagiv et al. (2015), andhence we discuss it here only briefly. Experimental findings in 2-D showed that, first, expanding the environment along one of the dimensions results in elongation of the place field along that dimension (O'Keefe & Burgess 1996). Second, larger place fields are observed in larger environments (Kjelstrup et al. 2008, O'Keefe & Burgess 1996). Such place-field expansion is also predicted from theoretical considerations of optimal tuning width, which suggest that the tuning width of neurons may depend both on the dimensionality of space (one, two, or three dimensions) and on its spatial scale (size) (Brown & Bäcker 2006, Zhang & Sejnowski 1999). For finite 3-D space, Fisher information calculations predicted an optimal tuning width of approximately 22.5% of the environment size in each dimension (**Figure 4c**, *red curve*)—which, interestingly, was close to the relative size of place fields that we found for bat 3-D place cells during volumetric navigation (Yartsev & Ulanovsky 2013). This theoretical consideration predicts huge place fields for kilometer-sized environments; however, we note that other theoretical studies suggested instead that larger environments should lead to multiple small fields, not to larger fields (Hedrick & Zhang 2013). Future experiments should test how 3-D place fields look under kilometer-scale natural navigation (Geva-Sagiv et al. 2015).

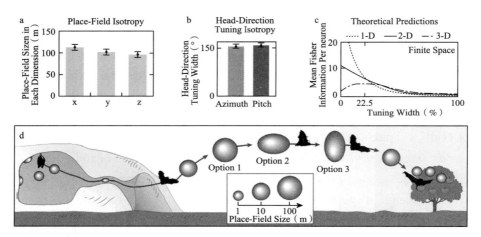

Figure 4

Isotropy and scale of 3-D neural representations. (*a*) 3-D place cells in bats during 3-D volumetric flight (measured as in **Figure 1*f***) exhibit spherical place fields with a very similar size in all three dimensions, suggesting that 3-D place fields are isotropic. (*b*) Head-direction cells in bats have similar tuning widths in azimuth and pitch, suggesting that 3-D head-direction tuning is isotropic.(*c*) Theoretical predictions for the positional information encoded by a population of neurons tuned to 1-D, 2-D, or 3-D for finite-sized space (Brown & Bäcker 2006). Note that for 3-D navigation, there is an optimal tuning width at approximately 22.5% of the total range, i.e., place-field size is predicted to be 22.5% of the room size. (d) There are three different possibilities for place-field representation during large-scale (kilometer-sized) navigation of an Egyptian fruit bat. In option 1, place fields would be isotropic. In option 2, place fields might be compressed vertically because the volume through which the bat typically navigates outdoors (the behavioral space) is compressed vertically, spanning 15–30 km horizontally but only about 700 m vertically (Tsoar et al. 2011). In option 3, place fields will be elongated vertically, reflecting the behavioral demands for long-range navigation. For all the options, we predict that place fields will be much larger during large-scale navigation. Panel *a* based on data from Yartsev & Ulanovsky (2013). Panel *b* based on data from Finkelstein et al. (2015). Panel *c* adapted with permission from Brown & Bäcker (2006).

Finally, the size and shape of 3-D place fields could be determined by complex combinations of the size and shape of the environment, as well as the degree of sensory isotropy and movement isotropy. For example, one possibility is that in open-field volumetric navigation, 3-D place cells would be isotropically tuned to all three dimensions but would have larger place fields compared to smaller environments (**Figure 4d**, option 1). However, even for infinite 3-D space, the behavioral space is anisotropic: For example, Egyptian fruit bats commute distances of 15–30 km horizontally, but their flight altitudes reach only up to about 700 m (Tsoar et al. 2011)—an approximately 30:1 ratio between the scales of horizontal and vertical movements, which can be viewed as a 2.5-D behavioral space. According to the aforementioned theoretical studies of optimal tuning (**Figure 4c**), in such a 2.5-D scenario, the place fields should shrink along the vertical dimension, maintaining a proportionally scaled optimal tuning width, yielding vertically compressed place fields (**Figure 4d**, option 2). Note that we assumed here that in outdoor open spaces, the relevant scale is the scale of the behavioral space of the animal and not the absolute environmental scale, which is practically infinite. A third possibility is that during volumetric navigation toward a distant target, the bat might not need to have a precise estimation of its altitude aboveground (as long as it is not hitting it), whereas estimating horizontal distance is more important. This might result hypothetically in elongation of place fields along the vertical dimension (**Figure 4d**, option 3), reflecting spatial uncertainty along the nonsalient dimension. Experiments in kilometer-sized environments are needed to test these contrasting possibilities.

FUNCTIONAL AND ANATOMICAL INTERACTIONS BETWEEN 3-D MAPS AND COMPASSES

Functional Anatomy of 3-D Spatial Circuits in the Hippocampal Formation

How are the map and compass systems in the brain wired anatomically to each other, and how does their wiring shape their function? We found a functional-anatomical gradient of head-direction cells along the transverse axis of the bat dorsal presubiculum, with 2-D azimuth cells located proximally in the presubiculum (close to the hippocampus) and 3-D azimuth × pitch cells located distally (close to the entorhinal cortex). This 2-D–to–3-D functional gradient might arise from a combination of azimuthal directional inputs from subcortical nuclei such as the anterodorsal thalamic nucleus (Taube 2007)—which presumably projects to all parts of the dorsal presubiculum (**Figure 5a**, *blue arrows*)—and a putative pitch input that might project predominantly to the distal part of the dorsal presubiculum (**Figure 5a**, *red arrows*). A convergence of azimuth and pitch inputs in the distal presubiculum may thus give rise to 3-D head-direction cells, whereas the proximal parts may receive only azimuthal input and therefore process mostly 2-D information.

This 2-D–to–3-D gradient of head-direction cells may affect other spatial cell types—in particular, grid cells in the medial entorhinal cortex (MEC), which were suggested to rely on head-direction inputs for computing the grid metric along each dimension (Burak & Fiete 2009). Interestingly, the transverse axis of the presubiculum, where we found the 2-D–to–3-D gradient of head-direction cells, projects in an inverted manner to the transverse axis of the MEC (**Figure 5a**) (Honda & Ishizuka 2004). Thus, distal MEC receives inputs from the proximal part of the dorsal presubiculum—which contains predominantly azimuthal head-directions cells—but does not receive pitch (vertical) head-direction inputs. If 3-D head-direction information is indeed crucial for grid formation in 3-D, then distal MEC grid cells would be insensitive to the vertical dimension and would discharge in 3-D space along vertically elongated hexagonal columns. In contrast, proximal MEC, which receives inputs predominantly from the distal presubiculum (that contains 3-D head-direction cells), is expected to contain 3-D grid cells tuned to all three axes, including altitude—namely, these 3-D grid cells should exhibit spherical firing fields, which could be arranged on a perfect 3-D lattice or might have a nonperfect 3-D arrangement.

Finally, we note that the entorhinal-hippocampal anatomical connectivity is very similar between bats and rats (Kleven et al. 2014), suggesting that the 2-D-to–3-D gradient of head-direction cells that we found in the presubiculum (and the corresponding predictions for grid cells in MEC) might be general across mammals. Because a 3-D head-direction tuning has never been studied in rodent presubiculum, this further highlights the need to record in the dorsal presubiculum of rats and mice during 3-D behaviors.

The Hippocampal Formation: A Neural Realization of a Kalman Filter?

A common problem encountered in human-made 3-D navigation systems, such as airplanes, spaceships, and submarines, is that directional and positional measurements by compasses and GPS can be noisy. In addition, similar information can also be computed from self-motion by an inertial navigation system (INS), which uses motion sensors and a set of navigational equations to compute the expected position, orientation, and velocity of a moving

Kalman filter:
an optimal recursive estimation algorithm; in human-made navigation systems, it combines noisy sensor

object via dead reckoning (path integration). Yet such estimation by an INS is very noisy and accumulates substantial errors over time. There is, however, a well-established engineering solution to this noisy-navigation problem: A recursive algorithm, known as the Kalman filter, can improve the navigational accuracy by comparing the GPS and compass measurements with those predicted by the INS and using both systems to generate a refined estimate of position and direction (Mohamed & Schwarz 1999).

In neuroscience, the Kalman filter was used to date mostly to model sensorimotor integration, such as tracking the hand position in space (Wolpert et al. 1995). In that case, researchers suggested that the brain generates an estimate of where the hand should be, based on an internal model of the arm kinematics (similar to INS), and compares it with the sensory-based estimation of the hand position (similar to GPS and compass). The error between the estimate of the internal model and the estimate based on sensory input is used by the Kalman filter to generate a more precise approximation of the hand position.

Here we hypothesize that, similar to the internal model that is needed to control 3-D arm movement (Wolpert et al. 1995), one would also need an internal model for controlling 3-D whole-body movement for navigation. Therefore, an intriguing possibility is that similar recursive interactions might exist in the hippocampal formation between the 3-D map and compass systems and the path-integration system (**Figure 5*b***). These systems could mutually correct each other, thus providing refined directional and positional estimates by implementing a neural analogue of a Kalman filter. According to our hypothesis, a sensory-based estimate of position and direction is computed in the hippocampus and the presubiculum by place cells and head-direction cells, respectively, which act as the neural analogue of a GPS and a compass. Additionally, we posit that the MEC forms an internal model of the animal's position and orientation, based on path integration (analogously to an INS).

The process of path integration (McNaughton et al. 2006) uses self-motion information to generate predictions of the animal's position and orientation—similarly to the function of internal models (forward models) in the motor system (Wolpert & Ghahramani 2000, Wolpert et al. 1995). The proposed internal model in the entorhinal cortex receives vestibular inputs (Jacob et al. 2014) that may update the model's prediction via navigational equations that are implemented by the entorhinal network, which includes grid cells (Hafting et al. 2005) and speed cells (Kropff et al. 2015), all of which were suggested to be important for path integration. The proposed involvement of grid cells and speed cells in path integration is one of the key reasons for our suggestion that the internal model for navigation resides in the MEC. We note that another component potentially important for path integration—angular velocity cells (Taube 2007)—was reported so far only in subcortical regions, and therefore it would be interesting to search for an angular-velocity signal in the MEC. Finally, we suggest that the sensory-based navigational measurements in the hippocampus and the presubiculum are compared with the prediction of the path-integration system in the MEC, and the resulting error is used by a neural analogue of a Kalman filter to weigh the relative contribution of each of the components and to refine the final positional and directional estimates, which are then sent back to the hippocampus and presubiculum (**Figure 5*b***).

The proposed putative Kalman filtering (or a similar recursive errorcorrection algorithm that compares sensory measurements with an estimate of an internal model) could be implemented by a network with attractor dynamics (Denève et al. 2007) in the MEC (Yoon et al. 2013) and might depend specifically on directionally tuned grid cells (Sargolini et al. 2006). These neurons are tuned to both position and head direction, and they thus contain the information required to compute a refined directional and positional estimate. Notably, such a Kalman-based error–correction mechanism might be relevant not only for 3-D but also for 2-D navigation (Bousquet et al. 1998). Moreover, this proposed notion could reconcile the arguments on whether spatially tuned neurons in the hippocampal formation reflect sensory information or path integration: We propose that this circuit combines both sensory information and path integration to guide navigation and that this is done via a recursive error–correction mechanism such as a Kalman filter.

What is the experimental evidence in favor of this hypothesis? First, both place cells and head-direction cells rely on a combination of external sensory processing and path integration (McNaughton et al. 2006). In the case of place cells, recordings in echolocating bats moving in darkness showed that place fields sharpen after each echolocation call (Ulanovsky & Moss 2011), as would be expected from a path-integration system that accumulates errors between the sensory inputs (between calls) and is then refined by each incoming sensory input. Similarly, head-direction representation accumulates errors in the absence of visual cues but can be corrected by positional cues such as the borders of the environment (Valerio & Taube 2012), indicating that head-direction computation also involves combined processing of external sensory inputs and path integration. Second, although the sensory-based positional and directional systems can function without each other, we predict that their estimate is refined via a Kalman filter (**Figure 5*b***), and hence the accuracy of one type of system (i.e., place cells or head-direction cells) is expected to degrade if the other is impaired. Indeed, in rats with lesions of the dorsal presubiculum, place cells remained active, but their tuning

a Functional Connections in the 3-D Navigation System

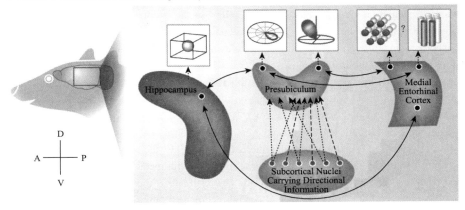

b The hippocampal formation as a Kalman Filter?

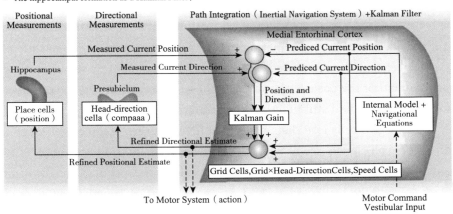

Figure 5

Functional-anatomical interactions between 3-D maps and 3-D compasses. (*a*) The mammalian navigation circuit and its different functional cell types. Known excitatory projections are shown as black arrows. This illustration shows the functional gradient of head-direction cells along the transverse axis of the bat's dorsal presubiculum from 2-D to 3-D (see Finkelstein et al. 2015); this gradient can be explained by putative differences between subcortical azimuthal inputs that may project to the entire dorsal presubiculum (*blue arrows*) versus hypothesized pitch inputs from subcortical nuclei that may project only to the distal part of dorsal presubiculum (*red arrows*). The 2-D-to-3-D gradient in head-direction tuning might in turn give rise to a 2-D-to-3-D gradient of grid cells along the transverse axis of the medial entorhinal cortex. The diagram below the animal's head shows the directions of the dorsal-ventral (D-V) axis and anterior-posterior (A-P) axis. (*b*) The hippocampal-presubicular-entorhinal loop could implement a neural analogue of a Kalman filter—a standard 3-D navigation system used in airplanes, spaceships, and submarines—that is comprised of positional and directional sensors (e.g., GPS and compass); an inertial navigation system (INS) containing gyroscopes, accelerometers, and speedometers, which allow path integration; and a set of navigational equations that allow combining the positional and INS information to perform error-correction for computing a more accurate navigational estimate. We propose that, first, hippocampal place cells and presubicular head-direction cells, respectively, implement the positional system and compass system. Second, we posit that the entorhinal cortex contains an internal model that performs path integration via grid cells and speed cells—so the internal model functions similarly to an INS. Third, the error-correction is done by a neural analogue of a Kalman filter in the medial entorhinal cortex, where conjunctive grid × head-direction cells might carry the refined navigational information. According to this view, the internal model is initialized by the previous location and orientation of the animal and is updated by self-motion cues (vestibular inputs as well as motor commands) to generate a prediction of the current position and orientation of the animal via path integration. This estimate is then compared in the medial entorhinal cortex to the measured current position and direction signals arising from the hippocampus and the presubiculum. The resulting error is used by the Kalman gain to weigh the sensory-based estimate versus the internal estimate (i.e., the signal from the hippocampus and presubiculum versus the prediction of the internal model in the medial entorhinal cortex) to generate a refined positional and directional estimate. This new estimate is then sent back to the hippocampus and presubiculum to refine the place-cell and head-direction–cell tuning; it also leaves the hippocampal formation toward the motor system to guide action..

became broader (Calton et al. 2003). Similarly, without the hippocampus, head-direction cells still provided direction information but became less stable (Golob & Taube 1999). Third, we expect that the components of a Kalman filter should be strongly affected in the absence of either the positional or directional systems; conversely, the positional and directional systems would still function (albeit more noisily) without the Kalman filter. Indeed, whereas grid cells lose their hexagonal firing pattern when either place-cell or head-direction–cell signals are disrupted (Bonnevie et al. 2013, Winter et al. 2015), the place and head-direction cells are affected only mildly by the loss of grid cells (Brun et al. 2008, Clark & Taube 2011, Hales et al. 2014, Koenig et al. 2011). Fourth, the Kalman model makes a specific prediction that if the positional input from the hippocampus is turned off, then the directional tuning in the entorhinal cortex should become sharper (because the Kalman filter gives higher weight to the more reliable input-channel)—which is exactly what was observed in experiments in which the hippocampus was inactivated, and medial entorhinal neurons were then shown to sharpen their head-direction tuning (Bonnevie et al. 2013). Fifth, if grid cells are implementing a Kalman filter, they should develop during ontogeny at a later stage than the positional and directional signals (i.e., place cells and head-direction cells). This is indeed the case (Langston et al. 2010, Wills et al. 2010). Sixth, the anatomy of the hippocampal-presubicular-entorhinal loop seems to suit particularly well a Kalman-filter implementation: compare **Figure 5*a*** and *b*. Taken together, it seems possible that the navigational function of the hippocampal formation is consistent with a biological realization of a Kalman filter—an algorithm that is used heavily for 3-D navigation in human-made vehicles. It remains to be seen in future experiments whether biology and technology indeed evolved similar solutions to the problem of 3-D navigation.

FUTURE OUTLOOK

Natural 3-D navigation behaviors are complex and consist of different navigation modes—planar, multilayered, and volumetric navigation—with most animals being able to switch flexibly between these different modes. In this review, we proposed that the properties of neural maps and compasses in the hippocampal formation depend strongly on behavior, being influenced both by pup behavior during ontogeny and by the adult behavioral repertoire. For example, the dimensionality of movement through space may determine the dimensionality of the map, and different map fragments will be formed consequently for 2-D and 3-D navigation. Similarly, the statistics and topology of movement through space define the topology of the neural representation, as exemplified by the toroidal nature of the 3-D neural compass in bats. An important future challenge will therefore be to test systematically in multiple species how the properties of 3-D behavior affect the nature of 3-D maps and compasses.

We next suggested that hippocampal place cells form a mosaic of 2-D and 3-D spatial maps, and we proposed that a global 3-D compass may stitch these map fragments into a global 3-D spatial representation that allows novel 3-D shortcuts. We argued that current data suggest that 3-D place cells and 3-D head-direction cells embody a metric representation of 3-D volumetric space (as in GPS). An important future test for this idea would be to record from 3-D grid cells: Do they support the notion of a metric representation of 3-D space?

We also proposed that the hippocampal-presubicular-entorhinal loop may implement a Kalman filter recursive-prediction algorithm for 2-D and 3-D navigation. Wolpert et al. (1995) proposed similar ideas for the interaction between the body and the world in the motor system. We speculate here that, more broadly, a similar process of recursive prediction might form the basis for all two-way interactions between an organism and its environment.

Finally, a major challenge would be to elucidate the properties of 2-D and 3-D maps and compasses during natural 3-D navigation. What are the neural codes underlying real-life navigation in complex 3-D environments and over large naturalistic scales (kilometers)? Answering this question will be one of the key future challenges for the field of behavioral neuroscience.

DISCLOSURE STATEMENT

The authors are not aware of any affiliations, memberships, funding, or financial

holdings that might be perceived as affecting the objectivity of this review.

ACKNOWLEDGMENTS

We thank Tamir Eliav, Maya Geva-Sagiv, Slava Kerner, Emily Mackevicius, Ron Meir, David Omer, Ayelet Sarel, Alessandro Treves, and Menno Witter for helpful discussions and comments on the manuscript and Genia Brodsky for assistance with graphics. The research of N.U. was supported by grants from the European Research Council (ERC-NEUROBAT), Human Frontiers Science Program (HFSP RGP0062/2009-C), Israel Science Foundation (ISF 1017/08 and ISF 1319/13), Minerva Foundation, Abe and Kathryn Selsky Foundation, Harold I. and Faye B. Liss Foundation, Mr. and Mrs. Steven Harowitz, and the Lulu P. and David J. Levidow Fund for Alzheimer's Disease and Neuroscience Research. A.F. was supported by a Clore predoctoral excellence fellowship.

LITERATURE CITED

Alme CB, Miao C, Jezek K, Treves A, Moser EI, Moser M-B. 2014. Place cells in the hippocampus: eleven maps for eleven rooms. *PNAS* 111:18428–35

Bjerknes TL, Langston RF, Kruge IU, Moser EI, Moser M-B. 2015. Coherence among head direction cells before eye opening in rat pups. *Curr. Biol.* 25:103–8

Blohm G, Keith GP, Crawford JD. 2009. Decoding the cortical transformations for visually guided reaching in 3D space. *Cereb. Cortex* 19:1372–93

Bonnevie T, Dunn B, Fyhn M, Hafting T, Derdikman D, et al. 2013. Grid cells require excitatory drive from the hippocampus. *Nat. Neurosci.* 16:309–17

Bousquet O, Balakrishnan K, Honavar V. 1998. Is the hippocampus a Kalman filter? *Proc. Pac. Symp. Biocomput.* 1998:657–68

Brandt T, Huber M, Schramm H, Kugler G, Dieterich M, Glasauer S. 2015. "Taller and shorter": Human 3-D spatial memory distorts familiar multilevel buildings. *PLOS ONE* 10:e0141257

Breveglieri R, Hadjidimitrakis K, Bosco A, Sabatini SP, Galletti C, Fattori P. 2012. Eye position encoding in three-dimensional space: integration of version and vergence signals in the medial posterior parietal cortex. *J. Neurosci.* 32:159–69

Brown W, Bäcker A. 2006. Optimal neuronal tuning for finite stimulus spaces. *Neural Comput.* 18:1511–26

Brun VH, Leutgeb S, Wu H-Q, Schwarcz R, Witter MP, et al. 2008. Impaired spatial representation in CA1 after lesion of direct input from entorhinal cortex. *Neuron* 57:290–302

Bueti D, Walsh V. 2009. The parietal cortex and the representation of time, space, number and other magnitudes. *Philos. Trans. R. Soc. B* 364:1831–40

Buneo CA, Jarvis MR, Batista AP, Andersen RA. 2002. Direct visuomotor transformations for reaching. *Nature* 416:632–36

Burak Y, Fiete IR. 2009. Accurate path integration in continuous attractor network models of grid cells. *PLOS Comput. Biol.* 5:e1000291

Calton JL, Stackman RW, Goodridge JP, Archey WB, Dudchenko PA, Taube JS. 2003. Hippocampal place cell instability after lesions of the head direction cell network. *J. Neurosci.* 23:9719–31

Calton JL, Taube JS. 2005. Degradation of head direction cell activity during inverted locomotion. *J. Neurosci.* 25:2420–28

Carpenter F, Manson D, Jeffery K, Burgess N, Barry C. 2015. Grid cells form a global representation of connected environments. *Curr. Biol.* 25:1176–82

Channon AJ, Crompton RH, Günther MM, D'Août K, Vereecke EE. 2010. The biomechanics of leaping in gibbons. *Am. J. Phys. Anthropol.* 143:403–16

Chen X, DeAngelis GC, Angelaki DE. 2013. Diverse spatial reference frames of vestibular signals in parietal cortex. *Neuron* 80:1310–21

Chen Z, Kloosterman F, Brown EN, Wilson MA. 2012. Uncovering spatial topology represented by rat hippocampal population neuronal codes. *J. Comput. Neurosci.* 33:227–55

Childs SB, Buchler ER. 1981. Perception of simulated stars by *Eptesicus fuscus* (Vespertilionidae): a potential navigational mechanism. *Anim. Behav.* 29:1028–35

Clark BJ, Taube JS. 2011. Intact landmark control and angular path integration by head direction cells in the anterodorsal thalamus after lesions of the medial entorhinal cortex. *Hippocampus* 21:767–82

Cohen RG, Rosenbaum DA. 2004. Where grasps are made reveals how grasps are planned: generation and recall of motor plans. Exp. *Brain Res.* 157:486–95

Cowen SL, Nitz DA. 2014. Repeating firing fields of CA1 neurons shift forward in response to increasing angular velocity. *J. Neurosci.* 34:232–41

Curto C, Itskov V. 2008. Cell groups reveal structure of stimulus space. *PLOS Comput. Biol.* 4:e1000205

Dabaghian Y, Brandt VL, Frank LM. 2014. Reconceiving the hippocampal map as a topological template. *eLife* 3:e03476

Dan Y, Poo M-M. 2004. Spike timing-dependent plasticity of neural circuits. *Neuron* 44:23–30

Davis VA, Holbrook RI, Schumacher S, Guilford T, de Perera TB. 2014. Three-dimensional spatial cognition in a benthic fish, *Corydoras aeneus. Behav. Process.* 109:151–56

Denève S, Duhamel J-R, Pouget A. 2007. Optimal sensorimotor integration in recurrent cortical networks: a neural implementation of Kalman filters. *J. Neurosci.* 27:5744–56

Derdikman D, Whitlock JR, Tsao A, Fyhn M, Hafting T, et al. 2009. Fragmentation of grid cell maps in a multicompartment environment. *Nat. Neurosci.* 12:1325–32

Domnisoru C, Kinkhabwala AA, Tank DW. 2013. Membrane potential dynamics of grid cells. *Nature* 495:199–204

Dunbar DC, Macpherson JM, Simmons RW, Zarcades A. 2008. Stabilization and mobility of the head, neck and trunk in horses during overground locomotion: comparisons with humans and other primates. *J. Exp. Biol.* 211:3889–907

Finkelstein A, Derdikman D, Rubin A, Foerster JN, Las L, Ulanovsky N. 2015. Three-dimensional head-direction coding in the bat brain. *Nature* 517:159–64

Foley JM. 1980. Binocular distance perception. *Psychol. Rev.* 87:411–34

Geva-Sagiv M, Las L, Yovel Y, Ulanovsky N. 2015. Spatial cognition in bats and rats: from sensory acquisition to multiscale maps and navigation. *Nat. Rev. Neurosci.* 16:94–108

Golob EJ, Taube JS. 1999. Head direction cells in rats with hippocampal or overlying neocortical lesions: evidence for impaired angular path integration. *J. Neurosci.* 19:7198–211

Grobéty M-C, Schenk F. 1992. Spatial learning in a three-dimensional maze. *Anim. Behav.* 43:1011–20

Hafting T, Fyhn M, Molden S, Moser M-B, Moser EI. 2005. Microstructure of a spatial map in the entorhinalcortex. *Nature* 436:801–6

Hales JB, Schlesiger MI, Leutgeb JK, Squire LR, Leutgeb S, Clark RE. 2014. Medial entorhinal cortex lesions only partially disrupt hippocampal place cells and hippocampus-dependent place memory. *Cell Rep.* 9:893–901

Hayman RMA, Casali G, Wilson JJ, Jeffery KJ. 2015. Grid cells on steeply sloping terrain: evidence for planar rather than volumetric encoding. *Front. Psychol.* 6:925

Hayman RMA, Verriotis MA, Jovalekic A, Fenton AA, Jeffery KJ. 2011. Anisotropic encoding of three-dimensional space by place cells and grid cells. *Nat. Neurosci.* 14:1182–88

Hedrick K, Zhang K. 2013. *Megamap: Continuous attractor model for place cells representing large environments.* Program No. 578.01. Presented at Soc. Neurosci. Annu. Meet., Nov. 12, San Diego

Honda Y, Ishizuka N. 2004. Organization of connectivity of the rat presubiculum: I. Efferent projections to the medial entorhinal cortex. *J. Comp. Neurol.* 473:463–84

Horiuchi TK, Moss CF. 2015. Grid cells in 3-D: reconciling data and models. *Hippocampus* 25:1489–1500

Iriarte-Díaz J, Swartz SM. 2008. Kinematics of slow turn maneuvering in the fruit bat *Cynopterus brachyotis.J. Exp. Biol.* 211:3478–89

Jacob P-Y, Poucet B, Liberge M, Save E, Sargolini F. 2014. Vestibular control of entorhinal cortex activity in spatial navigation. *Front. Integr. Neurosci.* 8:38

Jeffery KJ, Anand RL, Anderson MI. 2005. A role for terrain slope in orienting hippocampal place fields. *Exp.Brain Res.* 169:218–25

Jeffery KJ, Jovalekic A, Verriotis M, Hayman R. 2013. Navigating in a three-dimensional world. *Behav. Brain Sci.* 36:523–87

Jeffery KJ, Wilson JJ, Casali G, Hayman RM. 2015. Neural encoding of large-scale three-dimensional space–properties and constraints. *Front. Psychol.* 6:927

Jovalekic A, Hayman R, Becares N, Reid H, Thomas G, et al. 2011. Horizontal biases in rats' use of three-dimensional space. *Behav. Brain Res.* 222:279–88

Kaplan ED, Hegarty C. 2005. *Understanding GPS: Principles and Applications.* Boston: Artech House. 2nd ed.

Kjelstrup KB, Solstad T, Brun VH, Hafting T, Leutgeb S, et al. 2008. Finite scale of spatial representation in the hippocampus. *Science* 321:140–43

Klatzky RL, Giudice NA. 2013. The planar mosaic fails to account for spatially directed action. *Behav. Brain Sci.* 36:554–55

Kleven H, Gatome W, Las L, Ulanovsky N, Witter MP. 2014. *Organization of entorhinal-hippocampal projections in the Egyptian fruit bat.* FENS-3642, Poster No. F045. Presented at FENS Forum Neurosci., July 6, Milan

Knierim JJ, Hamilton DA. 2011. Framing spatial cognition: neural representations of proximal and distalrames of reference and their roles in navigation. *Physiol. Rev.* 91:1245–79

Knierim JJ, McNaughton BL. 2001. Hippocampal place-cell firing during movement in three-dimensional space. *J. Neurophysiol.* 85:105–16

Knierim JJ, McNaughton BL, Poe GR. 2000. Three-dimensional spatial selectivity of hippocampal neurons during space flight. *Nat. Neurosci.* 3:209–10

Knierim JJ, Rao G. 2003. Distal landmarks and hippocampal place cells: effects of relative translation versus rotation. *Hippocampus* 13:604–17

Ko H, Cossell L, Baragli C, Antolik J, Clopath C, et al. 2013. The emergence of functional microcircuits in visual cortex. *Nature* 496:96–100

Koenig J, Linder AN, Leutgeb JK, Leutgeb S. 2011. The spatial periodicity of grid cells is not sustained during reduced theta oscillations. *Science* 332:592–95

Kress D, Egelhaaf M. 2012. Head and body stabilization in blowflies walking on differently structured substrates. *J. Exp. Biol.* 215:1523–32

Kropff E, Carmichael JE, Moser M-B, Moser EI. 2015. Speed cells in the medial entorhinal cortex. Nature523:419–424

Kruge IU, Wernle T, Moser EI, Moser M-B. 2013. *Grid cells of animals raised in spherical environments.* Program No. 769.14. Presented at Soc. Neurosci. Annu. Meet., Nov. 13, San Diego

Langston RF, Ainge JA, Couey JJ, Canto CB, Bjerknes TL, et al. 2010. Development of the spatial representation system in the rat. *Science* 328:1576–80

Laurens J, Meng H, Angelaki DE. 2013. Neural representation of orientation relative to gravity in the macaque cerebellum. *Neuron* 80:1508–18

Lever C, Burton S, Jeewajee A, O'Keefe J, Burgess N. 2009. Boundary vector cells in the subiculum of the hippocampal formation. *J. Neurosci.* 29:9771–77

Ludvig N, Tang HM, Gohil BC, Botero JM. 2004. Detecting location-specific neuronal firing rate increases in the hippocampus of freely-moving monkeys. *Brain Res.* 1014:97–109

Mathis A, Stemmler MB, Herz AVM. 2015. Probable nature of higher-dimensional symmetries underlying mammalian grid-cell activity patterns. *eLife* 4:e05979

McNaughton BL, Battaglia FP, Jensen O, Moser EI, Moser M-B. 2006. Path integration and the neural basis of the 'cognitive map.' *Nat. Rev. Neurosci.* 7:663–78

Mohamed AH, Schwarz KP. 1999. Adaptive Kalman filtering for INS/GPS. *J. Geodesy* 73:193–203

Montello DR, Pick HL. 1993. Integrating knowledge of vertically aligned large-scale spaces. *Environ. Behav.* 25:457–84

Moser EI, Kropff E, Moser M-B. 2008. Place cells, grid cells, and the brain's spatial representation system. *Annu. Rev. Neurosci.* 31:69–89

Nitz DA. 2012. Spaces within spaces: rat parietal cortex neurons register position across three reference frames. *Nat. Neurosci.* 15:1365–67

O'Keefe J, Burgess N. 1996. Geometric determinants of the place fields of hippocampal neurons. *Nature* 381:425–28

O'Keefe J, Dostrovsky J. 1971. The hippocampus as a spatial map: preliminary evidence from unit activity in the freely-moving rat. *Brain Res.* 34:171–75

O'Keefe J, Nadel L. 1978. *The Hippocampus as a Cognitive Map.* Oxford, UK: Clarendon

Poucet B. 1993. Spatial cognitive maps in animals: new hypotheses on their structure and neural mechanisms. *Psychol. Rev.* 100:163–82

Pozzo T, Berthoz A, Lefort L. 1990. Head stabilization during various locomotor tasks in humans. *Exp. Brain Res.* 82:97–106

Ranck JB Jr. 1984. Head-direction cells in the deep cell layers of dorsal presubiculum in freely moving rats. *Soc. Neurosci. Abstr.* 10:599

Rosenberg A, Cowan NJ, Angelaki DE. 2013. The visual representation of 3D object orientation in parietal cortex. *J. Neurosci.* 33:19352–61

Sargolini F, Fyhn M, Hafting T, McNaughton BL, Witter MP, et al. 2006. Conjunctive representation of position, direction, and velocity in entorhinal cortex. *Science* 312:758–62

Singer AC, Karlsson MP, Nathe AR, Carr MF, Frank LM. 2010. Experience-dependent development of coordinated hippocampal spatial activity representing the similarity of related locations. *J. Neurosci.* 30:11586–604

Solstad T, Boccara CN, Kropff E, Moser M-B, Moser EI. 2008. Representation of geometric borders in the entorhinal cortex. *Science* 322:1865–68

Spiers HJ, Hayman RMA, Jovalekic A, Marozzi E, Jeffery KJ. 2015. Place field repetition and purely local remapping in a multicompartment environment. *Cereb. Cortex* 25:10–25

Stackman RW, Taube JS. 1998. Firing properties of rat lateral mammillary single units: head direction, head pitch, and angular head velocity. *J. Neurosci.* 18:9020–37

Stackman RW, Tullman ML, Taube JS. 2000. Maintenance of rat head direction cell firing during locomotion in the vertical plane. *J. Neurophysiol.* 83:393–405

Stella F, Cerasti E, Treves A. 2013. Unveiling the metric structure of internal representations of space. *Front.Neural Circuits* 7:81

Stella F, Treves A. 2015. The self-organization of grid cells in 3D. *eLife* 4:e05913

Stensola T, Stensola H, Moser M-B, Moser EI. 2015. Shearing-induced asymmetry in entorhinal grid cells. *Nature* 518:207–12

Tan HM, Bassett JP, O'Keefe J, Cacucci F, Wills TJ. 2015. The development of the head direction system before eye opening in the rat. *Curr. Biol.* 25:479–83

Taube JS. 2007. The head direction signal: origins and sensory-motor integration. *Annu. Rev. Neurosci.* 30:181–207

Taube JS, Muller RU, Ranck JB Jr. 1990a. Head-direction cells recorded from the postsubiculum in freely moving rats. I. Description and quantitative analysis. *J. Neurosci.* 10:420–35

Taube JS, Muller RU, Ranck JB Jr. 1990b. Head-direction cells recorded from the postsubiculum in freelymoving rats. II. Effects of environmental manipulations. *J. Neurosci.* 10:436–47

Taube JS, Stackman RW, Calton JL, Oman CM. 2004. Rat head direction cell responses in zero-gravity parabolic flight. *J. Neurophysiol.* 92:2887–997

Taube JS, Wang SS, Kim SY, Frohardt RJ. 2013. Updating of the spatial reference frame of head direction cells in response to locomotion in the vertical plane. *J. Neurophysiol.* 109:873–88

Thibault G, Pasqualotto A, Vidal M, Droulez J, Berthoz A. 2012. How does horizontal and vertical navigation influence spatial memory of multifloored environments? *Atten. Percept. Psychophys.* 75:10–15

Tolman EC. 1948. Cognitive maps in rats and men. *Psychol. Rev.* 55:189–208

Tsoar A, Nathan R, Bartan Y, Vyssotski A, Dell'Omo G, Ulanovsky N. 2011. Large-scale navigational map in a mammal. *PNAS* 108:E718–24

Ulanovsky N. 2011. Neuroscience: How is three-dimensional space encoded in the brain? *Curr. Biol.* 21:R886–88

Ulanovsky N, Finkelstein A. 2013. *Hippocampal representation of multiple spatial scales during 2D versus 3D navigation in bats.* Program No. 863.02. Presented at Soc. Neurosci. Annu. Meet., Nov. 13, San Diego

Ulanovsky N, Moss CF. 2007. Hippocampal cellular and network activity in freely moving echolocating bats. *Nat. Neurosci.* 10:224–33

Ulanovsky N, Moss CF. 2011. Dynamics of hippocampal spatial representation in echolocating bats. *Hippocampus* 21:150–61

Urdapilleta E, Troiani F, Stella F, Treves A. 2015. Can rodents conceive hyperbolic spaces? *J. R. Soc. Interface* 12:20141214

Valerio S, Clark BJ, Chan JHM, Frost CP, Harris MJ, Taube JS. 2010. Directional learning, but no spatial mapping by rats performing a navigational task in an inverted orientation. *Neurobiol. Learn. Mem.* 93:495–505

Valerio S, Taube JS. 2012. Path integration: how the head direction signal maintains and corrects spatialorientation. *Nat. Neurosci.* 15:1445–53

Viollet S, Zeil J. 2013. Feed-forward and visual feedback control of head roll orientation in wasps (*Polistes humilis,* Vespidae, Hymenoptera). *J. Exp. Biol.* 216:1280–91

Wallace DJ, Greenberg DS, Sawinski J, Rulla S, Notaro G, Kerr JND. 2013. Rats maintain an overhead binocular field at the expense of constant fusion. *Nature* 498:65–69

Wallraff HG. 2005. *Avian Navigation: Pigeon Homing as a Paradigm.* Berlin/Heidelberg, Ger.: Springer-Verlag

Wang RF. 2012. Theories of spatial representations and reference frames: What can configuration errors tellus? *Psychon. Bull. Rev.* 19:575–87

Whitlock JR, Derdikman D. 2012. Head direction maps remain stable despite grid map fragmentation. *Front.Neural Circuits* 6:9

Whitlock JR, Pfuhl G, Dagslott N, Moser M-B, Moser EI. 2012. Functional split between parietal and entorhinal cortices in the rat. *Neuron* 73:789–802

Whitlock JR, Sutherland RJ, Witter MP, Moser M-B, Moser EI. 2008. Navigating from hippocampus toparietal cortex. *PNAS* 105:14755–62

Wills TJ, Cacucci F, Burgess N, O'Keefe J. 2010. Development of the hippocampal cognitive map in preweanling rats. *Science* 328:1573–76

Wilson JJ, Harding E, Fortier M, James B, Donnett M, et al. 2015. Spatial learning by mice in three dimensions. *Behav.Brain Res.* 289:125–32

Wilson PN, Foreman N, Stanton D, Duffy H. 2004. Memory for targets in a multilevel simulated environment: evidence for vertical asymmetry in spatial memory. *Mem. Cogn.* 32:283–97

Winter SS, Clark BJ, Taube JS. 2015. Disruption of the head direction cell network impairs the parahip-pocampal grid cell signal. *Science* 347:870–74

Wolpert DM, Ghahramani Z. 2000. Computational principles of movement neuroscience. *Nat. Neurosci.* 3:1212–17

Wolpert DM, Ghahramani Z, Jordan MI. 1995. An internal model for sensorimotor integration. *Science* 269:1880–82

Yartsev MM, Ulanovsky N. 2013. Representation of three-dimensional space in the hippocampus of flying bats. *Science* 340:367–72

Yoder RM, Clark BJ, Brown JE, Lamia MV, Valerio S, et al. 2011. Both visual and idiothetic cues contribute to head direction cell stability during navigation along complex routes. *J. Neurophysiol.* 105:2989–3001

Yoon K, Buice MA, Barry C, Hayman R, Burgess N, Fiete IR. 2013. Specific evidence of low-dimensional continuous attractor dynamics in grid cells. *Nat. Neurosci.* 16:1077–84

Yoon K, Lewallen S, Kinkhabwala AA, Tank DW, Fiete IR. 2016. Grid cell responses in 1D environments assessed as slices through a 2D lattice. *Neuron* 89:1086–89

Zhang K, Sejnowski TJ. 1999. Neuronal tuning: to sharpen or broaden? *Neural Comput.* 11:75–84

Zugaro MB, Berthoz A, Wiener SI. 2001. Background, but not foreground, spatial cues are taken as references for head direction responses by rat anterodorsal thalamus neurons. *J. Neurosci.* 21:RC154

Neuronal Mechanisms of Visual Categorization: An Abstract View on Decision Making

David J. Freedman[1,2] *and John A. Assad*[3,4]

[1]Department of Neurobiology, University of Chicago, Chicago, Illinois 60637;
email: dfreedman@uchicago.edu
[2]The Grossman Institute for Neuroscience, Quantitative Biology, and Human Behavior, University of Chicago, Chicago, Illinois 60637
[3]Department of Neurobiology, Harvard Medical School, Boston, Massachusetts 02115;
email: jassad@hms.harvard.edu
[4] Istituto Italiano di Tecnologia, 16163 Genova, Italy

Key Words

categorization, decision making, vision, parietal cortex, prefrontal cortex, learning and memory, recognition, perception

Abstract

Categorization is our ability to flexibly assign sensory stimuli into discrete, behaviorally relevant groupings. Categorical decisions can be used to study decision making more generally by dissociating category identity of stimuli from the actions subjects use to signal their decisions. Here we discuss the evidence for such abstract categorical encoding in the primate brain and consider the relationship with other perceptual decision paradigms. Recent work on visual categorization has examined neuronal activity across a hierarchically organized network of cortical areas in monkeys trained to group visual stimuli into arbitrary categories. This has revealed a transformation of visual-feature encoding in early visual cortical areas into more flexible categorical representations in downstream parietal and prefrontal areas. These neuronal category representations are encoded as abstract internal cognitive states because they are not rigidly linked with either specific sensory stimuli or the actions that the monkeys use to signal their categorical choices.

The primate brain contains a hierarchically organized network of brain areas for processing visual stimuli. Neurons in early structures in the hierarchy, such as the lateral geniculate nucleus and primary visual cortex, encode simple visual features in a fairly stereotyped manner. As one ascends the cortical hierarchy, neuronal responses become less stereotyped and more flexible, reflecting nonvisual influences related to immediate or learned behavioral demands. Flexible neuronal responses presumably contribute to the enormous yet nuanced capability of higher organisms to modulate, select, and learn new behaviors and internal states in response to sensory input.

Nowhere is this flexibility more apparent than in decision making. All organisms continually face a plethora of options: We decide the characteristics or identity of objects in our sensory world; we choose actions to make and when to make them; and we select goods based on their subjective value. For centuries, decision making has fascinated and perplexed thinkers in a broad variety of disciplines, from philosophy to psychology to economics, with neuroscience entering the fray only recently (Glimcher 2003). From the neuroscientific perspective, the most-studied decisions have been of the perceptual variety, in which subjects (commonly macaque monkeys) discriminate the features of a visual stimulus. In an influential perceptual decision paradigm developed more than two decades ago by Newsome and colleagues (1989), monkeys observe a patch of randomly arrayed dots placed in the receptive field (RF) of the neuron under study. Some percentage of the dots move together in one or the opposite direction while the remaining dots meander about randomly. The animals discriminate the net direction of the dots and signal that direction by making a saccade to one of two target spots. The difficulty of the discrimination can be controlled by simply adjusting the percentage of dots that move coherently: The discrimination is easy if a large percentage of the dots move coherently and difficult if a small percentage (or none) of the dots move coherently. Neuronal responses in several brain regions have been found to reflect the animal's direction choice. For example, trial-by-trial variations in spike counts of single neurons in the middle temporal area (MT), an important motion-processing area, correlate with the animal's decision when the motion coherence is low (Britten et al. 1996).

Shadlen and colleagues exploited a variant of the noisy direction–discrimination task to study the neuronal mechanisms for perceptual decisions in more detail, focusing on neurons in the lateral intraparietal area (LIP). LIP was a candidate for mediating the noisy motion task because it receives inputs from visual cortical areas and projects to oculomotor structures (Andersen 1989, Lewis & Van Essen 2000). Moreover, consistent with its hybrid anatomical standing, many LIP neurons display activity that is neither strictly sensory nor motor, such as sustained delay activity that outlasts visual stimulation when animals are planning saccades in specific directions (Gnadt & Andersen 1988). In the version of the noisy motion task adapted for LIP neurons, one choice target is placed inside the RF, with the other target placed across the fixation point in the opposite visual hemifield. The moving dots are located outside the neuron's RF. In this configuration, many LIP neurons reveal a fascinating pattern of activity: During the presentation of the noisy motion, the average activity ramps up (or down) depending on whether the animal ultimately chooses the target inside (or outside) the RF. Moreover, the steepness of the ramping depends on the strength of the motion stimulus (Shadlen & Newsome 2001). Observations of this sort led to an appealing model for how LIP neurons (and presumably other neurons) might contribute to decisions (Gold & Shadlen 2007). The firing rate of the neurons was taken as an explicit representation of a decision variable that develops by integrating momentary sensory evidence. When that decision variable reaches a critical system bound corresponding to one or the other outcome (much like a race), the decision is triggered in favor of that outcome. The model can explain a host of observations, including the overall distribution of choices as a function of motion strength and the time to react. The model can even incorporate knotty cognitive factors, such as the animal's confidence in its choice (Shadlen & Kiani 2013).

Although the noisy motion task has contributed much to our understanding of perceptual decisions, it leaves open the question of the level of the decision. For example, the decision could be implemented ultimately as a competition between the two possible directions of eye movement used to signal the discrimination. Alternatively, the decision could be more abstract, signaling the percept of the direction of the moving dots independently of the concrete motor action the animal uses to report its percept (Freedman & Assad 2011). The basic noisy motion task cannot easily dissociate an action-based decision from a more abstract, nonmotor-based decision because the direction of the moving dots is yoked to the direction of the eye movement (see the sidebar, Intentional Versus Abstract Frameworks for Decision Making). This problem is more than semantics; the level of the decision bears on where and how in the brain the decision is adjudicated.

A limitation of the classic noisy motion task is not that it involves an eye movement per se but that it employs eye movements to different locations in the visual field. The information conveyed in responses of LIP neurons has been debated briskly (e.g., visual response, saccade planning, spatial attention), but there is nonetheless broad consensus that LIP neurons have spatially selective RFs. Spatial selectivity complicates interpretation of decision-related neural activity because (notwithstanding that LIP activity ramps up or down during the motion-observation period) there will assuredly be differences in LIP activity related to planning eye movements to locations either inside or outside the RF. The spatial concern applies to other neural representations that have been reported in LIP. For example, Kiani & Shadlen (2009) explored neural representations of decision confidence by using a task in which animals could opt out of difficult discriminations by making a saccade to a third, sure-reward target outside the RF. The authors found that LIP activity was intermediate between that elicited when the animals chose one of the two conventional targets, inside or outside the RF, suggesting that LIP neurons encode the animal's degree of confidence in the direction percept explicitly. However, there are alternative explanations to consider, such as the animal's dividing its attention or rapidly vacillating its saccade plan between the two risky targets, precisely because the animal lacks confidence in its decision on difficult trials. This example illustrates the difficulty in assessing cognitive signals independently in a spatially selective area such as LIP.

INTENTIONAL VERSUS ABSTRACT FRAMEWORKD FOR DECISION MAKING

In a hypothetical intentional framework, the brain encodes decisions according to the actions that subjects use to report their decisions. For example, if a subject makes an eye movement to the left or right to signal the outcome of a binary decision, the decision could be adjudicated by the brain's oculomotor circuits that move the eyes to the left or right. In contrast, in an abstract framework, the decision would be encoded independently of the action used to report the decision. For example, in perceptual decisions based on discriminating a feature of a sensory cue, the decision could be encoded in a sensory space (for example, a noisy visual stimulus is perceived as moving to the right or to the left) regardless of how the subject signals that percept.

The two frameworks are not mutually exclusive. However, if a perceptual decision task links different perceptual outcomes rigidly to particular actions (e.g., if a stimulus is perceived as moving to the left, always saccade to the left), it is inherently difficult to dissociate the representation of the percept from the representation of the action. If instead the particular action needed to signal the percept is unknown until some time after the stimulus presentation, one could reveal an abstract representation of the decision outcome, uncontaminated by action planning..

One way to reconcile these arguments is to assume that the decision process is inextricably linked to the neurons' spatially selective motor function. That is, decisions could be encoded in an action-based or intentional framework (Shadlen et al. 2008). In this view, an LIP neuron participates in a particular decision precisely because it is well suited to discriminate between saccade targets inside versus outside its RF. If the targets were positioned in some other place in the visual field—or if the decision involved a report other than an eye movement—different neurons or different brain areas would presumably mediate the decision.

One could imagine potential advantages of an intentional framework for decision making, such as shortening the time between decisions and actions. However, the intentional framework seems complex (if not clumsy) from a computational perspective because different neuronal populations would presumably have to mediate the decision for even small changes in spatiomotor contingencies. A nonintentional decision process would seem more efficient, or at least would complement intention-based decision making in specific contexts. This issue extends beyond perceptual decisions: Similar questions have been raised concerning intentional versus goods-based frameworks for economic decisions (Padoa-Schioppa 2011).

Regardless of computational considerations, it is a safe statement that the noisy motion task was not designed to test the idea of intentional frameworks for decision making rigorously because decisions were mapped rigidly to specific actions. It is thus difficult to ascertain whether a neuronal response represents a specific motor plan or a judgment of the relevant stimulus attribute (i.e., motion direction), independent of the motor plan. A similar challenge is faced by studies of other behavioral and cognitive functions, such as reward and attention (Maunsell 2004; Leathers & Olson 2012, 2013). To take an extreme example, recordings from brainstem neurons that innervate oculomotor muscles would correlate perfectly with an animal's perceptual decision. Of course, oculomotor neurons carry only motor signals. Higher brain regions, by contrast, often convey complex mixtures of sensory, cognitive, and motor signals, necessitating an experimental approach that can dissociate these factors.

An additional consideration is that in all such published experiments in LIP using the noisy motion task, the motion stimulus was placed outside the RF of the neuron under study. In this configuration, it is not obvious how visual motion information could be relayed to the neurons that were recorded. In contrast, LIP neurons are driven strongly by moving stimuli inside their RFs (Eskandar & Assad 1999, 2002; Herrington & Assad 2009, 2010) and in a direction-selective manner (Fanini & Assad 2009, Ibos & Freedman 2014, Sarma et al. 2016). However, these responses have not been examined in the published papers on the noisy motion task in LIP. Moreover, LIP neurons that are driven directly by the stimulus likely receive the most direct input from direction-selective cells in lower motion areas. It seems highly unlikely that spikes from these LIP neurons would be ignored or irrelevant to the decision process, motivating experiments that examine the role of stimulus-driven activity in LIP during perceptual decisions.

Our goal in this review is to focus on another behavioral paradigm that extends our understanding of the neuronal mechanisms of decision making. In a series of experiments, we trained animals to make decisions concerning learned visual categories, but with the task explicitly constructed to avoid visual-spatial and movement-related confounds. The experiments also placed the visual stimuli in the RF of the neuron under study to examine how the neurons that are driven directly by the stimuli might contribute to the categorical decision. These experiments were not intended to test frameworks for decision making originally, but they nonetheless challenge the hypothesis that decisions are encoded exclusively in an intentional framework.

VISUAL CATEGORIZATION AND THE DELAYED MATCH-TO-CATEGORY PARADIGM

Categorization is our ability to divide stimuli into distinct, and often arbitrary, groups (Barsalou 1992; Ashby & Maddox 2005, 2011). Categories are often taken as synonymous with the semantic relationships between objects. For example, we classify visual stimuli easily and almost automatically into familiar groups such as cars and trucks or fruits and vegetables. However, we are not born with a built-in library of thousands of categories that we are preprogrammed to recognize; rather, we acquire most categories through experience, presumably by forming learned associations between stimuli. Researchers assume that signals about visual features encoded in early visual areas are bound together by higher brain areas to engender categorical divisions between stimuli.

How can one study the brain's representation of learned categories? Exactly as in perceptual decision tasks, if an animal were trained to report category identity by a rigid mapping between stimulus and action, distinguishing category-related activity from motor-related activity would be difficult. The need to decouple subjects' categorical decisions from motor responses was a primary motivation for using the delayed match-to-sample (DMS) paradigm in neuronal categorization studies. The DMS approach was applied decades ago in neurophysiological studies of short-term working memory (Fuster & Jervey 1981; Fuster et al. 1981, 1982). In the DMS task, a subject is presented with a sample stimulus followed by a brief (usually one or more seconds) delay period. The delay is followed by a test stimulus, chosen randomly to be either a match (identical to the sample) or a nonmatch (different from the sample). Subjects must report, usually with a manual response or saccade, whether each test stimulus is a match or nonmatch to the sample. Successful task performance requires the subject to remember information about the sample during the delay in order to identify a matching test stimulus. To study categorization, the DMS paradigm can be modified so that a matching test stimulus is defined as belonging to the same category as the sample but is not necessarily an identical match (e.g., an apple and banana are both fruit). We refer to this task as a delayed match-to-category (DMC) task (**Figure 1**). Importantly, DMC tasks decouple information about the sample stimulus from the subjects' motor plan: The subject cannot predict whether a test stimulus will be a match or nonmatch to the sample, so the subject cannot plan their behavioral report until the test appears after the delay period.

NEURONAL CORRELATES OF VISUAL CATEGORIZATION

The DMC task was first employed more than a decade ago to examine the impact of categorization training on neuronal encoding in prefrontal cortex (PFC) and inferotemporal cortex (ITC)—cortical areas known to be engaged in higher cognitive functions (PFC) (Miller & Cohen 2001, Wallis 2007, Squire et al. 2013) and visual form and object encoding (ITC) (Logothetis & Sheinberg 1996, Tanaka 1996, Tsao & Livingstone 2008, Kourtzi & Connor 2011). This work has been reviewed recently (Freedman & Miller 2008; Seger & Miller 2010); thus, we discuss only the most relevant points here. In these studies, monkeys were trained to group a large, continuously varying set of animal-like stimuli into two categories, cats and dogs, defined by an arbitrary category boundary (Freedman et al. 2001, 2002). Visually similar stimuli could belong to different categories if they were close to, but on opposite sides of, the boundary, whereas stimuli in the same category could be visually quite dissimilar. This allows a dissociation of neuronal encoding of stimulus features (or similarity) from category membership—and category membership itself is dissociated from the animals' motor responses by the design of the DMC task.

Neurophysiological recordings during the DMC task revealed that PFC contained a population of category-selective neurons that encoded the learned cat and dog categories explicitly. Many such neurons showed binary-like categorical tuning, which mirrored the animals' behavior because both behavioral categorization and neuronal firing rates were similar for stimuli in the same category and different between categories. In contrast, ITC activity was dominated by strong feature selectivity, which tracked the perceptual similarity, but not category membership, of stimuli (Freedman et al. 2003, 2006) and was not impacted strongly by changes in task context (McKee et al. 2014). This suggests that categorization performance is mediated by brain areas associated with higher behavioral or cognitive functions (e.g., PFC) rather than areas involved in visual-feature encoding (e.g., ITC). Studies showing encoding of mnemonic, decision, and contextual factors in PFC during DMC tasks supported this further (Cromer et al. 2010, 2011; Swaminathan & Freedman 2012; McKee et al. 2014; Roy et al. 2014).

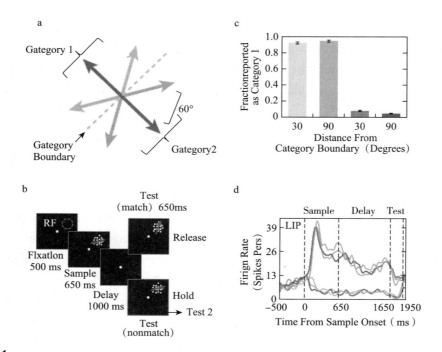

Figure 1

Delayed match-to-category (DMC) task. (*a*) Monkeys grouped six motion directions into two categories [corresponding to the *blue* and *red arrows* for category 1 (C1) and category 2 (C2), respectively] separated by a learned category boundary (*dashed line*). (*b*) Monkeys performed a DMC task and had to indicate (by releasing a lever) whether sample and test stimuli were in the same category. RF indicates the position of a neuron's receptive field. (*c*) The monkeys' average categorization performance (proportion of directions classified as C1) during lateral intraparietal area (LIP) recordings is shown as a function of distance from the category boundary. (*d*) Six traces indicate the average spike rate of a sample category-selective LIP neuron to the three directions in C1 (*blue*) and C2 (*red*). The dark traces represent directions in the center of each category (same color scheme as in panel *a*), and the pale blue and red traces represent directions closest to the category boundary. The neuron's activity reflects the category membership of motion directions as activity was similar within each category but differed sharply between categories. Selectivity for the sample category for this neuron is evident during the sample, delay, and test epochs. Data are shown only for correct trials. Figure adapted with permission from Swaminathan & Freedman (2012).

VISUAL CATEGORIZATION IN THE DORSAL STREAM

Visual shape categorization was a natural starting point for studying the categorization process, but the object-recognition system presents major challenges for studying mechanisms underlying categorization. Foremost among these is that visual object space is huge, making it impossible to estimate a neuron's response to more than a tiny fraction of possible stimuli. Natural stimuli are defined by multiple features (e.g., birds have wings, feathers, and beaks), and those component features are themselves defined by multiple features. It remains unknown how local-feature representations are integrated into holistic object representations or the degree to which ITC activity reflects local-feature versus more global object encoding. Furthermore, stimuli within the same category (e.g., plants) often have features in common and are more perceptually similar to one another than to stimuli in other categories. This poses a challenge in studies with natural stimuli, as similar neuronal responses within a category could be due to either visual similarity or their shared category membership (Vogels 1999a,b; Kriegeskorte et al. 2008; Baldassi et al. 2013; Liu et al. 2013; Long et al. 2016).

These challenges motivated an ongoing series of categorization studies using simpler visual stimuli—visual motion. Compared with the enormous space of visual shapes, visual motion is well parameterized based on speed and the 360°range of direction, making it much easier to characterize neuronal direction than shape selectivity. Work from many laboratories has provided a rich understanding of neuronal motion processing, from simple local motion representations in early visual cortex (Hubel & Wiesel 1959, 1968) to more global motion encoding in the MT (Born & Bradley 2005), culminating in complex motion representations (e.g., optic flow patterns) in the medial superior temporal (MST) area (Saito et al. 1986, Duffy & Wurtz 1991). Rich theoretical frameworks have also emerged that can account for the computation of global motion from local image features (Movshon et al. 1986) and the relationship between neuronal direction encoding and perception and behavior (Parker & Newsome 1998). This provides a promising platform for studying mechanisms of categorization in the motion-processing system.

We trained monkeys to perform a DMC task in which they categorized visual motion directions into two categories. Similar to the cat versus dog studies, monkeys learned to classify stimuli based on a category boundary that divided 360°of motion directions into two arbitrary categories (**Figure 1**) (Freedman & Assad, 2006). Directions could be visually similar (e.g., 30°apart) but belong to different categories if they were on opposite sides of the boundary (**Figure 1*a***). Likewise, stimuli in the same category could differ greatly (e.g., 150°apart) in their direction. As in the DMC studies described above, monkeys had to indicate (by releasing a lever) whether a test stimulus was a categorical match to a previously presented sample (**Figure 1*b***). Monkeys learned the two categories during one month or more of daily DMC training sessions, after which they could categorize stimuli with high accuracy (approximately 90% correct on average; **Figure 1*c***).

Our initial studies using the motion-DMC task focused on MT and LIP. MT was an obvious choice because it is an important visual motion–processing stage and projects anatomically to areas that are candidates for mediating the computation of direction categories (Lewis & Van Essen 2000, Born & Bradley 2005). We focused on LIP because of previous work showing that LIP neurons convey diverse sensory, cognitive, and motor signals, including direction selectivity (Eskandar & Assad 1999, Fanini & Assad 2009, Fitzgerald et al. 2012, Ibos & Freedman 2014), attentional modulation (Colby et al. 1996; Herrington & Assad 2009, 2010; Bisley & Goldberg 2010), action planning (Gnadt & Andersen 1988, Andersen et al. 1997, Snyder et al. 1997), behavioral context or rules (Stoet & Snyder 2004, Oristaglio et al. 2006), and decision-related activity (Britten et al. 1996, Shadlen & Newsome 2001, Gold & Shadlen 2007). Furthermore, previous work comparing activity across MT, MST, and LIP suggested a transition from sensory processing to a representation that was more aligned with subjective perception and more strongly influenced by nonvisual, task-related factors (Eskandar & Assad 1999, Williams et al. 2003, Maimon & Assad 2006a). Thus, LIP seemed a reasonable place to look for more abstract, categorical signals in the context of the DMC task.

Indeed, neuronal recordings from MT and LIP revealed that the two areas play highly distinct roles during the DMC task (Freedman & Assad 2006). Nearly all MT neurons showed strong direction selectivity, but no obvious relationship existed between the pattern of direction selectivity and the learned categories, suggesting that MT activity was not directly related to the animals' categorical decisions. In contrast, many LIP neurons showed strong category selectivity, analogous to that observed in PFC during the cat versus dog categorization task: similar activity for stimuli within each category and sharply different activity between categories (**Figure 1d**). Moreover, LIP category representations shifted when the animals were retrained with a new direction-category boundary. Thus, LIP neurons were strongly influenced by categorization training. Importantly, category selectivity in LIP was not due to the way the animals reported their categorical decisions because the DMC task dissociated category identity from the animals' motor action (releasing or holding the lever). The category identity could not be represented in an intentional framework because strong neuronal category selectivity was observed at a time when the animals did not even know yet what movement they would make to indicate their categorical decision. Category signals could also not be confused with differential spatial encoding (e.g., spatial attention) because all stimuli were presented in the same location, in the RF of the neuron under study.

FEEDFORWARD VERSUS FEEDBACK CATEGORY COMPUTATIONS

An appealing hypothesis is that the transformation of direction selectivity in MT to abstract categorical encoding in LIP is mediated by their direct anatomical interconnection (Lewis & Van Essen 2000). As we had speculated in our original report, the category learning process could have driven plasticity in the interconnections between MT and LIP (Ferrera & Grinband 2006, Freedman & Assad 2006). In this scheme, LIP neurons would gradually become category selective during categorization training. This scheme was made more explicit in a recent theoretical model examining the interplay between direction encoding, intercortical plasticity, and choice-correlated neuronal activity fluctuations that can account for encoding of learned categories in LIP (Engel et al. 2015). This model predicts that categorical representations in LIP should be relatively slow to develop during learning because the process would require reorganization of synaptic inputs from direction-selective areas such as MT. This is consistent with the long training durations (weeks to months) required for the animals to become proficient in the DMC task (Sarma et al. 2016).

Although the hypothesis that LIP plays a central role in generating category encoding from MT inputs is appealing, the experiments described so far do not exclude the possibility that category selectivity in LIP arises via feedback from other brain areas that might be more directly involved in category computation. An obvious candidate is the PFC, which encodes learned visual shape categories (Freedman et al. 2001, 2002, 2003) and is directly interconnected with posterior parietal cortex (PPC) (Petrides & Pandya 1984, Cavada & Goldman-Rakic 1989, Felleman & Van Essen 1991). Further evidence comes from a recent study that found that the frontal eye field of the PFC encoded visual motion categories during a speed categorization (e.g., fast versus slow) task (Ferrera et al. 2009).

A recent study in the Freedman lab (Swaminathan & Freedman 2012) compared PFC and LIP, providing evidence that LIP is more involved than PFC in mediating the motion-DMC task. First, LIP showed stronger category encoding than PFC. Second, category encoding emerged in the course of a trial with a shorter latency in LIP than PFC, consistent with a feedforward projection of category encoding from LIP to PFC, but not vice versa. Finally, the study included directions positioned on the category boundary, which had ambiguous category membership. The monkeys' performance for on-boundary stimuli was near 50%, providing an opportunity to examine the trial-by-trial relationship between activity in LIP and PFC and the monkeys' category decisions. This revealed that LIP activity was more strongly correlated with monkeys' behavior than PFC, suggesting that LIP is more directly involved in categorization performance during this task.

UNDERSTANDING CONTRIBUTIONS OF MULTIPLE POSTERIOR PARIETAL CORTEX AREAS

A line of work from Chafee's laboratory examined frontal and parietal cortices in a spatial categorization task in which monkeys categorized visual targets according to their spatial position (Goodwin et al. 2012, Crowe et al. 2013). The task required greater behavioral flexibility than the motion-DMC task because the orientation of the spatial category boundary varied between horizontal (creating above/below categories) and vertical (left/right categories) on different trials. This required the monkeys to keep track of the relevant category rule. PFC and PPC recordings revealed that both areas encoded the task-relevant category rules (horizontal or vertical boundary) as well as the spatial categories. Interestingly, rule encoding and rule-based modulation of category encoding was stronger and appeared with a shorter latency in PFC than PPC, suggesting a greater involvement of PFC in mediating this task—the opposite pattern of results seen in the motion-DMC study described above. In addition to testing different visual features (motion versus space), the motion- and spatial-categorization studies differed in several key aspects. In particular, the flexible category boundary in the spatial task places greater cognitive demands on the monkey, perhaps necessitating a greater involvement of PFC. The two studies also examined different parietal regions (LIP and 7a), which may differ in their function.

The observation of categorical encoding in 7a and LIP raises a question about the relative roles of PPC subdivisions. Although LIP and 7a have not yet been directly compared during visual categorization, a recent study compared LIP and another neighboring parietal region, the medial intraparietal (MIP) area (Brodmann area 5). MIP and LIP are anatomically interconnected (Lewis & Van Essen 2000) but show different patterns of interconnections with other areas. LIP is more strongly connected with visual areas, whereas MIP is more strongly connected with somatosensory and motor areas. In contrast to LIP, MIP is strongly engaged in behaviors involving hand and arm movements but not eye movements (Snyder et al. 1997, Eskandar & Assad 2002, Maimon & Assad 2006b). Because the monkeys report their decisions in the DMC task with an arm movement, we reasoned that MIP might play a role in the DMC decision process. Indeed, we found that neuronal category encoding was evident in both MIP and LIP during the motion-DMC task (Swaminathan et al. 2013). However, category encoding in LIP was stronger and had a shorter latency than in MIP, suggesting a leading role for LIP in the categorization process. Furthermore, this study employed a population-decoding approach to measure the strength of direction selectivity (independent of category selectivity) and category selectivity (independent of direction selectivity). This revealed that direction and category were both encoded by LIP, whereas MIP contained only category but not direction information. This provides additional evidence for LIP's involvement in transforming direction into category encoding; by contrast, MIP's position in the categorization process could be postdecisional—perhaps a recipient of category encoding from LIP.

GENERALITY OF CATEGORICAL AND ABSTRACT COGNITIVE ENCODING IN POSTERIOR PARIETAL CORTEX

The work discussed so far supports the idea that LIP is a candidate for transforming motion encoding in visual cortex into learning-dependent categorical representations. However, the generality of LIP's role in categorization is unclear. One possibility is that LIP is specialized for categorizing visual motion or other visual-spatial features that are encoded in parietal cortex. However, parietal cortex also shows selectivity for a variety of nonspatial features, including visual shapes (Sereno & Maunsell 1998), raising the possibility that it could play a more general role in categorical decision making. This was tested in our recent study, which trained monkeys on a shape-association task based on the DMC paradigm (Fitzgerald et al. 2011). In this pair-association task, six geometric shapes were grouped into three associated pairs. The pairs were similar to the categories in the DMC task, in that stimuli within a pair were treated as matches, whereas stimuli in different pairs were nonmatches. The shape study was conducted in monkeys that had also been trained on the motion-DMC task and could alternate flexibly between the shape and motion tasks. This allowed a direct comparison of LIP activity between the shape and motion tasks. This study found a striking generality of category-related encoding in LIP, with strong selectivity observed for both the shape pairs and the direction categories. Most interestingly, strong shape and motion category encoding were often observed in the same neurons, suggesting that the two tasks likely tap into common underlying mechanisms.

The generalized categorical encoding in PPC suggests it is involved in a wide variety of abstract cognitive functions beyond categorization. For example, several groups have reported diverse cognitive signals in PPC, including task rules (Stoet & Snyder 2004, Oristaglio et al. 2006), numerosity (Nieder & Miller 2004, Nieder et al. 2006, Tudusciuc & Nieder 2007), and salience (Leathers & Olson 2012). All these effects could be captured under the broad rubric of categories. Similar cognitive signals are also present in PFC (Wallis et al. 2001, Nieder et al. 2002, Padoa-Schioppa & Assad 2006) and other areas (Wallis & Miller 2003, Muhammad et al. 2006, Antzoulatos & Miller 2011, Cromer et al. 2011, Vallentin et al. 2012, Eiselt & Nieder 2014), suggesting that PPC is most likely a node in the network mediating abstract cognitive computations.

RELATIONSHIP BETWEEN SPATIAL AND NONSPATIAL ENCODING IN LATERAL INTRAPARIETAL AREA

Given the spatial framework dominating most theories regarding PPC (Andersen et al. 1997, Gold & Shadlen 2007, Bisley & Goldberg 2010), the function of nonspatial cognitive signals (such as category selectivity) and their relationship with spatial functions is uncertain. This was addressed by work comparing spatial and cognitive encoding in LIP. One study ran the DMC task with stimuli shown either inside or outside neurons' RFs (Freedman & Assad 2009). As expected, LIP neurons were strongly spatially selective, with much stronger activity when stimuli were shown in the RF (by definition). However, category selectivity was still evident for stimuli outside neurons' RFs, despite weaker overall activity. This is consistent with visual-spatial and category encoding arising from distinct inputs—a strong visual input driving high firing rates for stimuli shown in the RF, with a modulatory input (because it did not itself drive large firing-rate fluctuations) conveying category information. The findings also argue strongly against the possibility that the category encoding in LIP is some sort of spatial artifact, such as the animals' using a spatial mnemonic (e.g., a differential covert eye-movement plan) to represent or remember the learned categories.

A second study examined the relationship between category and spatiomotor encoding in LIP by examining the impact of saccades on LIP category encoding (Rishel et al. 2013). Monkeys were trained on a variant of the motion-DMC task, called the DMC+saccade task, which assessed the monkeys' category decisions and saccade plans independently (**Figure 2a**). The DMC+saccade task required the animal to make a visually cued saccade during the delay period. In some trials, the saccade was directed toward the neuron's RF, whereas in other trials, the saccade was directed away from the RF. The purpose of the saccade was to modulate LIP activity, as saccades toward the RF elicit much stronger firing than saccades away from the RF. Importantly, the saccade itself was unrelated to the animals' category or match/nonmatch decisions, and the monkey still had to remember sample-category information during the delay period. Category and saccade-related encoding in LIP were found to be independent, with each factor exerting an independent influence on neuronal firing rates (**Figure 2b–d**). In fact, on average, category-related selectivity was still present even during the saccade, with category selectivity superimposed on top of the saccade-related activity (**Figure 2c**).

Together, this work reveals that LIP activity reflects diverse sensory, cognitive, and motor signals. This is evident in both the DMC+saccade task—which revealed independent category and saccade encoding—and the study that showed spatially independent category encoding. This idea is supported further by a recent study that found multiplexed representations in LIP during the noisy motion decision task (Meister et al. 2013).

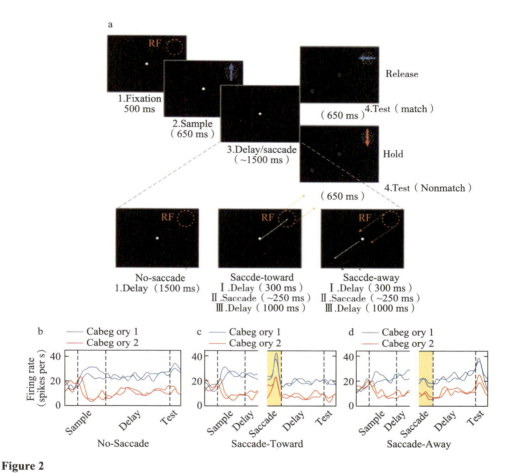

Figure 2

Delayed match-to-category (DMC)+saccade task. Monkeys performed a variant of the motion-DMC task (with four rather than six motion directions) in which they made a saccade during the delay period. (*a*) Schematic of the DMC+saccade task. The task was similar to that in **Figure 1*b*** except that in some trials, a saccade was required during the early delay period (300 ms after sample offset), directed either toward or away from the neuron's receptive field (RF). After the saccade, the monkey maintained its gaze at the new fixation location for the remainder of the trial. The fixation point is indicated by the white spot on each panel, and the dotted-outline spots indicate the test-period panels (indicating the three possible postsaccade fixation locations depending on the saccade condition). (*b–d*) Example lateral intraparietal area (LIP) neuron during the DMC+saccade task that showed independent encoding of direction categories and saccade direction. Neuronal activity is shown in the no-saccade (*b*), saccade-toward the RF (*c*), and saccade-away from the RF (*d*) conditions. The orange areas on the saccade-condition plots indicate the saccade periods. Note that in the saccade conditions (panels *c* and *d*), the neuron shows selectivity for both category and saccade direction during the saccade period, indicating distinct category and spatial encoding in LIP. In the no-saccade condition (*b*), the three vertical dotted lines indicate, from left to right, the sample onset, sample offset, and test onset. In the saccade conditions (*c, d*), the left portion of the plot (prior to the saccade period) is aligned on sample onset, whereas the right portion is aligned on test onset. The two vertical dotted lines on the left portion of the plot indicate sample onset and saccade cue. The two vertical dotted lines on the right portion indicate stable fixation after the saccade and test onset. Figure adapted with permission from Rishel et al. (2013).

INDEPENDENT COGNITIVE AND SPATIAL ENCODING— IMPLICATIONS FOR MODELS OF DECISION MAKING

We are now in a position to consider the category encoding of LIP neurons in the broader framework of decision making, as outlined in the introductory comments. The categorization studies relate back to the question of whether visual decisions are carried out in an intentional framework in the brain. In the context of the DMC task, the category selectivity in LIP could not be attributed to planned movements, because the animals did not know whether or not they would move until the test stimulus was presented, nor to spatial selectivity of the neurons (such as a differential eye-movement plan), because all stimuli were presented in the same location. Thus, at least in the DMC task, categories were represented in a nonintentional framework. This suggests that abstract (i.e., nonintentional) frameworks may be a basis for perceptual decision processes.

However, the DMC task required that the category decision be made in a nonintentional framework because the movement plan had to be deferred. In perceptual decision tasks in which the motor plan is known in advance (such as the noisy motion task), an intentional framework might be a natural strategy for performing the task. But the design of the noisy motion experiments cannot rule out that the decision might be encoded as an abstract representation of stimulus direction because the stimulus direction and eye-movement direction were confounded.

Interestingly, a recent study from the laboratory of Gold, a major contributor to the earlier work in LIP using the noisy motion task, helps to bridge the gap between the DMC and noisy motion approaches (Bennur & Gold 2011). Instead of training monkeys to saccade left or right to report leftward or rightward motion, Bennur and Gold trained the animals to saccade to either a red or green target to report the two directions of motion. Moreover, the positions of the targets were randomized from trial to trial. Crucially, in some trials, the targets were not shown to the monkeys until after the presentation of the motion stimulus, making it impossible for the monkeys to plan their eye movement until after they had formed their decision about the direction of the stimulus. Because the direction of the stimulus and the eye movement were dissociated in this task, the authors were able to identify signals in LIP neurons that reflected the stimulus direction, independent of the direction of the eye movement. These signals were very reminiscent of the categorical signals that we found in LIP using the DMC task.

One could argue that Bennur & Gold (2011) found the representation of stimulus direction precisely because the animals had to defer their eye-movement plan, as we had also required in the DMC task. However, they also included trials in which the red and green saccade targets were presented simultaneously with the noisy motion stimulus. In this case, the animals could plan their eye movement as quickly as they perceived the stimulus direction, yet the cells still showed selectivity for the stimulus direction in this context. This indicates that an abstract perceptual interpretation of stimulus direction may be present all the time in LIP and may form the basis for perceptual decisions, rather than a competition between eye-movement plans.

Although we have found strong evidence for abstract categorical selectivity in parietal cortex, we have not observed the ramping-up or ramping-down of activity that could reflect an accumulation of momentary evidence toward a decision bound. Rather, we have found the categorical signals present early after stimulus presentation. In a sense, we have observed the outcome of the decision process in parietal neuronal responses, but not necessarily the deliberation toward that decision. However, we have always used suprathreshold motion stimuli that, according to the accumulate-to-bound model, should provide strong momentary evidence driving a quick decision. It would be interesting to examine whether parietal neurons show evidence of ramping activity in the DMC task with more noisy stimuli that require more integration to discriminate. Of course, the decision process could also evolve in very different ways (or in different brain areas) between the DMC and noisy motion tasks.

LOOMING CHALLENGES AND A WAY FORWARD

In light of experiments that dissociated categorical representations explicitly from movement plans, the evidence for abstract categorical representations is now firmly established. Moreover, a parsimonious working hypothesis is that categorical representations could underlie other signals, including decision variables and a host of other reported cognitive representations. In parietal cortex, categorization may be a general feature of neural circuitry in low-dimensional decision tasks involving only a few options or outcomes (Fitzgerald et al. 2013).

Moving forward, we believe one of the most important and challenging issues is to understand the neuronal mechanisms underlying categorization. In considering mechanisms, one of the most salient issues is learning. It is inescapable that categorical representations must develop over time, on account of training in the behavioral tasks. How and where does this learning occur?

A recent study from the Freedman lab has given an intriguing look at the impact of learning and task demands on decision-related encoding by recording from LIP in the same monkeys before and after they learned the motion-DMC task (**Figure 3a**; Sarma et al. 2016). Before learning the DMC task, monkeys performed a DMS task in which they had to identify test stimuli that were exact matches of the sample. Although we knew to expect strong category encoding in LIP after (but not before) learning the DMC task, we did not know precisely how LIP would be affected by DMC training. To our surprise, learning the DMC task had a dramatic impact on the patterns of delay-period activity in LIP. Before DMC training, while the animals were performing the DMS task, we did not detect encoding of the sample direction during the delay (**Figure 3b**)—a surprise because the monkey had to remember the sample direction to solve the task, and PPC has been implicated in short-term memory (Gnadt & Andersen 1988, Chafee & Goldman-Rakic 1998). This was in sharp contrast to LIP responses after DMC training, which showed strong category selectivity during the delay, consistent with previous work (**Figure 3c**). This shows that learning the DMC task causes a new representation (delay-period selectivity) to emerge. Furthermore, LIP differed strikingly from PFC in this study, as strong delay-period selectivity was evident in PFC both before and after learning the DMC task, suggesting a more general role for PFC in delay-period representation of task-relevant information (**Figure 3d**). Thus, something about learning the DMC task alters encoding in LIP specifically, providing an intriguing clue that it may be uniquely involved in learning to perform abstract visual decision tasks.

Finally, another challenge is to reconcile the cognitive signals that we and others have identified in parietal neurons with the more classic view of parietal cortex mediating spatiomotor behavior. We presented evidence that cognitive and spatial signals may be largely independent influences in parietal neurons, but many questions remain. What are the origins of these influences? Is there a straightforward way for the signals to be read out independently by downstream areas? Moreover, the confluence of sensory, cognitive, and motor signals in a single cortical area suggests that the entire sensorimotor decision arc could, in principle, be played out within a fairly local network. This will be fertile ground for future investigations.

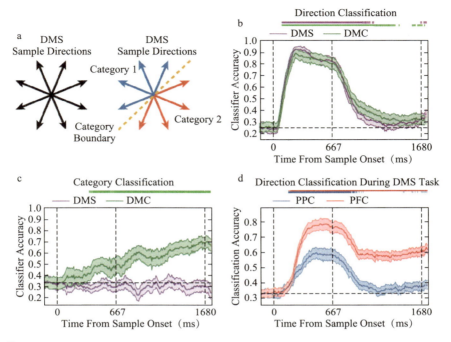

Figure 3

Direction and category selectivity in parietal and prefrontal cortices before and after categorization training. (*a*) Lateral intraparietal area (LIP) recordings were conducted in the same two monkeys during a delayed match-to-sample (DMS) task (*left*) prior to categorization training and a delayed match-to-category (DMC) task (*right*) after categorization training. (*b*) The time course of sample direction selectivity (assessed via a population-decoding analysis) in LIP is shown during the DMS task before categorization training (*purple*) and DMC task after categorization training (*green*). From left to right, the three dashed, vertical lines represent the start of the sample epoch, the end of the sample epoch, and the end of the delay epoch. The shaded areas indicate SEM. The horizontal dashed line indicates chance classification performance. The colored bars at the top of the panel indicate times at which classification accuracy was significantly above chance (*light*, $P < 0.05$; *dark*, $P < 0.01$; bootstrap). (*c*) The time course of sample category selectivity in LIP is shown during the DMS task before categorization training (*purple*) and DMC task after categorization training (*green*). Conventions are the same as in panel *b*. (*d*) Neuronal direction selectivity is shown in prefrontal cortex (PFC) (*red*) and posterior parietal cortex (PPC) (*blue*) from two monkeys performing the direction DMS task (prior to categorization training). Similar to panel *b*, PPC shows direction selectivity in the sample but not delay period. Notably, PFC shows direction selectivity during the sample and delay, suggesting a greater involvement in working memory for direction than PPC. Figure adapted with permission from Sarma et al. (2016).

DISCLOSURE STATEMENT

The authors are not aware of any affiliations, memberships, funding, or financial holdings that might be perceived as affecting the objectivity of this review.

LITERATURE CITED

Andersen RA. 1989. Visual and eye movement functions of the posterior parietal cortex. *Annu. Rev. Neurosci.* 12:377–403

Andersen RA, Snyder LH, Bradley DC, Xing J. 1997. Multimodal representation of space in the posterior parietal cortex and its use in planning movements. *Annu. Rev. Neurosci.* 20:303–30

Antzoulatos EG, Miller EK. 2011. Differences between neural activity in prefrontal cortex and striatum during learning of novel abstract categories. *Neuron* 71:243–49

Ashby FG, Maddox WT. 2005. Human category learning. *Annu. Rev. Psychol.* 56:149–78

Ashby FG, Maddox WT. 2011. Human category learning 2.0. *Ann. N.Y. Acad. Sci.* 1224:147–61

Baldassi C, Alemi-Neissi A, Pagan M, Dicarlo JJ, Zecchina R, Zoccolan D. 2013. Shape similarity, better than semantic membership, accounts for the structure of visual object representations in a population of monkey inferotemporal neurons. *PLOS Comput. Biol.* 9:e1003167

Barsalou L. 1992. *Cognitive Psychology: An Overview for Cognitive Scientists*. Hillsdale, NJ: Erlbaum

Bennur S, Gold JI. 2011. Distinct representations of a perceptual decision and the associated oculomotor plan in the monkey lateral intraparietal area. *J. Neurosci.* 31:913–21

Bisley JW, Goldberg ME. 2010. Attention, intention, and priority in the parietal lobe. *Annu. Rev. Neurosci.* 33:1–21

Born RT, Bradley DC. 2005. Structure and function of visual area MT. *Annu. Rev. Neurosci.* 28:157–89

Britten KH, Newsome WT, Shadlen MN, Celebrini S, Movshon JA. 1996. A relationship between behavioral choice and the visual responses of neurons in macaque MT. *Vis. Neurosci.* 13:87–100

Cavada C, Goldman-Rakic PS. 1989. Posterior parietal cortex in rhesus monkey: II. Evidence for segregated corticocortical networks linking sensory and limbic areas with the frontal lobe. *J. Comp. Neurol.* 287:422–45

Chafee MV, Goldman-Rakic PS. 1998. Matching patterns of activity in primate prefrontal area 8a and parietal area 7ip neurons during a spatial working memory task. *J. Neurophysiol.* 79:2919–40

Colby CL, Duhamel JR, Goldberg ME. 1996. Visual, presaccadic, and cognitive activation of single neurons in monkey lateral intraparietal area. *J. Neurophysiol.* 76:2841–52

Cromer JA, Roy JE, Buschman TJ, Miller EK. 2011. Comparison of primate prefrontal and premotor cortex neuronal activity during visual categorization. *J. Cogn. Neurosci.* 23:3355–65

Cromer JA, Roy JE, Miller EK. 2010. Representation of multiple, independent categories in the primate prefrontal cortex. *Neuron* 66:796–807

Crowe DA, Goodwin SJ, Blackman RK, Sakellaridi S, Sponheim SR, et al. 2013. Prefrontal neurons transmit signals to parietal neurons that reflect executive control of cognition. *Nat. Neurosci.* 16:1484–91

Duffy CJ, Wurtz RH. 1991. Sensitivity of MST neurons to optic flow stimuli. I. A continuum of response selectivity to large-field stimuli. *J. Neurophysiol.* 65:1329–45

Eiselt A-K, Nieder A. 2014. Rule activity related to spatial and numerical magnitudes: comparison of prefrontal, premotor, and cingulate motor cortices. *J. Cogn. Neurosci.* 26:1000–12

Engel TA, Chaisangmongkon W, Freedman DJ, Wang X-J. 2015. Choice-correlated activity fluctuations underlie learning of neuronal category representation. *Nat. Commun.* 6:6454

Eskandar EN, Assad JA. 1999. Dissociation of visual, motor and predictive signals in parietal cortex during visual guidance. *Nat. Neurosci.* 2:88–93

Eskandar EN, Assad JA. 2002. Distinct nature of directional signals among parietal cortical areas during visual guidance. *J. Neurophysiol.* 88:1777–90

Fanini A, Assad JA. 2009. Direction selectivity of neurons in the macaque lateral intraparietal area. *J. Neurophysiol.* 101:289–305

Felleman DJ, Van Essen DC. 1991. Distributed hierarchical processing in the primate cerebral cortex. *Cereb.Cortex* 1991 1:1–47

Ferrera VP, Grinband J. 2006. Walk the line: Parietal neurons respect category boundaries. *Nat. Neurosci.* 9:1207–8

Ferrera VP, Yanike M, Cassanello C. 2009. Frontal eye field neurons signal changes in decision criteria. *Nat.Neurosci.* 12:1458–62

Fitzgerald JK, Freedman DJ, Assad JA. 2011. Generalized associative representations in parietal cortex. *Nat.Neurosci.* 14:1075–79

Fitzgerald JK, Freedman DJ, Fanini A, Bennur S, Gold JI, Assad JA. 2013. Biased associative representations in parietal cortex. *Neuron* 77:180–91

Fitzgerald JK, Swaminathan SK, Freedman DJ. 2012. Visual categorization and the parietal cortex. *Front. Integr. Neurosci.* 6:18

Freedman DJ, Assad JA. 2006. Experience-dependent representation of visual categories in parietal cortex. *Nature* 443:85–88

Freedman DJ, Assad JA. 2009. Distinct encoding of spatial and nonspatial visual information in parietal cortex. *J. Neurosci.* 29:5671–80

Freedman DJ, Assad JA. 2011. A proposed common neural mechanism for categorization and perceptual decisions. *Nat. Neurosci.* 14:143–46

Freedman DJ, Miller EK. 2008. Neural mechanisms of visual categorization: insights from neurophysiology. *Neurosci. Biobehav. Rev.* 32:311–29

Freedman DJ, Riesenhuber M, Poggio T, Miller EK. 2001. Categorical representation of visual stimuli in the primate prefrontal cortex. *Science* 291:312–16

Freedman DJ, Riesenhuber M, Poggio T, Miller EK. 2002. Visual categorization and the primate prefrontal cortex: neurophysiology and behavior. *J. Neurophysiol.* 88:929–41

Freedman DJ, Riesenhuber M, Poggio T, Miller EK. 2003. A comparison of primate prefrontal and inferior temporal cortices during visual categorization. *J. Neurosci.* 23:5235–46

Freedman DJ, Riesenhuber M, Poggio T, Miller EK. 2006. Experience-dependent sharpening of visual shape selectivity in inferior temporal cortex. *Cereb. Cortex* 16:1631–44

Fuster JM, Bauer RH, Jervey JP. 1981. Effects of cooling inferotemporal cortex on performance of visual memory tasks. *Exp. Neurol.* 71:398–409

Fuster JM, Bauer RH, Jervey JP. 1982. Cellular discharge in the dorsolateral prefrontal cortex of the monkey in cognitive tasks. *Exp. Neurol.* 77:679–94

Fuster JM, Jervey JP. 1981. Inferotemporal neurons distinguish and retain behaviorally relevant features of visual stimuli. *Science* 212:952–55

Glimcher PW. 2003. *Decisions, Uncertainty, and the Brain: The Science of Neuroeconomics.* Cambridge, MA: MIT Press

Gnadt JW, Andersen RA. 1988. Memory related motor planning activity in posterior parietal cortex of macaque. *Exp. Brain Res.* 70:216–20

Gold JI, Shadlen MN. 2007. The neural basis of decision making. *Annu. Rev. Neurosci.* 30:535–74

Goodwin SJ, Blackman RK, Sakellaridi S, Chafee MV. 2012. Executive control over cognition: stronger and earlier rule-based modulation of spatial category signals in prefrontal cortex relative to parietal cortex. *J. Neurosci.* 32:3499–515

Herrington TM, Assad JA. 2009. Neural activity in the middle temporal area and lateral intraparietal area during endogenously cued shifts of attention. *J. Neurosci.* 29:14160–76

Herrington TM, Assad JA. 2010. Temporal sequence of attentional modulation in the lateral intraparietal area and middle temporal area during rapid covert shifts of attention. *J. Neurosci.* 30:3287–96

Hubel DH, Wiesel TN. 1959. Receptive fields of single neurones in the cat's striate cortex. *J. Physiol.* 148:574–91

Hubel DH, Wiesel TN. 1968. Receptive fields and functional architecture of monkey striate cortex. *J. Physiol.* 195:215–43

Ibos G, Freedman DJ. 2014. Dynamic integration of task-relevant visual features in posterior parietal cortex. *Neuron* 83:1468–80

Kiani R, Shadlen MN. 2009. Representation of confidence associated with a decision by neurons in the parietal cortex. *Science* 324:759–64

Kourtzi Z, Connor CE. 2011. Neural representations for object perception: structure, category, and adaptivecoding. *Annu. Rev. Neurosci.* 34:45–67

Kriegeskorte N, Mur M, Ruff DA, Kiani R, Bodurka J, et al. 2008. Matching categorical object representations in inferior temporal cortex of man and monkey. *Neuron* 60:1126–41

Leathers ML, Olson CR. 2012. In monkeys making value-based decisions, LIP neurons encode cue salience and not action value. *Science* 338:132–35

Leathers ML, Olson CR. 2013. Response to comment on "In monkeys making value-based decisions, LIP neurons encode cue salience and not action value." *Science* 340:430

Lewis JW, Van Essen DC. 2000. Corticocortical connections of visual, sensorimotor, and multimodal processing areas in the parietal lobe of the macaque monkey. *J. Comp. Neurol.* 428:112–37

Liu N, Kriegeskorte N, Mur M, Hadj-Bouziane F, Luh W-M, et al. 2013. Intrinsic structure of visual exemplar and category representations in macaque brain. *J. Neurosci.* 33:11346–60

Logothetis NK, Sheinberg DL. 1996. Visual object recognition. *Annu. Rev. Neurosci.* 19:577–621

Long B, Konkle T, Cohen MA, Alvarez GA. 2016. Mid-level perceptual features distinguish objects of different real-world sizes. *J. Exp. Psychol. Gen.* 145:95–109

Maimon G, Assad JA. 2006a. A cognitive signal for the proactive timing of action in macaque LIP. *Nat. Neurosci.* 9:948–55

Maimon G, Assad JA. 2006b. Parietal area 5 and the initiation of self-timed movements versus simple reactions. *J. Neurosci.* 26:2487–98

Maunsell JHR. 2004. Neuronal representations of cognitive state: reward or attention? *Trends Cogn. Sci.* 8:261–65

McKee JL, Riesenhuber M, Miller EK, Freedman DJ. 2014. Task dependence of visual and category representations in prefrontal and inferior temporal cortices. *J. Neurosci.* 34:16065–75

Meister MLR, Hennig JA, Huk AC. 2013. Signal multiplexing and single-neuron computations in lateral intraparietal area during decision-making. *J. Neurosci.* 33:2254–67

Miller EK, Cohen JD. 2001. An integrative theory of prefrontal cortex function. *Annu. Rev. Neurosci.* 24:167–202

Movshon JA, Adelson EH, Gizzi MS, Newsome WT. 1986. The analysis of moving visual patterns. *Exp. BrainRes. Suppl.* 11:117–51

Muhammad R, Wallis JD, Miller EK. 2006. A comparison of abstract rules in the prefrontal cortex, premotor cortex, inferior temporal cortex, and striatum. *J. Cogn. Neurosci.* 18:974–89

Newsome WT, Britten KH, Movshon JA. 1989. Neuronal correlates of a perceptual decision. *Nature* 341:52–54

Nieder A, Diester I, Tudusciuc O. 2006. Temporal and spatial enumeration processes in the primate parietal cortex. *Science* 313:1431–35

Nieder A, Freedman DJ, Miller EK. 2002. Representation of the quantity of visual items in the primate prefrontal cortex. *Science* 297:1708–11

Nieder A, Miller EK. 2004. A parieto-frontal network for visual numerical information in the monkey. *PNAS* 101:7457–62

Oristaglio J, Schneider DM, Balan PF, Gottlieb J. 2006. Integration of visuospatial and effector information during symbolically cued limb movements in monkey lateral intraparietal area. *J. Neurosci.* 26:8310–19

Padoa-Schioppa C. 2011. Neurobiology of economic choice: a good-based model. *Annu. Rev. Neurosci.* 34:333–59

Padoa-Schioppa C, Assad JA. 2006. Neurons in the orbitofrontal cortex encode economic value. *Nature* 441:223–26

Parker AJ, Newsome WT. 1998. Sense and the single neuron: probing the physiology of perception. *Annu. Rev. Neurosci.* 21:227–77

Petrides M, Pandya DN. 1984. Projections to the frontal cortex from the posterior parietal region in the rhesus monkey. *J. Comp. Neurol.* 228:105–16

Rishel CA, Huang G, Freedman DJ. 2013. Independent category and spatial encoding in parietal cortex. *Neuron* 77:969–79

Roy JE, Buschman TJ, Miller EK. 2014. PFC neurons reflect categorical decisions about ambiguous stimuli. *J. Cogn. Neurosci.* 26:1283–91

Saito H, Yukie M, Tanaka K, Hikosaka K, Fukada Y, Iwai E. 1986. Integration of direction signals of image motion in the superior temporal sulcus of the macaque monkey. *J. Neurosci.* 6:145–57

Sarma A, Masse NY, Wang X-J, Freedman DJ. 2016. Task-specific versus generalized mnemonic representations in parietal and prefrontal cortices. *Nat. Neurosci.* 19:143–49

Seger CA, Miller EK. 2010. Category learning in the brain. *Annu. Rev. Neurosci.* 33:203–19

Sereno AB, Maunsell JHR. 1998. Shape selectivity in primate lateral intraparietal cortex. *Nature* 395:500–3 Shadlen MN, Kiani R. 2013. Decision making as a window on cognition. *Neuron* 80:791–806

Shadlen MN, Kiani R, Hanks TD, Churchland AK. 2008. Neurobiology of decision making: an intentional framework. In *Better Than Conscious? Decision Making, the Human Mind, and Implications for Institutions,* ed. C Engel, W Singer, pp. 71–102. Cambridge, MA: MIT Press

Shadlen MN, Newsome WT. 2001. Neural basis of a perceptual decision in the parietal cortex (area LIP) of the rhesus monkey. *J. Neurophysiol.* 86:1916–36

Snyder LH, Batista AP, Andersen RA. 1997. Coding of intention in the posterior parietal cortex. *Nature*

386:167–70

Squire RF, Noudoost B, Schafer RJ, Moore T. 2013. Prefrontal contributions to visual selective attention. *Annu. Rev. Neurosci.* 36:451–66

Stoet G, Snyder LH. 2004. Single neurons in posterior parietal cortex of monkeys encode cognitive set. *Neuron* 42:1003–12

Swaminathan SK, Freedman DJ. 2012. Preferential encoding of visual categories in parietal cortex compared with prefrontal cortex. *Nat. Neurosci.* 15:315–20

Swaminathan SK, Masse NY, Freedman DJ. 2013. A comparison of lateral and medial intraparietal areas during a visual categorization task. *J. Neurosci.* 33:13157–70

Tanaka K. 1996. Inferotemporal cortex and object vision. *Annu. Rev. Neurosci.* 19:109–39

Tsao DY, Livingstone MS. 2008. Mechanisms of face perception. *Annu. Rev. Neurosci.* 31:411–37

Tudusciuc O, Nieder A. 2007. Neuronal population coding of continuous and discrete quantity in the primate posterior parietal cortex. *PNAS* 104:14513–18

Vallentin D, Bongard S, Nieder A. 2012. Numerical rule coding in the prefrontal, premotor, and posterior parietal cortices of macaques. *J. Neurosci.* 32:6621–30

Vogels R. 1999a. Categorization of complex visual images by rhesus monkeys. Part 1: behavioural study. *Eur. J. Neurosci.* 11:1223–38

Vogels R. 1999b. Categorization of complex visual images by rhesus monkeys. Part 2: single-cell study. *Eur. J. Neurosci.* 11:1239–55

Wallis JD. 2007. Orbitofrontal cortex and its contribution to decision-making. *Annu. Rev. Neurosci.* 30:31–56

Wallis JD, Anderso KC, Miller EK. 2001. Single neurons in prefrontal cortex encode abstract rules. *Nature* 411:953–56

Wallis JD, Miller EK. 2003. From rule to response: neuronal processes in the premotor and prefrontal cortex. *J. Neurophysiol.* 90:1790–806

Williams ZM, Elfar JC, Eskandar EN, Toth LJ, Assad JA. 2003. Parietal activity and the perceived direction of ambiguous apparent motion. *Nat. Neurosci.* 6:616–23

The Role of the Lateral Intraparietal Area in (the Study of) Decision Making

Alexander C. Huk, Leor N. Katz, and Jacob L. Yates

Center for Perceptual System, Department of Neuroscience and Psychology, The University of Texas at Austin, Austin, Texas 78712; email:huk@utexas.edu, leor.katz@nih.gov, jyates7@ur,rochester.edu

Key Words

decision making, visual motion, visual perception, parietal, lateral intraparietal cortex

Abstract

Over the past two decades, neurophysiological responses in the lateral intraparietal area (LIP) have received extensive study for insight into decision making. In a parallel manner, inferred cognitive processes have enriched interpretations of LIP activity. Because of this bidirectional interplay between physiology and cognition, LIP has served as fertile ground for developing quantitative models that link neural activity with decision making. These models stand as some of the most important frameworks for linking brain and mind, and they are now mature enough to be evaluated in finer detail and integrated with other lines of investigation of LIP function. Here, we focus on the relationship between LIP responses and known sensory and motor events in perceptual decision-making tasks, as assessed by correlative and causal methods. The resulting sensorimotor-focused approach offers an account of LIP activity as a multiplexed amalgam of sensory, cognitive, and motor-related activity, with a complex and often indirect relationship to decision processes. Our data-driven focus on multiplexing (and de-multiplexing) of various response components can complement decision-focused models and provides more detailed insight into how neural signals might relate to cognitive processes such as decision making.

INTRODUCTION: THE INTERPLAY BETWEEN LIP PHYSIOLOGY AND DECISION MAKING

In this review, we focus on LIP neural responses during various versions of a perceptual decision-making task that involves discriminating the direction of visual motion

(Newsome & Paré 1988, Shadlen & Newsome 1996) (see the sidebar titled Direction Discrimination Protocols for Studying LIP and Decision Making). LIP activity has long been regarded as an explicit neural representation of evidence accumulation during these decisions (Shadlen & Kiani 2013). Here, we put forth an analysis that considers how numerous sensory, cognitive, and motor events are related to patterns of LIP activity. While sensory and motor events are typically controlled or measured by the experimenter, cognitive components are almost by definition inferred without direct observation. We therefore adopt a sensorimotor perspective to explore how much of LIP responses can be explained by simple sensory and motor factors. After reviewing a number of correlational and causal investigations, we propose that LIP dynamics during decision making are very well explained by a mixture of visual and premotor signals, with some remaining response components dissociable as decision related. In explicating this perspective, we hope to offer a general framework for interpreting neural correlates of cognitive processes in LIP and elsewhere.

Perceptual decision making is easily conceived of as a two-stage process in which a decision variable evolves until a stopping mechanism commits the process to a particular choice (Schall 2001). The motivation for recording from LIP is guided by the assumption that somewhere in the brain, the decision variable is represented explicitly in the spike rates of single neurons, and that recording from those neurons would effectively let biology constrain models of decision making. LIP has been taken as a likely candidate, and the spike rates of single LIP neurons have been interrogated as direct neural correlates of an evolving decision variable (Gold & Shadlen 2007). Later progress has sometimes reversed the direction of inference, relying on particular decision-making models to constrain (or at least focus) the interpretation of responses in LIP. This work stands as a critical domain for interpreting neural correlates of cognition, in which models of decision making can constrain the interpretation of biology and biology can constrain models of decision making. At times, this bidirectionality can pose a challenge for refining theories, if and when physiology and behavior provide ambiguous or contradictory descriptions of the decision process. Thus, to complement decision-focused explanations of what LIP activity encodes during perceptual decision making, we propose the adoption of a firm sensorimotor starting point. This simplifying perspective derives from the long history of interpreting neural activity in many brain areas (including LIP) in terms of known variables such as visual and oculomotor factors (Colby et al. 1996,Gnadt & Andersen 1988).

DIRECTION DISCRIMINATION PROTOCOLS FOR STUDYING LIP AND DECISION MAKING

The classical paradigms used to study evidence accumulation are the response-time (RT) and fixed-duration (FD) tasks. In the FD task, the sensory evidence is presented for a fixed amount of time regardless of whether the subject has reached a decision or not. The RT task is different in that the sensory evidence is presented for an indefinite amount of time and it is up to the observer to communicate their decision at any given time (Laming 1968). Thus, whereas the FD task produces one data point for every trial—accuracy—the RT task produces two: accuracy and response time.

The response time on every trial in the RT task is thought to reflect the internal process of accumulation and decision formation. The richness of the RT data has appealed to LIP researchers who want to gain more traction on single trials. Although the RT task provides a temporal upper bound on how long a decision took, there is no direct knowledge of all components that constitute the total reaction time. Although some portion of the reaction time is indeed related to the accumulation of sensory evidence, the remainder (termed non-decision time) includes a number of other components (Palmer et al. 2005). The study of neurons in the RT task is challenging because the link between neural activity and evidence accumulation could be mixed with neural responses linked to response preparation.

The FD task thus complements the RT paradigm in this regard, as the motor response is prompted only after the stimulus has been extinguished, such that neural activity related to accumulation is at least separated in time from response preparation, at least to a greater degree than in RT tasks. Of course, the stimulus duration can be varied across trials in a variable-duration (VD) version of the task such that subject accuracy may be probed at different times, providing the researcher with the average state of the evolving decision at any probed time. However, both the FD and VD approaches, in which the experimenter (and not the subject) determines the motion viewing period, are more limited in providing constrained insight into the time frame of individual decisions and might themselves change the strategic weighting of evidence over time without providing a direct means of assessing these effects.

Perceptual decisions can also be probed in a reverse correlation framework. In reverse correlation analysis, it is possible to compute the (average) influence of each stimulus epoch in order to discriminate between different temporal weighting schemes and infer subject strategy. Recent years have seen an increased use of stimuli that are directly amenable to this type of analysis (Brunton et al. 2013, Katz et al. 2016, Raposo et al. 2014, Wyart et al. 2012).

Such reverse-correlation stimuli contrast with the classic random dot motion stimulus, for which reverse correlation calculations must be done post hoc by filtering the moving dot stimulus (Kiani et al. 2008). These analyses should be viewed with caution, as such filtering exercises typically work with simplified simulations of the sensory signals, and any analysis based purely on behavior can miss the temporal dynamics actually present in neural representations of the stimulus (see main text for more discussion of this point). Missing some of the realistic dynamics, sensory filtering could in turn complicate estimates of the time course of temporal integration. More generally, although these new tasks offer a finer level of detail in mapping the relationship between the stimulus and behavior, it is important to note that the resulting kernels are the result of averaging over trials and do not necessarily by themselves indicate the strategy employed on each individual trial.

In the following sections, we focus on studies of LIP activity in the brains of rhesus monkeys while they perform a visual motion direction discrimination task (**Figure 1**). This review is not comprehensive, even within the LIP motion discrimination literature, but we have attempted to provide a thorough treatment of studies most relevant to disentangling sensory, cognitive, and motor factors. Similarly, although we discuss specific insights from complementary studies in other brain areas, other task frameworks, and other species, each of these topics is deserving of longer consideration and has recently been the focus of excellent reviews (Ding & Gold 2013, Freedman & Assad 2016, Hanks & Summerfield 2017).

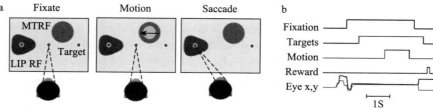

Figure 1

Generic direction discrimination task and neural recording geometry. (*a*) In a standard version of the direction discrimination task, monkeys are trained to discriminate the net direction of visual motion and communicate their decision with a saccadic eye movement to one of two diametrically opposite choice targets. When electrophysiological measurements are taken, task elements are positioned in reference to the receptive field (RF) of the neuron under study. For lateral intraparietal area (LIP) recordings, one of the saccade targets is typically placed in the LIP RF (*blue patch*). For middle temporal (MT) recordings, the motion stimulus is placed in the MT RF (*green patch*). (*b*) Sequence of task events. In this example, the motion epoch is a fixed duration (FD) of approximately 1 s. Timings of individual events can be temporally jittered to decorrelate components. Other task permutations may include motion that has a variable duration (VD) between trials or with a duration determined by the animal's response time (RT).

WHAT WE LEARN ABOUT LIP FROM DECISION MAKING

The first characterization of LIP activity during visual motion decision making was performed in the Newsome lab, following their classic and extensive studies of the middle temporal (MT) area's role in the discrimination of motion direction (Britten et al. 1992, 1993, 1996; Newsome & Paré 1988, Salzman et al. 1992). Having established MT's sensory role in the discrimination of a noisy moving dot stimulus, Shadlen & Newsome (1996, 2001) ventured beyond this key sensory representation to record from macaque LIP during the direction discrimination task. They laid out a compelling set of reasons why LIP was well poised to integrate noisy sensory signals from MT in favor of an oculomotor decision: LIP receives anatomical input from MT (Lewis & Van Essen 2000a, b; Maunsell & Van Essen 1983; Ungerleider & Desimone 1986), projects to oculomotor centers such as the frontal eye fields and superior colliculus (Andersen et al. 1985, 1990; Ferraina et al. 2002; Pare & Wurtz 1997; Wurtz et al. 2001), and displays persistent activity during sensorimotor delay periods (Gnadt & Andersen 1988), reminiscent of the sustained activity in prefrontal cortex during working memory (Goldman-Rakic 1995). Thus, based on both anatomy and physiology, area LIP stood as a potential sensorimotor nexus for the physiological linkage of sensory to motor events (Felleman & Van Essen 1991, Lennie 1998).

LIP was studied using the same motion direction discrimination task used to study MT, in which monkeys view a random dot motion stimulus and communicate their choice about motion direction with a saccadic eye movement to one of two choice targets. For LIP, instead of placing the motion stimulus in the response field (RF) of the cell, the experimenters placed one of the two choice targets in the RF of the LIP neuron under study (**Figure 1**). When the monkey communicated its choice with a saccade into the RF of the neuron, neural activity gradually increased, long before the actual saccade was made. Motor buildup signals are often observed in many oculomotor structures (including LIP) during saccade preparation (Janssen & Shadlen 2005, Wurtz et al. 2001), but this ramping activity had a unique characteristic: The slope of the ramping response was proportional to the strength of the motion stimulus. Thus, the dynamics of the LIP response mimicked what one would expect from the accumulation of motion evidence (i.e., the time integral). These early observations were made in a fixed-duration (FD) version of the random dot task (Shadlen & Newsome 2001), but related observations were made in a response-time (RT) version of the task (Roitman & Shadlen 2002). (The two paradigms are complementary, and we discuss their tradeoffs in the sidebar titled Direction Discrimination Protocols for Studying LIP and Decision Making.)

LIP's ramping responses led to the hypothesis that this brain area reflects the accumulation of evidence, and a quantitative model showed that LIP responses were qualitatively similar to the time integral of differential MT activity (Mazurek et al. 2003). Such temporal integration could be the neural basis of evidence accumulation, as formulated in a variety of sequential sampling models in statistical and psychological work, most notably, the drift-diffusion model (Palmer et al. 2005, Ratcliff 1978). This was a powerful instantiation of connecting physiology to an existing mathematical model of decision making, broadly analogous to efforts put forth by others in more rapidly performed oculomotor tasks (Carpenter & Williams 1995; Hanes & Schall 1995, 1996). To test this apparent correspondence, Huk & Shadlen (2005) used brief pulses of motion to test whether LIP firing rates matched the predictions of an integrator. They predicted that if LIP reflected the accumulated motion evidence, then these additional pulses of evidence should affect LIP firing in a sustained (as opposed to transient) manner. In line with this prediction, LIP responses reflected the brief added pulses for many hundreds of milliseconds after the motion pulse had ended—and the length of these effects corresponded quantitatively to the ending of the decision process, as might be implemented by a decision bound. This result reinforced the notion that the accumulation of motion evidence believed to underlie decisions was reflected explicitly in the spike rates of individual LIP neurons. Because the quantitative relation between motion stimuli and the time-varying LIP responses is at the heart of what computations LIP might reflect during decision making, we revisit repeatedly this apparent mapping between LIP activity and the temporal accumulation of motion evidence.

Although these foundational studies bolstered interest in LIP responses as a neural correlate of evidence accumulation, it has long been appreciated that LIP activity during decision making comprises a number of non-decision-related response components as well. One such component is the onset of the choice targets preceding the motion stimulus (one of which is placed within the LIP RF), which elicit strong responses in LIP. Strong, time-locked responses to sensory stimuli in the LIP RF have been well documented (Bisley et al. 2004, Blatt et al. 1990, Robinson et al. 1978), but it is not clear whether the sensory-driven responses are punctate (and effectively over by the time of motion onset) or whether a more persistent, target-driven response component bleeds into the motion viewing epoch in which the decision about motion direction is made. A second component is the saccade used to communicate the decision, which is likely preceded by a premotor response that may emerge during the motion viewing epoch as well. Although initial analyses assumed that only the final 75–100 ms of LIP activity during decisions are affected by a premotor burst (Roitman & Shadlen 2002), pre-saccadic activity often builds up gradually over hundreds of milliseconds across multiple areas of the oculomotor system in simpler tasks (Mountcastle et al. 1975, Wurtz et al. 2001). Despite these considerations, early experimental paradigms and analyses were not designed to isolate either persistent responses to the visual targets or premotor buildup responses to the saccade, both of which could temporally overlap and complexify the interpretation of responses that occur during motion evidence accumulation.

For LIP activity to provide clear insight into the decision process not available from the behavioral responses themselves, it is important to characterize and quantify how these secondary sensory and motor signals are encoded in LIP spike rate and, ideally, to isolate them from the decision formation signals. This is a challenging endeavor because in the standard motion discrimination paradigm, the spatial and motor aspects of the task are often inextricably linked to the decision about the sensory stimulus (Gold & Shadlen 2007, Roitman & Shadlen 2002, Shadlen & Newsome 1996). Simply put, in standard versions of the task, there are correlations between the motion stimulus, the decision about motion direction, and the motor response used to indicate the decision. This structure calls for a more targeted dissection of the response components in order to isolate decision-making signals; we next describe both empirical and analytic attempts to do so.

Dissection of LIP's Correlations with Decision Making and Identification of Sensorimotor Multiplexing

The isolation of sensory and motor responses from decision-correlated activity has recently received renewed consideration. In one especially powerful variant of the motion discrimination task, the impending direction of the saccade was decoupled from the directional decision using a color-coded mapping between motion direction and saccade targets, revealed at different times relative to decision formation (Bennur & Gold 2011). LIP neurons were found to distinctly represent the onset of targets, the direction of motion, and the premotor plan to saccade, mixed in their spike rates. Importantly, the signals associated with saccadic choices (and even the color coding instruction) were substantial in magnitude relative to the direction-selective motion response, and their time course appeared to begin to emerge as soon as the stimulus-response mapping was revealed to the subject. Furthermore, the direction-selective motion response could be present irrespective of whether or not the mapping to the saccade target locations had been revealed (and could even be inconsistent with the preferred direction of the choice-related response), raising the question as to whether motion-driven responses (dissociable from saccadic choice signals) might be a sensory phenomenon, akin to observations of direction selectivity in other LIP studies (Fanini & Assad 2009).

In another empirical sensorimotor dissociation, Meister et al. (2013) manipulated the presence of the choice targets to isolate the target-driven visual response from the decision-correlated activity. Monkeys performed the standard direction discrimination task, but on a fraction of trials—instead of leaving the visual targets on for the course of the trials—the targets appeared for only 100 ms and then quickly extinguished. This simple manipulation revealed that the visual drive from the target exerted prolonged effects on firing rates throughout the trial, including the decision-making epoch. Thus, LIP spikes during motion viewing were, at the least, a function of both the decision process and the simple presence of a target in the RF. Meister et al. (2013) additionally highlighted substantial heterogeneity in the response dynamics and sensorimotor multiplexing of individual LIP neurons. A similar multiplexed encoding of task and saccade components in LIP has been observed in a perceptual categorization task (Rishel et al. 2013).

These sensorimotor manipulations paint a picture of multiple, temporally overlapping response components that contribute to the neural response during the putative decision phase of the trial. Although such empirical investigations were valuable, these first steps involved coarse analyses, such as comparison of the mean responses across a small number of differential experimental conditions. In focusing on such simple comparisons, analyses do not access all the factors that might be driving neurons, nor do they provide insight into single-trial response dynamics. It struck us as potentially useful to characterize LIP's multiple response components by leveraging the temporal dissociation between events on individual trials that are typically present in experimental designs (i.e., jittered timings between experimenter-controlled events, and variable onsets of behavioral responses) to provide detailed and data-driven characterizations of the sensory and motor components, and to quantify how such response components might interact.

To dissect LIP's multiplexed encoding in such a general-purpose analytic framework, Park et al. (2014) applied a generalized linear model (GLM) to statistically characterize the multiple factors that drive spiking in LIP during decision making. The GLM is a form of the linear-nonlinear-Poisson model often used to describe sensory and motor responses as a function of external variables (Brown et al. 1998; Fernandes et al. 2014; Jacobs et al. 2009; Mayo et al. 2015, 2016; Paninski et al. 2004; Pillow et al. 2008; Ramkumar et al. 2016; Yu et al. 2009). These encoding models describe the probability of spiking for a single LIP neuron at a particular time as a nonlinear function of a weighted combination of the stimulus and task components, whose contributions can overlap in time or space (**Figure 2a**) (Pillow et al. 2005, Truccolo et al. 2005).

Applied to LIP, the GLM analysis revealed that activity throughout the trial reflected responses to the choice targets, the visual motion, and the impending oculomotor choice, each with their own shape and time course (**Figure 2b**). These multiple components overlapped over long timescales, meaning that LIP spikes, even during the period of motion viewing (and hence, during decision making), could be described by a combination of factors (including a decaying visual response to the onset of the choice targets and a growing response corresponding to the impending saccade), as opposed to a pure neural correlate of evidence accumulation. Within this framework, these contributions were characterized as the output of linear filters (convolved with the time course of corresponding events). Importantly, although these components are summed

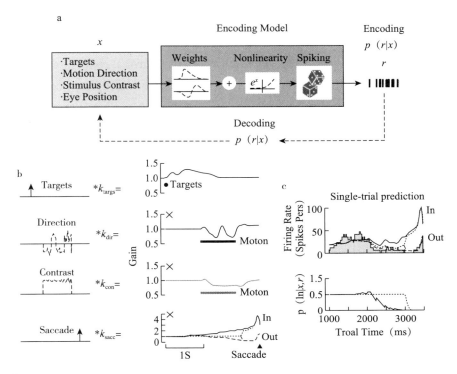

Figure 2

Generalized linear model (GLM) applied to lateral intraparietal area (LIP). (*a*) The GLM describes the probability of a spike train *r* given external variables *x*. This relationship, *p*(*r*|*x*), is given by a Poisson process with a rate that is generated by filtering the external variables linearly and then passing the summed output through a static nonlinearity. The conventional exponential nonlinearity implies that all linear terms interact multiplicatively. (*b*) The separate contribution of the targets, direction of motion, contrast of the stimulus, and saccade of the animal (*left column*) to spike rate are depicted as spike rate gains. The individual component gains (*right column*) are produced by convolving the stimulus covariates with their respective linear filters and exponentiating them. The × indicates that these gains are multiplied together to produce the spike rate for the neuron, a result of the exponential nonlinearity. The saccade kernel exerts a choice-dependent effect on spike rate with rate increasing for choices in the response field (RF) (*blue*) and decreasing for choices out of the RF (*red*). The dashed lines represent the effect the saccade would have if it affected spike rate only up to 500 ms before the saccade. (*c*) An example single-trial prediction for an LIP neuron (*top*). The predicted rates for a choice into the RF (*blue*) and out of the RF (*red*) are overlaid with the binned spike count for this neuron (*gray*). The probability of a choice into the RF is derived from the two predicted rates (*black*). Predicted responses using more punctate (truncated) saccade kernels are shown by the dashed lines, and clearly fail to account for the observed response. Figure modified with permission from Park et al. (2014) using data from Katz et al. (2016) and Yates et al. (2017).

before being passed through a static nonlinearity, the particular form of the nonlinearity can capture multiplicative interactions between them (**Figure 2*b***). In LIP, many of the neurons were better described by multiplicative interactions between different sensory and motor components than by linear combinations (Park et al. 2014). These nonlinear interactions between task and stimulus variables, combined with substantial encoding heterogeneity across the population of LIP neurons, reflects an instance of mixed selectivity, for which theoretical treatments have highlighted computational benefits (Fusi et al. 2016, Pagan & Rust 2014, Raposo et al. 2014, Rigotti et al. 2013). Thus, the multiplexed nature of LIP responses may be purely accidental but could also carry information processing benefits, wherein the heterogeneous population can still be read out with a simple mechanism (Park et al. 2014).

One limitation of Park et al.'s (2014) analysis was that the motion stimulus was modeled with a single fixed coherence value throughout each trial. It therefore characterized LIP's motion-related response component in reference to the idealized stimulus (in the statistical sense, the expectation of the stimulus), as opposed to the actual noisy output of MT neurons, which Newsome's work has compellingly demonstrated provides a critical source of sensory data to decision-making stages. This simplified formulation obscures the degree to which LIP reflected a particular computation as opposed to inheriting it from MT's sensory processing stages, especially at the scale of single trials. This motivated us to characterize the responses of both LIP and MT neurons using a reverse correlation stimulus (see the sidebar titled Direction Discrimination Protocols for Studying LIP and Decision Making) while simultaneously recording from neurons in both MT and LIP (Yates et al. 2017).

The first notable observation from these multiarea recordings was that MT responses to motion exhibited direction selectivity that was initially strong and decreased over the course of the motion epoch. Such dynamics have important implications for interpreting the time-varying effects of motion in later cortical stages such as LIP and in the psychophysical behavior. By analyzing LIP with respect to MT's time-varying output (instead of with respect to the stimulus), the oft-observed decay of motion-driven responses in LIP could largely be explained by inheritance of temporally decaying weighting already evident in MT. Isolating this stimulus-related term then revealed the second most notable observation: That LIP's ramping responses were more closely linked to the upcoming saccade than to the integration of motion. The motion integration response was small, and a larger saccade-related response buildup was necessary to quantitatively explain the underlying steepness of the response ramps (**Figure 2c**). These motion-integration and saccade linked responses were distinguishable by virtue of each trial having a distinctly variable (and known) time course of motion, thus improving the statistical dissection of decision from saccadic response. In fact, many single LIP units exhibited only the saccade-linked response component, further supporting the notion that the majority of LIP's ramping is dissociable from motion integration.

Regardless of the source of the motion response in LIP, the primary take-home message from this analysis is simply that LIP responses may be based on a correlation with the impending saccade more than with the accumulated evidence per se. When viewed through a sensorimotor lens, LIP's responses do not require an appeal to a primary status as a direct neural correlate of decision formation. Instead, LIP is well explained as a combination of decision-irrelevant sensory responses, a large premotor component, and a substantially smaller component related to the integrated motion signal, which itself inherits important dynamics from the MT stage of processing. The isolated motion response (i.e., the representation of the putative decision variable) is certainly deserving of more study. Based on the sensorimotor multiplexing perspective, however, one might suspect that the small motion responses in LIP—despite having been isolated from sensory and motor events—could at least in part result from the motion stimulus grazing the RF of some LIP neurons. This conjecture is consistent with two established aspects of LIP responses: (*a*) that LIP RFs are quite large (Blatt et al. 1990) and (*b*) that when a motion stimulus is intentionally placed within the RF, LIP neurons exhibit direction-selective responses (Fanini & Assad 2009).

Causal Interrogations of LIP and Distributed Processing for Decision Making

Although LIP has received extensive focus, decision-related ramping activity has been observed in many brain areas, even during motion direction discrimination tasks. These areas include the frontal eye fields (Ding & Gold 2012a, Gold & Shadlen 2000, Kim & Shadlen 1999), the parietal reach region (de Lafuente et al. 2015), and subcortical structures such as the superior colliculus (Horwitz & Newsome 1999) and the caudate nucleus (Ding & Gold 2010, 2012b). Decision-related activity has even been observed outside the central nervous system, in muscle stretch reflexes (Selen et al. 2012) and inferred from overt arm movements (Song & Nakayama 2009, Spivey et al. 2005). Even saccades, which exhibit considerably more ballistic dynamics, can reflect evidence accumulation during a motion discrimination task (Joo et al. 2016). In light of the complexities in interpreting LIP's ramping responses (described in the previous section), this constellation of results is a reminder that thorough physiological insight into the mechanisms of decision making will likely require recording from areas beyond LIP and also calls for more nuanced dissections of their respective decision-correlated activity patterns.

To explore the functional relevance of one node of this larger network, we recently tested LIP's role in decisions by inactivating it reversibly during a direction discrimination task (Katz et al. 2016). The musicmol infusions targeted the very clusters of LIP cells that exhibited substantial decision-related activity in preceding electrophysiological recordings, and silencing of these clusters was confirmed electrophysiologically on every session. Despite having exhibited strong decision-related activity prior to muscimol infusion, inactivation of neurons in LIP resulted in no clear impact on the measured decision-making performance, indicated by virtually indistinguishable behavior between the baseline and LIP inactivation conditions (**Figure 3**). This held for both a reverse-correlation direction discrimination task and an FD version of the classic moving dots task. Behavior in a free-choice control task, in contrast, was significantly disrupted by the LIP in activations. In this control task, monkeys chose freely between two saccade targets in the absence of a motion stimulus. Inactivation in LIP shifted choices away from the target in the contralesional hemifield, consistent with previous inactivation studies in monkeys, rodents, and humans (Balan & Gottlieb 2009, Erlich et al. 2015, Kerkhoff 2001, Kubanek et al. 2015, Wardak et al. 2002, Wilke et al. 2012, Zirnsak et al. 2015), and indicating that neural activity in LIP—although necessary for guiding free choices—is not necessary for the type of perceptual decision traditionally used to elicit decision-correlated activity in LIP.

Do these results mean that LIP is not used for decision making? Interpreting the results of a causal manipulation is challenging when the manipulation produces a positive result but is all the more so for a null result. A lack of measurable impact of inactivation is inconsistent with the notion that clusters of decision-correlated neurons in LIP play some sort of large and stable role in decision making. However, it could also be the case that these LIP neurons are in fact usually relied upon, but that other structures are recruited in compensation. This is an intriguing, but we think unlikely, hypothesis, given that inactivation of LIP in other cognitive tasks (e.g., attentional selection, visual search, free choices) does result in reliable behavioral deficits (Balan & Gottlieb 2009, Kubanek et al. 2015, Liu et al. 2010, Wardak et al. 2004, Wilke et al. 2012). Such an account would require that LIP be compensated for by other structures in a decision-making task but not in these related tasks—including the free-choice control task in which behavioral disruption was observed (in the same experimental sessions as the motion discrimination task inactivations).

A related issue is the unilateral nature of the inactivations performed. Could bilateral inactivation of LIP have revealed an effect that is unobservable in the unilateral condition? Along similar lines of reasoning, it seems unlikely that unilateral inactivation in LIP is sufficient to elicit clear behavioral deficits in the studies mentioned above but not in a direction discrimination task. Additionally, Katz et al. (2016) tested for this possibility by placing both saccadic choice targets within contralesional space and still

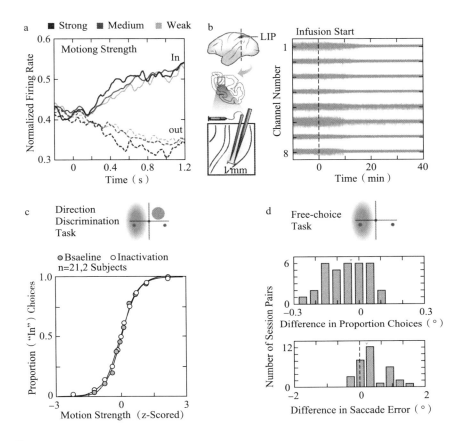

Figure 3

Inactivation of decision-correlated activity in lateral intraparietal area (LIP) does not have a significant effect on decision making.(*a*)Average response of 113 LIP neurons as a function of motion strength and direction (in versus out of the cell's response field, *solid* and *dashed lines,* respectively), aligned to motion onset. (*b*) Schematic of the inactivation protocol (*left*). A multielectrode array was lowered alongside the infusion cannula to identify the targeted cortical location, verify neural selectivity before infusion, and confirm neural silencing after. On the right, continuous voltage traces are shown from an example inactivation session in which neural silencing is evident approximately 10 min after infusion start. (*c*) Psychophysical data for the direction discrimination task, averaged over pairs of baseline and muscimol treatment sessions in area LIP. Upper panel shows the experimental geometry along with the estimated inactivated field (*gray cloud*). (*d*) Distribution of behavioral data in the free-choice task in which the animal chose between two targets that flashed in variable locations in the absence of a motion stimulus. Histograms show baseline and inactivation differences in proportion contralateral choices (*top*) and saccade error (*bottom*); positive numbers indicate an increase following inactivation. Figure modified with permission from Katz et al. (2016).

found no effect (i.e., instead of bilateralizing the inactivations, they unilateralized the task geometry). Thus, although LIP is critical for other cognitive functions, it is unlikely to be a critical node, and definitely not a fixed bottleneck for the behavior under study in the decision-making task. Of course, it remains possible that more nimble perturbation techniques (e.g., temporally precise and trial-interleaved optogenetic silencing) stand to reveal more nuanced contributions of LIP to decisions in the future (Dai et al. 2014, Li et al. 2016). If such techniques generate results not easily reconcilable with those of pharmacological inactivation, that would be particularly exciting, because they would contrast starkly with the history of causal perturbations in MT—for which slow, coarse techniques (lesion, pharmacological inactivation) and fast, localized techniques (microstimulation) paint a fairly cohesive picture with little need to appeal to fast or middle time frames of compensation (Chowdhury & DeAngelis 2008; Fetsch et al. 2014; Katz et al. 2016; Murasugi et al. 1993; Newsome & Paré 1988; Salzman et al. 1990, 1992).

These LIP inactivation results weigh against a critical role for LIP in the accumulation of sensory evidence, but a previous study may appear superficially at odds: Hanks et al. (2006) performed microstimulation of LIP neurons during the motion viewing epoch of the direction discrimination task. They found that microstimulation shortened response times in both monkeys, and subtly shifted the proportions of choices in favor of the target in the microstimulated LIP RF. Reduction in RT and increased choices to the target in the RF of the microstimulated LIP neurons is indeed consistent with an offset of the integrated signal. However, the results from this experiment should be interpreted with an open mind for several reasons. First, microstimulation may recruit areas other than the targeted area (LIP) via antidromic stimulation (Histed et al. 2009, Tehovnik et al. 2006), consistent with a distributed network. Second, the reduction in RT and subtle increase in choices can be explained in multiple forms outside the evidence accumulation framework, for example, as an increase in attention to one choice target over the other. In studies that focus on target selection in the presence of distractor targets, both microstimulation and inactivation in LIP produce behavioral deficits consistent with the results reported by Hanks et al. (2006) (Balan & Gottlieb 2009, Cutrell & Marrocco 2002, Dai et al. 2014, Kubanek et al. 2015, Wardak et al. 2002, Zirnsak et al. 2015).

Attempts to elucidate the differential role of decision-related activity have recently emerged in the rat model system, with a particular focus on posterior parietal cortex (PPC) (Erlich et al. 2015, Hanks et al. 2015, Harvey et al. 2013, Licata et al. 2016, Raposo et al. 2014). Despite an incomplete case for homology between rodent and primate PPC (Belmonte et al. 2015, Cooke et al. 2014), one clear result is that inactivations of rat PPC did not affect behavior in an auditory evidence accumulation task (Erlich et al. 2015), whereas similar inactivations in PPC by a different group did have an impact in a visual decision-making task (Licata et al. 2016, Raposo et al. 2014). Although one way to reconcile these results with the macaque results reported above is to posit that rodent

PPC and macaque LIP are fundamentally different, another is that the visual deficits observed by the Churchland group may be more sensory in nature and correspond to the motion-driven responses observed in macaques when motion is placed in the LIP RF (Fanini & Assad 2009, Freedman & Assad 2006, Licata et al. 2016).

In light of the above correlational and causal observations, our perspective is that the neural basis of evidence accumulation requires considerable additional investigation. Temporal accumulation could be performed in a single cortical area, but it could just as well be performed in a subcortical region or within a more distributed network. Regardless of how one might map a cognitive process onto a brain area or multiarea circuit, LIP's complex and perhaps indirect relation to decision formation may soften focus on recording from this area to draw inferences about nuanced neural mechanisms of decision formation, a topic we turn to in the following section.

APPLYING THE SENSORIMOTOR MULTIPLEXING PERSPECTIVE TO OTHER ASPECTS OF LIP RESPONSES

In the previous section, we focused on how LIP responses can be described in terms of experimenter-controlled or experimenter-observed external variables, such as visual stimuli or saccadic eye movements. This is not the approach that has generated so much interest in LIP, however. Instead, the appeal of LIP recordings is the prospect of access to internal (cognitive) dynamics that are not directly experimentally controlled or observed. These dynamics are traditionally interpreted under the umbrella of drift-diffusion models (Ratcliff et al. 2016), where a number of model parameters describe the relationship between LIP responses and decision formation. In contrast, our statistical dissections and causal interrogations of LIP support the notion of LIP as carrying a combination of sensory and motor signals that are related to the task but less directly correlated with decision making. In the current section, to test the viability of our complementary focus on sensorimotor multiplexing, we ask whether simple sensory and motor components might also account for neural activity that has typically been taken as an impetus to append the drift-diffusion model of LIP.

Perhaps the most obvious mismatch between LIP responses and generic forms of drift-diffusion models of evidence accumulation is the initial phase of the neural responses. LIP motion-related responses often begin with a nonselective phase that lasts ~200 ms and then evolve into the classic ramping dynamics, which follow the motion stimulus with the same latency. Within those first ~200 ms, LIP responses often follow a dip-and-rise pattern that is not an inherent part of abstract drift-diffusion models (Churchland et al. 2008; de Lafuente et al. 2015; Huk & Shadlen 2005; Katz et al. 2016; Kiani & Shadlen 2009; Kiani et al. 2008; Meister et al. 2013; Roitman & Shadlen 2002; Shadlen & Newsome 1996, 2001). In the past, it has been posited that the dip and rise reflects the brain's way of resetting the integration process (Huk & Shadlen 2005), a transient attentional shift to the motion stimulus (Wong 2007), or both, although neither cognitive explanation has direct empirical support. Alternatively, a parsimonious account of the dip and rise derives from a sensory account. In an experiment in which we presented the motion stimulus very soon (~200 ms) after the onset of the response targets (Meister et al. 2013), we found that the dip and rise now exhibited a coherence dependence that was nondirectional: The stronger the motion, the lower the LIP response dipped, independent of motion direction. This would be quite perplexing if one were to take the LIP response as a neural correlate of a decision variable, because a decision-framework explanation would posit that the animal had an early hunch about the strength of motion and then lowered the starting point of evidence accumulation on easier trials.

Instead, the nondirectional coherence-dependent dip-and-rise pattern of responses can be explained parsimoniously as an instance of widefield divisive normalization, as suggested in other LIP studies (Falkner et al. 2010, Louie et al. 2014). In this sensory-based explanation, the response of the LIP neuron under study (which, in the idealized case, has a single response target in its RF) is divided by the net response of other neurons in the area. If the response of other LIP neurons (which have the motion stimulus in their RFs) is larger for higher motion strengths [as it is in many other areas, including MT and medial superior temporal area (MST)], then the normalization term would be stronger and the response would be reduced more. This idea is further supported given the temporal match between maximal dipping time of the LIP dip and rise (when the target is in the RF, ~200 ms) and the time when directional selectivity of LIP neurons is maximal (when motion is in the RF, also ~200 ms) (Fanini & Assad 2009). The only difference between this scheme and the normalization that is better understood in earlier visual areas is that it would require broad spatial tuning of the normalization pool. This seems reasonable given the large RF sizes and messy retinotopy in LIP (Ben Hamed et al. 2001, 2002; Blatt et al. 1990; Janssen et al. 2008; Platt & Glimcher 1997; but see Patel et al. 2010) and the flexible tuning observed in LIP during learning (Sarma et al. 2015), but more direct analyses are required.

In other experiments, the number of targets has been manipulated to change the decision-making context. In one important study, the number of directions to be judged was increased to four, and so was the number of choice targets (Churchland et al. 2008). Responses in LIP started lower in the dip-and-rise phase when more choice alternates were allowed (and when more targets were present). This response baseline could be interpreted cognitively, as a lower starting point for a drift-diffusion process when more options are contemplated. Alternatively, the sensorimotor framework expanded above highlights that more targets would drive the full-field normalization pool more strongly and thus reduce the level of activity in a particular LIP neuron being recorded from. Similar target-number effects have been observed in other oculomotor structures (Basso & Wurtz 1997, 1998). Such normalization could also contribute to response dynamics when a third target is presented in variants of the perceptual decision-making task to study confidence (Kiani & Shadlen 2009). Although such studies have included some controls, the strongest isolation of sensorimotor/normalization contributions would involve always presenting the same number (and geometric layout) of targets while changing only their instructive meaning (by virtue of color coding or similar manipulations). Such controls are difficult to implement in practice (Churchland et al. 2008), but the sensorimotor framework will hopefully motivate such considerations in related future work.

Under some conditions, modulations of the dip and rise not easily ascribed to simple stimulus components have been observed. When there are asymmetric rewards or prior probabilities associated with one direction relative to the other, the dip and rise exhibits modest but sometimes measurable offsets in response—slightly higher activity in neurons that correspond to the more preferred target (Hanks et al. 2011, Rao et al. 2012, Rorie et al. 2010). This could be interpreted in terms of a changing starting point of evidence accumulation. However, because decisions and motor responses were linked in these studies, it is not possible to identify whether the observed bias is an offset in the decision variable or an increased level of motor preparation. In our work, the statistical relation between LIP activity at one point in time and the eventual saccade spans a wide time range. Thus, when manipulations of bias affect behavior, they are likely to affect the neural responses in commensurate ways, and the exact form (i.e., whether they offset the starting level or have more complex time-varying effects) likely depends on the timing of the saccade (or the saccade planning) relative to the motion viewing period (Rao et al. 2012). However, it is certainly possible that other types of decisionrelated variables (e.g., payoffs, rewards) could be directly reflected in LIP activity, and the sorts of dissections performed in some studies (Park et al. 2014, Rorie et al. 2010) are likely to continue to provide insights on this issue.

Analogous to the difficulties in interpreting the beginning of LIP responses as the start of evidence accumulation, the final level of activity is equally difficult to interpret as reflecting the final level of accumulated evidence. The sensorimotor approach taken by Meister et al. (2013), for example, showed that the level of LIP response around the time of the saccade—even though it was separated from the motion viewing by a delay period—was lower in trials in which the visual targets had been extinguished at the beginning of the trial compared to those in which the targets stayed on (despite no obvious difference in psychophysical performance). This indicates that the final level of LIP activity cannot be interpreted as a singular correlate (e.g., a decision bound). A visual manipulation occurring at the start of a trial affecting the response level at the end of the trial highlights the limitations in interpreting any phase of the LIP response in uniquely decision-related terms.

LIP responses sometimes exhibit a global upward trend that is not consistent with a pure diffusion to bound mechanism. This pattern of residuals has been cast in decision-related terms as an added urgency signal that serves to push the accumulated evidence closer to the bound (Carland et al. 2016, Churchland et al. 2008, Cisek et al. 2009, Hanks et al. 2014). In LIP experiments, this urgency signal was originally defined as the residuals between the observed LIP response and the predicted time course based on symmetric drift-diffusion. Although this upward trend could indeed be an added cognitive signal, the fact that the urgency signal is the mathematical difference between LIP responses and pure accumulation could also be couched in a sensorimotor manner: The same premotor buildup terms we have described can capture this overall positive tendency, because the saccade component related to saccades into the RF is larger than the corresponding term for out-of-RF saccades. It would therefore be interesting to see manipulations of urgency that do not manipulate the timing of motor responses, although such a dissociation is challenging to envision in practice.

One other feature of LIP responses that has recently garnered interest is the dynamics of single trial spike trains during decision making. A recent application of advanced statistical modeling argued that discrete transitions (steps) provided a better description of single-trial LIP activity than noisy accumulation trajectories (ramps), the latter being the core component of drift-diffusion models (Latimer et al. 2015). Although these conclusions have been followed by extensive and ongoing debate, the sensorimotor multiplexing hypothesis highlights that single-trial responses are a function of multiple overlapping features with their own dynamics. Just as this perspective tempers attempts to treat averaged response ramps as direct correlations of decisions, it seems only fair to also apply this caveat to single-trial assessments of either stepping or ramping. We propose that if isolatable decision-related dynamics exist at the single-neuron or single-trial level, a more definitive analysis scheme will need to be built upon single-trial isolation of both sensory and motor factors. Otherwise, such finer-grained analyses of the statistics of LIP responses may reflect differential dynamics that are in part due to sensory responses, presaccadic activity, or both (see also Churchland et al. 2011). If we are to draw a constructive take-home message from the step-ramp debate at this time, it is simply that the time is right to specify models of decisions and their relation to physiology with greater precision, and that this level of precision requires simultaneous consideration of a multitude of sensory, cognitive, and motor signals. Of course, given the lack of a clear causal role of LIP in motion decisions, recordings from areas (or circuits) more clearly involved in decisions may provide more direct insight into the dynamics of the decision-making process (Brody & Hanks 2016, Pesaran & Freedman 2016, Pisupati et al. 2016).

Taken together, dynamics in LIP activity during decision formation could still potentially map onto evidence accumulation, though in a complex and indirect manner. We hope to have highlighted the challenges in doing so, with a constructive emphasis on sensory and motor factors. We put forth that renewed interest in multiplexed sensorimotor factors will allow for less controversial and more powerful isolation of cognitive signals.

CONCLUSIONS

In this review, we have highlighted a complementary framework for interpreting LIP activity with respect to perceptual decision making. The two basic ideas behind the sensorimotor multiplexing framework are that (*a*) LIP (and many other higher brain areas) responds to a multiplicity of sensory, cognitive, and motor factors, and (*b*) in the motion discrimination task (and essentially every other perceptual decision-making task), there are a multitude of sensory and motor task elements present. These sensorimotor task elements can have varying degrees of relation to the decision process; at one end, saccadic buildup is typically tightly related to the decisions, whereas at the other, visual responses to the choice targets are an indirect by-product of the way the task is implemented. Such signals, even during simple tasks, are multiplexed, distributed, and not necessarily isomorphic with psychological theories. At an extreme, the close correspondence between LIP spike rates and theories of decision making could result primarily from an amalgam of partially correlated factors overlapping. Based on the evidence summarized above, we believe that a moderate form of this implication is warranted—that significant components of LIP activity are sensorimotor correlates of task performance, and that direct correlates of decision formation should be isolated only after taking sensorimotor multiplexing into account, either in experimental design or in data analysis.

Although the overlap and possible mixture of decision and motor signals are always a theoretical possibility, this review has suggested that in practice, a substantial part of LIP activity is easily thought of as premotor buildup activity, as distinct from the temporal accumulation of evidence. Although a small component of LIP activity does indeed carry the signature of the motion time course on each trial, the majority of decision-correlated activity can be described as a response component that builds up (or not) if the animal is forming a decision that corresponds to an eye movement into (or out of) the RF of the neuron under study. More direct tests of this point are called for, especially in light of category-specific responses in LIP that also appear to be dissociable from eye movement signals (Rishel et al. 2013). We propose that LIP activity will continue to be inextricably linked with choices across a wide range of tasks whenever the decisions are communicated with geometrically consistent saccades (Kira et al. 2015, Yang & Shadlen 2007), and that many instances of decision-correlated LIP dynamics will remain ambiguous until experimental and analytic approaches facilitate the disentangling of its multiplexed response components.

The sensorimotor framework we have proposed emphasizes an aspect inherent to many studies and thus initially aligns with an intentional framework: that decision-making signals and premotor activity are often commensurate and mixed together in many experimental designs and brain areas. However, the framework also comprises both clever experimental designs and sensorimotor de multiplexing analyses, which suggest that such signals are potentially dissociable (Chowdhury & DeAngelis 2008, Freedman & Assad 2011, Shadlen et al. 2008). Decision and premotor signals are multiplexed and intertwined to the extent that decision-making signals may even be observed outside the central nervous system, in the motor periphery (Joo et al. 2016, Selen et al. 2012, Song & Nakayama 2009, Spivey et al. 2005). However, not all task variants produce this multiplexing phenomenon (Bennur & Gold 2011, Gold & Shadlen 2003), presumably because the mixed nature of decision and premotor responses may only be present once the mapping between abstract decision and appropriate motor action is predictable and has been learned (Law & Gold 2008, Sarma et al. 2015). Thus, our perspective is that rather than study particular tasks in which LIP resembles a neural correlate of decisions, it will be particularly illuminating to seek and identify when and how these multiplexed signals are (and are not) present to understand how the brain computes decisions and plans corresponding actions.

It is intuitive to raise a philosophical objection here—that presaccadic buildup in and of itself is a neural correlate of the evolving decision. However, the sensorimotor (de-)

multiplexing focus on dissecting multiple decision-correlated response components is important to appreciate. If LIP primarily represented a decision variable, its activity would be well explained by the transformation of the motion stimulus, whereas responses related to the saccade would be of smaller amplitude and precede the saccade by a brief period of time (as is commonly posited by analyses that simply censor the ~75 ms before a saccade). Instead, statistical analyses suggest that the motion-related response is relatively tiny, and that the dominating aspect of LIP response is a slow buildup of saccade-related activity. Of course, the response buildup is also correlated with the passage of time preceding a saccade, which LIP activity is known to track (in a choice-selective manner) even without ongoing perceptual decisions (Ipata et al. 2006, Janssen & Shadlen 2005, Jazayeri & Shadlen 2015, Leon & Shadlen 2003).

We have hypothesized that saccade-correlated activity emerges before decisions are complete, and that this premotor component interacts with a small stimulus-related signal. If one drops the assumption that LIP signals must map onto decisions explicitly, the potentially unintuitive nature of this sensorimotor interaction dissolves, as these signals may in fact be irrelevant to aspects of task performance upon which we focus, or even epiphenomenal. Of course, what we have called a presaccadic signal might not be inherently motoric but could reflect computations and signals only correlated with saccade planning (e.g., attention) and that are difficult to dissociate in the tasks we have discussed. Finally, we also note that our statistical identification of this presaccadic signal was performed within a GLM framework that included multiplicative interactions. As described above, this form of interaction is a basic component of all linear-nonlinear models of neural processing (Chichilnisky 2001). If such interactions are reflected in LIP's computations, this would imply that as the visual target response fades and the presaccadic signal increases, any true motion-driven responses will have changing gains. This last point also relates our perspective to models that assert that a time-varying signal gates a low-passed version of the stimulus (Carland et al. 2016, Cisek et al. 2009).

Although we have focused on the relations between decision formation and motor preparation, the sensorimotor framework also calls for renewed focus on the sensory representations that feed the decision mechanism. In particular, a detailed understanding of the sensory stimulus and neural representation of that stimulus have important implications for inferences about evidence accumulation derived from behavior. The classical motion discrimination experiments have used an algorithm for modulating motion strength of a moving dot display with strong stochastic elements (Newsome & Paré 1988). This means that although each trial is designed to have a particular coherence and direction, the actual motion strength presented can be quite different from the statistical expectation, both in the moment-by-moment fluctuations within a trial and in the sense of the cumulative net motion the subject is supposed to discriminate. Thus, accuracy in the dots task is limited by a combination of both internal neural noise as well as external noise (i.e., variability of the noisy stimulus as presented on a particular trial, point in time within a trial, or both). The discrepancy between motion expectation and cumulative net motion presented in a particular trial is mitigated for longer viewing durations (as the net motion approaches the expectation in the limit of time). Improvements in accuracy as a function of duration thus could reflect both internal evidence accumulation and the external process of the stimulus coming closer to its statistical expectation. Both types of temporal integration could play out similarly, and simultaneously, during the time course of decisions. Given that MT's encoding of motion is sensitive to variability in the motion stimulus and has significant temporal dynamics, it is important to frame decision-related signals in higher brain regions with respect to the MT response, as opposed to with respect to the stimulus, and especially not with respect to the expected or average stimulus. This is why some of our recent studies have used a stimulus and generation algorithm with more direct control over the stochastic elements—an approach motivated by elegant designs in other lines of decision-making research (Katz et al. 2016, Raposo et al. 2014, Brunton et al. 2013, Wyart et al. 2012).

Importantly, LIP does not function in isolation, and the sensorimotor perspective offered here is similarly applicable across the entire relevant network. In the case of motion decisions, we have already argued that the sensory filtering reflected in MT outputs is an important element in the proper interpretation of how later brain areas and decisions process motion information (which is not the stimulus, but the relevant neural signals encoding the stimulus with only partial veracity). Tasks that involve changing the urgency of responses (such as manipulations of speed accuracy tradeoffs) could also change the MT representation via the same attentional mechanisms observed in other tasks (Cook & Maunsell 2004, Ghose & Bearl 2010, Ghose & Maunsell 2002). Direct measurements of the sensory signals will continue to be important. Likewise, many other brain areas have been implicated in carrying ramping responses during motion discrimination (de Lafuente et al. 2015, Ding & Gold 2010, Horwitz & Newsome 1999, Kim & Shadlen 1999, Mante et al. 2013, Shadlen et al. 1996) and are also likely to contain a mixture of over lapping, decision-relevant and irrelevant sensory and motor response components. Partitioning out these elements may reveal greater functional distinctions across this complex and distributed network.

SUMMARY

The sensorimotor framework is simple but still speculative, and we hope it evolves to complement and enrich interpretation of neural recordings during decision making. Ideally, it will provide a useful basis for interpreting complex patterns of response, clarify apparent neuronal or cross-area heterogeneity, and allow the remaining dynamics of decision-related activity to directly inform models of how the brain approximates mathematical theories of decision making. Perhaps most importantly, the sensorimotor multiplexing framework should allow decision-making studies of LIP to connect more tightly with other lines of work investigating LIP function. Similarly, this approach could facilitate comparisons across brain areas (Ding & Gold 2012a, 2013; Horwitz et al. 2004) and allow for the isolation of factors of focus in various lines of work (e.g., attention, motor intention, categorization, and abstract decisions) (Freedman & Assad 2016, Gottlieb & Balan 2010, Snyder et al. 2000). Complementing experimental designs that explicitly try to dissociate one or two factors that drive LIP, we have argued for the value of comprehensive regression-based analyses that decompose all possible impacts of external variables on the response of an LIP neuron. This approach is far from a deep functional theory but, as an exploratory tool, would provide an organizational scheme for allowing the data to speak to the formation of new ideas. Finally, sensorimotor multiplexing reminds us that many aspects of a response might not be decision relevant or decision critical and thus calls for increasingly nimble and powerful techniques for perturbing these signals. It would be exciting if some of the simpler conclusions we have put forth were finetuned by insights from future methodologies.

Whether or not LIP proves a strongly informative neural correlate of decision formation, it is absolutely the case that recording from LIP has generated many important ideas for relating neural activity and decisions (Shadlen & Kiani 2013). These ideas remain incontrovertibly historically groundbreaking for understanding how neural signal, noise, and correlation could relate to cognition and behavior. A core assumption of these exercises is that the brain must somehow accumulate evidence. However, this need not imply that a particular brain area or a special set of neurons must integrate sensory evidence in a manner that is directly instantiated in the firing rates of individual neurons on individual trials. The sensorimotor multiplexing perspective on LIP thus motivates a broader search for the cellular and network-level factors that contribute to such fundamental cognitive computations. This is likely to remain a topic that the entire field remains deeply enthusiastic about, even if the weight of evidence ultimately shifts neurophysiological recording choices away from LIP.

FUTURE ISSUES

1. How does LIP interact with other areas that exhibit similar decision-correlated activity (e.g., frontal eye fields, superior colliculus, basal ganglia)? Can we arrive at a neural circuit model of decision making that takes new data into account and moves beyond MT-LIP? What, if anything, is LIP's causal role in the decision-making circuit? Can additional permutations of tasks be used to isolate or expose a functional contribution to well-studied tasks? Can additional techniques that allow for more precise and nuanced manipulations of neural activity provide greater insight?

2. How do neurons in LIP come to reflect temporally integrated representations with time constants considerably longer than those evident in single neurons in sensory areas? Are there unique cellular properties? What are the network-scale contributions to integration, and do these arise within a brain area, across areas, or both?

3. Can behavioral paradigms be refined to glean more direct insight into the underlying strategies of subjects at the scale of individual trials, or at least experimental sessions? As analytic tools become more adept at interrogating single-trial responses, it becomes critical to have parallel single-trial insight into the behavior that does not rely on averaging or on using the physiological responses to make inferences.

4. Can behavioral paradigms be extended to directly test the posited extensions of the drift diffusion model that have been hinted at by LIP physiology? Can urgency be separated from motor preparation—i.e., are the dynamics of LIP that are not directly linked to the stimulus, but that do evolve over time in a decision-correlated manner, the reflection of decision formation per se, or are they simply an evolving motor plan that is downstream of a critical decision-making process?

5. What are the functional roles of LIP neurons that are driven by visual motion, as opposed to saccade targets, in their RFs? This has been explored in a categorization task with suprathreshold motion stimuli, but it is unknown if such conditions might reveal LIP's contributions to temporal integration.

6. What is the functional structure of LIP? Our understanding of its basic functional architecture is nascent. More precise knowledge of multiple subregions, the nature and extent of maps, and cell types should allow for more detailed insights into what LIP does.

7. What is the homology between macaque LIP and regions in PPC in other species, many of which would allow for more immediate deployment of cutting-edge neurophysiological tools? Is rodent PPC functionally similar to LIP, or is it better thought of as a more sensory visual area? Does the marmoset model system have an LIP that might be a complementary sweet spot between highly trainable macaques and cutting-edge tool availability?

8. Can these studies benefit from current efforts to standardize research procedures? Would broader and less focused sampling of LIP neurons help or hurt our ability to infer its functional contributions to decision making? Would preregistering numbers of trials or cells paint a different picture than those from more flexible procedures? Can standard training procedures and pre hoc criteria for behavior be shared and adapted broadly? Can a full set of models be developed and then used to decide on specific experiments to select between these established options? Would these lead to conclusions different than more exploratory practices that interpret deviations from models after data collection?

DISCLOSURE STATEMENT

The authors are not aware of any affiliations, memberships, funding, or financial holdings that might be perceived as affecting the objectivity of this review.

ACKNOWLEDGMENTS

This research was supported by a Howard Hughes Medical Institute International Student Research Fellowship to L.N.K., a National Eye Institute grant (R01-EY017366) to A.C.H. and J. Pillow, a Ruth L. Kirschstein National Research Service Award (T32DA018926) from the National Institute on Drug Abuse, and grant number T32EY021462 from the National Eye Institute.

LITERATURE CITED

Andersen RA, Asanuma C, Cowan WM. 1985. Callosal and prefrontal associational projecting cell populations in area 7A of the macaque monkey: a study using retrogradely transported fluorescent dyes. *J. Comp. Neurol.* 232(4):443–55

Andersen RA, Asanuma C, Essick G, Siegel RM. 1990. Corticocortical connections of anatomically and physiologically defined subdivisions within the inferior parietal lobule. *J. Comp. Neurol.* 296(1):65–113

Balan PF, Gottlieb J. 2009. Functional significance of nonspatial information in monkey lateral intraparietal area. *J. Neurosci.* 29(25):8166–76

Basso MA, Wurtz RH. 1998. Modulation of neuronal activity in superior colliculus by changes in target probability. *J. Neurosci.* 18(18):7519–34

Basso MA, Wurtz RH. 1997. Modulation of neuronal activity by target uncertainty. *Nature* 389(6646):66–69

Belmonte JCI, Callaway EM, Churchland P, Caddick SJ, Feng G, et al. 2015. Brains, genes, and primates. *Neuron* 86(3):617–31

Ben Hamed S, Duhamel JR, Bremmer F, Graf W. 2001. Representation of the visual field in the lateral intraparietal area of macaque monkeys: a quantitative receptive field analysis. *Exp. Brain Res.* 140(2):127–44

Ben Hamed S, Duhamel JR, Bremmer F, Graf W. 2002. Visual receptive field modulation in the lateral intraparietal area during attentive fixation and free gaze. *Cereb. Cortex* 12(3):234–45

Bennur S, Gold JI. 2011. Distinct representations of a perceptual decision and the associated oculomotor plan in the monkey lateral intraparietal area. *J. Neurosci.* 31(3):913–21

Bisley JW, Krishna BS, Goldberg ME. 2004. A rapid and precise on-response in posterior parietal cortex. *J. Neurosci.* 24(8):1833–38

Blatt GJ, Andersen RA, Stoner GR. 1990. Visual receptive field organization and corticocortical connections of the lateral intraparietal area (area LIP) in the macaque. *J. Comp. Neurol.* 299(4):421–45

Britten KH, Newsome WT, Shadlen MN, Celebrini S, Movshon JA. 1996. A relationship between behavioral choice and the visual responses of neurons in macaque MT. *Vis. Neurosci.* 13:87–100

Britten KH, Shadlen MN, Newsome WT, Movshon JA. 1992. The analysis of visual motion: a comparison of neuronal and psychophysical performance. *J. Neurosci.* 12(12):4745–65

Britten KH, Shadlen MN, Newsome WT, Movshon JA. 1993. Responses of neurons in macaque MT to stochastic motion signals. *Vis. Neurosci.* 10(6):1157–69

Brody CD, Hanks TD. 2016. Neural underpinnings of the evidence accumulator. *Curr. Opin. Neurobiol.* 37:149–57

Brown EN, Frank LM, Tang D, Quirk MC, Wilson MA, Wilson MA. 1998. A statistical paradigm for neural spike train decoding applied to position prediction from ensemble firing patterns of rat hippocampal place cells. *J. Neurosci.* 18(18):7411–25

Brunton BW, Botvinick MM, Brody CD. 2013. Rats and humans can optimally accumulate evidence for decision-making. *Science* 340(6128):95–98

Carland MA, Marcos E, Thura D. 2016. Evidence against perfect integration of sensory information during perceptual decision making. *J. Neurophysiol.* 115(2):915–30

Carpenter RH, Williams ML. 1995. Neural computation of log likelihood in control of saccadic eye movements. *Nature* 377(6544):59–62

Chichilnisky EJ. 2001. A simple white noise analysis of neuronal light responses. *Network* 12(2):199–213

Chowdhury SA, DeAngelis GC. 2008. Fine discrimination training alters the causal contribution of macaque area MT to depth perception. *Neuron* 60(2):367–77

Churchland AK, Kiani R, Chaudhuri R, Wang XJ, Pouget A, Shadlen MN. 2011. Variance as a signature of neural computations during decision making. *Neuron* 69(4):818–31

Churchland AK, Kiani R, Shadlen MN. 2008. Decision-making with multiple alternatives. *Nat. Neurosci.* 11(6):693–702

Cisek P, Puskas GA, ElMurr S. 2009. Decisions in changing conditions: the urgency-gating model. *J. Neurosci.* 29(37):11560–71

Colby CL, Duhamel JR, Goldberg ME. 1996. Visual, presaccadic, and cognitive activation of single neurons in monkey lateral intraparietal area. *J. Neurophysiol.* 76(5):2841–52

Cook EP, Maunsell JHR. 2004. Attentional modulation of motion integration of individual neurons in the middle temporal visual area. *J. Neurosci.* 24(36):7964–77

Cooke DF, Goldring A, Recanzone GH, Krubitzer L. 2014. The evolution of parietal areas associated with visuomanual behavior: from grasping to tool use. In *The New Visual Neurosciences,* ed. JS Werner, LM Chalupa, pp. 1049–63. Cambridge, MA: MIT Press

Cutrell EB, Marrocco RT. 2002. Electrical microstimulation of primate posterior parietal cortex initiates orienting and alerting components of covert attention. *Exp. Brain Res.* 144(1):103–13

Dai J, Brooks DI, Sheinberg DL. 2014. Optogenetic and electrical microstimulation systematically bias visuospatial choice in primates. *Curr. Biol.* 24(1):63–69

de Lafuente V, Jazayeri M, Shadlen MN. 2015. Representation of accumulating evidence for a decision in two parietal areas. *J. Neurosci.* 35(10):4306–18

Ding L, Gold JI. 2010. Caudate encodes multiple computations for perceptual cecisions. *J. Neurosci.* 30(47):15747–59

Ding L, Gold JI. 2012a. Neural correlates of perceptual decision making before, during, and after decision commitment in monkey frontal eye field. *Cereb. Cortex* 22(5):1052–67

Ding L, Gold JI. 2012b. Separate, causal roles of the caudate in saccadic choice and execution in a perceptual decision task. *Neuron* 75(5):865–74

Ding L, Gold JI. 2013. The basal ganglia's contributions to perceptual decision making. *Neuron* 79(4):640–49

Erlich JC, Brunton BW, Duan CA, Hanks TD, Brody CD. 2015. Distinct effects of prefrontal and parietal cortex inactivations on an accumulation of evidence task in the rat. *eLife* 4:e05457

Falkner AL, Krishna BS, Goldberg ME. 2010. Surround suppression sharpens the priority map in the lateral intraparietal area. *J. Neurosci.* 30(38):12787–97

Fanini A, Assad JA. 2009. Direction selectivity of neurons in the macaque lateral intraparietal area. J. Neuro physiol. 101(1):289–305 Felleman DJ, Van Essen DC. 1991. Distributed hierarchical processing in the primate cerebral cortex. *Cereb. Cortex* 1(1):1–47

Fernandes HL, Stevenson IH, Phillips AN, Segraves MA, Körding KP. 2014. Saliency and saccade encoding in the frontal eye field during natural scene search. *Cereb. Cortex* 24(12):3232–45

Ferraina S, Paré M, Wurtz RH. 2002. Comparison of corticocortical and corticocollicular signals for the generation of saccadic eye movements. *J. Neurophysiol.* 87(2):845–58

Fetsch CR, Kiani R, Newsome WT, Shadlen MN. 2014. Effects of cortical microstimulation on confidencein a perceptual decision. *Neuron* 83(4):797–804

Freedman DJ, Assad JA. 2006. Experience-dependent representation of visual categories in parietal cortex. *Nature* 443(7107):85–88

Freedman DJ, Assad JA. 2011. A proposed common neural mechanism for categorization and perceptual decisions. *Nat. Neurosci.* 14(2):143–46

Freedman DJ, Assad JA. 2016. Neuronal mechanisms of visual categorization: an abstract view on decision making. *Annu. Rev. Neurosci.* 39:129–47

Fusi S, Miller EK, Rigotti M. 2016. Why neurons mix: high dimensionality for higher cognition. *Curr. Opin. Neurobiol.* 37:66–74

Ghose GM, Bearl DW. 2010. Attention directed by expectations enhances receptive fields in cortical area MT. *Vis. Res.* 50(4):441–51

Ghose GM, Maunsell J. 2002. Attentional modulation in visual cortex depends on task timing. *Nature* 419(6907):616–20

Gnadt JW, Andersen RA. 1988. Memory related motor planning activity in posterior parietal cortex of macaque. *Exp. Brain Res.* 70(1):216–20

Gold JI, Shadlen MN. 2000. Representation of a perceptual decision in developing oculomotor commands. *Nature* 404(6776):390–94

Gold JI, Shadlen MN. 2003. The influence of behavioral context on the representation of a perceptual decision in developing oculomotor commands. *J. Neurosci.* 23(2):632–51

Gold JI, Shadlen MN. 2007. The neural basis of decision making. *Annu. Rev. Neurosci.* 30:535–74

GoldmanRakic PS. 1995. Cellular basis of working memory. *Neuron* 14(3):477–85

Gottlieb J, Balan P. 2010. Attention as a decision in information space. *Trends Cogn. Sci.* 14(6):240–48

Hanes DP, Schall JD. 1995. Countermanding saccades in macaque. *Vis. Neurosci.* 12(5):929–37

Hanes DP, Schall JD. 1996. Neural control of voluntary movement initiation. *Science* 274(5286):427–30

Hanks TD, Ditterich J, Shadlen MN. 2006. Microstimulation of macaque area LIP affects decisionmaking in a motion discrimination task. *Nat. Neurosci.* 9(5):682–89

Hanks TD, Kiani R, Shadlen MN. 2014. A neural mechanism of speedaccuracy tradeoff in macaque area

LIP. *eLife* 3:e02260

Hanks TD, Kopec CD, Brunton BW, Duan CA, Erlich JC, Brody CD. 2015. Distinct relationships of parietal and prefrontal cortices to evidence accumulation. *Nature* 520(7546):220–23

Hanks TD, Mazurek ME, Kiani R, Hopp E, Shadlen MN. 2011. Elapsed decision time affects the weighting of prior probability in a perceptual decision task. *J. Neurosci.* 31(17):6339–52

Hanks TD, Summerfield C. 2017. Perceptual decision making in rodents, monkeys, and humans. *Neuron* 93(1):15–31

Harvey CD, Coen P, Tank DW. 2013. Choicespecific sequences in parietal cortex during a virtualnavigation decision task. *Nature* 484(7392):62–68

Histed MH, Bonin V, Reid RC. 2009. Direct activation of sparse, distributed populations of cortical neurons by electrical microstimulation. *Neuron* 63(4):508–22

Horwitz GD, Batista AP, Newsome WT. 2004. Representation of an abstract perceptual decision in macaque superior colliculus. *J. Neurophysiol.* 91(5):2281–96

Horwitz GD, Newsome WT. 1999. Separate signals for target selection and movement specification in the superior colliculus. *Science* 284(5417):1158–61

Huk AC, Shadlen MN. 2005. Neural activity in macaque parietal cortex reflects temporal integration of visual motion signals during perceptual decision making. *J. Neurosci.* 25(45):10420–36

Ipata AE, Gee AL, Goldberg ME, Bisley JW. 2006. Activity in the lateral intraparietal area predicts the goal and latency of saccades in a freeviewing visual search task. *J. Neurosci.* 26(14):3656–61

Jacobs AL, Fridman G, Douglas RM, Alam NM, Latham PE, et al. 2009. Ruling out and ruling in neural codes. *PNAS* 106(14):5936–41

Janssen P, Shadlen MN. 2005. A representation of the hazard rate of elapsed time in macaque area LIP. *Nat. Neurosci.* 8(2):234–41

Janssen P, Srivastava S, Ombelet S, Orban GA. 2008. Coding of shape and position in macaque lateral intraparietal area. *J. Neurosci.* 28(26):6679–90

Jazayeri M, Shadlen MN. 2015. A neural mechanism for sensing and reproducing a time interval. *Curr. Biol.* 25(20):2599–609

Joo SJ, Katz LN, Huk AC. 2016. Decisionrelated perturbations of decision-irrelevant eye movements. *PNAS* 113(7):1925–30

Katz LN, Yates JL, Pillow JW, Huk AC. 2016. Dissociated functional significance of decisionrelated activity in the primate dorsal stream. *Nature* 535(7611):285–88

Kerkhoff G. 2001. Spatial hemineglect in humans. *Prog. Neurobiol.* 63(1):1–27

Kiani R, Hanks TD, Shadlen MN. 2008. Bounded integration in parietal cortex underlies decisions even when viewing duration is dictated by the environment. *J. Neurosci.* 28(12):3017–29

Kiani R, Shadlen MN. 2009. Representation of confidence associated with a decision by neurons in the parietal cortex. *Science* 324(5928):759–64

Kim JN, Shadlen MN. 1999. Neural correlates of a decision in the dorsolateral prefrontal cortex of the macaque. *Nat. Neurosci.* 2:176–85

Kira S, Yang T, Shadlen MN. 2015. A neural implementation of Wald's sequential probability ratio test. *Neuron* 85(4):861–73

Kubanek J, Li JM, Snyder LH. 2015. Motor role of parietal cortex in a monkey model of hemispatial neglect. *PNAS* 112(16):E2067–72

Laming DRJ. 1968. *Information Theory of Choice-Reaction Times*. New York: Acad. Press

Latimer KW, Yates JL, Meister MLR, Huk AC, Pillow JW. 2015. Singletrial spike trains in parietal cortex reveal discrete steps during decision-making. *Science* 349(6244):184–87

Law CT, Gold JI. 2008. Neural correlates of perceptual learning in a sensory-motor, but not a sensory, cortical area. *Nat. Neurosci.* 11(4):505–13

Lennie P. 1998. Single units and visual cortical organization. *Perception* 27(8):889–935

Leon MI, Shadlen MN. 2003. Representation of time by neurons in the posterior parietal cortex of the macaque. *Neuron* 38(2):317–27

Lewis JW, Van Essen DC. 2000a. Corticocortical connections of visual, sensorimotor, and multimodal processing areas in the parietal lobe of the macaque monkey. *J. Comp. Neurol.* 428(1):112–37

Lewis JW, Van Essen DC. 2000b. Mapping of architectonic subdivisions in the macaque monkey, with emphasis on parieto-occipital cortex. *J. Comp. Neurol.* 428(1):79–111

Li N, Daie K, Svoboda K, Druckmann S. 2016. Robust neuronal dynamics in premotor cortex during motor planning. *Nature* 532(7600):459–64

Licata AM, Kaufman MT, Raposo D, Ryan MB. 2016. Posterior parietal cortex guides visual decisions in rats. bioRxiv 066639. https://dx.doi.org/10.1101/066639

Liu Y, Yttri EA, Snyder LH. 2010. Intention and attention: different functional roles for LIPd and LIPv. Nat. *Neurosci.* 13(4):495–500

Louie K, LoFaro T, Webb R, Glimcher PW. 2014. Dynamic divisive normalization predicts time-varying value coding in decision-related circuits. *J. Neurosci.* 34(48):16046–57

Mante V, Sussillo D, Shenoy KV, Newsome WT. 2013. Context-dependent computation by recurrent dynamics in prefrontal cortex. *Nature* 503(7474):78–84

Maunsell JH, Van Essen DC. 1983. Functional properties of neurons in middle temporal visual area of the macaque monkey. I. Selectivity for stimulus direction, speed, and orientation. *J. Neurophysiol.* 49(5):1127–47

Mayo JP, DiTomasso AR, Sommer MA, Smith MA. 2015. Dynamics of visual receptive fields in the macaque frontal eye field. *J. Neurophysiol.* 114(6):3201–10

Mayo JP, Morrison RM, Smith MA. 2016. A probabilistic approach to receptive field mapping in the frontal eye fields. *Front. Syst. Neurosci.* 10:25

Mazurek ME, Roitman JD, Ditterich J, Shadlen MN. 2003. A role for neural integrators in perceptual decision making. *Cereb. Cortex* 13(11):1257–69

Meister MLR, Hennig JA, Huk AC. 2013. Signal multiplexing and singleneuron computations in lateral intraparietal area during decision-making. *J. Neurosci.* 33(6):2254–67

Mountcastle VB, Lynch JC, Georgopoulos A, Sakata H, Acuna C. 1975. Posterior parietal association cortex of the monkey: command functions for operations within extrapersonal space. *J. Neurophysiol.* 38(4):871–908

Murasugi CM, Salzman CD, Newsome WT. 1993. Microstimulation in visual area MT: effects of varying pulse amplitude and frequency. *J. Neurosci.* 13(4):1719–29

Newsome WT, Paré EB. 1988. A selective impairment of motion perception following lesions of the middle temporal visual area (MT). *J. Neurosci.* 8(6):2201–11

Pagan M, Rust NC. 2014. Quantifying the signals contained in heterogeneous neural responses and determining their relationships with task performance. *J. Neurophysiol.* 112(6):1584–98

Palmer J, Huk AC, Shadlen MN. 2005. The effect of stimulus strength on the speed and accuracy of a perceptual decision. *J. Vis.* 5(5):376–404

Paninski L, Shoham S, Fellows MR, Hatsopoulos NG, Donoghue JP. 2004. Superlinear population encoding of dynamic hand trajectory in primary motor cortex. *J. Neurosci.* 24(39):8551–61

Pare M, Wurtz RH. 1997. Monkey posterior parietal cortex neurons antidromically activated from superior colliculus. *J. Neurophysiol.* 78(6):3493–97

Park IM, Meister MLR, Huk AC, Pillow JW. 2014. Encoding and decoding in parietal cortex during sensorimotor decision-making. *Nat. Neurosci.* 17(10):1395–403

Patel GH, Shulman GL, Baker JT, Akbudak E, Snyder AZ, et al. 2010. Topographic organization of macaque area LIP. *PNAS* 107(10):4728–33

Pesaran B, Freedman DJ. 2016. Where are perceptual decisions made in the brain? *Trends Neurosci.* 39(10):642–44

Pillow JW, Paninski L, Uzzell VJ, Simoncelli EP, Chichilnisky EJ. 2005. Prediction and decoding of retinal ganglion cell responses with a probabilistic spiking model. *J. Neurosci.* 25(47):11003–13

Pillow JW, Shlens J, Paninski L, Sher A, Litke AM, et al. 2008. Spatio-temporal correlations and visual signalling in a complete neuronal population. *Nature* 454(7207):995–99

Pisupati S, Chartarifsky L, Churchland AK. 2016. Decision activity in parietal cortex-leader or follower? *Trends Cogn. Sci.* 20(11):788–89

Platt ML, Glimcher PW. 1997. Responses of intraparietal neurons to saccadic targets and visual distractors. *J. Neurophysiol.* 78(3):1574–89

Ramkumar P, Dekleva B, Cooler S, Miller L, Kording K. 2016. Premotor and motor cortices encode reward. *PLOS ONE* 11(8):e0160851

Rao V, DeAngelis GC, Snyder LH. 2012. Neural correlates of prior expectations of motion in the lateral intraparietal and middle temporal areas. *J. Neurosci.* 32(29):10063–74

Raposo D, Kaufman MT, Churchland AK. 2014. A category-free neural population supports evolving demands during decision-making. *Nat. Neurosci.* 17(12):1784–92

Ratcliff R. 1978. A theory of memory retrieval. *Psychol. Rev.* 85(2):59–108

Ratcliff R, Smith PL, Brown SD, McKoon G. 2016. Diffusion decision model: current issues and history. *Trends Cogn. Sci.* 20(4):260–81

Rigotti M, Barak O, Warden MR, Wang XJ, Daw ND, et al. 2013. The importance of mixed selectivity in complex cognitive tasks. *Nature* 497(7451):585–90

Rishel CA, Huang G, Freedman DJ. 2013. Independent category and spatial encoding in parietal cortex. *Neuron* 77(5):969–79

Robinson DL, Goldberg ME, Stanton GB. 1978. Parietal association cortex in the primate: sensory mechanisms and behavioral modulations. *J. Neurophysiol.* 41(4):910–32

Roitman JD, Shadlen MN. 2002. Response of neurons in the lateral intraparietal area during a combined visual discrimination reaction time task. *J. Neurosci.* 22(21):9475–89

Rorie AE, Gao J, McClelland JL, Newsome WT. 2010. Integration of sensory and reward information during perceptual decision-making in lateral intraparietal cortex (LIP) of the macaque monkey. *PLOS ONE* 5(2):e9308

Salzman CD, Britten KH, Newsome WT. 1990. Cortical microstimulation influences perceptual judgements of motion direction. *Nature* 346:174–77

Salzman CD, Murasugi CM, Britten KH, Newsome WT. 1992. Microstimulation in visual area MT: effects on direction discrimination performance. *J. Neurosci.* 12(6):2331–55

Sarma A, Masse NY, Wang XJ, Freedman DJ. 2015. Task-specific versus generalized mnemonic representations in parietal and prefrontal cortices. *Nat. Neurosci.* 19(1):143–49

Schall JD. 2001. Neural basis of deciding, choosing and acting. *Nat. Rev. Neurosci.* 2(1):33–42

Selen LPJ, Shadlen MN, Wolpert DM. 2012. Deliberation in the motor system: reflex gains track evolving evidence leading to a decision. *J. Neurosci.* 32(7):2276–86

Shadlen MN, Britten KH, Newsome WT, Movshon JA. 1996. A computational analysis of the relationship between neuronal and behavioral responses to visual motion. *J. Neurosci.* 16(4):1486–510

Shadlen MN, Kiani R. 2013. Decision making as a window on cognition. *Neuron* 80(3):791–806

Shadlen MN, Kiani R, Hanks TD, Churchland AK. 2008. Neurobiology of decision making: an intentional framework. In *Better Than Conscious? Decision Making, the Human Mind, and Implications for Institutions,* ed. C Engel, W Singer, pp. 71–101. Cambridge, MA: MIT Press

Shadlen MN, Newsome WT. 1996. Motion perception: seeing and deciding. *PNAS* 93(2):628–33

Shadlen MN, Newsome WT. 2001. Neural basis of a perceptual decision in the parietal cortex (area LIP) of the rhesus monkey. *J. Neurophysiol.* 86(4):1916–36

Snyder LH, Batista AP, Andersen RA. 2000. Intention-related activity in the posterior parietal cortex: a review. *Vis. Res.* 40(10–12):1433–41

Song JH, Nakayama K. 2009. Hidden cognitive states revealed in choice reaching tasks. *Trends Cogn. Sci.* 13(8):360–66

Spivey MJ, Grosjean M, Knoblich G. 2005. Continuous attraction toward phonological competitors. *PNAS* 102(29):10393–98

Tehovnik EJ, Tolias AS, Slocum WM, Logothetis NK. 2006. Direct and indirect activation of cortical neurons by electrical microstimulation. *J. Neurophysiol.* 96(2):512–21

Truccolo W, Eden UT, Fellows MR, Donoghue JP, Brown EN. 2005. A point process framework for relating neural spiking activity to spiking history, neural ensemble, and extrinsic covariate effects. *J. Neurophysiol.* 93(2):1074–89

Ungerleider LG, Desimone R. 1986. Projections to the superior temporal sulcus from the central and peripheral field representations of V1 and V2. *J. Comp. Neurol.* 248(2):147–63

Wardak C, Olivier E, Duhamel JR. 2002. Saccadic target selection deficits after lateral intraparietal area inactivation in monkeys. *J. Neurosci.* 22(22):9877–84

Wardak C, Olivier E, Duhamel JR. 2004. A deficit in covert attention after parietal cortex inactivation in the monkey. *Neuron* 42(3):501–8

Wilke M, Kagan I, Andersen RA. 2012. Functional imaging reveals rapid reorganization of cortical activity after parietal inactivation in monkeys. *PNAS* 109(21):8274–79

Wong KF. 2007. Neural circuit dynamics underlying accumulation of timevarying evidence during perceptual decision making. *Front. Comput. Neurosci.* 1:6

Wurtz RH, Sommer MA, Pare M, Ferraina S. 2001. Signal transformations from cerebral cortex to superior colliculus for the generation of saccades. *Vis. Res.* 41(25–26):3399–412

Wyart V, De Gardelle V, Scholl J, Summerfield C. 2012. Rhythmic fluctuations in evidence accumulation during decision making in the human brain. *Neuron* 76(4):847–58

Yang T, Shadlen MN. 2007. Probabilistic reasoning by neurons. *Nature* 447(7148):1075–80

Yates JL, Park IM, Katz LN, Pillow JP, Huk AC. 2017. Functional dissection of signal and noise in MT and LIP during decision making. *Nat. Neurosci.* In press

Yu BM, Cunningham JP, Santhanam G, Ryu SI, Shenoy KV, Sahani M. 2009. Gaussian-process factor analysis for low-dimensional single-trial analysis of neural population activity. *J. Neurophysiol.* 102(1):1881–88

Zirnsak M, Chen X, Lomber SG, Moore T. 2015. *Effects of reversible inactivation of parietal cortex on the processing of visual salience in the frontal eye field.* Presented at Soc. Neurosci., Prog. No. 747.06, Oct. 21, Chicago

II

Socialization of Nervous System:
From Neuroculture to Socialized Cognition

神经系统的社会化：从神经文化到社会化认知

Habits, Rituals, and the Evaluative Brain

Ann M. Graybiel

Department of Brain and Cognitive Science and the McGovern Institute for Brain Research,
Massachusetts Institute of Technology, Cambridge, Massachusetts 02139; email:Graybie@mit.edu

Key Words

striatum, reinforcement learning, stereotypy, procedural learning, addiction, automatization, obsessive compulsive disorder

Abstract

Scientists in many different fields have been attracted to the study of habits because of the power habits have over behavior and because they invoke a dichotomy between the conscious, voluntary control over behavior, considered the essence of higher order deliberative behavioral control, and lower order behavioral control that is scarcely available to consciousness. A broad spectrum of behavioral routines and rituals can become habitual and stereotyped through learning. Others have a strong innate basis. Repetitive behaviors can also appear as cardinal symptoms in a broad range of neurological and neuropsychiatric illness and in addictive states. This review suggests that many of these behaviors could emerge as a result of experience dependent plasticity in basal ganglia–based circuits that can influence not only overt behaviors but also cognitive activity. Culturally based rituals may reflect privileged interactions between the basal ganglia and cortically based circuits that influence social, emotional, and action functions of the brain.

Habit is the most effective teacher of all things.

> —Pliny

We are what we repeatedly do. Excellence, then, is not an act, but a habit.

> —Aristotle

Habit is second nature, or rather, ten times nature.

> —William James

For in truth habit is a violent and treacherous schoolmistress. She establishes in us, little by little, stealthily, the foothold of her authority; but having by this mild and humble beginning settled and planted it with the help of time, she soon uncovers to us a furious and tyrannical face against which we no longer have the liberty of even raising our eyes.

—*Montaigne*

INTRODUCTION

Habit, to most of us, has multiple connotations. On the one hand, a habit is a behavior that we do often, almost without thinking. Some habits we strive for, and work hard to make part of our general behavior. And still other habits are burdensome behaviors that we want to abolish but often cannot, so powerfully do they control our behavior. Viewed from this broad and intuitive perspective, habits can be evaluated as relatively neutral, or as "good" (desirable) or as "bad" (undesirable). Yet during much of our waking lives, we act according to our habits, from the time we rise and go through our morning routines until we fall asleep after evening routines. Taken in this way, habits have long attracted the interest of philosophers and psychologists, and they have been alternatively praised and cursed.

Whether good, bad, or neutral, habits can have great power over our behavior. When deeply enstated, they can block some alternate behaviors and pull others into the habitual repertoire. In early accounts, habits were broadly defined. Mannerisms, customs, and rituals were all considered together with simple daily habits, and habituation or sensitization (the lessening or increase in impact of stimuli and events with repetition) were included. Much current work on habit learning in neuroscience has pulled away from this broad view in an effort to define habit in a way that makes it accessible to scientific study. Much insight can also be gained by extending such constructs of habit and habit learning to include the rich array of behaviors considered by ethologists, neuropharmacologists, neurologists, and psychiatrists, as well as by students of motor control. Below, I review some of the definitions of habit that have developed in cognitive neuroscience and psychology and how these views have been formalized in computational theories. I then point to work on extreme habits and compulsions, ritualistic behaviors and mannerisms, stereotypies, and social and cultural "habits" and suggest that these are critical behaviors to consider in a neuroscience of habit formation.

This proposal is based on mounting evidence that this broad array of behaviors can engage neural circuits interconnecting the neocortex with the striatum and related regions of the basal ganglia. Different basal ganglia–based circuits appear to operate predominantly in relation to different types of cognitive and motor actions, for example, in intensely social behaviors such as mating and in the performance of practiced motor skills. Remarkably, however, evidence suggests that many of these basal ganglia-based subcircuits participate during the acquisition of habits, procedures, and repetitive behaviors, and these may be reactivated or misactivated in disorders producing repetitive thoughts and overt behaviors.

A starting point is to consider defining characteristics of habits. First, habits (mannerisms, customs, rituals) are largely learned; in current terminology, they are acquired via experience dependent plasticity. Second, habitual behaviors occur repeatedly over the course of days or years, and they can become remarkably fixed. Third, fully acquired

FAPs: fixed action patterns
OCD: obsessive compulsive disorder

habits are performed almost automatically, virtually nonconsciously, allowing attention to be focused elsewhere. Fourth, habits tend to involve an ordered, structured action sequence that is prone to being elicited by a particular context or stimulus. And finally, habits can comprise cognitive expressions of routine (habits of thought) as well as motor expressions of routine. These characteristics suggest that habits are sequential, repetitive, motor, or cognitive behaviors elicited by external or internal triggers that, once released, can go to completion without constant conscious oversight.

This description is familiar to many who study animal behavior and observe complex repetitive behaviors [fixed action patterns (FAPs)]. Some of these appear to be largely innate, such as some mating behaviors, but

S-R: stimulus-response

others are learned, such as the songs of some orcene birds. Repetitive behaviors and thoughts are also major presenting features in human disorders such as Tourette syndrome and obsessive compulsive disorder (OCD). Stereotypies and repetitive behaviors appear in a range of other clinical disorders including schizophrenia and Huntington's disease, as well as in addictive states. I suggest that there may well be a common theme across these behavioral domains. Many of these repetitive behaviors, whether motor or cognitive, are built up in part through the action of basal ganglia-based neural circuits that can iteratively evaluate contexts and select actions and can then form chunked representations of action sequences that can influence both cortical and subcortical brain structures (**Figure 1**). Both experimental evidence and computational analysis suggest that a shift from largely evaluation-driven circuits to those engaged in performance is a critical feature of habit learning. Chronic multielectrode recordings suggest that within the habit production system, as habits are acquired, neural activity patterns change dynamically and eventually settle into specific chunked patterns. This shift in neural activity from variable to repetitive matches the explore exploit transition in behavioral output from a testing, exploratory mode to a focused, exploitive mode as habitual behaviors crystallize. This process may be critical to allow the emergence of habitual behaviors as entire structured entities once they are learned.

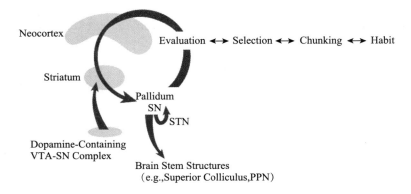

Figure 1
Schematic representation of the development of habits through iterative action of cortico-basal ganglia circuits. Circuits mediating evaluation of actions gradually lead to selection of particular behaviors that, through the chunking process, become habits. PPN, pedunculopontine nucleus; SN, substantia nigra; STN, subthalamic nucleus; VTA, ventral tegmental area.

DEFINITIONS OF HABIT LEARNING IN COGNITIVE NEUROSCIENCE AND EXPERIMENTAL PSYCHOLOGY

Classic studies of habit learning distinguished this form of learning as a product of a procedural learning brain system that is differentiable from declarative learning brain systems for encoding facts and episodes. These definitions rest on findings suggesting that these two systems have different brain substrates (Knowlton et al. 1996, Packard & Knowlton 2002, Packard & McGaugh 1996). Deficits in learning facts contrast vividly with the preserved habits, daily routines, and procedural capabilities of patients with medial temporal lobe damage (Salat et al. 2006). By contrast, patients with basal ganglia disorders exhibit, in testing, procedural learning deficits and deficits in implicit (nonconsciously recognized) learning such as performance on mazes and probabilistic learning tasks in which the subject learns the probabilities of particular stimulus-response (S-R) associations without full awareness (Knowlton et al. 1996, Poldrack et al. 2001). The nonconscious acquisitions of S-R habits by amnesic patients has been documented most clearly by the performance of a patient who learned a probabilistic task with an apparent total lack of awareness of the acquired habit (Bayley et al. 2005).

Despite these distinctions, human imaging experiments suggest that both the basal ganglia (striatum) and the medial temporal lobe are active in such probabilistic learning tasks. When task conditions favor implicit learning, however, activity in the medial temporal lobe decreases as striatal activity increases, and when conditions favor explicit learning, the reverse is true (Foerde et al. 2006, Poldrack et al. 2001, Willingham et al. 2002). Moreover, in disease states involving dysfunction of the basal ganglia, medial temporal lobe activity can appear under conditions in which striatal activity normally would dominate (Moody et al. 2004, Rauch et al. 2006, Voermans et al. 2004). These findings demonstrate conjoint but differentiable contributions of both the declarative and the procedural memory systems to behaviors, as well as interactions between these two.

Comparable distinctions have been drawn for memory systems in experimental animals. The striatum is required for repetitive S-R or win stay behaviors (for example, always turning right in a maze to obtain reward) as opposed to behaviors that can be flexibly adjusted when the context or rules change (for example, not just turning right, but turning toward the rewarded side even if it is now on the left). By contrast, the hippocampus is required for flexible (win-shift) behaviors (Packard & Knowlton 2002, Packard & McGaugh 1996). Nevertheless, the control systems for these behaviors cannot be simply divided into hippocampal and basal ganglia systems because both types of behavior can be supported by the striatum, depending on the hippocampal and sensorimotor connections of the striatal regions in question (Devan & White 1999, Yin & Knowlton 2004). Moreover, as conditional procedures are learned, neural activities in the striatum and hippocampus can become highly coordinated in the frequency domain (DeCoteau et al. 2007a).

In an effort to promote clearly interpretable experimentation on habit formation, Dickinson and his collaborators developed an operational definition of habits using characteristics of reward-based learning in rodents (Adams & Dickinson 1981, Balleine & Dickinson 1998, Colwill & Rescorla 1985). In the initial stages of habit learning, behaviors are not automatic. They are goal directed, as in an animal working to obtain a food reward. But with extended training or training with interval schedules of reward, animals typically come to perform the behaviors repeatedly, on cue, even if the value of the reward to be received is reduced so that it is no longer rewarding (for example, if the animal is tested when it is sated or if its food reward has been repetitively paired with a noxious outcome). Dickinson defined the goal-oriented, purposeful, nonhabitual behaviors as action-outcome (A-O) behaviors and labeled the habitual behaviors occurring despite reward devaluation as S-R behaviors. Thus, in addition to habits being learned, repetitive, sequential, context-triggered behaviors, habits can be defined experimentally as being performed not in relation to a current or future goal but rather in relation to a previous goal and the antecedent behavior that most successfully led to achieving that goal.

The central finding from lesion work based on the reward-devaluation paradigm is that the transition from goal-oriented A-O to habitual S-R modes of behavior involves transitions in the neural circuits predominantly controlling the behaviors (**Figure 2**). Specifically, experiments suggest that different regions of the

A-O: action outcome

prefrontal cortex, the striatum, and the amygdala and other limbic sites critically influence these two different behavioral modes.

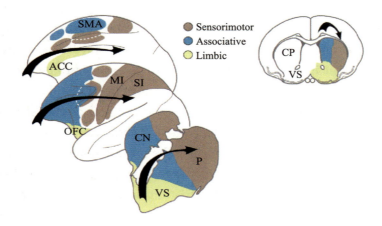

Figure 2

Dynamic shifts in activity in cortical and striatal regions as habits and procedures are learned. Sensorimotor, associative, and limbic regions of the frontal cortex (medial and lateral views) and striatum (single hemisphere) are shown for the monkey (*left*), and corresponding striatal regions are indicated for the rat (*right*). These functional designations are only approximate and are shown in highly schematic form.ACC, anterior cingulate cortex; CN, caudate nucleus; CP, caudoputamen; MI, primary motor cortex;OFC, orbitofrontal cortex; P, putamen; SI, primary somatosensory cortex; SMA, supplementary motor area; VS, ventral striatum.

In rats, lesions in either the sensorimotor striatum (dorsolateral caudoputamen) or the infralimbic prefrontal cortex reduce the insensitivity to reward devaluation that defines habitual behavior in this paradigm. With such lesions, the animals exhibit sensitivity to reward value (A-O behavior) rather than habitual (S-R) behavior, even after overtraining (Killcross & Coutureau 2003, Yin & Knowlton 2004). By contrast, lesions of either the caudomedial (associative) striatum or the prelimbic prefrontal cortex reduce the sensitivity to reward devaluation that defines goal-oriented behavior in this paradigm; the animals are habit driven (Killcross & Coutureau 2003, Yin et al. 2005). The fact that lesions in either the striatum or the frontal cortex are disruptive suggests that the controlling systems represent neural circuits that have both cortical and subcortical components. In macaque monkeys, the basolateral amygdala and the orbitofrontal cortex are also required for sensitivity to reward devaluation (Izquierdo et al. 2004, Wellman et al. 2005). Thus multiple components of the goal-oriented system have been demonstrated across species, and these include regions strongly linked with the limbic system (Balleine et al. 2003, Corbit & Balleine 2005, Gottfried et al. 2003, Wellman et al. 2005).

Like the declarative vs. habit system distinction made in studies on humans, the distinction based on these experiments between action-outcome vs. stimulus-response systems is not absolute (Faure et al. 2005). Evidence suggests that these are not independent "systems." For example, after training that produces habitual behavior in rats, goal-oriented behavior can be reinstated if the infralimbic prefrontal cortex is inactivated (Coutureau & Killcross 2003). This finding suggests that the circuits controlling goal-directed behavior may be actively suppressed when behavior becomes habitual (Coutureau & Killcross 2003). The dichotomy between A-O and S-R behaviors also does not reflect the richness of behavior outside the narrow boundaries of their definitions in reward-devaluation paradigms (for example, when multiple choices are available or different reward schedules are used). The idea that there is a dynamic balance between control systems governing flexible cognitive control and more nearly automatic control of behavioral responses supports the long-standing view from clinical studies that frontal cortical inhibitory zones can suppress lower-order behaviors. This view has become important in models of such system-level interactions (Daw et al. 2005).

Most of these studies have been based on the effects of permanent lesions made in parts of either the dorsal striatum or the neocortex. The use of reversible inactivation procedures suggests that during early stages of instrumental learning, activity in the ventral striatum (nucleus accumbens) is necessary for acquisition of the behavior (Atallah et al. 2007, Hernandez et al. 2002, Hernandez et al. 2006, Smith Roe & Kelley 2000). This requirement for the nucleus accumbens is apparently transitory: After learning, inactivation of the nucleus accumbens has less or no effect. Notably, inactivating the dorsolateral striatum during the very early stages of conditioning does not block learning and can even improve performance. This last result at first glance seems to conflict with the many reports concluding that the dorsolateral striatum is necessary for habit learning. However, these results fit well with the view, encouraged here, that the learning process is highly dynamic and engages in parallel, not simply in series, sets of neural circuits ranging from those most tightly connected with limbic and midbrain-ventral striatal reward systems to circuits engaging the dorsal striatum, neocortex, and motor structures such as the cerebellum.

Several groups have suggested that eventually the "engram" of the habit shifts to regions outside the basal ganglia, including the neocortex (Atallah et al. 2007, Djurfeldt et al. 2001, Graybiel 1998, Houk & Wise 1995, O'Reilly & Frank 2006). Evidence to settle this point is still lacking. There could be a competition between the early-learning ventral striatal system and the late-learning dorsal striatal system (Hernandez et al. 2002), an idea parallel to the proposal that, in maze training protocols that eventually produce habitual behavior, the hippocampus is required for learning early on, whereas later the dorsal striatum is required (Packard & McGaugh 1996). However, things are not likely to be so simple. The dorsal striatum can be engaged very early in the learning process (Barnes et al. 2005, Jog et al. 1999). And "the striatum" and "the hippocampus" each actually comprise a composite of regions that are interconnected with different functional networks.

COMPUTATIONAL APPROACHES TO HABIT LEARNING: HABIT LEARNING AND VALUE FUNCTIONS

Work on habit learning has been powerfully invigorated by computational neuroscience. A critical impetus for this effort came from the pioneering work of Sutton & Barto (1998), which explicitly outlined the essential characteristics of reinforcement learning (RL) and summarized a series of alternative models to account for such learning (RL models). For experimental neuroscientists, this work is of remarkable interest because neural signals and activity patterns are being identified that fit well with the essential elements of RL models (Daw et al. 2005, Daw & Doya 2006). The key characteristics of these models are that an agent (animal, machine, algorithm) undergoing learning starts with a goal and senses and explores the environment by making choices (selecting behaviors) in order to reach that goal optimally. The agent's actions are made in the context of uncertainty about the environment. The agent must explore the environment to reduce the uncertainty, but it must also exploit (for example, by selecting or deselecting an action) to attain the goal. Sequences of behaviors are seen as guided by subgoals, and the learning involves determining the immediate value of the state or state action set (a reward function), the estimated (predicted) future value of the state in terms of that reward (a value function). To make this value estimate, the agent needs some

RL: reinforcement learning
VTA: ventral tegmental area

representation of future actions (a policy). Then the choice can be guided by the estimated value of taking a given action in a given state with that policy (the action value). These value estimates are principal drivers of behavior. Most behaviors do not immediately yield primary reward, and so ordinarily they involve the generation of a model of the action space (environment) to guide future actions (planning) in the sense of optimal control. Thus the control of behavior crucially depends on value estimates learned through experience.

A pivotal convergence of RL models and traditional learning experiments came with two sets of findings based on conditioning experiments in monkeys (**Figure 3**). First, dopamine-containing neurons of the midbrain substantia nigra pars compacta and the ventral tegmental area (VTA) can fire in patterns that correspond remarkably closely to the properties of a positive reward prediction error of RL models such as in the temporal difference model (Montague et al. 1996, Romo & Schultz 1990, Schultz et al. 1997). Second, during such conditioning tasks, striatal neurons gradually acquire a response to the conditioning stimulus, and this acquired response depends on dopamine signaling in the striatum (Aosaki et al. 1994a,b). These two sets of findings suggested a teacher (dopamine)–student (striatum) sequence in which dopamine-containing nigral neurons, by coding reward-prediction errors, teach learning-related circuits in the striatum (Graybiel et al. 1994). The actor-critic architecture and its variants, in which the critic sup plies value predictions to guide action selection by the actor, have been used to model these relationships (Schultz et al. 1997). Many studies have now focused on identifying signals corresponding to the parameters in models of this learning process.

The firing characteristics of midbrain dopamine-containing neurons suggest that they can signal expected reward value (reward probability and magnitude including negative reward prediction error) and motivational state in a context-dependent manner (Bayer & Glimcher 2005, Morris et al. 2004, Nakahara et al. 2004, Satoh et al. 2003, Tobler et al. 2005, Waelti et al. 2001), that they are specialized to respond in relation to positive but not aversive reinforcements, and that they may code uncertainty (Fiorillo et al. 2003, Hsu et al. 2005, Niv et al. 2005, Ungless et al. 2004) or salience (Redgrave & Gurney 2006). These characteristics may, among others, account for the remarkable capacity for placebo treatments to elicit dopamine release in the striatum (de la Fuente Fernandez et al. 2001). Action value encoding was not detected by the original experimental paradigms used for recording from the dopaminergic neurons, which focused mostly on noninstrumental learning. Morris et al. (2006), using a decision task with a block design, have now shown that the action value of a future action can be coded in the firing of these neurons. This result is important in favoring computational models that take into account the value of a given action in a given state (the Q value). Remarkably, the dopaminergic neurons can signal which of two alternate actions will subsequently be taken in a given experimental task with a latency of less than 200 ms after the start of a given trial. This fast response suggests that another brain region has coded the decision and sent the information about the forth coming action to the nigral neurons (Morris et al. 2006; compare Dommett et al. 2005). Models that incorporate the value of chosen actions in a particular state include those known as the state-action-reward-state-action or SARSA models and advantage learning models.

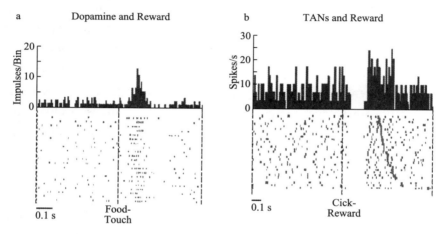

Figure 3

Reward related activity of dopamine-containing neurons of nigral and striatal neurons. (*a*) Activity of nigral-VTA complex neurons (from Romo & Schultz 1990). (*b*) Activity of tonically active neurons in the striatum (from Aosaki et al. 1994b) Spike rasters (*below*) and histograms of those spikes (*above*) are aligned (*vertical lines*) at touch of the food reward (*a*) and at the conditional stimulus click sound indicating reward.

Ironically, a main candidate for a neural structure that could deliver the action value signal to the midbrain dopamine-containing neurons is the striatum, the region originally thought to be the student of the dopaminergic substantia nigra. Many projection neurons in the striatum encode action value when monkeys perform in block design paradigms in which action values are experimentally manipulated (Samejima et al. 2005). Other structures projecting to the nigral dopamine-containing neurons are also candidates, including the pedunculopontine nucleus (one of the brain stem regions noted in **Figures 1** and **5**), the raphe nuclei including the dorsal raphe nucleus, the lateral habenular nucleus and forebrain regions including the amygdala and limbic related cortex, and also the striatum itself, including the striosomal system.

These findings highlight the difficulty of as signing an exclusive teaching function to any one node in interconnected circuits such as those linking the dopamine-containing midbrain neurons, the basal ganglia, and the cerebral cortex. Reinforcement-related signals of different sorts have been found in all of these brain regions (e.g., Glimcher 2003, Padoa-Schioppa & Assad 2006, Paton et al. 2006, Platt & Glimcher 1999, Sugrue et al. 2004), suggesting that signals related to reinforcement and motivation are widely distributed and can be used to modulate distributed neural representations guiding action. Reward-related activity has even been identified in the primary visual cortex (Shuler & Bear 2006) and the hippocampus (Suzuki 2007), neither of which is part of traditional reinforcement learning circuits. How these distributed mechanisms are coordinated is not yet clear.

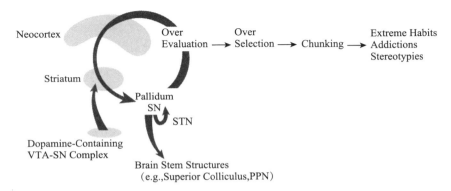

Figure 4

Schematic diagram suggesting the progression of functional activation in cortico-basal ganglia circuits as highly repetitive habits, addiction, and stereotypies emerge behaviorally. Note that in contrast to normal everyday habit learning (**Figure 1**), even the early stages of extreme habit formation involve steps that tend not to be readily reversible. PPN, pedunculopontine nucleus; SN, substantia nigra; STN, subthalamic nucleus; VTA, ventral tegmental area

Many of the ideas in reinforcement learning models and their close allies in neuroeconomics are now central to any consideration of habit learning. Experiments on goal-directed behavior in animals, including some with reward devaluation protocols, are increasingly being interpreted within the general framework of reinforcement learning (Daw & Doya 2006, Niv et al. 2006). For example, Daw et al. (2005) have proposed a model with two behavioral controllers. One (identified with the prefrontal cortex) uses a step-by-step, model-based reinforcement learning system to explore alternatives and make outcome predictions (their "tree search" learning system for goal-oriented behaviors). The second, identified with the dorsolateral striatum, is a nonmodel-based cache system for determining a fixed value for an action or context that can be stored but that then is inflexible, corresponding to the habit system. The transition between behavioral control between the tree-search and cache systems is determined by the relative Bayesian uncertainty of the two systems. Allowing for interactions between these two systems would bring them into correspondence with the goal-oriented and habit systems of the reward devaluation literature. It seems unlikely, however, that there are only two learning systems or that these are dissociable as being exclusively cortical and subcortical.

The shift from ventral striatal to dorsal striatal activation during habit learning is also being incorporated explicitly into modified RL models. For example, evidence from human brain-scanning experiments has implicated the ventral striatum—active during the initial stages of learning—in reward prediction error encoding whether or not responses by the subject are required. By contrast, reward prediction error signaling that occurs during instrumental responding is differentially associated with activity in the dorsal striatum, especially dorsomedially (O'Doherty et al. 2004). These findings favor modified RL actor-critic models including advantage learning (Dayan 2002), in which the critic (ventral striatum) is influenced by motivational state (e.g., hunger) as well as by the ongoing evaluation of the particular state, whereas the dorsal striatum (actor) is engaged when actions are instrumental in bringing reward (Dayan & Balleine 2002).

This reformulation of the actor-critic model adds the critical feature of motivation to RL treatments. Haruna & Kawato (2006) have drawn a contrast between the caudate nucleus (and ventral striatum) as being associated with reward-prediction error and the putamen as being associated with a stimulus-action-dependent reward-prediction error. The early requirement for ventral striatal activity in procedural learning has prompted the view that the ventral striatum is the director, guiding the dorsal striatal actor (Atallah et al. 2007). Imaging technologies cannot yet detect the neurochemically defined striosome and matrix compartments of the dorsal striatum, but the limbic-associated striosomes likely share some characteristics of the ventral striatum (Graybiel 1990, 1998). In line with this possibility, striosomes may represent state value in actor-critic architectures (Doya 2000, 2002) and thus may be critical components of striatum-based learning circuits.

The convergence of computational approaches with neuroimaging in humans has also led to new cognitive neuroscience models in which cognitive evaluation and choice depend on current reinforcement and expected future reinforcement and also on the value of behaviors not chosen so that choices themseves can be evaluated by comparing alternatives (Montague et al. 2006). The ventral-to-dorsal gradients found in these and other experiments (e.g., Tanaka et al. 2004, Zald et al. 2004) may in part reflect the predominance of immediate rewards (ventral striatum) and future rewards (dorsal striatum) in influencing behavior (Tanaka et al. 2004). This view emphasizes what I, below, call mental exploration.

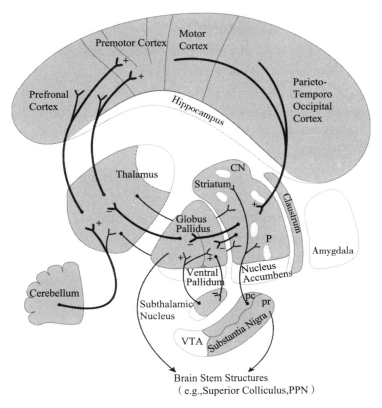

Figure 5
Basal ganglia circuit anatomy shown in simplified form with motor control pathways (*gray*) and regions related to limbic-system function (*yellow*). Descending pathways to brain stem structures (such as the superior colliculus and pedunculopontine nucleus) are not shown in detail. CN, caudate nucleus; P, putamen; pc, substantia nigra pars compacta; pr, substantia nigra pars reticulata; VTA, ventral tegmental area.

EXTREME HABITS

Investigations of normal habit learning have immediate relevance for the field of addiction research. Habits induced by exposure to drugs of abuse can dominate behavior and produce states of craving and drug-seeking that persist for years. The dopamine-containing VTA and its striatal target region (the ventral striatum including the nucleus accumbens) have long been identified as forming a reward circuit essential for the initiation and expression of addictive behavior. Experiments on addiction have led to conclusions that are strikingly similar to those emerging from study of the normal transition from goal-directed to habit-driven behavior (**Figure 4**) (Everitt et al. 2001, Hyman et al. 2006, Ito et al. 2002, Kalivas & Volkow 2005, Porrino et al. 2004, Wise 2004). (*a*) The addictive process involves plasticity in neural circuits, not just in a single brain region. (*b*) The circuits critical for addiction include midbrain dopamine-containing cell groups and the striatum (in particular, the ventral striatum) and the neocortex (especially the anterior cingulate cortex and orbitofrontal cortex), as well as limbic parts of the pallidum (ventral pallidum), the thalamus (mediodorsal nucleus), and the amygdala and extended amygdala (Everitt & Robbins 2005, Kalivas et al. 2005, Kalivas & Volkow 2005). (*c*) The neocortex (especially the anterior cingulate and orbitofrontal cortex) power fully influences this circuit at multiple levels. (*d*) Within this cortico-subcortical circuitry, different subcircuits appear to be essential at different stages of acquisition: the VTA and shell of the nucleus accumbens for the initial learning process in addiction, and the core of the nucleus accumbens and neocortical regions for the expression of the learned behaviors. Thus, the predominant activity in a given learning context shifts over time as the addiction becomes fixed (Sellings & Clarke 2003).

A striking example of this dynamic patterning comes from imaging studies in which radiolabeled dopamine-receptor agonist ligands are given to addicted subjects. Given exposures to a drug (IV methylphenidate), increased dopamine-related signaling (displacement of the ligand) in cocaine addicts occurs in the ventral striatum (and anterior cingulate and or bitofrontal cortices). But when cocaine addicts view video tapes showing cues associated with cocaine use, it is the dorsal striatum (including the putamen) that exhibits the differentially heightened dopamine signal (Volkow et al. 2006, Wong et al. 2006).

These findings are supported by work on experimental animals. In macaque monkeys, cocaine Self-administration alters metabolic activity mainly in the ventral striatum after a brief period of self-administration, but with extended self-administration, such changes occur increasingly in the dorsal striatum (Porrinoet al. 2004). The gradients in these effects fit with anatomical evidence that the dopamine containing inputs to the striatum, via in directly recursive connections between the striatum and substantia nigra, follow such a ventral-to-dorsal gradient (Haber et al. 2000). Many other neurotransmitter-related compounds also follow such gradients, however, so that many aspects of striatal circuitry are modulated differently in ventral and dorsal striatum (and in medial and lateral striatal regions as well).

Differential release of dopamine in the nucleus accumbens in response to drug-associated cues has now been demonstrated directly in the rat by measuring extracellular dopamine by electrochemical detection with intrastriatal electrodes (Phillips et al.

2003). Cues produce dopamine release after the drug habit has been established (Ito et al. 2002). If the drug-taking behavior is extinguished and then is reinstated by exposing rodents to cues associated with the drug, as a model of relapse, dopamine release is also prominent in the neocortex and amygdala. These studies, in their entirety, suggest that as an individual moves from spaced, cognitively controlled experience with an addictive drug to increasingly repeated experience with the drug and then to the addicted state of compulsive drug use, major circuit-level changes occur at both cortical and subcortical levels. Within the striatum, there is a strong gradient from ventral to dorsal regions during the course of the addictive process.

Studies on normal habit learning and these studies on addictive habit learning both indicate a gradual change from ventral striatal control over the addictive behaviors to engagement of the dorsal striatum and neocortex (Everitt & Robbins 2005, Fuchs et al. 2006, Nelson & Killcross 2006). Such shared dynamic patterning (**Figure 2**) suggests that both in addiction and in the acquisition of nondrug related habits, a similar set of coordinated changes in basic activity occurs at the circuit level. These dynamic patterns are a major clue for future experimental and computational work (Redish 2004).

A particular advantage of studies on addiction is that they can be used to uncover cellular and molecular mechanisms that promote transition to the addicted state (Hyman et al. 2006, Robinson & Kolb 2004, Self 2004). Many of these mechanisms are likely conserved and influence the learning processes that lead to repetitive, compulsive habits that are not triggered directly by drugs (Graybiel et al. 2000, Graybiel & Rauch 2000, Hyman et al. 2006). Links between the neurobiology of addictive and nonaddictive habits have already begun to be established by considering commonalities and interactions between drug effects and learning (e.g., Everitt & Robbins 2005, Fuchs et al. 2006, Nelson & Killcross 2006, Nordquist et al. 2007, Willuhn & Steiner 2006, Wise 2004).

LEARNING HABITS AND LEARNING PROCEDURES

How does habit learning relate to procedural or skill learning? Both involve learning sequential behaviors, but there are obvious differences: Learning how to ride a bicycle is quite different from having the habit of biking every evening after work. The skilled procedural performance of the baseball player is distinct from the habits and rituals for which baseball players are famous. And a regularly practiced habit may actually not be very skilled. Nevertheless, the learning of sequential actions to the point at which they can be performed with little effort of attention is common to both. So is the fact that, once consolidated, most acquired habits and procedures can be retained for long periods of time.

This transition from novice to expert is called proceduralization in the literature on hu man skill learning, most of which lies outside the reinforcement learning framework. These studies typically use reaction times (either simple or serial) as performance measures, along with a gradual reduction in interference if other tasks must be performed (Logan 1988, Nissen& Bullemer 1987). A large literature summarizes the varieties of motor sequenceen coding in monkeys performing movement sequences (Georgopoulos & Stefanis 2007, Matsuzaka et al. 2007, Tanji & Hoshi 2007). This work demonstrates that neurons can have responses related to individual or multiple elements of movement sequences and that these sequence-selective activities can occur in widely distributed brain regions, both cortical and sub-cortical (**Figure 5**). Notably, imaging and other studies demonstrate dynamic changes in the patterns of neural activity in the neocortex and striatum that closely parallel those observed in studies of habit learning (**Figure 2**). With practice, the activity shifts from anterior and ventral cortical regions to more posterior zones of the neocortex, and from more ventral and anterior striatal regions to more caudal zones in the striatum (Doyon & Benali 2005, Graybiel 2005, Isoda & Hikosaka 2007, Poldrack et al. 2005). These shifts have been interpreted as representing a shift in the coordinate frames of the early and late representations (Hikosaka et al. 1999, Isoda & Hikosaka 2007). Doyon et al. (2003) propose a key function for cortico striatal circuits in this process, contrasting these with cerebellum-based learning processes that allow adjustments of motor behavior to imposed changes (Kawato & Gomi 1992).

A function in online correction has been ascribed to the striatum (Smith & Shadmehr 2005), which would be consistent with evidence for the basal ganglia–based song system in birds (Brainard & Doupe 2000). The experiments of Haruno & Kawato (2006) suggest that the differential engagements of the more anterior striatum (caudate nucleus) and more posterior striatum (putamen) at different points during learning reflect a fundamental difference in the representations of learning-related signals in the caudate nucleus and ventral striatum (representing reward-prediction error) and the putamen (representing stimulus-action-reward association). These and related studies (Samejima et al. 2005) indicate the value of extending reinforcement learning and neuroeconomics models of habit learning to the study of the proceduralization process itself and underscore the dynamic shifts in focus of neural activity as learning proceeds. However, these models will likely be revised as more is learned about the concurrent neural activities in the neocortex, striatum, and cerebellum. For example, cerebellar input can reach the putamen via a cerebello-thalamo-striatal pathway (Hoshi et al. 2005). Activity in the putamen may in part reflect input from a cerebellar circuit. Perhaps the striatum evaluates the forward models of the cerebellum.

These findings may help in understanding the progression of symptoms in Parkinson's disease patients. The advance of dopamine denervation in Parkinson's disease follows a posterior-to-anterior gradient, the reverse of the gradient found for procedural learning and habit learning. The predominant difficulty of Parkinson's patients to perform even well-known procedures, like rising from a chair or trying to do two things at once, may reflect learning gradients seen during proceduralization.

HABITS, RITUALS, AND FIXED-ACTION PATTERNS

Work on habits and habit learning has clear connectivity with the study of repetitive behaviors that includespecies-specific, apparently instinctual action sequences: the fixed action patterns described by ethologists. Like habits, FAPs are regularly contrasted with voluntary behaviors and relegated to low-level behavioral control schemes in which a particular stimulus/context elicits an entire behavioral sequence, for example, a chick pecking the orange spot on the beak of its parent (Tinbergen 1953). Here, value functions are estimated over evolutionary time.

Rituals are common across animal species and can be either remarkably complex, as in the nest building of the bowerbird (Diamond 1986), or relatively simple, as in grazing animals following set routes to a water source. Rituals related to territoriality, mating, and social interactions of many types seem to dominate the lives of animals in the wild. These behaviors share cardinal characteristics of habits. They are repetitive, sequential action streams and can be triggered by particular cues.

Nearly all so-called fixed action patterns exhibit some flexibility (vulnerability to experience-based adjustment), and many rituals are acquired. A particularly instructive example of ritualistic behavior in animals is the use of song in mating behavior and other social behaviors in orcine birds. Their songs are composed of fixed sequences and, though in some species these appear to be genetically determined, in many species the songs are acquired by trial-and-error learning aimed at matching a tutor's template song. Bird song learning critically depends on a forebrain circuit that corresponds to a cortico–basal ganglia loop in mammals. Lesions within this circuit (called the anterior forebrain pathway) prevent the development of a mature song matching the tutor's song. The early song becomes prematurely stereotyped. At adulthood, such lesions impair the corrective adjustments of song required when abnormalities in singing behavior are induced experimentally (Brainard & Doupe 2000, Kao & Brainard 2006, Olveczky et al. 2005), and in some species the lesions can seriously degrade the mature song (Kobayashi et al. 2001). The song system is likely strongly influenced by dopamine, as the central core of the song circuit, area X (corresponding to the mammalian striatum and pallidum), receives a dopamine containing fiber projection from the midbrain. Altogether, many aspects of avian song learning and song production bear a strong resemblance to the learning and performance of what we think of as habits and procedures in mammals: Both involve the acquisition of ordered sequences of behavior that are learned by balancing exploration and exploitation to attain a goal. And both require the activity of basal ganglia–based circuits as they are learned.

STEREOTYPIES

Very highly repetitive behavioral routines in animals are qualitatively distinguished as stereotypies on the basis of their apparent purposelessness and their great repetitiveness. In contrast to the habitual displays and rituals that are triggered in the course of normal behaviors, stereotypies (extremely rigid, repetitive sequences of behavior) are most prominent under aversive conditions, including stress, social isolation, and sensory deprivation (Ridley 1994). Exposure to psychomotor stimulants such as amphetamine also induces stereotypies, either of locomotion or of naturalistic behaviors such as sniffing, rearing, mouthing, and huddling. These categories can be likened to the route stereotypies and focused stereotypies of unmedicated animals. A distinction is often made between such highly repetitive, apparently purposeless behaviors and highly repetitive behaviors that appear goal directed, for example, repetitive grooming (barbering) of another animal or self-grooming.

Work on drug-induced stereotypies demonstrates that the basal ganglia are central to the development of these repetitive behaviors (Cooper & Dourish 1990). Low doses of drugs such as amphetamine (which releases dopamine and other biogenic amines) and cocaine (which blocks the reuptake of dopamine) induce prolonged and often repetitive bouts of locomotion. This behavior requires the functioning of the ventral striatum. The same drugs, given at higher doses, also induce bouts of highly stereotyped behavior, but these are more focused behaviors such as sniffing and grooming. These behaviors depend on the dorsal striatum (Joyce & Iversen 1984). Such dose-dependent drug effects ranging from locomotion to focused stereotypies follow the ventral striatal-to-dorsal striatal gradient that is thought to underlie both the acquisition of normal procedures and habits and the acquisition of addictive habits.

The strength of drug-induced stereotypic behaviors is correlated with differential activation of the striosomal compartment of the striatum, relative to activation of the matrix (**Figure 6**). This relationship has been demonstrated in rats, mice, and monkeys by the use of early-response gene assays (Canales 2005, Canales & Graybiel 2000, Graybiel et al. 2000, Saka et al. 2004). These findings tie the acquisition of at least one class of stereotypic behavior to a specialized set of striosome predominant basal ganglia–based circuits. Striosomes in the anterior part of the striatum receive inputs from the anterior cingulate cortex and posterior orbitofrontal cortex and project to the substantia nigra both directly and indirectly via the pallidum and the limbic lateral habenular nucleus (Graybiel 1997). Through these connections, the striosomal system likely integrates and evaluates limbic information and affects the firing of dopamine-containing neurons in the nigral complex or immediately neighboring neurons. The striosomal system could thus alter the set point of subsets of the dopaminergic neurons and be a driving force behind some classes of stereotypic behaviors. We do not yet know what signal is transmitted by striosomes. One possibility is that they gate the magnitude of positive reward-prediction error or signal negative reward-prediction error through a connection with the pars compacta or nearby nigral neurons. They may also indirectly influence the lateral habenula, itself a source of negative reward-prediction error (Matsumoto & Hikosaka 2007).

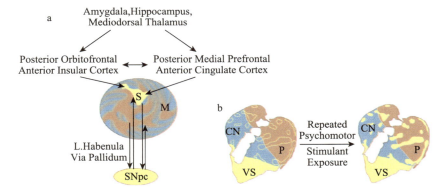

Figure 6

The striosomal system of the striatum. (*a*) Connections of striosomes with forebrain and brain stem regions including projections to nigral dopamine-containing neurons and/or nondopamine-containing nigral neurons near them. (*b*) Schematic illustration of the progressive accentuation of activity in striosomes that occurs as psychomotor stimulant-induced stereotypies develop. (Modified from Graybiel & Saka 2004). CN, caudate nucleus; M, matrix; P, putamen; S, striosome; SNpc, substantia nigra pars compacta; VS, ventral striatum.

A second line of evidence that the basal ganglia are required for some stereotypic behaviors comes from work on the highly repetitive movement sequences made by rodents during bouts of grooming called syntatic grooming (Berridge 1990). The grooming movements are strictly repetitive and temporally ordered, and lesion studies in the rat have demonstrated that a region in the mid-anterior dorsolateral caudoputamen must be intact for syntatic grooming to occur. Recording experiments suggest that some single striatal neurons in that region respond selectively to grooming movements that are part of a given syntatic grooming bout (Aldridge & Berridge 1998). The dopaminergic system can modulate the intensity of the grooming as well as the intensity of drug-induced stereotypies (Berridge et al. 2005).

How do stereotypies relate to habits? Evidently, they are different phenotypically. Stereotypies are more repetitive, are not as regularly elicited by triggers, and may not be governed in the same way by reinforcement learning. Yet many types of stereotypic behavior depend on basal ganglia–based circuits for their development and production, as do habits. Stereotypies, like addictions, seem to reflect an extreme state of functioning of these brain circuits, in which flexibility is minimal and repetitiveness is maximal. The well-learned song of a zebra finch, the product of a basal ganglia circuit analogue in the avian brain, is as stereotyped as a typical drug-induced stereotypy. The fact that one is naturally prompted and the other is not should not obscure a potential common neural basis of these behaviors in the operation of cortico–basal ganglia circuits and related brain networks.

HABITS, STEREOTYPIES, AND RITUALISTIC BEHAVIORS IN HUMANS

Habits and routines are woven into the fabric of our personal and social lives as humans. One can scarcely call to mind the events of a day without running up against them. In fact, "getting in a rut" is so easy to do for many of us that we must fight the tendency in order to get a fresh look at life. Yet, as William James and many others have argued, those ruts, fashioned carefully, are invaluable aids to making one's way through life and are critical in social order: famously, the flywheel of society (James 1890).

Helpful as habits can be in daily life, they can become dominant and intrusive in neuropsychiatric disorders, including obsessive-compulsive disorder (OCD), Tourette syndrome, and other so-called OC-spectrum disorders (Albin & Mink 2006, Graybiel & Rauch 2000, Leckman & Riddle 2000). They are major features of some autism spectrum disorders and can dominate in some of these conditions (for example, in Rett syndrome). Stereotypies are pronounced in unmedicated schizophrenics and in some patients with Huntington's disease or dystonia, and they appear in exaggerated form in some medicated Parkinson's disease patients. In these disorders—as in normal behavior—habits, stereotypies, and rituals can be cognitive as well as motor, and often, cognitive and motor acts and rituals are interrelated. This commonality across cognitive and motor domains is a crucial attribute that suggests that there may be commonality also in the neural mechanisms underlying repetitive thoughts and actions.

In OCD, classified as an anxiety disorder, the repetitive behaviors that often occur do so in response to compelling and disturbing repetitive thoughts and felt needs (obsessions) that drive the repetitive behaviors (compulsions). The repetitive thoughts are dominant despite the fact that the person usually is aware that they are happening and does not want them to happen. Notably, the most common obsessions and compulsions are strongly cross cultural. Checking ("Is it done? Did I do it just right?"), washing and cleaning due to obsessions about contamination, and repetitive ordering (lining up, straightening) are universal. OC-spectrum disorders include grooming and body-centered obsessions and compulsions, for example, extreme hair pulling or trichotillomania, for which barbering in animals is thought to be a model.

In Tourette syndrome, tics take the form of highly repetitive movements or vocalizations that range from abrupt single movements (twitches or sounds) to complex sequences of a behavior (including vocalizations) that appear as whole purposeful behaviors. They are exacerbated by stress, are sensitive to context, often occur in runs or bouts, and though suppressible for a time, often break through into overt expression. These actions are often prompted by an internal sensation or urge that builds up and produces stress; the urge can be relieved for a time by the expression of the tics. Nearly any piece of voluntary behavior can appear as a tic in Tourette syndrome. Sensory tricks, for example, touching a particular body part in a particular way, can also relieve the tics in some instances. It is as though a circuit goes into overdrive during a bout of tics but can be normalized by receiving new inputs.

At the behavioral level, obsessions and compulsions also can focus on a wide range of behaviors from grooming to eating to hoarding. Some specific genetic disorders involve compulsive engagement in one or more of these fundamental behaviors (for example, ritualistic hyperphagia and food-related obsessions in Prader-Willi syndrome). These disorders seem to accentuate one or another of a library of species-specific and culturally molded behaviors. This remarkable characteristic has led many investigators, from ethologists and neuroscientists to neurologists and psychiatrists, to make extensive comparisons between these human behaviors and the rituals, stereotypies, and FAPs of nonhuman species (Berridge et al. 2005, Graybiel 1997). These behaviors also have high relevance for the study of habit. Across the spectrum of OC disorders, the behaviors are repetitive and usually sequential; they are often performed as a whole after being triggered by external or internal stimuli or contexts (including thoughts); and they can be largely involuntary.

It is not yet known whether the brain states generated by these disorders are akin to the brain states generated as a result of habit learning. However, dysfunction of basal ganglia–based circuits appear to be centrally implicated in these disorders, from OCD-Tourette and other OC-spectrum disorders to eating disorders to the punding of overmedicated Parkinson's disease patients (e.g., Graybiel & Rauch 2000, Palmiter 2007, Voon et al. 2007). A disordered OCD circuit has been identified in imaging studies, and this circuit includes the regions most often implicated in drug addiction: the anterior cingulate cortex and the posterior orbitofrontal cortex and their thalamic correspondents, and the anterior part of the caudate nucleus and the ventral striatum. In Tourette syndrome, differential degeneration of parvalbumin-containing neurons in the striatum and pallidum has been reported (Kalanithi et al. 2005), suggesting that the basal ganglia also contribute to the symptomatology in this disorder. The focal nature of tics may reflect local abnormal activation of subdivisions in the matrix compartment of the striatum (matrisomes) (Mink 2001) or defects in particular cortico–basal ganglia subcircuits (Grabli et al. 2004). The efficacy of sensory tricks in some Tourette patients may also reflect this architecture, whereby inputs and outputs are matched in focal striatal domains (Flaherty & Graybiel 1994). Because of the limited spatial resolution of current imaging methods, it is not known whether the striosomal system is differentially involved; this would not be surprising, however. Differential vulnerability of striosomes has been reported for a subset of Huntington's disease patients in whom mood disorders, including OC-like symptoms, were predominant at disease onset (Tippett et al. 2007). Predominant striosomal degeneration has also been reported in X-linked dystonia parkinsonism (DYT-3, Lubag), in which many affected individuals exhibit OCD-like symptoms and suffer from severe depression (Goto et al. 2005). Sites being targeted for deep brain stimulation therapy in OCD and Tourette syndrome are also within these basal ganglia–based circuits; anterior sites near the nucleus accumbens are favored for OCD, and intralaminar and internal pallidal sites are favored for Tourette syndrome (Mink et al. 2006).

HABITS AND RITUALS: THE BASAL GANGLIA AS A COMMON THEME

Habits, whether they are reflected in motor or cognitive activity, typically entail a set of actions, and these action steps typically are released as an entire behavioral episode once the habit is well engrained. Here, I have reviewed evidence that this characteristic expression of an entire sequential behavior is also typical of well practiced procedures and extends to stereo typies and rituals, including personal and cultural rituals in humans. Neural mechanisms that could account for such extended, encapsulated behaviors are not yet understood, but clues are coming from experiments in which chronic electrophysiological recordings were made from ensembles of neurons of rats and monkeys performing repetitive tasks to receive reward (**Figure 7**) (Barnes et al. 2005; Fujii & Graybiel 2003, 2005; Graybiel et al. 2005; Joget al. 1999).

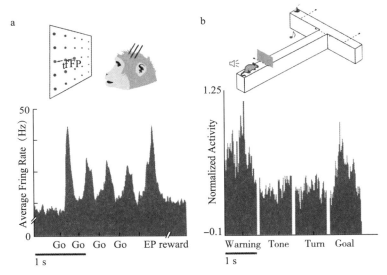

Figure 7

Heightened neural responses at action boundaries. Comparison between neural activity in prefrontal cortex accentuating the beginning and the end of saccade sequences in macaque monkeys performing sequential saccade tasks (*a*) and activity in sensorimotor striatum (dorsolateral caudoputamen) accentuating beginning and the end of maze runs in rats performing maze runs (*b*). Part a is modified from data in Fujii & Graybiel (2003), and part *b* is modified from data in Barnes et al. (2005).

In monkeys thoroughly trained to make sequences of saccadic eye movements, subsets of neurons in both the dorsolateral prefrontal cortex and the striatum have accentuated responses related to the first and last movements of the sequences, as though marking the boundaries of the action sequences (Fujii & Graybiel 2003, 2005). The accentuated beginning and end representations can be detected not only by looking at the task-related activity of the neurons, but also by looking at the temporal resolution of the neural representations (Graybiel et al. 2005). Time can be decoded with higher resolution at the beginning and the end of the movement sequences than during them. This last finding suggests that the neural representation of boundaries involves both action and time.

An explicit attempt has been made in rodents to track the neural activity patterns that occur in the striatum as habits are acquired (Barnes et al. 2005, Jog et al. 1999). Rats were trained to run a T-maze task in which they were cued to turn right or left to receive reward. The recordings were made in the dorsolateral caudoputamen, the region identified with S-R habits in reward devaluation studies and RL models. According to such studies, one would expect neural activity to be low during initial training but to increase during overtraining. The activity should be maintained during reward devaluation. This was not found. Instead, early in training, activity was strong throughout the maze runs. With extended overtraining, the activity did not increase. Instead, the activity changed in its pattern of distribution over the course of the maze runs. Task-related neural firing became concentrated at the beginning and end of the maze runs, in a pattern resembling the action boundary pattern found in the overtrained monkeys performing saccade sequences (**Figure 7**). Simultaneously, apparently nontask-responsive neural firing in the striatum was markedly reduced. When extinction training was then given (akin to reward devaluation), these acquired neural activity patterns were not maintained, but instead were gradually reversed. They were not lost or forgotten, however, because once reacquisition training began, the acquired activity patterns reappeared.

These findings suggest that one result of habit learning is to build in the sensorimotor striatum chunked, boundary-marked representations of the entire set of action steps that make up the behavioral habit. The changes in neural activity found in the maze experiments have been interpreted as representing the neural analog of explore-exploit behavior (Barnes et al. 2005, Graybiel 2005). During initial training, the animals shifted from an exploratory (variable) mode to a repetitive mode of running the maze. In parallel, the neural activity in the sensorimotor striatum measured across task time was initially variable (neural exploration), in that activity occurred throughout the maze runs. As the behavior became repetitive, the neural activity took on the accentuated beginning and end patterns (neural exploitation). This shift could represent part of the process by which action sequences are chunked for representation as a result of habit learning: When they are packaged as a unit ready for expression, the boundaries of the unit are marked and the behavioral steps unfold from the first to the last boundary marker (Graybiel 1998).

These shifts in neural activity during learning could be carried to the striatum by its input connections—notably, inputs from the cerebral cortex and the thalamus. Alternatively, they may be produced by intrastriatal network activity or by some combination of such circuit processing. Evidence favors a role for striatal interneurons in such circuit-level plasticity. For example, as monkeys are taken through successive bouts of conditioning and extinction of conditioned eyeblink training, the tonically active neurons of the striatum (thought to be the cholinergic interneurons) successively gain and extinguish conditioned responses in parallel with the shifts in eyeblink behavior (Blazquez et al. 2002).

Further findings from the maze-learning experiments suggest that the formulation of the beginning and end pattern may depend, in part, on the rats learning the association between the instruction cues and the actions they instruct (T. Barnes, D. Hu, M. Howe, A. Dreyer, E. Brown, A. Graybiel, unpublished findings). The chunking process in sensorimotor tasks may thus occur as a result of S-R learning and be a hallmark of activity in the part of the striatum most directly engaged in affecting motor output. Work in progress also suggests that, at least in rats, the associative (dorsomedial) stria tum exhibits different patterns during learning (Kubota et al. 2002a,b; C.A. Thorn, H. Atallah,Y. Kubota, A.M. Graybiel, unpublished findings). More work needs to be done to characterize these other learning-related striatal patterns along with those in other regions, including the neocortex; the locus coeruleus, identified as modulating exploration-exploitation balances (Aston-Jones & Cohen 2005); and serotonergic and cholinergic nuclei of the brainstem, which can modulate nigral neurons and reward functions (e.g., Doya 2002).

These results suggest that models epitomizing what physiological changes occur as habits and procedures are learned need to be revised to allow for repatterning of activity in different parts of the striatum and corresponding cortical and subcortical circuits. The gradients in activity traced from ventral to dorsal, anterior to posterior, and medial to lateral regions during habit learning do not necessarily mean that one region is transiently active and then becomes inactive as another region takes over, as if the habits are stored in just one site. The electrophysiological recordings suggest that we need dynamic models in which activity can occur simultaneously in multiple cortico-basal ganglia loops, not move in to from one site to another, and models in which, as the learning process occurs, activity patterns change at all these sites. We need to capture how such simultaneous, dynamically changing activity patterns can become coordinated over time through the actions of plasticity mechanisms that act on neurotransmitter signaling systems that themselves are expressed in differential gradients and compartmental patterns across the striatum. An interesting possibility is that oscillatory activity helps to coordinate these activities (DeCoteau et al. 2007b, Thorn & Graybiel 2007). For example, even though different patterns of neuronal activity are found with simultaneous recordings in dorsomedial and dorsolateral regions of the striatum during maze learning, these two regions exhibit coordination of oscillatory activity in particular frequency ranges (Thorn & Graybiel 2007).

If this dynamic repatterning is a general function of cortico–basal ganglia loops, then it should occur for cognitive activity patterns as well, with a shift from mental exploration to mental exploitation as habits of thought are developed. Some evidence suggests that the striatum and associated cortico–basal ganglia loops are involved in such processing and chunking in human language (e.g., Crinion et al. 2006, Lieberman et al. 2004, Liegeois et al. 2003, Tettamanti et al. 2005). A system of executive control using such start and end states has been proposed for activity in Broca's area in the human, supporting the idea of boundary markers in hierarchically organized domains (Fujii & Graybiel 2003, Koechlin & Jubault 2006).

Could the chunking process also relate to the low levels of attention that we typically need to pay to a familiar behavior when performing it as a habit? The deaccentuation of neural activity in between the accentuated beginning and end activities in cortico–basal ganglia circuits could reflect this attribute of habits. The eventual chunking of action repertoires during habit learning is an endpoint of successive shifting of neural activity from regions more closely related to the limbic system to regions more closely related to motor and cognitive output. These shifts represent stages of evaluation, and if the stages are successfully met—if the behavior is evaluated sufficiently positively—it is rerepresented in a chunked, readily releasable form. Thus, the relation between habits and the evaluative brain is that habits are an endpoint of the valuation process. Altogether, this process may engage a range of different cortico–basal ganglia loops and other neural circuits, potentially influencing different types of habit, from seemingly innocent mannerisms and rituals to dominating addictions. Studying this process should help investigators identify the neural systems underlying the shift from deliberative behavior controls to the nearly automatic, scarcely conscious control that we associate with acting through habit. Tracking this process may help us to understand the conscious state itself.

The power of social rituals may in part reflect an endpoint of this progressive evaluation process (**Figure 8**). The basal ganglia are strongly tied to the control and modulation of social behaviors. Human brain-imaging experiments have demonstrated strong activation of the dorsal striatum in experiments tracking activation for maternal love and romantic love (Aron et al. 2005, Bartels & Zeki 2004) and in social situations mimicked by interactive games and cost-benefit protocols (de Quervain et al. 2004, Elliott et al. 2004, Harbaugh et al. 2007, King-Casas et al. 2005, Montague & Berns 2002, O'Doherty et al. 2004, Tricomi et al. 2004, Zink et al. 2004). The nucleus accumbens and its dopamine receptors are necessary for monogamous pair bonding and for the maintenance of these bonds in the prairie vole (Aragona et al. 2006). Both language and song, with strong cortico–basal ganglia neural bases, serve social communication and are self-generated. They have the characteristic of agency, which heightens activation in the striatum in instrumental tasks relative to striatal activity in passive but otherwise corresponding tasks (Harbaugh et al. 2007, Zink et al. 2004). Finally, many of the rituals encountered in normal societies, and many of the ritualistic behaviors in neuropsychiatric disorders, have a strong social element, both in their content and in their likelihood for expression, and rituals and stereotypies in animals can also be strongly influenced by social context. Neural processing in circuits related to the basal ganglia, with their widespread interconnections with both limbic and sensorimotor systems, provides a common mechanistic theme across this large array of behaviors.

Figure 8
A ritual in humans (bull jumping in ancient Greece). Fresco from the East Wing of Knossos Palace,~1500 B.C., Herakleion Museum, Crete.

DISCLOSURE STATEMENT

The author is not aware of any biases that might be perceived as affecting the objectivity of this review.

ACKNOWLEDGMENTS

The author acknowledges the support for her laboratory from the National Institute of Mental Health MH60379, the National Institute of Neurological Disorders and Stroke NS25529, the National Eye Institute EY12848, the National Parkinson Foundation, and the Office of Naval Research N00014-04-1-0208. The author thanks H.F. Hall and Emily Romano for help with the figures; Clark Brayton for manuscript processing; and colleagues who read an earlier draft of this manuscript, including Russell Poldrack, Peter Dayan, Kyle Smith, and Hisham Atallah.

LITERATURE CITED

Adams CD, Dickinson A. 1981. Instrumental responding following reinforcer devaluation. *Q. J. Exp. Psychol.* 33:109–21

Albin RL, Mink JW. 2006. Recent advances in Tourette syndrome research. *Trends Neurosci.* 29:175–82

Aldridge JW, Berridge KC. 1998. Coding of serial order by neostriatal neurons: a "natural action" approach to movement sequence. *J. Neurosci.* 18:2777–87

Aosaki T, Graybiel AM, Kimura M. 1994a. Effects of the nigrostriatal dopamine system on acquired neural responses in the striatum of behaving monkeys. *Science* 265:412–15

Aosaki T, Tsubokawa H, Ishida A, Watanabe K, Graybiel AM, Kimura M. 1994b. Responses of tonically active neurons in the primate's striatum undergo systematic changes during behavioral sensorimotor conditioning. *J. Neurosci.* 14:3969–84

Aragona BJ, Liu Y, Yu YJ, Curtis JT, Detwiler JM, et al. 2006. Nucleus accumbens dopamine differentially mediates the formation and maintenance of monogamous pair bonds. *Nat. Neurosci.* 9:133–39

Aron A, Fisher H, Mashek DJ, Strong G, Li H, Brown LL. 2005. Reward, motivation, and emotion systems associated with early-stage intense romantic love. *J. Neurophysiol.* 94:327–37

Aston-Jones G, Cohen JD. 2005. An integrative theory of locus coeruleus-norepinephrine function: adaptive gain and optimal performance. *Annu. Rev. Neurosci.* 28:403–50

Atallah HE, Lopez-Paniagua D, Rudy JW, O'Reilly RC. 2007. Separate neural substrates for skill learning and performance in the ventral and dorsal striatum. *Nat. Neurosci.* 10:126–31

Balleine BW, Dickinson A. 1998. Goal-directed instrumental action: contingency and incentive learning and their cortical substrates. *Neuropharmacology* 37:407–19

Balleine BW, Killcross AS, Dickinson A. 2003. The effect of lesions of the basolateral amygdala on instrumental conditioning. *J. Neurosci.* 23:666–75

Barnes T, Kubota Y, Hu D, Jin DZ, Graybiel AM. 2005. Activity of striatal neurons reflects dynamic encoding and recoding of procedural memories. *Nature* 437:1158–61

Bartels A, Zeki S. 2004. The neural correlates of maternal and romantic love. *Neuroimage* 21:1155–66

Bayer HM, Glimcher PW. 2005. Midbrain dopamine neurons encode a quantitative reward prediction error signal. *Neuron* 47:129–41

Bayley PJ, Frascino JC, Squire LR. 2005. Robust habit learning in the absence of awareness and independent of the medial temporal lobe. *Nature* 436:550–53

Berridge KC. 1990. Comparative fine structure of action: rules of form and sequence in the grooming patterns of six rodent species. *Behaviour* 113:21–56

Berridge KC, Aldridge JW, Houchard KR, Zhuang X. 2005. Sequential superstereotypy of an instinctive fixed action pattern in hyperdopaminergic mutant mice: a model of obsessive compulsive disorder and Tourette's. *BMC Biol.* 3:4

Blazquez P, Fujii N, Kojima J, Graybiel AM. 2002. A network representation of response probability in the striatum. *Neuron* 33:973–82

Brainard MS, Doupe AJ. 2000. Interruption of a basal ganglia-forebrain circuit prevents plasticity of learned vocalizations. *Nature* 404:762–66

Canales JJ. 2005. Stimulant-induced adaptations in neostriatal matrix and striosome systems: transiting from instrumental responding to habitual behavior in drug addiction. *Neurobiol. Learn. Mem.* 83:93–103

Canales JJ, Graybiel AM. 2000. A measure of striatal function predicts motor stereotypy. *Nat.Neurosci.* 3:377–83

Colwill RM, Rescorla RA. 1985. Postconditioning devaluation of a reinforcer affects instrumental responding. *J. Exp. Psychol. Anim. Behav. Process.* 11:120–32

Cooper SJ, Dourish CT, eds. 1990. *Neurobiology of Stereotyped Behaviour.* Oxford, UK: Clarendon

Corbit LH, Balleine BW. 2005. Double dissociation of basolateral and central amygdala lesions on the general and outcome-specific forms of pavlovian-instrumental transfer. *J. Neurosci.*25:962–70

Coutureau E, Killcross S. 2003. Inactivation of the infralimbic prefrontal cortex reinstates goal directed responding in overtrained rats. *Behav. Brain Res.* 146:167–74

Crinion J, Turner R, Grogan A, Hanakawa T, Noppeney U, et al. 2006. Language control in the bilingual brain. *Science* 312:1537–40

Daw ND, Doya K. 2006. The computational neurobiology of learning and reward. *Curr. Opin. Neurobiol.* 16:199–204

Daw ND, Niv Y, Dayan P. 2005. Uncertainty-based competition between prefrontal and dorsolateral striatal systems for behavioral control. *Nat. Neurosci.* 8:1704–11

Dayan P. 2002. Motivated reinforcement learning. In *Advances in Neural Information Processing Systems,* ed. TG Dietterich, S Becker, Z Ghahramans, pp. 11–18. San Mateo, CA: Morgan Kaufmann

Dayan P, Balleine BW. 2002. Reward, motivation, and reinforcement learning. *Neuron* 36:285–98

DeCoteau WE, Thorn CA, Gibson DJ, Courtemanche R, Mitra P, et al. 2007a. Learning-related coordination of striatal and hippocampal theta rhythms during acquisition of a procedural maze task. *Proc. Natl. Acad. Sci. USA* 104:5644–49

DeCoteau WE, Thorn C, Gibson DJ, Courtemanche R, Mitra P, et al. 2007b. Oscillations of local field potentials in the rat dorsal striatum during spontaneous and instructed behaviors. *J. Neurophysiol.* 97:3800–5

de la Fuente-Fernandez R, Ruth TJ, Sossi V, Schulzer M, Calne DB, Stoessl AJ. 2001. Expectation and dopamine release: mechanism of the placebo effect in Parkinson's disease. *Science* 293:1164–66

de Quervain DJ, Fischbacher U, Treyer V, Schellhammer M, Schnyder U, et al. 2004. The neural basis of altruistic punishment. *Science* 305:1254–58

Devan BD, White NM. 1999. Parallel information processing in the dorsal striatum: relation to hippocampal function. *J. Neurosci.* 19:2789–98

Diamond J. 1986. Animal art: variation in bower decorating style among male bowerbirds *Ambly ornis inornatus. Proc. Natl. Acad. Sci. USA* 83:3042–46

Djurfeldt M, Ekeberg Ö, Graybiel AM. 2001. Cortex-basal ganglia interaction and attractor states. *Neurocomputing* 38–40:573–79

Dommett E, Coizet V, Blaha CD, Martindale J, Lefebvre V, et al. 2005. How visual stimuli activate dopaminergic neurons at short latency. *Science* 307:1476–79

Doya K. 2000. Complementary roles of basal ganglia and cerebellum in learning and motor control. *Curr. Opin. Neurobiol.* 10:732–39

Doya K. 2002. Metalearning and neuromodulation. *Neural. Netw.* 15:495–506

Doyon J, Benali H. 2005. Reorganization and plasticity in the adult brain during learning of motor skills. *Curr. Opin. Neurobiol.* 15:161–67

Doyon J, Penhune V, Ungerleider LG. 2003. Distinct contribution of the cortico-striatal and cortico-cerebellar systems to motor skill learning. *Neuropsychologia* 41:252–62

Elliott R, Newman JL, Longe OA, Deakin JFW. 2004. Instrumental responding for rewards is associated with enhanced neuronal response in subcortical reward systems. *Neuroimage* 21:984–90

Everitt BJ, Dickinson A, Robbins TW. 2001. The neuropsychological basis of addictive behaviour. *Brain Res. Brain Res. Rev.* 36:129–38

Everitt BJ, Robbins TW. 2005. Neural systems of reinforcement for drug addiction: from actions to habits to compulsion. *Nat. Neurosci.* 8:1481–89

Faure A, Haberland U, Conde F, El Massioui N. 2005. Lesion to the nigrostriatal dopamine system disrupts stimulus-response habit formation. *J. Neurosci.* 25:2771–80

Fiorillo CD, Tobler PN, Schultz W. 2003. Discrete coding of reward probability and uncertainty by dopamine neurons. *Science* 299:1898–902

Flaherty AW, Graybiel AM. 1994. Input-output organization of the sensorimotor striatum in the squirrel monkey. *J. Neurosci.* 14:599–610

Foerde K, Knowlton BJ, Poldrack RA. 2006. Modulation of competing memory systems by distraction. *Proc. Natl. Acad. Sci. USA* 103:11778–83

Fuchs RA, Branham RK, See RE. 2006. Different neural substrates mediate cocaine seeking after abstinence versus extinction training: a critical role for the dorsolateral caudate-putamen. *J. Neurosci.* 26:3584–88

Fujii N, Graybiel A. 2003. Representation of action sequence boundaries by macaque prefrontal cortical neurons. *Science* 301:1246–49

Fujii N, Graybiel A. 2005. Time-varying covariance of neural activities recorded in striatum and frontal cortex as monkeys perform sequential-saccade tasks. *Proc. Natl. Acad. Sci. USA* 102:9032–37

Georgopoulos AP, Stefanis CN. 2007. Local shaping of function in the motor cortex: motor contrast, directional tuning. *Brain Res. Rev.* 55:383–89

Glimcher PW. 2003. *Decisions, Uncertainty, and the Brain: The Science of Neuroeconomics.* Cambridge, MA: MIT Press

Goto S, Lee LV, Munoz EL, Tooyama I, Tamiya G, et al. 2005. Functional anatomy of the basal ganglia in X-linked recessive dystonia-parkinsonism. *Ann. Neurol.* 58:7–17

Gottfried JA, O'Doherty J, Dolan RJ. 2003. Encoding predictive reward value in human amygdala and orbitofrontal cortex. *Science* 301:1104–7

Grabli D, McCairn K, Hirsch EC, Agid Y, Feger J, et al. 2004. Behavioural disorders induced by external globus pallidus dysfunction in primates: I. Behavioural study. *Brain* 127:2039–54

Graybiel AM. 1990. Neurotransmitters and neuromodulators in the basal ganglia. *Trends Neurosci.* 13:244–54

Graybiel AM. 1997. The basal ganglia and cognitive pattern generators. *Schizophr. Bull.* 23:459–69

Graybiel AM. 1998. The basal ganglia and chunking of action repertoires. *Neurobiol. Learn. Mem.* 70:119–36

Graybiel AM. 2005. The basal ganglia: learning new tricks and loving it. *Curr. Opin. Neurobiol.* 15:638–44

Graybiel AM, Aosaki T, Flaherty AW, Kimura M. 1994. The basal ganglia and adaptive motor control. *Science* 265:1826–31

Graybiel AM, Canales JJ, Capper-Loup C. 2000. Levodopa-induced dyskinesias and dopamine dependent stereotypies: a new hypothesis. *Trends Neurosci.* 23:S71–77

Graybiel AM, Fujii N, Jin DZ. 2005. Representation of time and states in the macaque prefrontal cortex and striatum during sequential saccade tasks. *Soc. Neurosci. Abstr. Viewer/Itiner.* 400.7

Graybiel AM, Rauch SL. 2000. Toward a neurobiology of obsessive-compulsive disorder. *Neuron.* 28:343–47

Graybiel AM, Saka E. 2004. The basal ganglia and the control of action. In *The New Cognitive Neurosciences,* ed. MS Gazzaniga, pp. 495–510. Cambridge, MA: MIT Press. 3rd ed.

Haber SN, Fudge JL, McFarland NR. 2000. Striatonigrostriatal pathways in primates form an ascending spiral from the shell to the dorsolateral striatum. *J. Neurosci.* 20:2369–82

Harbaugh WT, Mayr U, Burghart DR. 2007. Neural responses to taxation and voluntary giving reveal motives for charitable donations. *Science* 316:1622–25

Haruno M, Kawato M. 2006. Different neural correlates of reward expectation and reward expectation error in the putamen and caudate nucleus during stimulus-action-reward association learning. *J. Neurophysiol.* 95:948–59

Hernandez PJ, Sadeghian K, Kelley AE. 2002. Early consolidation of instrumental learning requires protein synthesis in the nucleus accumbens. *Nat. Neurosci.* 5:1327–31

Hernandez PJ, Schiltz CA, Kelley AE. 2006. Dynamic shifts in corticostriatal expression patterns of the immediate early genes Homer 1a and Zif268 during early and late phases of instrumental training. *Learn. Mem.* 13:599–608

Hikosaka O, Nakahara H, Rand MK, Sakai K, Lu X, et al. 1999. Parallel neural networks for learning sequential procedures. *Trends Neurosci.* 22:464–71

Hoshi E, Tremblay L, Feger J, Carras PL, Strick PL. 2005. The cerebellum communicates with the basal ganglia. *Nat. Neurosci.* 8:1491–93

Houk JC, Wise SP. 1995. Distributed modular architectures linking basal ganglia, cerebellum, and cerebral cortex: their role in planning and controlling action. *Cereb. Cortex* 5:95–110

Hsu M, Bhatt M, Adolphs R, Tranel D, Camerer CF. 2005. Neural systems responding to degrees of uncertainty in human decision-making. *Science* 310:1680–83

Hyman SE, Malenka RC, Nestler EJ. 2006. Neural mechanisms of addiction: the role of reward related learning and memory. *Annu. Rev. Neurosci.* 29:565–98

Isoda M, Hikosaka O. 2007. Switching from automatic to controlled action by monkey medial frontal cortex. *Nat. Neurosci.* 10:240–48

Ito R, Dalley JW, Robbins TW, Everitt BJ. 2002. Dopamine release in the dorsal striatum during cocaine-seeking behavior under the control of a drug-associated cue. *J. Neurosci.* 22:6247–53

Izquierdo A, Suda RK, Murray EA. 2004. Bilateral orbital prefrontal cortex lesions in rhesus monkeys disrupt choices guided by both reward value and reward contingency. *J. Neurosci.* 24:7540–48

James W. 1950 [1890]. *The Principles of Psychology.* New York: Dover

Jog M, Kubota Y, Connolly CI, Hillegaart V, Graybiel AM. 1999. Building neural representations of habits. *Science* 286:1745–49

Joyce EM, Iversen SD. 1984. Dissociable effects of 6-OHDA-induced lesions of neostriatum on anorexia, locomotor activity and stereotypy: the role of behavioural competition. *Psychopharmacology* (*Berl.*) 83:363–66

Kalanithi PS, Zheng W, Kataoka Y, DiFiglia M, Grantz H, et al. 2005. Altered parvalbumin positive neuron distribution in basal ganglia of individuals with Tourette syndrome. *Proc. Natl. Acad. Sci. USA* 102:13307–12

Kalivas PW, Volkow N, Seamans J. 2005. Unmanageable motivation in addiction: a pathology in prefrontal-accumbens glutamate transmission. *Neuron* 45:647–50

Kalivas PW, Volkow ND. 2005. The neural basis of addiction: a pathology of motivation and choice. *Am. J. Psychiatry* 162:1403–13

Kao MH, Brainard MS. 2006. Lesions of an avian basal ganglia circuit prevent context-dependent changes to song variability. *J. Neurophysiol.* 96:1441–55

Kawato M, Gomi H. 1992. A computational model of four regions of the cerebellum based on feedback-error learning. *Biol. Cybern.* 68:95–103

Killcross S, Coutureau E. 2003. Coordination of actions and habits in the medial prefrontal cortex of rats. *Cereb. Cortex* 13:400–8

King-Casas B, Tomlin D, Anen C, Camerer CF, Quartz SR, Montague PR. 2005. Getting to know you: reputation and trust in a two-person economic exchange. *Science* 308:78–83

Knowlton BJ, Mangels JA, Squire LR. 1996. A neostriatal habit learning system in humans. *Science* 273:1399–402

Kobayashi K, Uno H, Okanoya K. 2001. Partial lesions in the anterior forebrain pathway affect song production in adult Bengalese finches. *NeuroReport* 12:353–58

Koechlin E, Jubault T. 2006. Broca's area and the hierarchical organization of human behavior. *Neuron* 50:963–74

Kubota Y, DeCoteau WE, Liu J, Graybiel AM. 2002a. Task-related activity in the medial striatum during performance of a conditional T-maze task. *Soc. Neurosci. Abstr. Viewer/Itiner.* 765.7

Kubota Y, DeCoteau WE, Liu J, Graybiel AM. 2002b. Task-related activity in the medial striatum during performance of a conditional T-maze task. *Soc. Neurosci. Abstr. Viewer/Itiner.* 765.7

Leckman JF, Riddle MA. 2000. Tourette's syndrome: when habit-forming systems form habits of their own. *Neuron* 28:349–54

Lieberman MD, Chang GY, Chiao J, Bookheimer SY, Knowlton BJ. 2004. An event-related fMRI study of artificial grammar learning in a balanced chunk strength design. *J. Cogn. Neurosci.* 16:427–38

Liegeois F, Baldeweg T, Connelly A, Gadian DG, Mishkin M, Vargha-Khadem F. 2003. Language fMRI abnormalities associated with *FOXP2* gene mutation. *Nat. Neurosci.* 6:1230–37

Logan GD. 1988. Toward an instance theory of automatization. *Psychol. Rev.* 95:492–527

Matsumoto M, Hikosaka O. 2007. Lateral habenula as a source of negative reward signals in dopamine neurons. *Nature* 447:1111–15

Matsuzaka Y, Picard N, Strick PL. 2007. Skill representation in the primary motor cortex after long-term practice. *J. Neurophysiol.* 97:1819–32

Mink JW. 2001. Basal ganglia dysfunction in Tourette's syndrome: a new hypothesis. *Pediatr. Neurol.* 25:190–98

Mink JW, Walkup J, Frey KA, Como P, Cath D, et al. 2006. Patient selection and assessment recommendations for deep brain stimulation in Tourette syndrome. *Mov. Disord.* 21:1831–38

Montague PR, Berns GS. 2002. Neural economics and the biological substrates of valuation. *Neuron* 36:265–84

Montague PR, Dayan P, Sejnowski TJ. 1996. A framework for mesencephalic dopamine systems based on predictive Hebbian learning. *J. Neurosci.* 16:1936–47

Montague PR, King-Casas B, Cohen JD. 2006. Imaging valuation models in human choice. *Annu. Rev. Neurosci.* 29:417–48

Moody TD, Bookheimer SY, Vanek Z, Knowlton BJ. 2004. An implicit learning task activates medial temporal lobe in patients with Parkinson's disease. *Behav. Neurosci.* 118:438–42

Morris G, Arkadir D, Nevet A, Vaadia E, Bergman H. 2004. Coincident but distinct messages of midbrain dopamine and striatal tonically active neurons. *Neuron* 43:133–43

Morris G, Nevet A, Arkadir D, Vaadia E, Bergman H. 2006. Midbrain dopamine neurons encode decisions for future action. *Nat. Neurosci.* 9:1057–63

Nakahara H, Itoh H, Kawagoe R, Takikawa Y, Hikosaka O. 2004. Dopamine neurons can represent context-dependent prediction error. *Neuron* 41:269–80

Nelson A, Killcross S. 2006. Amphetamine exposure enhances habit formation. *J. Neurosci.* 26:3805–12

Nissen MJ, Bullemer P. 1987. Attentional requirements of learning: evidence from performance measures. *Cognit. Psychol.* 19:1–32

Niv Y, Duff MO, Dayan P. 2005. Dopamine, uncertainty and TD learning. *Behav. Brain Funct.* 1:6

Niv Y, Joel D, Dayan P. 2006. A normative perspective on motivation. *Trends Cogn. Sci.* 10:375–81

Nordquist RE, Voorn P, de Mooij-van Malsen JG, Joosten RN, Pennartz CM, Vanderschuren LJ. 2007. Augmented reinforcer value and accelerated habit formation after repeated amphetamine treatment. *Eur. Neuropsychopharmacol.* 17:532–40

O'Doherty J, Dayan P, Schultz J, Deichmann R, Friston K, Dolan RJ. 2004. Dissociable roles of ventral and dorsal striatum in instrumental conditioning. *Science* 304:452–54

Olveczky BP, Andalman AS, Fee MS. 2005. Vocal experimentation in the juvenile songbird requires a basal ganglia circuit. *PLoS Biol.* 3:e153

O'Reilly RC, Frank MJ. 2006. Making working memory work: a computational model of learning in the prefrontal cortex and basal ganglia. *Neural Comput.* 18:283–328

Packard MG, Knowlton BJ. 2002. Learning and memory functions of the basal ganglia. *Annu. Rev. Neurosci.* 25:563–93

Packard MG, McGaugh JL. 1996. Inactivation of hippocampus or caudate nucleus with lidocaine differentially affects expression of place and response learning. *Neurobiol. Learn. Mem.* 65:65–72

Padoa-Schioppa C, Assad JA. 2006. Neurons in the orbitofrontal cortex encode economic value. *Nature* 441:223–26

Palmiter RD. 2007. Is dopamine a physiologically relevant mediator of feeding behavior? *Trends Neurosci.*

30:375-81

Paton JJ, Belova MA, Morrison SE, Salzman CD. 2006. The primate amygdala represents the positive and negative value of visual stimuli during learning. *Nature* 439:865–70

Phillips PE, Stuber GD, Heien ML, Wightman RM, Carelli RM. 2003. Subsecond dopamine release promotes cocaine seeking. *Nature* 422:614–18

Platt ML, Glimcher PW. 1999. Neural correlates of decision variables in parietal cortex. *Nature* 400:233–38

Poldrack RA, Clark J, Pare Blagoev EJ, Shohamy D, Creso Moyano J, et al. 2001. Interactive memory systems in the human brain. *Nature* 414:546–50

Poldrack RA, Sabb FW, Foerde K, Tom SM, Asarnow RF, et al. 2005. The neural correlates of motor skill automaticity. *J. Neurosci.* 25:5356–64

Porrino LJ, Lyons D, Smith HR, Daunais JB, Nader MA. 2004. Cocaine self-administration produces a progressive involvement of limbic, association, and sensorimotor striatal domains. *J. Neurosci.* 24:3554–62

Rauch SL, Wedig MM, Wright CI, Martis B, McMullin KG, et al. 2006. Functional magnetic resonance imaging study of regional brain activation during implicit sequence learning in obsessive-compulsive disorder. *Biol. Psychiatry* 61:330–36

Redgrave P, Gurney K. 2006. The short-latency dopamine signal: a role in discovering novel actions? *Nat. Rev. Neurosci.* 7:967–75

Redish AD. 2004. Addiction as a computational process gone awry. *Science* 306:1944–47

Ridley RM. 1994. The psychology of perseverative and stereotyped behaviour. *Prog. Neurobiol.* 44:221–31

Robinson TE, Kolb B. 2004. Structural plasticity associated with exposure to drugs of abuse. *Neuropharmacology* 47(Suppl. 1):33–46

Romo R, Schultz W. 1990. Dopamine neurons of the monkey midbrain: contingencies of response to active touch during self-initiated arm movements. *J. Neurophysiol.* 63:592–606

Saka E, Goodrich C, Harlan P, Madras BK, Graybiel AM. 2004. Repetitive behaviors in monkeys are linked to specific striatal activation patterns. *J. Neurosci.* 24:7557–65

Salat DH, van der Kouwe AJ, Tuch DS, Quinn BT, Fischl B, et al. 2006. Neuroimaging H.M.: a 10-year follow-up examination. *Hippocampus* 16:936–45

Samejima K, Ueda Y, Doya K, Kimura M. 2005. Representation of action-specific reward values in the striatum. *Science* 310:1337–40

Satoh T, Nakai S, Sato T, Kimura M. 2003. Correlated coding of motivation and outcome of decision by dopamine neurons. *J. Neurosci.* 23:9913–23

Schultz W, Dayan P, Montague PR. 1997. A neural substrate of prediction and reward. *Science* 275:1593–99

Self DW. 2004. Regulation of drug-taking and -seeking behaviors by neuroadaptations in the mesolimbic dopamine system. *Neuropharmacology* 47(Suppl. 1):242–55

Sellings LH, Clarke PB. 2003. Segregation of amphetamine reward and locomotor stimulation between nucleus accumbens medial shell and core. *J. Neurosci.* 23:6295–303

Shuler MG, Bear MF. 2006. Reward timing in the primary visual cortex. *Science* 311:1606–9

Smith MA, Shadmehr R. 2005. Intact ability to learn internal models of arm dynamics in Huntington's disease but not cerebellar degeneration. *J. Neurophysiol.* 93:2809–21

Smith-Roe SL, Kelley AE. 2000. Coincident activation of NMDA and dopamine D1 receptors within the nucleus accumbens core is required for appetitive instrumental learning. *J. Neurosci.* 20:7737–42

Sugrue LP, Corrado GS, Newsome WT. 2004. Matching behavior and the representation of value in the parietal cortex. *Science* 304:1782–87

Sutton RS, Barto AG. 1998. *Reinforcement Learning: An Introduction.* Cambridge, MA: MIT Press

Suzuki WA. 2007. Associative learning signals in the monkey medial temporal lobe. Presented at *Cosyne* 2007, Salt Lake City, Utah

Tanaka SC, Doya K, Okada G, Ueda K, Okamoto Y, Yamawaki S. 2004. Prediction of immediate and future rewards differentially recruits cortico-basal ganglia loops. *Nat. Neurosci.* 7:887–93

Tanji J, Hoshi E. 2007. Role of the lateral prefrontal cortex in executive behavioral control. *Physiol. Rev.* 88:37–57

Tettamanti M, Moro A, Messa C, Moresco RM, Rizzo G, et al. 2005. Basal ganglia and language: phonology modulates dopaminergic release. *NeuroReport* 16:397–401

Thorn CA, Graybiel AM. 2007. Medial and lateral striatal LFPs exhibit task-dependent patterns of coherence in multiple frequency bands. *Soc. Neurosci. Abstr. Viewer/Itiner.* 622.14

Tinbergen N. 1953. *The Herring Gull's World: A Study of the Social Behaviour of Birds.* London:Collins

Tippett LJ, Waldvogel HJ, Thomas SJ, Hogg VM, van Roon-Mom W, et al. 2007. Striosomes and mood dysfunction in Huntington's disease. *Brain* 130:206–21

Tobler PN, Fiorillo CD, Schultz W. 2005. Adaptive coding of reward value by dopamine neurons. *Science* 307:1642–45

Tricomi EM, Delgado MR, Fiez JA. 2004. Modulation of caudate activity by action contingency. *Neuron* 41:281–92

Ungless MA, Magill PJ, Bolam JP. 2004. Uniform inhibition of dopamine neurons in the ventral tegmental area by aversive stimuli. *Science* 303:2040–42

Voermans NC, Petersson KM, Daudey L, Weber B, Van Spaendonck KP, et al. 2004. Interaction between the human hippocampus and the caudate nucleus during route recognition. *Neuron* 43:427–35

Volkow ND, Wang GJ, Telang F, Fowler JS, Logan J, et al. 2006. Cocaine cues and dopamine in dorsal striatum: mechanism of craving in cocaine addiction. *J. Neurosci.* 26:6583–88

Voon V, Potenza MN, Thomsen T. 2007. Medication-related impulse control and repetitive behaviors in Parkinson's disease. *Curr. Opin. Neurol.* 20:484–92

Waelti P, Dickinson A, Schultz W. 2001. Dopamine responses comply with basic assumptions of formal learning theory. *Nature* 412:43–48

Wellman LL, Gale K, Malkova L. 2005. GABA$_A$-mediated inhibition of basolateral amygdala blocks reward devaluation in macaques. *J. Neurosci.* 25:4577–86

Willingham DB, Salidis J, Gabrieli JD. 2002. Direct comparison of neural systems mediating conscious and unconscious skill learning. *J. Neurophysiol.* 88:1451–60

Willuhn I, Steiner H. 2006. Motor-skill learning-associated gene regulation in the striatum: effects of cocaine. *Neuropsychopharmacology* 31:2669–82

Wise RA. 2004. Dopamine, learning and motivation. *Nat. Rev. Neurosci.* 5:483–94

Wong DF, Kuwabara H, Schretlen DJ, Bonson KR, Zhou Y, et al. 2006. Increased occupancy of dopamine receptors in human striatum during cue-elicited cocaine craving. *Neuropsychophar macology* 31:2716–

27

Yin HH, Knowlton BJ. 2004. Contributions of striatal subregions to place and response learning. *Learn. Mem.* 11:459–63

Yin HH, Knowlton BJ, Balleine BW. 2005. Blockade of NMDA receptors in the dorsomedial striatum prevents action-outcome learning in instrumental conditioning. *Eur. J. Neurosci.* 22:505–12

Zald DH, Boileau I, El-Dearedy W, Gunn R, McGlone F, et al. 2004. Dopamine transmission in the human striatum during monetary reward tasks. *J. Neurosci.* 24:4105–12

Zink CF, Pagnoni G, Martin-Skurski ME, Chappelow JC, Berns GS. 2004. Human striatal responses to monetary reward depend on saliency. *Neuron* 42:509–17

Crittenden J, Sauvage M, Cepeda C, Andre V, Costa C, et al. 2007. *CalDAG-GEFI modulates behavioral sensitization to psycho motor stimulants and is required for cortico-striatal long-term potentiation.* Presented at Annu. Meet. Am. Coll. Neuropsychopharmacol., 46th, Boca Raton, FL

Welch J, Lu J, Rodriguez RM, Trotta NC, Peca J, et al. 2007. Cortico-striatal synaptic defects and OCD-like behaviours in Sapap3-mutant mice. *Nature* 448:894–900

NOTE ADDED IN PROOF

Valuable new mouse models of disorders involving action since this review was submitted include emerging Sapap-3 mutant mice and CalDAG-GEF I mutant mice (Crittenden et al. 2007, Welch et al. 2007).

Brain Plasticity Through the Life Span: Learning to Learn and Action Video Games

Daphne Bavelier,[1,2] C. Shawn Green,[3] Alexandre Pouget,[1,2] and Paul Schrater[4]

[1] Department of Psychology and Education Sciences, University of Geneva, 1211 Geneva 4, Switzerland

[2] Department of Brain and Cognitive Sciences, University of Rochester, Rochester, New York 14627-0268, USA; email: daphne@bcs.rochester.edu, alex@bcs.rochester.edu

[3] Department of Psychology, Eye Research Institute, University of Wisconsin, Madison, Wisconsin 53706, USA: email: csgreen2@wisc.edu

[4] Departments of Psychology and Computer Science, University of Minnesota, Minnesota 55455, USA; email: schrater@umn.edu

Key Words

transfer, generalization, probabilistic inference, cognitive control, resource allocation, knowledge, hierarchy, learning rules

Abstract

The ability of the human brain to learn is exceptional. Yet, learning is typically quite specific to the exact task used during training, a limiting factor for practical applications such as rehabilitation, workforce training, or education. The possibility of identifying training regimens that have a broad enough impact to transfer to a variety of tasks is thus highly appealing. This work reviews how complex training environments such as action video game play may actually foster brain plasticity and learning. This enhanced learning capacity, termed learning to learn, is considered in light of its computational requirements and putative neural mechanisms.

INTRODUCTION

Human beings have a tremendous capacity to learn. And although there are unquestionably gradients in the ability to learn that arise as a function of intrinsic factors such as age and individual genetic propensities, nearly all humans demonstrate the ability to acquire new skills and to alter behavior given appropriate training. One common finding, however, is that the learning that emerges as a result of training is often quite specific to the trained stimuli, context, and task. Such specificity has been documented in every subfield of neuroscience that focuses on learning—e.g., motor learning (Shapiro & Schmidt 1982), expertise (Chase & Simon 1973), or memory (Godden & Baddeley 1975). In addition to being of theoretical interest, this fact is of great practical relevance in areas such as rehabilitation and education in which the end goal necessarily requires that the benefits of learning extend beyond the confines of the particular training regimen. For instance, it is of little use for a patient with damage to the motor cortex to improve on reaching movements in therapy if it does not also improve their ability to reach for a bottle of milk in their refrigerator at home (Frey et al. 2011). Similarly, it is of limited benefit if pupils trained on one type of math problem are unable to solve similar problems when presented outside the classroom. Thus, for all practical purposes, the specificity that typically accompanies learning is a curse.

A crucial issue in the field of learning, therefore, concerns training conditions that result in observable benefits for untrained skills and tasks. A handful of behavioral interventions have recently been noted to induce more general learning than that typically documented in the learning literature. These learning paradigms are usually more complex than standard laboratory manipulations, reminiscent of the "enriched environment effects" seen in animal rearing (Renner & Rosenzweig 1987), and they typically correspond to real-life experiences, such as aerobic activity (Hillman et al. 2008), athletic training (Erickson & Kramer 2009), musical training (Schellenberg 2004), mind-body training (Lutz et al. 2008, Tang & Posner 2009), working memory training (Jaeggi et al. 2008, Klingberg 2010), and, the focus of this article, action video game training (Green & Bavelier 2003, Spence & Feng 2010). Here we first review the variety of tasks on which performance is enhanced as a result of action video game experiences. We then ask why action games might produce benefits on such a wide range of tasks. In particular, rather than hypothesizing that video games teach myriad individual specific skills (i.e., one for each laboratory test during which they have been shown to produce benefit), we instead consider the possibility that what video games teach is the capability to quickly learn to perform new tasks—a capability that has been dubbed "learning to learn" (Harlow 1949, Kemp et al. 2010).

The idea that the ultimate goal of training and education might be to enable learning to learn is not a new one. On the basis of his work on pairing tasks, in which training on task A may impact training on new task B, Thorndike & Woodworth (1901) observed,

"It might be that improvement in one function might fail to give in another improved ability, but succeed in giving ability to improve faster " (p. 248). Around the same time, the educational psychologist Alfred Binet [1984 (1909)] described his classroom instructional goals by saying, "[O]our first job was not to teach [the students] the things which seemed to us the most useful to them, but to teach them how to learn" (p. 111). To this end, Binet asked students to play games such as "statue," wherein the students learned to stay focused, quiet, and still for long periods of time. Although no new academic concepts or facts were taught by these games, they allowed the children to develop skills, such as attention and control, that underlie the ability to learn in school. In the 1940s, Harlow (1949) outlined what he called "the formation of learning sets" wherein animals learned general rules, such as "win-stay, lose-switch," that allowed them to quickly master new tasks that abided by this general rule. This work led Harlow to state, "The learning of primary importance to the primates is the learning how to learn efficiently in the situations the animal frequently encounters" (p. 51). Despite this early interest in the topic, the factors that facilitate learning to learn and the computational principles on which it relies remain largely unexplored. Here we discuss these issues by anchoring our discussion in recent advances in the field of cognitive training through action video game play.

BENEFITS OF ACTION VIDEO GAME EXPERIENCE

An overview of the existing literature on action video game (one specific subgenre of video game) play indicates performance benefits after action game play in many domains typically thought of as distinct. This breadth is particularly notable and runs counter to the predominant pattern of results in the learning literature, wherein training facilitates the trained task but typically leads to little benefit to other, even related tasks. In the case of video game training, the transfer of learning takes place between tasks and environments that have different goals and feel. Most laboratory tasks on which avid action video game players (VGPs) are tested involve simple stimuli and highly repetitive choices, making for a rather dull environment. In sharp contrast, action video games include complex visual scenes, an enthralling feel, and a wide variety of goals at different timescales (from momentary goals such as "jump over the rock" to game-long goals such as "rescue the princess") (see **Figure 1**). Although it seems reasonable to expect VGPs to show transfer across a broad range of similar games (e.g., first-person shooters or third-

VGP: action video game player
NVGP: nonaction-video game players

person shooters), the fact that VGPs best their nonaction game–playing peers (NVGPs) on standard laboratory tests that are quite dissimilar to video games begs further investigation.

Vision

Action video game play enhances the spatial and temporal resolution of vision as well as its sensitivity (Green et al. 2010). For example, when asked to determine the orientation of a T that is flanked by distracting shapes above and below VGPs can tolerate the distractors being nearer to the T shape while still maintaining a high level of accuracy (Green & Bavelier 2007). Such capacity to resolve small details in the context of clutter, also called crowding acuity, is thought of as a limiting property of spatial vision and is often compromised in low-vision patients who complain that the small print of newspapers is unreadable because letters are unstable and mingle into one another (Legge et al. 1985). In addition, the enhancements noted in this ability as a result of action video game play occur both within and outside the typical eccentricities of game play, a finding that stands in contrast to the bulk of the perceptual learning literature, which finds that learning is typically specific to the trained retinal location.

Along with this improvement in spatial resolution, enhanced temporal resolution of vision as a result of action game experience has also been noted. VGPs show, for example, significant reductions in backward masking (wherein a display of interest is more difficult to see if it is quickly followed by another display) as compared with NVGPs. This difference suggests a change in the dynamics of the visual system: VGPs can resolve events at a higher temporal frequency (Li et al. 2010). Although backward masking, which is believed to reflect limitations on cortical processing, is improved in VGPs, this is not true of forward masking (wherein the masking display precedes the display of interest), which is believed to reflect mostly limitations of early, retinal factors. As we discuss below, these results are consistent with the view that action video game play retrains cortical networks such that each layer of the processing hierarchy makes better use of the information it receives from earlier layers, rather than changing the nature of the early sensory input (e.g., altering the optics of the eye).

Finally, a third aspect of vision found to be enhanced in VGPs is contrast sensitivity, or the ability to detect small changes in levels of gray. Although this effect appears across spatial frequencies, it is particularly pronounced at intermediate spatial frequencies (Li et al. 2009). Thus, contrary to the folk belief that screen time is bad for eyesight, action video game play appears to enhance how well one sees. Video game training, therefore, may become a powerful tool to improve eyesight in situations in which the optics of the eye are not implicated in producing the poor vision (e.g., amblyopia; Bavelier et al. 2010, Li et al. 2011).

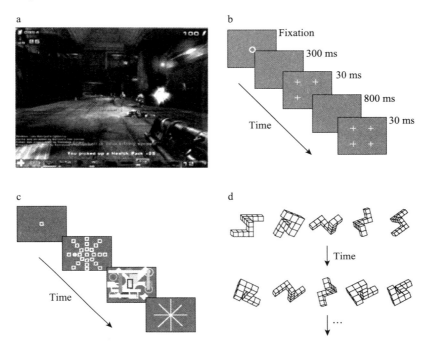

Figure 1

Action video game play (*a*) bears little resemblance in terms of stimuli and goals to perceptual, attentional, or cognitive tasks at which VGPs are found to excel in the laboratory (*b*: contrast sensitivity, *c*: visual search, *d*: mental rotation). This raises the question of why action video game play enhances such a varied set of skills.

Cognitive Functions

VGPs have also been documented to better their nongamer peers on several aspects of cognition such as visual short-term memory (Anderson et al. 2011, Boot et al. 2008), spatial cognition (Greenfield 2009), multitasking (Green & Bavelier 2006a), and some aspects of executive function (Anderson & Bavelier 2011, Chisholm & Kingstone 2011, Colzato et al. 2010, Karle et al. 2010; but see Bailey et al. 2010 for a different view). For example, whereas NVGPs were markedly slower and less accurate when asked to perform both a peripheral visual search task and a demanding central identification task concurrently [relative to performing the peripheral search task alone, VGPs showed no falloff in performance (Green & Bavelier 2006b)]. Several studies have also documented enhanced task-switching abilities in VGPs, meaning they pay less of a price for switching from one task to another (Andrews & Murphy 2006, Boot et al.

2008, Cain et al. 2012, Colzato et al. 2010, Green et al. 2012, Karle et al. 2010, Strobach et al. 2012). These results are all the more surprising given that such task-switching skills appear to be negatively affected by the extensive multitasking seen in many young adults who are heavy users of a variety of forms of technology (Ophir et al. 2009). This contrast highlights the need to assess separately the effects of different technology use on brain function.

Another cognitive domain enhanced by action games is spatial cognition. The beneficial aspect of video game play on spatial cognition was noted in the early days of video game development (McClurg & Chaille 1987, Okagaki & Frensch 1994, Subrahmanyam & Greenfield 1994). More recently, action game play has been shown to enhance mental rotation abilities (Feng et al. 2007). These results have received much attention because spatial skills are typically positively correlated with mathematical achievement in school (Halpern et al. 2007, Spelke 2005). Whether action game play may indeed foster mathematical ability is currently being investigated.

Decision Making

Benefits are also noted in decision making. In one perceptual decision-making task, for instance, VGPs and NVGPs were asked to determine whether the main flow of motion within a random dot kinematogram was to the left or to the right (**Figure 2*a***). Unlike most previous tasks used to study the effect of action gaming, in this task, participants are in control of when to terminate display presentation (by making a decision and pressing the corresponding key). It thereby provides a measure of how participants accumulate information over time in the service of decision making. Indeed, these types of task are well understood in terms of one's ability to first extract and integrate information from the environment and then to stop the integration and to select an action on the basis of the accumulated information (Palmer et al. 2005, Ratcliff & McKoon 2008). By presenting trials with variable signal-to-noise levels, the task allows the full chronometric and psychometric curves to be mapped, providing a unique description of how information about motion direction accrues over time (**Figure 2*b***). We found that action game play enhances the rate at which information accumulates over time by ~20% as compared with control participants (**Figure 2*c***). The net result is that VGPs can make more correct decisions per units of time. This is of practical relevance as illustrated by the recruitment by the Royal Air Force of young gamers to pilot unmanned drones or by findings indicating that young laparoscopic surgeons who are gamers outperform more seasoned surgeons, executing surgery procedure faster and as accurately if not more accurately (Daily Mail 2009, McKinley et al. 2011, Rosser et al. 2007, Schlickum et al. 2009).

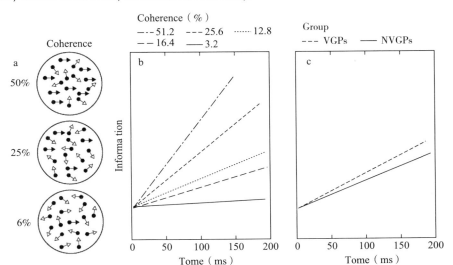

Figure 2

(*a*) In this perceptual decision-making task, participants are asked to judge whether the motion is mostly leftward or rightward. The number of dots moving coherently in one direction or the other can vary from none to many, allowing the investigator perfect control over the task difficulty. (*b*) As the level of coherence goes from low to high, the task varies from extremely difficult to easy. Accordingly, electrophysiological recordings in monkeys show that the amount of information about motion direction accumulated in the lateral intraparietal cortex (LIP), a neuronal structure associated with such decisions, increases (adapted from Beck et al. 2008, figure 3d).(*c*) Another way to increase the rate at which information accumulates is to enhance the decision-making characteristics of the viewer. This was shown to be the case in action video game players (VGPs) or after training on action video games: VGPs accumulated on average 20% more information per unit of time as compared with nonaction–video game players (NVGPs) (Green et al. 2010).

Reaction Time and Speed-Accuracy Trade-Off

More generally, a fourth domain of improvement concerns the ability to make quick and accurate decisions in response to changes in the environment, such as when one must brake suddenly while driving. Such improvement was noted in the decision-making tasks described above. Action video game play sped up reaction times, and this was true whether the task was in the visual or the auditory modality. Such tasks entail concurrently collection reaction times and accuracy along a wide range of task difficulties, easily revealing a speed–accuracy trade-off if present. Despite faster reaction times, accuracies were left unchanged, establishing that action game play does not result in trading speed for accuracy. A meta-analysis of more than 80 experimental conditions in which gamers and nongamers have been tested confirmed that VGPs are on average 12% faster than NVGPs, yet VGPs make no more errors (Dye et al. 2009a). This relationship held from simple decision tasks leading to reaction times as short as 250 ms, all the way to demanding, serial, visual search tasks eliciting reaction times as long as 1.5 s. The fact that the relationship was purely multiplicative, with no additive component, demonstrates that the faster reaction times seen in VGPs throughout the literature are not attributable to postdecisional factors such as faster motor execution time. Indeed, if the only difference between groups was in their ability to map a decision that had already been made into a button press on a keyboard, the difference in reaction time between groups should not depend on how long it took for the decision to be reached (i.e., would be only additive rather than multiplicative). Furthermore, the fact that there was no difference in accuracy in any of these tasks suggests that the differences also cannot be attributed to differences in criteria, or VGPs being "trigger-happy" or willing to trade reductions in accuracy for increases in speed. Whether this pattern of greater speed with matched accuracy will remain when using complex tasks with longer timescales and/or multiple task components is currently being investigated (A.F. Anderson, C.S. Green, and D. Bavelier, manuscript in preparation).

Attention

Several different facets of attention are improved following action game play. For instance, many aspects of top-down attentional control such as selective attention, divided attention, and sustained attention are enhanced in VGPs. In contrast, however, when attention is driven in a bottom-up fashion (i.e., by exogenous cueing) no differences have been found between VGPs and NVGPs, despite the fact that orienting to abrupt events is a key component of game play (Castel et al. 2005, Dye et al. 2009b, Hubert-Wallander et al. 2011). This pattern of results clearly illustrates the need for a careful investigation of those aspects of performance modified by action game play. Simply because a process is required during game play does not guarantee changes in that process. In a later section, we review in greater detail those aspects of attention and executive control that are modified by action game experience (see Resources, Knowledge, and Learning Rules in Action Video Game Play).

Causality

A key question concerns whether the effects of action video game play are causal or are instead reflective of population bias, wherein action gaming tends to attract individuals with inherently superior skills. Because our interest is in learning rather than in identifying individuals with extraordinary skills, causality is a crucial factor. The only way to establish firmly that the relationship is indeed causal is via well-controlled training studies (see Green & Bavelier 2012, box 1 for further discussion). In such studies, individuals who do not naturally play fast-paced, action video games are recruited and pretested on the task(s) of interest. A randomly selected sample of half of these subjects is then assigned to play an action game (e.g., Medal of Honor, Call of Duty, Unreal Tournament), whereas the other half is assigned to play an equally entertaining, but nonaction video game (e.g., Tetris, The Sims, Restaurant Empire). Note that in addition to controlling for simple test/retest effects (subjects may improve at posttest simply because they have experienced the task a second time) the presence of an active control condition guarantees that any effects observed in the action game training group are truly the result of action game play rather than simply a reflection of the power of an intervention per se. Indeed, individuals often feel special upon being included in an active intervention study and, as a result of receiving more attention, may perform better independently of the content of the intervention, an effect also known as the Hawthorne effect (Benson 2001).

Causality has been established through training studies, whereby both groups of participants come to the laboratory regularly to play 1–2 h per day over a period of 2–10 weeks, depending on the duration of the training (e.g., 10, 30, or 50 h of training were used in our work) (**Figure 3**). At the end of training, subjects are tested again on the same tasks of interest as those during pretest. However, it is important to note that in all cases we require that participants come back in the following days for posttesting, with no less than 24 h elapsing between the cessation of training and the beginning of posttesting. Indeed, our focus is on the durable learning effects of action game play and not on any transient effects of playing games (e.g., changes in arousal) that could contaminate posttest performance if assessed shortly after the end of gaming. If action game play truly has a causal effect on the skill under study, we expect greater pre- to posttest improvement in the action trainees than in the control trainees. Such causal effects have been established for vision (Green & Bavelier 2007; Li et al. 2009, 2010), some cognitive functions such as multitasking (Green et al. 2012) and mental rotation (Feng et al. 2007, Spence et al. 2009), many aspects of attention (Cohen et al. 2007; Feng et al. 2007; Green & Bavelier 2003, 2006a,b; Spence et al. 2009), and decision making (Green et al. 2010).

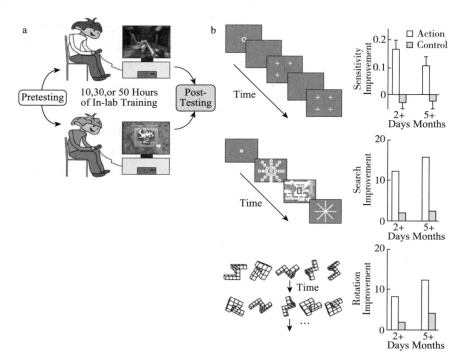

Figure 3

(*a*) Design of training studies to evaluate causality. Participants are randomly assigned to either the experimental group or a control group before being evaluated on perceptual, attentional, or cognitive laboratory tests. Both groups are then required to play commercially developed entertainment games. All experimental games are of the action genre (Unreal Tournament, Medal of Honor, Call of Duty), whereas control games are of other genres (Tetris, The Sims, Restaurant Empire). Following the end of training (and a 24-h waiting period) subjects are retested on the original measures. If the effect of action game play is truly causal, those assigned to the experimental games should display greater improvements between pre- and posttest than those assigned to the control games. (*b*) Such causal effects have been established for a variety of perceptual, attention, and cognitive tasks; some studies showed that the effects are quite long-lasting because robust effects were observed not only 1–2 days after the cessation of training (*red bars*) but also as long as 5+ months posttraining (*blue bars*) (panel b adapted from Feng et al. 2007, Li et al. 2009).

A COMMON MECHANISM: LEARNING TO LEARN

As illustrated above (see Benefits of Action Video Game Experience), the range of tasks seen to improve after action game play is quite atypical in the field of learning. VGPs perform better than NVGPs do on tasks neither group had previously experienced and that are, as noted earlier, quite different in nature from action game play. This finding begs the question of what exactly action video games teach that results in better performance. Rather than hypothesizing distinct mechanisms for each task improved, it seems more parsimonious to consider one common cause: learning to learn.

To understand what is meant by learning to learn, it is important first to realize that all the tasks on which VGPs have been found to improve share one fundamental computational principle: All require subjects to make a decision based on a limited amount of noisy data. Note that most everyday decisions fall into this category. Consider a radiologist reading computed tomography (CT) scans to determine the presence of bone fractures. Some scans will be clear-cut, leading to the immediate conclusion that a fracture is present. Other scans may present more subtle evidence, such that the radiologist may conclude a fracture has occurred, but his/her level of confidence will be much lower. In making these decisions, the radiologist is performing what is known as probabilistic inference. In essence, the radiologist computes the probability that each choice (fracture or no fracture) is correct given the evidence present in the scan(s). This quantity is known as the posterior distribution over choices, which we denote $p(c|e)$ where c are the choices, and e is the evidence (i.e., c = "fracture" or "no fracture", e = the evidence, which in this case is the scan itself). The key computational question then concerns how to compute the most accurate posterior distribution, $p(c|e)$, so that the best decision can be made. The main goal of learning is indeed to improve the precision of such probabilistic inference. The very fact that trained radiologists are more competent at diagnosing subtle fractures and identifying them with more confidence exemplifies that proper training does lead to more accurate posterior distributions.

How then does this process occur in practice? During training, a young radiologist sees many scans and is explicitly told the state of the patient that generated the image (i.e., whether a fracture was present). Radiologists are trained on the statistics of the evidence $p(e|c)$ or the probability of the evidence (how the scan looks) given a known state of the world (fracture is present/absent). Through Bayes' rule, this value, $p(e|c)$, is proportional to the posterior probability $p(c|e)$ on which the decision will be based. This set of computations is termed probabilistic inference (Knill & Richards 1996). One of the major advances of systems and computational neuroscience over the past 10 years has been to show how such probabilistic computations may be implemented in networks of spiking neurons (Deneve 2008, Ma et al. 2006, Rao 2004).

Participants who first encounter a new task in the lab are not very different from young radiologists in training. Initially, there is no way for them to have perfect knowledge of the statistics of the evidence, which in turn means that the calculated posterior distribution over choices will be suboptimal. It is only through repeated exposure to the task that subjects can learn these statistics and, as a result, make decisions on the basis of a more accurate posterior distribution. By using a well-studied perceptual decision-making task (as described above in Benefits of Action Video Game Experience), we more directly tested whether action video game experience indeed results in improvements in probabilistic inference. In this task, subjects were asked to determine whether the main flow of motion within a random dot kinematogram was to the left or to the right. Using a neural model of the task (Beck et al. 2008), we showed that the pattern of behavior after game play is captured via a single change in the model: an increase in the connectivity between the model's sensory layer, where neurons code for direction of motion, and its integration layer, where neurons accumulate the information they receive over time from the sensory layer until a criterion is reached for decision and response. This increase in connectivity can be shown to be mathematically equivalent to improved statistical inference; game experience resulted in more accurate knowledge of the statistics of the evidence for the task or more accurate knowledge of $p(e|c)$ (**Figure 2c**) (Green et al. 2010). This work ruled out simpler explanations for the variety of benefits observed after action game play, including simple speed–accuracy trade-offs or faster motor execution times in VGPs.

The modeling demonstrates that VGPs were making choices on the basis of a more accurate posterior distribution. Because these statistics could not have been directly taught by any action video games, but instead must have been learned through the course of the experiment, action video games thus taught these participants to learn the appropriate statistics for new tasks more quickly and more accurately than is possible for NVGPs.

The idea that VGPs have learned to learn the statistics of the task (or what we call later the generative model of the data) can be generalized beyond the specific example of perceptual decision making. Indeed, most of the behavioral tasks on which VGPs have been shown to excel can be formalized as instances of probabilistic inference. This includes tasks that are more classically categorized as attentional, such as multiple object tracking or locating a target among distractors, both of which have recently been modeled within the probabilistic framework, as well as more cognitive and perceptual tasks (Ma et al. 2011, Ma & Huang 2009, Vul et al. 2009).

COMPUTATIONAL PRINCIPLES OF LEARNING TO LEARN

Learning is commonly defined as a long-term improvement in performance due to training, a process that is often modeled as the tuning of parameters within a fixed architecture. The literature has countless examples wherein learning algorithms, such as gradient descent, are implemented within neural networks to tune weights to enhance performance on the trained task (Rumelhart et al. 1987, but see Gallistel & King 2009 and Gallistel 2008 for a different view on learning). For example, in classic orientation-discrimination training tasks, subjects have to learn through experience whether the image of a Gabor patch is oriented clockwise or counterclockwise from some reference angle (e.g., 45°). Human behavior on such tasks can be easily captured with two-layer networks, in which an input layer containing oriented filters projects to an output layer that produces a clockwise or counterclockwise decision (Dosher & Lu 1998, Pouget & Thorpe 1991). Learning in such models is typically produced via changes in the synaptic weights between the two layers. The pattern of synaptic weights that optimizes performance on this task is a simple instantiation of a discriminant function and thus is completely specific to the trained reference angle. Learning at one orientation does not transfer to other orientations. These models naturally capture the high degree of specificity typically seen in perceptual learning. However, such an implementation cannot be responsible for the types of benefits associated with learning to learn.

Transfer of learning is enabled by implementations that recognize the importance of resources, knowledge, and learning rules. These core computational elements are well illustrated by real-world tasks such as playing soccer, where players must keep track of an array of complex moving objects, including the ball and other players (**Figure 4a**). As the player moves and scans the scene, the color, shape, and motion information they receive must be converted into the three-dimensional trajectories of objects while maintaining their identities and roles (opponent, teammate, ball, referee, spectator). This complex problem can be decomposed into a set of core skills, many of which have been psychophysically studied in isolation, including multiple object tracking, object identification, resource allocation, and eye-movement planning. Whereas the novice player will only partially achieve the coordination of these different demands, expert players have learned through experience how best to represent game play as a function of the different identities and roles of the players and how best to allocate resources among these different demands. Accordingly, top athletes can keep track of all the players on both teams, including those out of view (Cavanagh & Alvarez 2005).

Skill acquisition in such complex domains requires more than instantiating a simple discriminant function of the type discussed above. A neural architecture that incorporates knowledge critical for the given task is needed. In the case of soccer, this knowledge encompasses tracking objects, including features important for maintaining object identity and role, developing representations that are invariant to irrelevant internal limb motions, and representing the statistical motion behavior typical of different object types (motion of the ball versus that of players). Learning the proper knowledge for

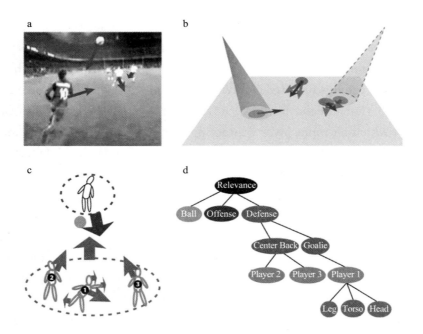

Figure 4

(*a*) In real-world tasks such as playing soccer, players must keep track of a complex array of moving objects, including the ball and all other players on the field. An expert player will readily recover the three-dimensional trajectories of various objects on the field, while maintaining their proper identities and roles at all times, allowing them to predict next game moves. This is possible as players gain experience with soccer play and build a rich internal model with the proper hierachical architecture to facilitate computation. (*b*)Schematic illustration of how resources can facilitate object tracking over time. Resources, shown as beams of light, are believed to be limited in number and require flexible allocation across objects as the number of objects increases. Recruiting additional resources, illustrated as the dashed beam, provides improved motion processing and tracking for more objects. (*c, d*) The dynamic of game play in a soccer game is best captured as the top node of a hierachical architecture, whereby motion information is pooled across different levels of representations, from low level such as limb motion within a player to the motion of individual players to assemblies of players (offense versus defense). An appropriate abstraction allows hierarchical architectures to capture regularity across variations in task and stimuli, a key feature of learning transfer.

the given task is critically dependent not only on the chosen neural architecture but also on the learning rules that adapt the neural architecture with experience. Finally, the ability to adapt swiftly to the different demands of such complex tasks greatly benefits from improvement in the number or allocation of core resources such as attention. Changes in any or all of these three—knowledge, learning rules, and resources—could be produced by action video game experience and could account for widespread benefits observed in VGPs. As illustrated below (see Resources, Knowledge, and Learning Rules in Action Video Game Play), there is much evidence that action video game play improves resource allocation, and there are some pointers toward knowledge and/or learning rule changes as well. Before we discuss these, we lay out a general framework for the type of probabilistic inference at the core of learning to learn and discuss how resources, knowledge, and learning rules may come to account for the widespread benefits associated with learning to learn.

Resources

VGPs have increased attentional resources in multiple-object-tracking tasks, allowing for more accurate representation of motion and features, which in turn enables more accurate tracking and identification (Boot et al. 2008, Dye & Bavelier 2010, Green & Bavelier 2006b, Trick et al. 2005). As illustrated in **Figure 4***b*, this ability should be advantageous when playing soccer. However, such additional resources will do more than just improve performance on a single soccer action; they are also a key determinant of learning. Greater resources allow for finer-grain distinctions to be made, be they perceptual, conceptual, or in motor output. An increase in resources may therefore enable learners to achieve greater asymptotic performance (more learning capacity), or even faster learning, because critical distinctions will be more accessible to those learners with greater resources. The concept of resources we use here is closely related to that defined in the field of attention and executive control under the terms goal-directed attention or top-down control (Corbetta & Shulman 2002, Dosenbach et al. 2008, Koechlin & Summerfield 2007, Rougier et al. 2005). The cholinergic basal forebrain system may orchestrate such resource allocation. More specifically, it appears that the release of acetylcholine regulates the extent to which targets are processed and distractors are ignored (Baluch & Itti 2011, Herrero et al. 2008, Sarter et al. 2005) as well as enhances the precision of sensory representations (Goard & Dan 2009).

The role of acetylcholine in selective attention and resource allocation goes hand in hand with its recognized role in cortical map plasticity (Ji et al. 2001, Kilgard & Merzenich 1998). In a seminal study, Kilgard & Merzenich (1998) established that adult rats exposed to tones exhibited larger cortical map plasticity in the presence of episodic stimulation of the nucleus basalis. Early studies in humans also suggest a possible link between learning and the cholinergic system. For example, intake of a choline agonist leads to improved learning when participants are asked to learn lists of words. In contrast, a choline antagonist leads to reduced learning (Sitaram et al. 1978). In this view, resource allocation acts as a key gateway to learning by refining the distinction between signal and noise and enhancing the quality and precision of the to-be-learned information. The role that the distribution of attentional resources plays in guiding learning is well exemplified by a study by Polley et al. (2006), whereby rats were trained to discriminate tones on the basis of either their frequency or their intensity. When animals performed the frequency task, changes in frequency maps were observed that correlated with behavior. In contrast, no such changes were noted in neurons coding for intensity. The opposite pattern was observed when the task focused the animals' attention on the intensity task. A similar link is also supported in humans (see Saffell & Matthews 2003). In line with this view, Li et al. (2009) have shown that prefrontal areas, known to be involved in resource allocation, code for different decision criteria as a result of learning episodes that emphasized the relevant criterion. This neural signature was present only when participants were engaged in the trained task and not just exposed to the learned stimuli. More recently, Baldassare et al. (2012) demonstrated that resting state connectivity between prefrontal areas and the visual cortex was predictive of the amount of learning in an upcoming perceptual learning task. This study is a direct illustration of how efficiency in resource allocation comes to guide learning abilities at the individual level. As we highlight below (see Resources, Knowledge, and Learning Rules in Action Video Game Play), much evidence indicates that action video game play improves top-down resource allocation, thus leading to better selective attention and enhanced precision in representations. A direct link to the acetylcholine system, however, remains to be elucidated.

More computational and memory resources can also foster learning to learn by providing the capacity to explore alternative hypotheses and architectures to find those that best account for the knowledge being acquired. The burgeoning literature on the training of executive function testifies to this link (Klingberg 2010, Lustig et al. 2009). For example, as participants train on an N-back task and can handle more demanding updating in memory, they also show enhanced fluid intelligence as measured by the Raven's Advanced Progressive Matrices (RAPM) (Jaeggi et al. 2008). The RAPM presents visual patterns with one missing component that participants must complete. This test relies on the ability to search swiftly through many hypotheses to find the best match. Enhanced executive resources will allow such searches to be more efficient. Although executive functions appear trainable across the life span, some evidence indicates that transfer effects may be more limited in older age (Buschkuehl et al. 2008, Dahlin et al. 2008). At the neural level, working-memory training results in enhanced recruitment of the frontoparietal network associated with top-down attentional control, possibly increased myelination among related areas, as well as a decrease in cortical dopamine D1 binding potential in large parts of frontal and posterior areas (McNab et al. 2009, Olesen et al. 2004, Takeuchi et al. 2010). Whether action video game play effects are mediated by similar mechanisms remains to be assessed. Finally, having greater resources does not systematically guarantee more efficient learning; resource allocation needs to be guided by the presence of structured knowledge that helps select where useful information lies, as discussed below.

Knowledge

Knowledge here refers to the representational structure used to guide behavior. The realization that complex behaviors cannot be appropriately captured by a chain of stimulus-response associations is far from new (Lashley 1951) and has led to a keen interest in the role of hierarchical architecture in human actionand cognition (see Botvinick 2008 for a review). Hierarchical behavior models capture the intuitive notion that tasks such as executing a play in soccer are organized into subtasks involving individual player moves, which are themselves decomposed into component actions of running, turning, kicking, and passing. In turn, knowing soccer makes learning a game such as hockey easier because the similarities between the games allow one to import skills and knowledge from one game to bootstrap learning of the other. Note that it is not just because both games share the category "goalies" that such generalization is possible. Rather, recent work in machine learning suggests that hierarchical architectures allow the decomposition of computations into multiple layers using greater and greater abstraction, a critical step for generalizable learning (Bengio 2009, Hinton 2007). In these architectures, categories such as "goalie" are distributed across multiple levels of abstraction, from sensory input to characteristic features to body plans to player types to the concept of a soccer player. In contrast, shallow architectures eschew abstraction and construct the category "goalie" by treating it as a classification problem that involves finding the right set of diagnostic features such as limb motion, jersey color or pattern, or position relative to the goal post. Although few examples of such deep coding in neuroscience exist as of yet, studies of categorization and object recognition point to the psychological grounding of deep architectures. For example, hierarchical architectures naturally account for the fact that categories appear organized around prototypes or most representative exemplars of a category, even when the prototypes do not reflect the frequency of diagnostic features (Rosch 1975). As work on the neural bases of object recognition progresses, characterizing how deep knowledge may constrain neural coding will become increasingly more relevant (DiCarlo & Cox 2007, Lee & Mumford 2003, Paupathy & Connor 2002, Yamane et al. 2008).

The shallow/deep distinction can be expressed in terms of Bayesian decision theory as the difference between learning a rich generative model that captures hidden structure in the data versus learning a discriminative model (a discriminant function) specific to a classification problem. Learning theory has shown that discriminative models require less data to learn the observed relationships but are task specific. For example, discriminative classifiers such as support vector machines learn boundaries in feature space. The fact that these boundaries can be quite complex does not guarantee generalizable knowledge. In fact, in the case of support vector machines, the learned boundaries typically depend intimately on the characteristics of the training set. Simple changes in team uniforms or players will require relearning a new boundary to classify them. Such implementations therefore appear ill adapted when considering learning to learn. In contrast, appropriate abstraction allows hierarchical architectures to capture regularity across variations in task and stimuli. **Figure 4c,d** illustrates a hierarchy for relevant motion in a soccer game. The tree structure shows how broader categories can be constructed by pooling across categories of motion at different levels, pooling limb motions to construct player motion, pooling player motions within a player type (such as goalie), pooling player-type motions within a team type (offense or defense), and pooling all game-relevant motions, including the ball, coaches, and referees, in opposition to nonrelevant motions such as spectators and billboards.

Abstract organization admits generalization to the extent that new tasks overlap and share structure and representations. For example, despite the fact that players use skates, hockey shares many of the same motion characteristics of the teams' offense and defense, roles of players (such as goalie), and requirements for passing and controlling the puck as does soccer. The generalization expected from a given architecture is not infinite, but will be determined instead by the level of abstraction at which knowledge is shared by the various tasks considered. Thus, for action video games to provide players with knowledge applicable to standard laboratory tasks, it must necessarily be the case that games and the laboratory tasks share structure at some level of abstraction. And perhaps, more importantly, training paradigms must necessarily share structure at some level of abstraction for them to affect performance in real-life tasks.

The need for structured representation to support more complex task performance has led to an interest into the neural bases of hierarchical tasks that embed goals and subgoals. Although this area of research is still fairly new, the available literature points to a crucial role of prefrontal cortex structures to enable hierarchical representations and coordinate resource allocation dynamically among these as task demands vary (Badre 2008, Botvinick 2008, Koechlin 2008, Sakai 2008). Indeed, the ability to enhance neural processing selectively by allocating more resources is central to behavior optimization, but it depends fundamentally on an internal understanding of which neural pool is important to augment at each point in time. The prefrontal cortex is a crucial candidate in the representation of these more complex behaviors. Prefrontal structures are key to maintaining context and goals over time with increasingly higher levels of behavioral abstraction being coded as one moves from caudal to rostral locations along the prefrontal cortex (Christoff & Keramatain 2007, Fuster 2004). There is certainly debate about the exact nature of the abstraction, with higher planning units, greater relational complexity, more branching control, or larger contextual integration being documented as one moves rostrally within the prefrontal cortex (see figure 2 in Badre 2008 for a summary). Yet, there is common agreement about the importance of hierarchical knowledge when considering how the brain learns from its successes and failures. Neural signatures for such an architecture are emerging, e.g., in the work of Ribas-Fernandes et al. (2011) who recently demonstrated that subgoals lead to reward-prediction signals even if they are not associated with primary reinforcers. Such an outcome is predicted under the assumption that the task is represented hierarchically but not otherwise. This research should be an important avenue for future research on learning, as the nature of the architecture fundamentally constrains the type of learning neural structures should implement.

Learning Algorithms

A common observation is that as an architecture becomes more complex, it becomes more critical that the learning algorithm can modify the representations in ways that are aligned with the nature of the problems to be solved. Simply put, knowledge and learning rules are intimately interdependent; different architectures typically call for different learning rules. A key question then concerns how improvement in learning algorithms and their related architecture may foster learning to learn. To facilitate generalization, learning algorithms require shared structure, a fact that constrains how learning rules should modify representations. Violating these constraints destroys transferrable knowledge because changes in representation that improve performance in one task can disrupt performance in another.

Over the past 20 years, this field has witnessed dramatic improvements in the basic set of machine-learning algorithms. For instance, algorithms such as those employed in Hinton's deep-belief networks have provided evidence that there is enough shared structure in the types of learning problems encountered by humans beings to allow for the use of flexible and generalizable learning algorithms (Hinton 2007, Hinton et al. 2006, Larochelle et al. 2007, Lee et al. 2009). In this case, speech recognition and handwritten character recognition can be achieved using initially similar architectures and the same learning rule. Of note, performance after learning rivals the best algorithms specifically developed for speech recognition or character recognition respectively (Mohamed et al. 2011). This approach is appealing because the lack of structural differentiation among cortical areas suggests a broadly shared learning algorithm, at least at the cortical level.

Fine-tuning of learning algorithms may occur through improvement in error detection, error computation, reward computation, and, more generally, synaptic learning rules. This fine tuning may involve neuromodulators, which are believed to be involved in the modulation of synaptic plasticity (Zhang et al. 2009) and in the computation of error signals (Schultz et al. 1997). Changes in the responses of the neurons synthesizing these neuromodulators could easily generalize across tasks and sensory modalities because these neurons tend to connect very wide cortical regions. Although few studies are available on this topic, action video game play likely alters patterns of neuromodulator release. For instance, an early positron emission tomography study indicated that a large release of striatal dopamine occurs when subjects play a basic shooting video game (Koepp et al. 1998, though see the call for further replication by Egerton et al. 2009). This should be a fruitful area of research for future studies.

RESOURCES, KNOWLEDGE, AND LEARNING RULES IN ACTION VIDEO GAME PLAY

All three factors that can contribute to learning to learn may, of course, be involved in the effects of video game training, but the experimental evidence so far has focused mainly on changes in resources and improvements in resource allocation. However, although no experiments have directly tested differences in knowledge or in learning rules, some data points do appear more consistent with these changes than does a resource-based account.

Attentional Resources

Enhanced resources. The proposal that VGPs benefit from greater attentional resources is supported by the observation that, in simple flanker compatibility tasks, VGPs' reaction times show a greater sensitivity to the identity of the flanker (Dye et al. 2009b; Green & Bavelier 2003, 2006a). Thus, reaction times on the main target task showed greater speeding (respectively, slowing down) when flankers were response compatible (respectively, response incompatible) in VGPs as compared with NVGPs. According to the work of Lavie and others (2005), in such flanker paradigms the resources utilized by the main target task determine the amount of processing the flanker receives and thus the size of the flanker compatibility effect. As the main target task exhausts more resources, the identity of the flanker affects reaction times to a lesser extent, leading to increasingly smaller compatibility effects. Green & Bavelier (2003) showed that although this pattern of results held in NVGPs, VGPs continued to show effects of flanker identity even for relatively high-load target tasks, thus suggesting the presence of greater attentional resources.

Selective attention. Action game play results in improved selective attention over space, over time, and to objects. VGPs exhibit more accurate spatial localization of a target whether presented in isolation as in Goldman perimetry (Buckley et al. 2010) or among distracting, irrelevant information as in the Useful Field of View task (Feng et al. 2007; Green & Bavelier 2003, 2006a; Spence et al. 2009; West et al. 2008). Action game play also enhances the ability to select relevant information over time. When viewing a stream of letters presented rapidly, one after the other, participants can typically identify the one white letter among black letters if asked. However, doing so creates a momentary blink in attention, leading the subject to be unaware of the next few items following the white letter. This effect, termed the attentional blink, is believed to measure a fundamental bottleneck in the dynamics of attention allocation (Shapiro et al. 1997). VGPs' blink is much less pronounced, with some VGPs failing to show any blink (at least at the rate of presentation tested; see Cohen et al. 2007, Green & Bavelier 2003). It is unlikely that action game play totally obliterates the attentional blink, but these data suggest a much faster rate of attentional recovery in VGPs. Such faster processing speed is in line with the reduced backward masking reviewed earlier and the report that VGPs perceive the timing of visual events more veridically than do NVGPs (Donohue et al. 2010, but see West et al. 2008 for a different view). A third aspect of selective attention documented to change for the better after action game play is attention to objects (Boot et al. 2008; Cohen et al. 2007; Green & Bavelier 2003, 2006b; Trick et al. 2005). Using the multiple-object-tracking task (Pylyshyn & Storm 1988), VGPs have been repeatedly shown to track more objects accurately than have NVGPs (Boot et al. 2008, Green & Bavelier 2006b, Trick et al. 2005). This skill requires efficient allocation of attentional resources as objects move and cross over each other in the display. The advantage noted in VGPs in this task is consistent with either large resources or a swifter allocation of resources as a result of game play.

The ability to focus attention on a target location and ignore irrelevant distracting information is a key determinant of selective attention. Converging evidence indicates that selective attention is a strong predictor of academic achievement in young children (Stevens & Bavelier 2012), illustrating the general benefits this skill may confer. Although seldom considered in conjunction with action game play, meditative activities such as mind-body training and eastern relaxation technique also act in part by enhancing selective attention (Lutz et al. 2008, Tang & Posner 2009). Whether these two rather different treatments achieve their effects through comparable neural modulation may be a fruitful avenue of investigation in the future.

Divided attention. The ability to divide attention between tasks or locations was one of the first attentional benefits noted in the early days of computer games (see Greenfield 2009 for a review). For example, when asked to report the appearance of a target in a high probability, low probability, or neutral location, VGPs did not show the expected decrement (slower reaction time compared with neutral) for the low probability location, suggesting a better strategy in spreading attention due to game expertise (Greenfield et al. 1994). More recently, VGPs have shown an enhanced ability at dividing attention between a peripheral localization task and a central identification task (Dye & Bavelier 2010, Green & Bavelier 2006a). VGPs also excelled when asked to detect a prespecified target in rapid sequences of Gabor stimuli presented in both visual fields. Concurrent event-related potential recordings indicated a larger occipital P1 component (latency 100–160ms) in VGPs during this divided-attention task, consistent with the proposal of greater attentional modulation in early stages of processing such as the extrastriate visual cortex in VGPs (Khoe et al. 2010).

Sustained attention and impulsivity. Some evidence also supports the notion that sustained attention benefits from action video game play. Using the Test of Variables of Attention (T.O.V.A.), a computerized test often used to screen for attention deficit disorder, Dye et al. (2009a) found that VGPs responded faster and made no more mistakes than did NVGPs on this test. Briefly, this test requires participants to respond as fast as possible to shapes appearing at the target location, while ignoring the same shapes if they appear at another location. By manipulating the frequency of targets, the T.O.V.A. offers a measure of both impulsivity (is the observer able to withhold a response to a nontarget when most of the stimuli are targets?) and a measure of sustained attention (is the observer able to stay on task and respond quickly to a target when most of the stimuli are nontargets?). In all cases, VGPs were faster but no less accurate than were the NVGPs, indicating, if anything, enhanced performance on these aspects of attention as compared with NVGPs. It may be worth noting that VGPs were often so fast that the built-in data analysis software of the T.O.V.A. considered their reaction times to be anticipatory (200 ms or less). However, a close look at the reaction time distribution indicated that these anticipatory responses were nearly all correct responses (which would obviously be unlikely if they were driven by chance, rather than being visually evoked). In summary, VGPs are faster but not more impulsive than NVGPs and equally capable of sustaining their attention. Although some have argued for a link between technology use and attention-deficit disorder (Rosen 2007, but see Durkin 2010), in the case of playing action games this does not appear to be the case.

Resource allocation. The proposal that action game play enhances top-down aspects of attention by allowing VGPs to allocate their resources more flexibly where it matters for the given task is supported by several independent sources. First, some evidence indicates that VGPs may better ignore sources of distraction, a key skill for properly allocating resources. For example, VGPs have been shown to be more capable of overcoming attentional capture. Chisholm et al. (2010, Chisholm & Kingstone 2011) compared the performance of VGPs and NVGPs on a target-search task known to engage top-down attention, while concurrently manipulating whether a singleton distractor, known to capture attention automatically, was or was not present. When top-down and bottom-up attention interact in such a way, investigators noted that although the singleton distractor captured attention in both groups, it did so to a much lesser extent in VGPs than in NVGPs. Thus VGPs may better employ executive strategies to reduce the effects of distraction. More recently, Mishra et al. (2011) made use of the steady-state visual-evoked potentials technique to examine the neural bases of the attentional enhancements noted in VGPs. They found that VGPs more efficiently suppressed unattended, potentially distracting information when presented with highly taxing displays. Participants viewed four different streams of rapidly flashing alphanumeric characters. Each stream flashed at a distinct temporal frequency, allowing retrieval of the brain signals evoked by each stream independently at all times. Thus, brain activation evoked by the attended stream could be retrieved, as could brain activation evoked by each of the unattended and potentially distracting streams. VGPs suppressed irrelevant streams to a greater extent than did NVGPs, and the extent of the suppression predicted the speed of their responses. VGPs seem to focus better on the task at hand by ignoring other sources of potentially distracting information. This view is in line with the proposal in the literature of greater distractor suppression as one possible determinant of more efficient executive and attentional control (Clapp et al. 2011, Serences et al. 2004, Toepper et al. 2010).

VGPs may also benefit from greater flexibility in resource allocation. The finding that VGPs switch tasks more swiftly, whether or not switches are predictable, is consistent with this view (Green et al. 2012). A recent brain-imaging study involving VGPs and NVGPs also speaks to this issue. In this study, the recruitment of the frontoparietal network of areas hypothesized to control the flexible allocation of top-down attention was compared in VGPs and NVGPs. As attentional demands increased, NVGPs showed increased activation in this network as expected. In contrast, VGPs barely engaged this network despite a matched increase in attentional demands across groups. This reduced activity in the frontoparietal network is compatible with the proposal that resource allocation may be more automatic and swift in VGPs (Bavelier et al. 2011). Furthermore, many models of the multiple-object-tracking task, on which VGPs have been seen to excel, intimate a close relationship between the number of items that can be tracked and the efficiency of resource allocation (Ma & Huang 2009, Vul et al. 2010).

In sum, much evidence demonstrates that action video game play leads to not only enhanced resources, but also a more intelligent allocation of these resources given the goals at hand. This is one of the ways action game play may result in learning to learn. We now turn to other evidence for learning to learn after action game play that suggests changes in knowledge and/or learning algorithm.

Knowledge/Learning Rule

Direct evidence for changes in knowledge or learning rules is still lacking. However, the literature contains pointers suggestive of such changes. First, the observation that action game play leads to more accurate probabilistic inference, as discussed above (see A Common Mechanism: Learning to Learn), suggests the development of new connectivity and knowledge to enable a more efficient architecture for the given task. Although some of these benefits could be driven initially by changes in resources, the fact that they can last for five months or longer suggests more profound and long-lasting changes in representation than what is typically afforded by attentional resources. Second, a number of experiments point to improvement despite little need for resources. This is true, for example, of very basic psychophysical skills such as visual acuity or contrast sensitivity (Green & Bavelier 2007, Li et al. 2009). These paradigms displayed only one target in isolation, and no uncertainty was present on either the place and/or time of the target's arrival (fixed target location, fixed SOA). In contrast, resources such as attention are typically called on when a target needs to be selected in time or space from distractors or when the time and/or place of arrival of the stimulus is uncertain (Carrasco et al. 2002). Third, an attentional explanation is not always in line with the noted changes. For example, contrast-sensitivity improvements were not larger as spatial frequency increased (and thus stimulus size decreased) as would be predicted by attentional accounts. Rather, the beneficial effect of action game play was maximal at intermediate frequencies. These considerations point to changes in representations likely to be mediated by enhancement either in learning rule or in knowledge.

Action game playing may act by enabling more generalizable knowledge through various abstractions, including the extent to which nontask-relevant information should be suppressed (Clapp et al. 2011, Serences et al. 2004, Toepper et al. 2010), how to modify performance to maximize reward rate (Simen et al. 2009), how to combine data across feature dimensions (which depends on the presence/type of dependencies between the dimensions; Perfors & Tenenbaum 2009, Stilp et al. 2010), and how to set a proper learning rate (which depends on the belief about whether generating distributions are stationary or drifting; Behrens et al. 2007). Although these avenues are only beginning to be explored, some support is emerging for distraction suppression and reward-rate changes after action game play. First, several studies document that action video game play alters the extent to which distracting information is suppressed. It is not always the case that VGPs show greater suppression of distractors. Rather, distractor processing appears to be greater in VGPs than in NVGPs under conditions of relatively low load. Under such conditions, task difficulty is sufficiently low for VGPs to remain efficient on the primary task and at the same time still be able to process distractors (Dye et al. 2009b, Green & Bavelier 2003). In contrast, VGPs show greater suppression of distractors under extremely high load conditions (Mishra et al. 2011). As task difficulty increases, suppression of distractors becomes necessary for VGPs to maintain their high performance on the primary task. Such differential effects may be understood if distractor processing has been encoded at a proper level of abstraction that is contingent on primary task success. Second, a key step in fostering learning is to engage the reward system (Bao et al. 2001, Koepp et al. 1998, Roelfsema et al. 2010). A distinguishing feature of action games is the layering of events/actions at many different timescales, resulting in a complex pattern of reward in time. This feature may explain, in part, why VGPs seem to maximize reward rate in a variety of tasks (A.F. Anderson, C.S. Green, D. Bavelier, manuscript in preparation; Green et al. 2010). Further investigation of the impact of action game play on knowledge and learning rules should provide an interesting test case, as the roles of resource control, hierarchical knowledge, and learning rules are refined.

Finally, the enhanced attentional control documented earlier is not, in this view, the proximal cause of the superior performance of VGPs, i.e., an end in and of itself. Instead, it is a means to an end, with that end being the development of more generalizable knowledge as one is faced with new tasks or new environments. Such learning to learn predicts that reasonably equivalent performance between groups should be seen early on when performing a new task, with the action gaming advantage appearing and then increasing through experience with the task. Such outcomes are being investigated in the domain of perceptual learning and decision making.

CONCLUDING REMARKS

In sum, we propose that action game play does not teach any one particular skill but instead increases the ability to extract patterns or regularities in the environment. As exemplified by their superior attentional control skills, VGPs develop the ability to exploit task-relevant information more efficiently while better suppressing irrelevant, potentially distracting sources of information. This may stem from the ability to discover the underlying structure of the task they face more readily, perform more accurate statistical inference over the data they experience, and thus exhibit superior performance on a variety of tasks previously thought to tap unrelated capacities. In this sense, action video game play may be thought of as fostering learning to learn. The hypothesis that action video game play may improve learning to learn is appealing because it naturally accounts for the wide range of skills enhanced in VGPs. Yet, learning to learn does not guarantee better performance on all tasks. Instead, like all forms of learning, it comes with a number of constraints that determine the nature and extent of the generalization possible. For instance, changes in knowledge produce benefits only to the extent to which new tasks share structure with action video games. No benefits are expected in tasks that share no such structure. Furthermore, and again like all forms of learning, potential drawbacks exist. For instance, an increase in resources may predict a tendency to over explore: to test models that are more elaborate than what is required by the task at hand, which could actually result in poorer performance and slower learning. Learning to learn does not predict that VGPs will excel at all tasks, but rather provides a theoretical framework to further our understanding of key concepts underlying generalization in learning as well as better characterizing the architecture, resources, and learning algorithms that generate the remarkable power of the human brain.

DISCLOSURE STATEMENT

D.B., C.S.G., and A.P. have patents pending concerning the use of video games for learning. D.B. is a consultant for PureTech Ventures, a company that develops approaches for various health areas, including cognition. D.B., C.S.G., and A.P. have a patent pending on action-video-game-based mathematics training; D.B. has a patent pending on action-video-game-based vision training.

ACKNOWLEDGMENTS

We thank Ted Jacques for his invaluable help with manuscript preparation. This work was made possible thanks to the support of the Office of Naval Research through a Multi University Research Initiative (MURI) that supported all four authors. Projects described here also received support in part from grants from the National Institutes of Health (EY-O16880 to D.B.; EY-0285 to P.S.; NIDA-0346785 to A.P.), from the National Science Foundation (BCS0446730 to A.P.), and from the James S. McDonnell Foundation (to D.B. and to A.P.).

LITERATURE CITED

Anderson AF, Bavelier D. 2011. Action game play as a tool to enhance perception, attention and cognition. In *Computer Games and Instruction,* ed. S Tobias, D Fleycher, pp. 307–29. Charlotte, NC: Information Age

Anderson AF, Kludt R, Bavelier D. 2011. *Verbal versus visual working memory skills in action video game players.* Poster presented at the Psychonomics Soc. Meet., Seattle

Andrews G, Murphy K. 2006. Does video-game playing improve executive function? In *Frontiers in Cognitive Psychology,* ed. MA Vanchevsky, pp. 145–61. New York: Nova Sci.

Badre D. 2008. Cognitive control, hierarchy, and the rostro-caudal organization of the frontal lobes. *Trends Cogn. Sci.* 12:193–200

Bailey K, West R, Anderson CA. 2010. A negative association between video game experience and proactive cognitive control. *Psychophysiology* 47:34–42

Baldassare A, Lewis CM, Committeri G, Snyder AZ, Romani GL, Corbetta M. 2012. Individual variability in functional connectivity predicts performance of a perceptual task. *Proc. Natl. Acad. Sci. USA* 109:3516–21

Baluch F, Itti L. 2011. Mechanisms of top-down attention. *Trends Neurosci.* 34:210–24

Bao S, Chan VT, Merzenich MM. 2001. Cortical remodelling induced by activity of ventral tegmental dopamine neurons. *Nature* 412:79–83

Bavelier D, Achtman RA, Mani M, Foecker J. 2011. Neural bases of selective attention in action video game players. *Vis. Res. doi:* 10.1016/j.visres.2011.08.007

Bavelier D, Levi DM, Li RW, Dan Y, Hensch TK. 2010. Removing brakes on adult brain plasticity: from molecular to behavioral interventions. *J. Neurosci.* 30:14964–71

Beck JM, Ma WJ, Kiani R, Hanks T, Churchland AK, et al. 2008. Bayesian decision making with probabilistic population codes. *Neuron* 60:1142–52

Behrens TE, Woolrich MW, Waldon ME, Rushworth MF. 2007. Learning the value of information in an uncertain world. *Nat. Neurosci.* 10:1214–21

Bengio Y. 2009. Learning deep architectures for AI. *Found. Trends Mach. Learn.* 2:1–127

Benson PG. 2001. Hawthorne effect. In *The Corsini Encyclopedia of Psychology and Behavioral Science,* ed. WE Craighead, CB Nemeroff, pp. 667–68. New York: Wiley

Binet A. 1984 (1909). *Les idées modernes sur les enfants, transl.* S Heisler. Menlo Park, CA: S. Heisler

Boot WR, Kramer AF, Simons DJ, Fabiani M, Gratton G. 2008. The effects of video game playing on attention, memory, and executive control. *Acta Psychol.* 129:387–98

Botvinick MM. 2008. Hierarchical models of behavior and prefrontal function. *Trends Cogn. Sci.* 12:201–8

Buckley D, Codina C, Bhardwaj P, Pascalis O. 2010. Action video game players and deaf observers have larger Goldmann visual fields. *Vis. Res.* 50:548–56

Buschkuehl M, Jaeggi SM, Hutchison S, Perrig-Chiello P, Däpp C, et al. 2008. Impact of working memory training on memory performance in old-old adults. *Psychol. Aging* 23:743–53

Cain MS, Landau AN, Shimamura AP. 2012. Action video game experience reduces the cost of switching tasks. *Atten. Percept. Psychophysics* 74:641–47

Carrasco M, Williams PE, Yeshurun Y. 2002. Covert attention increases spatial resolution with or without masks: support for signal enhancement. *J. Vis.* 2:467–79

Castel AD, Pratt J, Drummond E. 2005. The effects of action video game experience on the time course of inhibition of return and the efficiency of visual search. *Acta Psychol.* 119:217–30

Cavanagh P, Alvarez GA. 2005. Tracking multiple targets with multifocal attention. *Trends Cogn. Sci.* 9:349–54

Chase WG, Simon HA. 1973. The mind's eye in chess. In *Visual Information Processing,* ed. WG Chase, pp. 215–81. New York: Academic

Chisholm JD, Hickey C, Theeuwes J, Kingstone A. 2010. Reduced attentional capture in action video game players. *Atten. Percept. Psychophys.* 72:667–71

Chisholm JD, J, Kingstone A. 2011. Improved top-down control reduces oculomotor capture: the case of action video game players. *Atten. Percept. Psychophys.* 74:257–62

Christoff K, Keramatain K. 2007. Abstraction of mental representations: theoretical considerations and neuroscientific evidence. In *Perspectives on Rule-Guided Behavior,* ed. SA Bunge, JD Wallis, pp. 107–27. New York: Oxford Univ. Press

Clapp WC, Rubens MT, Sabharwal J, Gazzaley A. 2011. Deficit in switching between functional brain networks underlies the impact of multitasking on working memory in older adults. *Proc. Natl. Acad. Sci. USA* 108:7212–17

Cohen JE, Green CS, Bavelier D. 2007. Training visual attention with video games: Not all games are created equal. In *Computer Games and Adult Learning,* ed. H O'Neil, R Perez, pp. 205–27. Oxford, UK: Elsevier

Colzato LS, van Leeuwen PJA, van den Wildenberg WPM, Hommel B. 2010. DOOM'd to switch: superior cognitive flexibility in players of first person shooter games. *Front. Psychol.* 1:1–5

Corbetta M, Shulman G. 2002. Control of goal-directed and stimulus-driven attention in the brain. *Nat. Rev. Neurosci.* 3:201–15

Dahlin E, Nyberg L, Bäckman L, Neely AS. 2008. Plasticity of executive functioning in young and older adults: immediate training gains, transfer, and long-term maintenance. *Psychol. Aging* 23:720–30

Daily Mail. 2009. RAF jettisons its top guns: drones to fly sensitive missions over Afghanistan. *Daily Mail Online* Feb.28:http://www.dailymail.co.uk/news/worldnews/article-1158084/RAF-jettisons-Top-Guns-Drones-fly-sensitive-missions-Afghanistan.html

Deneve S. 2008. Bayesian spiking neurons I: inference. *Neural Comput.* 20:91–117

DiCarlo JJ, Cox DD. 2007. Untangling invariant object recognition. *Trends Cogn. Sci.* 11:333–41

Donohue SE, Woldorff MG, Mitroff SR. 2010. Video game players show more precise multisensory temporal processing abilities. *Atten. Percept. Psychophys.* 72:1120–29

Dosenbach NUF, Fair DA, Cohen AL, Schlaggar BL, Petersen SE. 2008. A dual-networks architecture of top-down control. *Trends Cogn. Sci.* 12:99–105

Dosher BA, Lu Z-L. 1998. Perceptual learning reflects external noise filtering and internal noise reduction through channel reweighting. *Proc. Natl. Acad. Sci. USA* 95:13988–93

Durkin K. 2010. Videogames and young people with developmental disorders. *Rev. Gen. Psychol.* 14:122–40

Dye MW, Green CS, Bavelier D. 2009a. Increasing speed of processing with action video games. *Curr. Dir. Psychol. Sci.* 18:321–26

Dye MWG, Bavelier D. 2010. Differential development of visual attention skills in school-age children. *Vis. Res.* 50:452–59

Dye MWG, Green CS, Bavelier D. 2009b. The development of attention skills in action video game players. *Neuropsychologia* 47:1780–89

Egerton A, Mehta MA, Montgomery AJ, Lappin JM, Howes OD, et al. 2009. The dopaminergic basis of human behaviors: a review of molecular imaging studies. *Neurosci. Biobehav. Rev.* 33:1109–32

Erickson KI, Kramer AF. 2009. Aerobic exercise effects on cognitive and neural plasticity in older adults. *Br. J. Sports Med.* 43:22–24

Feng J, Spence I, Pratt J. 2007. Playing an action videogame reduces gender differences in spatial cognition. *Psychol. Sci.* 18:850–55

Frey SH, Fogassi L, Grafton S, Picard N, Rothwell JC, et al. 2011. Neurological principles and rehabilitation of action disorders: computation, anatomy, and physiology (CAP) model. *Neurorehabil. Neural Repair* 25(5):6S–20

Fuster JM. 2004. Upper processing stages of the perception-action cycle. *Trends Cogn. Sci.* 8:143–45

Gallistel CR. 2008. Learning and representation. In *Learning Theory and Behavior.* Vol. 1: *Learning and Memory: A Comprehensive Reference,* ed. R Menzel, pp. 529–48. Oxford: Elsevier

Gallistel CR, King AP. 2009. *Memory and the Computational Brain: Why Cognitive Science Will Transform Neuroscience.* New York: Wiley/Blackwell

Goard M, Dan Y. 2009. Basal forebrain activation enhances cortical coding of natural scenes. *Nat. Neurosci.* 12:1444–49

Godden D, Baddeley A. 1975. Context dependent memory in two natural environments. *Br. J. Psychol.* 66:325–31

Green CS, Bavelier D. 2003. Action video games modify visual selective attention. *Nature* 423:534–37

Green CS, Bavelier D. 2006a. Effects of action video game playing on the spatial distribution of visual selective attention. *J. Exp. Psychol.: Hum. Percept. Perform.* 32:1465–78

Green CS, Bavelier D. 2006b. Enumeration versus multiple object tracking: the case of action video game players. *Cognition* 101:217–45

Green CS, Bavelier D. 2007. Action video game experience alters the spatial resolution of vision. *Psychol. Sci.* 18:88–94

Green CS, Bavelier D. 2012. Learning, attentional control and action video games. *Curr. Biol.* 22:R167–

206

Green CS, Pouget A, Bavelier D. 2010. Improved probabilistic inference as a general learning mechanism with action video games. *Curr. Biol.* 20:1573–79

Green CS, Sugarman MA, Medford K, Klobusicky E, Bavelier D. 2012. The effect of action video games on task switching. *Comput. Hum. Behav.* 12:984–94

Greenfield PM. 2009. Technology and informal education: What is taught, what is learned. *Science* 323:69–71

Greenfield PM, DeWinstanley P, Kilpatrick H, Kaye D. 1994. Action video games and informal education: effects on strategies for dividing visual attention. *J. Appl. Dev. Psychol.* 15:105–23

Halpern DF, Benbow CP, Geary DC, Gur RC, Hyde JS, Gernsbacher MA. 2007. The science of sex differences in science and mathematics. *Psychol. Sci. Public Interest* 8:1–51

Harlow HF. 1949. The formation of learning sets. *Psychol. Rev.* 56:51–65

Herrero JL, Roberts MJ, Delicato LS, Gieselmann MA, Dayan P, Thiele A. 2008. Acetylcholine contributes through muscarinic receptors to attentional modulation in V1. *Nature* 454:1110–14

Hillman CH, Erickson KI, Kramer AF. 2008. Be smart, exercise your heart: exercise effects on brain and cognition. *Nat. Rev. Neurosci.* 9:58–65

Hinton GE. 2007. Learning multiple layers of representation. *Trends Cogn. Sci.* 11:428–34

Hinton GE, Osindero S, Tej YH. 2006. A fast learning algorithm for deep belief nets. *Neural Comput.* 18:1527–54

Hubert-Wallander BP, Green CS, Sugarman M, Bavelier D. 2011. Altering the rate of visual search through experience: the case of action video game players. *Atten. Percept. Psychophys.* 73:2399–412

Jaeggi SM, Buschkuehl M, Jonides J, Perrig WJ. 2008. Improving fluid intelligence with training on working memory. *Proc. Natl. Acad. Sci. USA* 105:6829–33

Ji W, Gao E, Suga N. 2001. Effects of acetylcholine and atropine on plasticity of central auditory neurons caused by conditioning in bats. *J. Neurophysiol.* 86:211–25

Karle JW, Watter S, Shedden JM. 2010. Task switching in video game players: benefits of selective attention but not resistance to proactive interference. *Acta Psychol.* 134:70–78

Kemp C, Goodman ND, Tenenbaum JB. 2010. Learning to learn causal models. *Cogn. Sci.* 23:1–59

Khoe WW, Bavelier D, Hillyard SA. 2010. Enhanced attentional modulation of early visual processing in video game players. *Soc. Neurosci. Meet. Plann.* Abstr. 399.1

Kilgard MP, Merzenich MM. 1998. Cortical map reorganization enabled by nucleus basalis activity. *Science* 279:1714–18

Klingberg T. 2010. Training and plasticity of working memory. *Trends Cogn. Sci.* 14:317–24

Knill DC, Richards W. 1996. *Perception as Bayesian Inference.* Cambridge, UK: Cambridge Univ. Press

Koechlin E. 2008. The cognitive architecture of the human lateral prefrontal cortex. In *Attention and Performance,* ed. P Haggard, Y Rosetti, M Kawato, pp. 483–509. New York: Oxford Univ. Press

Koechlin E, Summerfield C. 2007. An information theoretical approach to prefrontal executive function. *Trends Cogn. Sci.* 11:229–35

Koepp MJ, Gunn RN, Lawrence AD, Cunningham VJ, Dagher A, et al. 1998. Evidence for striatal dopamine release during a video game. *Nature* 393:266–68

Larochelle H, Erhan D, Courville A, Bergstra J, Bengio Y. 2007. *An empirical evaluation of deep architectures on problems with many factors of variation.* Presented at Proc. Int. Conf. Mach. Learn., 24th,

Corvallis, Or.

Lashley KS. 1951. The problem of serial order in behavior. In *Cerebral Mechanisms in Behavior*, ed. LA Jeffress, pp. 112–36. New York: Wiley

Lavie N. 2005. Distracted and confused? Selective attention under load. *Trends Cogn. Sci.* 9:75–82

Lee H, Grosse R, Ranganath R, Ng AY. 2009. *Convolutional deep belief networks for scalable unsupervised learning of hierarchical representations.* Presented at Proc. Int. Conf. Mach. Learn., 26th, Montreal

Lee TS, Mumford D. 2003. Hierarchical Bayesian inference in the visual cortex. *J. Opt. Soc. Am.* 20:1434–48

Legge GE, Rubin GS, Pelli DG, Schleske MM. 1985. Psychophysics of reading–II. Low vision. *Vis. Res.* 25:253–65

Li R, Ngo C, Nguyen J, Levi DM. 2011. Video game play induces plasticity in the visual system of adults with amblyiopia. *Public Libr. Sci.* 9(8):e1001135

Li R, Polat U, Makous W, Bavelier D. 2009. Enhancing the contrast sensitivity function through action video game training. *Nat. Neurosci.* 12:549–51

Li RW, Polat U, Scalzo F, Bavelier D. 2010. Reducing backward masking through action game training. *J. Vis.* 10:1–13

Li S, Mayhew SD, Kourtzi Z. 2009. Learning shapes the representation of behavioral choice in the human brain. *Neuron* 62:441–52

Lustig C, Shah P, Seidler R, Reuter PA. 2009. Aging, training, and the brain: a review and future directions. *Neuropsychol. Rev.* 19:504–22

Lutz A, Slagter HA, Dunne JD, Davidson RJ. 2008. Attention regulation and monitering in meditation. *Trends Cogn. Sci.* 12:163–69

Ma WJ, Beck JM, Latham PE, Pouget A. 2006. Bayesian inference with probabilistic population codes. *Nat. Neurosci.* 9:1432–38

Ma WJ, Huang W. 2009. No capacity limit in attentional tracking: evidence for probabilistic inference under a resource constraint. *J. Vis.* 9:1–30

McClurg PA, Chaille C. 1987. Computer games: environments for developing spatial cognition. *J. Educ. Comput. Res.* 3:95–111

McKinley RA, McIntire LK, Funke MA. 2011. Operator selection for unmanned aerial systems: comparing video game players and pilots. *Aviat. Space Environ. Med.* 82:635–42

McNab F, Varrone A, Farde L, Jucaite A, Bystritsky P, et al. 2009. Changes in cortical dopamine D1 receptor binding associated with cognitive training. *Science* 323:800–2

Mishra J, Zinni M, Bavelier D, Hillyard SA. 2011. Neural basis of superior performance of video-game players in an attention-demanding task. *J. Neurosci.* 31:992–98

Mohamed A, Dahl GE, Hinton G. 2011. Acoustic modeling using deep belief networks. IEEE *Trans. Audio, Speech, Lang. Proc.*

Okagaki L, Frensch PA. 1994. Effects of video game playing on measures of spatial performance: gender effects in late adolescence. *J. Appl. Dev. Psychol.* 15:33–58

Olesen P, Westerberg H, Klingberg T. 2004. Increased prefrontal and parietal activity after training of working memory. *Nat. Neurosci.* 7:75–79

Ophir E, Nass C, Wagner AD. 2009. Cognitive control in media multitaskers. *Proc. Natl. Acad. Sci. USA* 106:15583–87

Palmer J, Huk AC, Shadlen MN. 2005. The effect of stimulus strength on the speed and accuracy of a perceptual decision. *J. Vis.* 5:376–404

Paupathy A, Connor CE. 2002. Population coding of shape in area V4. *Nat. Neurosci.* 5:1332–38

Perfors AF, Tenenbaum JB. 2009. *Learning to learn categories.* Presented at Annu. Conf. Cogn. Sci. Soc., 31st, Amsterdam

Polley DB, Steinberg EE, Merzenich MM. 2006. Perceptual learning directs auditory cortical map reorganization through top-down influences. *J. Neurosci.* 26:4970–82

Pouget A, Thorpe SJ. 1991. Connectionist model of orientation identification. *Connect. Sci.* 3:127–42

Pylyshyn ZW, Storm RW. 1988. Tracking multiple independent targets: evidence for a parallel tracking mechanism. *Spat. Vis.* 3:179–97

Rao RP. 2004. Bayesian computation in recurrent neural circuits. *Neural Comput.* 16:1–38

Ratcliff R, McKoon G. 2008. The diffusion decision model: theory and data or two-choice decision tasks. *Neural Comput.* 20:873–922

Renner MJ, Rosenzweig MR. 1987. *Enriched and Impoverished Environments Effects on Brain and Behavior Recent Research in Psychology.* New York: Springer-Verlag

Ribas-Fernandes JJF, Solway A, Diuk C, McGuire JT, Barto AG, et al. 2011. A neural signature of hierarchical reinforcement learning. *Neuron* 71:370–79

Roelfsema PR, van Ooyen A, Watanabe T. 2010. Perceptual learning rules based on reinforcers and attention. *Trends Cogn. Sci.* 14:64–71

Rosch E. 1975. Cognitive representations of semantic categories. *J. Exp. Psychol. Gen.* 104:192–233

Rosen LD. 2007. *Me, MySpace, and I.* New York: Macmillan

Rosser JCJ, Lynch PJ, Cuddihy L, Gentile DA, Klonsky J, Merrell R. 2007. The impact of video games on training surgeons in the 21st century. *Arch. Surg.* 142:181–86

Rougier NP, Noelle DC, Braver TS, Cohen JD, O'Reilly RC. 2005. Prefrontal cortex and flexible cognitive control: rules without symbols. *Proc. Natl. Acad. Sci. USA* 102:7338–43

Rumelhart DE, McClelland JL, PDP Res. Group. 1987. *Parallel Distributed Processing–Vol. 1.* Boston, MA: MIT Press

Saffell T, Matthews N. 2003. Task-specific perceptual learning on speed and direction discrimination. *Vis. Res.* 43(12):1365–74

Sakai K. 2008. Task set and prefrontal cortex. *Annu. Rev. Neurosci.* 31:219–45

Sarter M, Hasselmo ME, Bruno JP, Givens B. 2005. Unraveling the attentional functions of cortical cholinergic inputs: interactions between signal-driven and cognitive modulation of signal detection. *Brain Res. Rev.* 48:98–111

Schellenberg EG. 2004. Music lessons enhance IQ. *Psychol. Sci.* 15:511–14

Schlickum MK, Hedman L, Enochsson L, Kjellin A, Fellander-Tsai L. 2009. Systematic video game training in surgical novices improves performance in virtual reality endoscopic surgical simulators: a prospective randomized study. *World J. Surg.* 33:2360–67

Schultz W, Dayan P, Montague PR. 1997. A neural substrate of prediction and reward. *Science* 275:1593–99

Serences JT, Yantis S, Culberson A, Awh E. 2004. Preparatory activity in visual cortex indexes distractor suppression during covert spatial orienting. *J. Neurophysiol.* 92:3538–45

Shapiro DC, Schmidt RA. 1982. The schema theory: recent evidence and developmental implications.

In *The Development of Movement Control and Co-ordination,* ed. JAS Kelso, JE Clark, pp. 113–50. New York: Wiley

Shapiro KL, Arnell KM, Raymond JE. 1997. The attentional blink. *Trends Cogn. Sci.* 1:291–96

Simen P, Contreras D, Buck C, Hu P, Holmes P, Cohen D. 2009. Reward rate optimization in two-alternative decision making: empirical tests of theoretical predictions. *J. Exp. Psychol.: Hum. Percept. Perform.* 35:1865–97

Sitaram N, Weingartner H, Gillin JC. 1978. Human serial learning: enhancement with arecholine and choline impairment with scopolamine. *Science* 201:274–76

Spelke ES. 2005. Sex differences in intrinsic aptitude for mathematics and science?: A critical review. *Am. Psychol.* 60:950–58

Spence I, Feng J. 2010. Video games and spatial cognition. *Rev. Gen. Psychol.* 14:92–104

Spence I, Yu JJ, Feng J, Marshman J. 2009. Women match men when learning a spatial skill. *J. Exp. Psychol.: Learn. Mem. Cogn.* 35:1097–103

Stevens C, Bavelier D. 2012. The role of selective attention on academic foundations: a cognitive neuroscience perspective. *Dev. Cogn. Neurosci.* 2:S30–48

Stilp CE, Rogers TT, Kluender KR. 2010. Rapid efficient coding of correlated complex acoustic properties. *Proc. Natl. Acad. Sci. USA* 107:21914–19

Strobach T, Frensch PA, Schubert T. 2012. Video game practice optimizes executive control skills in dual tasks and task switching situations. *Acta Psycholog.* 140:13–24

Subrahmanyam K, Greenfield PM. 1994. Effect of video game practice on spatial skills in girls and boys. *J. Appl. Dev. Psychol.* 15:13–32

Takeuchi H, Sekiguchi A, Taki Y, Yokoyama S, Yomogida Y, et al. 2010. Training of working memory impacts structural connectivity. *J. Neurosci.* 30:3297–303

Tang YY, Posner MI. 2009. Attention training and attention state training. *Trends Cogn. Sci.* 13:222–27

Thorndike EL, Woodworth RS. 1901. The influence of improvement in one mental function upon the efficiency of other functions. (I). *Psychol. Rev.* 8:247–61

Toepper M, Gebhardt H, Beblo T, Thomas C, Driessen M, et al. 2010. Functional correlates of distractor suppression during spatial working memory encoding. *Neuroscience* 165:1244–53

Trick LM, Jaspers-Fayer F, Sethi N. 2005. Multiple-object tracking in children: the "Catch the Spies" task. *Cogn. Dev.* 20:373–87

Vul E, Frank MC, Alvarez GA, Tenenbaum JB. 2010. Explaining human multiple object tracking as resource-constrained approximate inference in a dynamic probabilistic model. *Adv. Neural Inf. Proc. Syst.* 32:1955–63

Vul E, Hanus D, Kanwisher N. 2009. Attention as inference: Selection is probabilistic; responses are all-or-none samples. *J. Exp. Psychol. Gen.* 138(4):546–60

West GL, Stevens SA, Pun C, Pratt J. 2008. Visuospatial experience modulates attentional capture: evidence from action video game players. *J. Vis.* 8:1–9

Yamane Y, Carlson ET, Bowman KC, Wang Z, Connor CE. 2008. A neural code for three-dimensional object shape in macaque inferotemporal cortex. *Nat. Neurosci.* 11:1352–60

Zhang J-C, Lau P-M, Bi G-Q. 2009. Gain in sensitivity and loss in temporal contrast of STDP by dopaminergic modulation at hippocampal synapses. Proc. Natl. Acad. Sci. USA 106:13028–33

RELATED RESOURCES

Bavelier D, Levi DM, Li RW, Dan Y, Hensch TK. 2010. Removing brakes on adult plasticity: from molecular to behavioral interventions. *J. Neurosci.* 30(45):14964–71

Botvinick MM. 2008. Hierarchical models of behavior and prefrontal function. *Trends Cogn. Sci.* 12:201–8

Green CS, Pouget A, Bavelier D. 2010. Improved probabilistic inference, as a general learning mechanism with action video games. *Curr. Biol.* 20:1573–79

Hinton GE. 2007. Learning multiple layers of representation. *Trends Cogn. Sci.* 11:428–34

Spence I, Feng J. 2010. Video games and spatial cognition. *Rev. Gen. Psychol.* 14(2):92–104

Neural Basis of Reinforcement Learning and Decision Making

Daeyeol Lee[1,2], Hyojung Seo[1], and Min Whan Jung[3]

Daeyeol Lee: daeyeol.lee@yale.edu; Hyojung Seo: hyojun.seo@yale.edu; Min Whan Jung: min@ajou.ac.kr

[1]Department of Neurobiology, Kavli Institute for Neuroscience, Yale University School of Medicine, New Haven, Connecticut 06510, USA

[2]Department of Psychology, Yale University, New Haven, Connecticut 06520, USA

[3]Neuroscience Laboratory, Institute for Medical Sciences, Ajou University School of Medicine, Suwon 443-721, Republic of Korea

Key Words

prefrontal cortex; neuroeconomics; reward; striatum; uncertainty

Abstract

Reinforcement learning is an adaptive process in which an animal utilizes its previous experience to improve the outcomes of future choices. Computational theories of reinforcement learning play a central role in the newly emerging areas of neuroeconomics and decision neuroscience. In this framework, actions are chosen according to their value functions, which describe how much future reward is expected from each action. Value functions can be adjusted not only through reward and penalty, but also by the animal's knowledge of its current environment. Studies have revealed that a large proportion of the brain is involved in representing and updating value functions and using them to choose an action. However, how the nature of a behavioral task affects the neural mechanisms of reinforcement learning remains incompletely understood. Future studies should uncover the principles by which different computational elements of reinforcement learning are dynamically coordinated across the entire brain.

INTRODUCTION

Decision making refers to the process by which an organism chooses its actions, and has been studied in such diverse fields as mathematics, economics, psychology, and neuroscience. Traditionally, theories of decision making have fallen into two categories. On the one hand, normative theories in economics generate well-defined criteria for identifying best choices. Such theories, including expected utility theory (von Neumann and Morgenstern 1944), deal with choices in an idealized context, and often fail to account for actual choices made by humans and animals. On the other hand, descriptive psychological theories try to account for failures of normative theories by identifying a set of heuristic rules applied by decision makers. For example, prospect theory (Kahneman & Tversky 1979) can successfully account for the failures of expected utility theory in describing human decision making under uncertainty. These two complementary theoretical frameworks are essential for neurobiological studies of decision making. However, a fundamental question not commonly addressed by either approach is the role of learning.

How do humans and animals acquire their preference for different actions and outcomes in the first place?

Our goal in this paper is to review and organize recent findings about the functions of different brain areas that underlie experience-dependent changes in choice behaviors. Reinforcement learning theory (Sutton & Barto 1998) has been widely adopted as the main theoretical framework in designing experiments as well as interpreting empirical results. Since this topic has been frequently reviewed (Dayan & Niv 2008, van der Meer & Redish 2011, Ito & Doya 2011, Bornstein & Daw, 2011), we only briefly summarize the essential elements of reinforcement learning theory and focus on the following questions. First, where and how in the brain are the estimates of expected reward represented and updated by the animal's experience? Converging evidence from a number of recent studies suggests that many of these computations are carried out in multiple interconnected regions in the frontal cortex and basal ganglia. Second, how are the value signals for potential actions transformed to the final behavioral response? Competitive interactions among different pools of recurrently connected neurons are a likely mechanism for this selection process (Usher & McClelland 2001, Wang 2002), but their neuroanatomical substrates remain poorly understood (Wang 2008). Finally, how is model-based reinforcement learning implemented in the brain? Humans and animals can acquire new knowledge about their environment without directly experiencing reward or penalty, and this knowledge can be used to influence subsequent behaviors (Tolman 1948). This is referred to as model-based reinforcement learning, whereas reinforcement learning entirely relying on experienced reward and penalty is referred to as model-free. Recent studies have begun to shed some light on how these two different types of reinforcement learning are linked in the brain. We conclude with some suggestions for future research on the neurobiological mechanisms of reinforcement learning.

REINFORCEMENT LEARNING THEORIES OF DECISION MAKING

Economic Utilities and Value Functions

Economic theories of decision making focus on how numbers can be attached to alternative actions so that choices can be understood as selecting an action that has the maximum value among all possible actions. These hypothetical quantities are often referred to as utilities, and can be applied to all types of behaviors. By definition, behaviors chosen by an organism are those that maximize the organism's utility (Figure 1a). These theories are largely agnostic about how these utilities are determined, although they are presumably constrained by evolution and individual experience. By contrast, reinforcement learning theories describe how the animal's experience alters its value functions, which in turn influence subsequent choices (Figure 1b).

The goal of reinforcement learning is to maximize future rewards. Analogous to utilities in economic theories, value functions in reinforcement learning theory refer to the estimates for the sum of future rewards. However, since the animal cannot predict the future changes in its environment perfectly, value functions, unlike utilities, reflect the animal's empirical estimates for its future rewards. Rewards in the distant future are often temporally discounted so that more immediate rewards exert stronger influence on the animal's behavior. The reinforcement learning theory utilizes two different types of value functions. First, action value function refers to the sum of future rewards expected for taking a particular action in a particular state of the environment, and is often denoted by $Q(s,a)$, where s and a refer to the state of the environment and the animal's action, respectively. The term action is used formally: it can refer to not only a physical action, such as reaching to a coffee mug in a particular location with a specific limb, but also an abstract choice, such as buying a particular guitar. Second, state value function, often denoted by $V(s)$, refers to the sum of future rewards expected from a particular state of the animal's environment. If the animal always chooses only one action in a given state, then its action value function would be equal to the state value function. Otherwise, the state value function would correspond to the average of action value functions weighted by the probability of taking each action in a given state. State value functions can be used to evaluate the action outcomes, but in general, action value functions are required for selecting an action.

Model-free vs. Model-based Reinforcement Learning

Value functions can be updated according to two different types of information. First, they can be revised according to the reward or penalty received by the animal after each action. Value functions would not change and hence no learning would be necessary, if choice outcomes are always perfectly predicted from the current value functions. Otherwise, value functions must be modified to reduce errors in reward predictions. The signed difference between the actual reward and the reward expected by the current value functions is referred to as a reward prediction error (Sutton & Barto 1998). In a class of reinforcement learning algorithms, referred to as simple or model-free reinforcement learning, reward prediction error is the primary source of changes in value functions. More specifically, the value function for the action chosen by the animal or the state visited by the animal is updated according to the reward prediction error, while the value functions for all other actions and states remain unchanged or simply decay passively (Barraclough et al. 2004, Ito & Doya 2009).

In the second class of reinforcement learning, referred to as model-based reinforcement learning, value functions can be changed more flexibly. These algorithms can update the value functions on the basis of the animal's motivational state and its knowledge of the environment without direct reward or penalty. The use of cognitive models allows the animal to adjust its value functions immediately, whenever it acquires a new piece of information about its internal state or external environment. There are many lines of evidence that animals as well as humans are capable of model-based reinforcement learning. For example, when an animal is satiated for a particular type of reward, the subjective value of the same food would be diminished. However, if the animal relies entirely on simple reinforcement learning, the tendency to choose a given action would not change until it experiences the devalued reward through the same action. Previous work has shown that rats can change their behaviors immediately according to their current motivational states following the devaluation of specific food items. This is often used as a test for goal-directed behaviors, and indicates that animals are indeed capable of model-based reinforcement learning (Balleine & Dickinson 1998; Daw et al. 2005). Humans and animals can also simulate the consequences of potential actions that they could have chosen. This is referred to as counterfactual thinking (Roese & Olson 1995), and the information about hypothetical outcomes from unchosen actions can be incorporated into value functions when they are different from the outcomes predicted by the current value functions (Lee et al. 2005, Coricelli et al. 2005, Boorman et al. 2011). Analogous to reward prediction error, the difference between hypothetical and predicted outcomes is referred to as fictive or counterfactual reward prediction error (Lohrenz et al. 2007, Boorman et al. 2011).

Learning During Social Decision Making

During social interaction, the outcomes of actions are often jointly determined by the actions of multiple players. Such strategic situations are referred to as games (von Neumann & Morgenstern 1944). Decision makers can improve the outcomes of their choices during repeated games by applying a model-free reinforcement learning algorithm. In fact, for relatively simple games, such as two-player zero-sum games, humans and animals gradually approximate optimal strategies using model-free reinforcement learning algorithms (Mookherjee & Sopher 1994, Erev & Roth 1998, Camerer 2003, Lee et al. 2004). On the other hand, players equipped with a model-based reinforcement learning algorithm can adjust their strategies more flexibly according to the predicted behaviors of other players. The ability to predict the beliefs and intentions of other players is often referred to as the theory of mind (Premack & Woodruff 1978), and during social decision making, this may dramatically improve the efficiency of reinforcement learning. In game theory, this is also referred to as belief learning (Camerer 2003).

In pure belief learning, the outcomes of actual choices by the decision makers do not exert any additional influence on their choices, unless such outcomes can modify the beliefs or models of the decision makers about the other players. In other words, reward or penalty does not have any separate roles other than affecting the decision maker's belief about the other players. Results from studies in behavioral economics showed that such pure belief learning models do not account for human choice behaviors very well (Mookherjee & Sopher, 1997, Erev & Roth 1998, Feltovich 2000, Camerer 2003). Instead, human behaviors during repeated games are often consistent with hybrid learning models, such as the experience-weighted attraction model (Camerer & Ho 1999). In these hybrid models, value functions are adjusted by both real and fictive reward prediction errors. Learning rates for these different reward prediction errors can be set independently. Similar to the results from human studies, hybrid models also account for the behaviors of non-human primates performing a competitive game task against a computer better than either model-free reinforcement learning or belief learning model (Lee et al. 2005, Abe & Lee 2011).

NEURAL REPRESENTATION OF VALUE FUNCTIONS

Utilities and value functions are central to economic and reinforcement learning theories of decision making, respectively. In both theories, these quantities are assumed to capture all the relevant factors influencing choices. Thus, brain areas or neurons involved in decision making are expected to harbor signals related to utilities and value functions. In fact, neural activity related to reward expectancy has been found in many different brain areas (Schultz et al. 2000, Hikosaka et al. 2006; Wallis & Kennerley 2010), including sensory cortical areas (Shuler & Bear 2006, Serences 2008; Vickery et al. 2011). The fact that signals related to reward expectancy are widespread in the brain suggests that they are likely to subserve not only reinforcement learning, but also other related cognitive processes, such as attention (Maunsell 2004, Bromberg-Martin et al. 2010a, Litt et al. 2011). Neural signals related to reward expectancy can be divided into at least two different categories, depending on whether they are related to specific actions or states. Neural signals related to action value functions would be useful in choosing a particular action, especially if such signals are observed before the execution of a motor response. Neural activity related to state value functions may play more evaluative roles. In particular, during decision making, the state value function changes from the weighted average of action values for alternative choices to the action value function for the chosen action. The latter is often referred to as chosen value (Padoa-Schioppa & Assad 2006, Cai et al. 2011).

During experiments on decision making, choices can be made among alternative physical movements with different spatial trajectories, or among different objects regardless of the movements required to acquire them. Whereas these different options are all considered actions in the reinforcement learning theory, neural signals related to the corresponding action value functions may vary substantially according to the dimension in which choices are made. In most previous neurobiological studies, different properties of reward were linked to different physical actions. These studies have identified neural activity related to action value functions in numerous brain areas, including the posterior parietal cortex (Platt & Glimcher 1999, Sugrue et al. 2004, Dorris & Glimcher 2004, Seo et al. 2009), dorsolateral prefrontal cortex (Barraclough et al. 2004, Kim et al. 2008), premotor cortex (Pastor-Bernier& Cisek 2011), medial frontal cortex (Seo & Lee 2007, 2009, Sul et al. 2010, So & Stuphorn 2010), and striatum (Samejima et al. 2005, Lau & Glimcher 2008; Kim et al. 2009; Cai et al. 2011). Despite the limited spatial and temporal resolutions available in neuroimaging studies, metabolic activity related to action value functions has been also identified in the supplementary motor area (Wunderlich et al. 2009). In contrast, neurons in the primate orbitofrontal cortex are not sensitive to spatial locations of targets associated with specific rewards (Tremblay & Schultz 1999, Wallis & Miller 2003), suggesting that they encode action value functions related to specific objects or goals. When animals chose between two different flavors of juice, neurons in the primate orbitofrontal cortex indeed signaled the action value functions associated with specific juice flavors rather than the directions of eye movements used to indicate the animal's choices (Padoa-Schioppa & Assad, 2006).

Neurons in many different brain areas often combine value functions for alternative actions and other decision-related variables. Precisely how multiple types of signals are combined in the activity of individual neurons can therefore provide important clues about how such signals are computed and utilized. For example, likelihood of choosing one of two alternative choices is determined by the difference in their action value functions, and therefore neurons encoding such signals may be closely involved in the process of action selection. During a binary choice task, neurons in the primate posterior parietal cortex (Seo et al. 2009), dorsolateral prefrontal cortex (Kim et al. 2008), premotor cortex (Pastor-Bernier & Cisek 2011), supplementary eye field (Seo & Lee 2009), and dorsal striatum (Cai et al. 2011), as well as the rodent secondary motor cortex (Sul et al. 2011) and striatum (Ito & Doya 2009), encode the difference between the action value functions for two alternative actions.

Signals related to state value functions are also found in many different brain areas. During a binary choice task, the sum or average of the action value functions for two alternative choices corresponds to the state value function before a choice is made. In the posterior parietal cortex and dorsal striatum, signals related to such state value functions and action value functions coexist (Seo et al. 2009, Cai et al. 2011). Neurons encoding state value functions are also found in the ventral striatum (Cai et al. 2011), anterior cingulate cortex (Seo & Lee, 2007), and amygdala (Belova et al. 2008). Neural activity related to chosen values that correspond to post-decision state value functions is also widespread in the brain, and has been found in the orbitofrontal cortex (Padoa-Schioppa & Assad, 2006, Sul et al. 2010), medial frontal cortex (Sul et al. 2010), dorsolateral prefrontal cortex (Kim & Lee 2011), and striatum (Lau & Glimcher 2008, Kim et al. 2009, Cai et al. 2011). Since reward prediction error corresponds to the difference between the outcome of a choice and chosen value, neural activity related to chosen values might be utilized to compute reward prediction errors and update value functions. In some of these areas, such as the dorsolateral prefrontal cortex (Kim & Lee 2011) and dorsal striatum (Cai et al. 2011), activity related to chosen value signals emerges later than the signals related to the sum of the value functions for alternative actions, suggesting that action selection might take place during this delay (Figure 2).

NEURAL MECHANISMS OF ACTION SELECTION

During decision making, neural activity related to action value functions must be converted to the signals related to a particular action and transmitted to motor structures. The precise anatomical location playing a primary role in action selection may vary with the nature of a behavioral task. For example, actions selected by fixed stimulus-action associations or well-practiced motor sequences might rely more on the dorsolateral striatum compared to flexible goal-directed behaviors (Knowlton et al. 1996, Hikosaka et al. 1999, Yin & Knowlton 2006). Considering that spike trains of cortical neurons are stochastic (Softky & Koch 1993), the process of action selection is likely to rely on a network of neurons temporally integrating the activity related to difference in action value functions (Soltani & Wang 2006, Krajbich et al. 2010). An analogous process has been extensively studied for action selection based on noisy sensory stimulus during perceptual decision making. For example, psychophysical performance during a two-alternative forced choice task is well described by the so-called random-walk or drift-diffusion model in which a particular action is selected when the gradual accumulation of noisy evidence reaches a threshold for that action (Laming 1968, Roitman & Shadlen 2002, Smith & Ratcliff 2004).

Neurons in multiple brain areas involved in motor control often build up their activity gradually prior to specific movements, suggesting that these areas might also be involved in action selection. Execution of voluntary movements are tightly coupled with phasic neural activity in a number of brain areas, such as the primary motor cortex (Georgopoulos et al. 1986), premotor cortex (Churchland et al. 2006), frontal eye field (Hanes & Schall 1996), supplementary eye field (Schlag & Schlag-Rey 1987), posterior parietal cortex (Andersen et al. 1987), and superior colliculus (Schiller & Stryker 1972, Wurtz & Goldberg 1972). All of these structures are closely connected with motor nuclei in the brainstem and spinal cord. In addition, neurons in these areas display persistent activity related to the metrics of upcoming movements when the desired movement is indicated before a go signal, suggesting that they are also involved in motor planning and preparation (Smyrnis et al. 1992, Glimcher & Sparks 1992, Weinrich & Wise 1982, Bruce & Goldberg 1985, Schall 1991, Gnadt & Andersen 1988). Such persistent activity has been often associated with working memory (Funahashi et al. 1989, Wang 2001), but may also subserve the temporal integration of noisy inputs (Shadlen & Newsome 2001, Roitman & Shadlen, 2002, Wang, 2002, Curtis & Lee 2010). In fact, neural activity in accordance with gradual evidence accumulation has been found in the same brain areas that show persistent activity related to motor planning, such as the posterior parietal cortex (Roitman & Shadlen 2002), frontal eye field (Ding & Gold 2011), and superior colliculus (Horwitz & Newsome 2001).

Computational studies have demonstrated that a network of neurons with recurrent excitation and lateral inhibition can perform temporal integration of noisy sensory inputs and produce a signal corresponding to an optimal action (Wang 2002, Lo & Wang, 2006, Beck et al. 2008, Furman & Wang 2008). Most of these models have been developed to account for the pattern of activity observed in the lateral intraparietal (LIP) cortex during a perceptual decision making task. Nevertheless, value-dependent action selection might also involve attractor dynamics in a similar network of neurons, provided that their input synapses are adjusted in a reward-dependent manner (Soltani & Wang 2006, Soltani et al. 2006). Therefore, neurons involved in evaluation of unreliable sensory information may also contribute to value-based decision making. Consistent with this possibility, neurons in the LIP tend to change their activity according to the value of rewards expected from alternative actions (Platt & Glimcher 1999, Sugrue et al. 2004, Seo et al. 2009, Louie & Glimcher 2010).

In contrast to the brain areas involved in selecting a specific physical movement, other areas might be involved in more abstract decision making. For example, the orbitofrontal cortex might play a particularly important role in making choices among different objects or goods (Padoa-Schioppa 2011). On the other hand, an action selection process guided by internal cues rather than external sensory stimuli might rely more on the medial frontal cortex. Activity related to action value functions have been found in the supplementary and presupplementary motor areas (Sohn & Lee 2007, Wunderlich et al. 2009), as well as the supplementary eye field (Seo & Lee 2009, So & Stuphorn 2010). More importantly, neural activity related to an upcoming movement appears in the medial frontal cortex earlier than in other areas of the brain. For example, when human subjects are asked to initiate a movement voluntarily without any immediate sensory cue, scalp EEG displays the so-called readiness potential well before movement onset, and its source has been localized to the supplementary motor area (Haggard 2008, Nachev et al. 2008). The hypothesis that internally generated voluntary movements are selected in the medial frontal cortex is also consistent with the results from single-neuron recording and neuroimaging studies. Individual neurons in the primate supplementary motor area often begin to change their activity according to an upcoming limb movement earlier than those in the premotor cortex or primary motor cortex, especially when the animal is required to produce such movements voluntarily without immediate sensory cues (Tanji & Kurata 1985, Okano & Tanji 1987). Similarly, neurons in the supplementary eye field begin to modulate their activity according to the direction of an upcoming saccade earlier than similar activity recorded in the frontal eye field and LIP (Coe et al. 2002). In rodents performing a dynamic foraging task, signals related to the animal's choice appear in the medial motor cortex, presumably a homolog of the primate supplementary motor cortex, earlier than many other brain areas, including the primary motor cortex and basal ganglia (Sul et al. 2011). Furthermore, lesions in this area make the animal's choices less dependent on action value functions (Sul et al., 2011). Finally, analysis of BOLD activity patterns during a self-timed motor task has also identified signals related to an upcoming movement even several seconds before the movement onset in the human supplementary motor area (Soon et al. 2008).

In summary, neural activity potentially reflecting the process of action selection has been identified in multiple regions, including areas involved in motor control, orbitofrontal cortex, and medial frontal cortex. It would be therefore important to investigate in the future how these multiple areas interact cooperatively or competitively depending on the demands of specific behavioral tasks. The frame of reference in which different actions are represented varies across brain areas, and therefore, how the actions encoded in one frame of reference, for example, in objects space, are transformed to another, such as visual or joint space, needs to be investigated (Padoa-Schioppa 2011).

NEURAL MECHANISMS FOR UPDATING VALUE FUNCTIONS

Temporal Credit Assignment and Eligibility Trace

Reward resulting from a particular action is often revealed after a substantial delay, and an animal might carry out several other actions before collecting the reward resulting from a previous action. Therefore, it can be challenging to associate an action and its corresponding outcome correctly, and this is referred to as the problem of temporal credit assignment (Sutton & Barto 1998). Not surprisingly, loss of the ability to link specific outcomes to corresponding choices interferes with the process of updating value functions appropriately. While normal animals can easily alter their preferences between two objects when the probabilities of getting rewards from the two objects are switched, humans, monkeys, and rats with lesions in the orbitofrontal cortex are impaired in such reversal learning tasks (Iversen & Mishkin 1970, Schoenbaum et al. 2002; Fellows & Farah 2003, Murray et al. 2007). These deficits may arise due to failures in temporal credit assignment. For example, during a probabilistic reversal learning task, in which the probabilities of rewards from different objects were dynamically and unpredictably changed, deficits produced by the lesions in the orbitofrontal cortex were due to erroneous associations between the choices and their outcomes (Walton et al. 2010).

In reinforcement learning theory, the problem of temporal credit assignment can be resolved in at least two different ways. First, a series of intermediate states can be introduced during the interval between an action and a reward, so that they can propagate the information about the reward to the value function of the correct action (Montague et al. 1996). This basic temporal difference model was initially proposed to account for the reward prediction error signals conveyed by dopamine neurons, but was shown to be inconsistent with the actual temporal profiles of dopamine neuron signals (Pan et al. 2005). Another possibility is to utilize short-term memory signals related to the states or actions selected by the animal. Such memory signals are referred to as eligibility traces (Sutton & Barto 1998), and can facilitate action-outcome association even when the outcome is delayed. Therefore, eligibility traces can account for the temporally discontinuous shift in the phasic activity of dopamine neurons observed during classical conditioning (Pan et al. 2005). Signals related to the animal's previous choices have been observed in a number of brain areas, including the prefrontal cortex and posterior parietal cortex in monkeys (Barraclough et al. 2004, Seo & Lee 2009, Seo et al. 2009), as well as many regions in the rodent frontal cortex and striatum (Kim et al. 2007, 2009, Sul et al. 2010, 2011; Figure 3), and they might provide eligibility traces necessary to form associations between actions and their outcomes (Curtis & Lee 2010). In addition, neurons in many of these areas, including the orbitofrontal cortex, often encode specific conjunctions of chosen actions and their outcomes, for example, by increasing their activity when a positive outcome is obtained from a specific action (Barraclough et al. 2004, Seo & Lee 2009, Kim et al. 2009, Roesch et al. 2009, Sul et al. 2010, Abe & Lee 2011). Such action-outcome conjunction signals may also contribute to the resolution of the temporal credit assignment problem.

Integration of Chosen Value and Reward Prediction Error

In model-free reinforcement learning, the value function for a chosen action is revised according to reward prediction error. Therefore, signals related to chosen value and reward prediction error must be combined in the activity of individual neurons involved in updating value functions. Signals related to reward prediction error were first identified in the midbrain dopamine neurons (Schultz 2006), but later found to exist in many other areas, including the lateral habenula (Matsumoto & Hikosaka 2007), globus pallidus (Hong & Hikosaka 2008), anterior cingulate cortex (Matsumoto et al. 2007, Seo & Lee 2007), orbitofrontal cortex (Sul et al. 2010), and striatum (Kim et al. 2009, Oyama et al. 2010). Thus, the extraction of reward prediction error signals might be gradual and implemented through a distributed network of multiple brain areas. Dopamine neurons might then play an important role in relaying these error signals to update the value functions broadly represented in different brain areas. Signals related to chosen values are also distributed in multiple brain areas, including the medial frontal cortex, orbitofrontal cortex, and striatum (Padoa-Schioppa & Assad 2006, Lau & Glimcher 2008, Kim et al. 2009, Sul et al. 2010, Cai et al. 2011). The areas in which signals related to chosen value and reward prediction error converge, such as the orbitofrontal cortex and striatum, might therefore play an important role in updating value functions (Kim et al. 2009, Sul et al. 2010).

It is often hypothesized that the primary site for updating and storing action value functions is at the synapses between axons from cortical neurons and dendrites of medium spiny neurons in the striatum (Reynolds et al. 2001, Hikosaka et al. 2006, Lo & Wang 2006, Hong & Hikosaka 2011). Signals related to reward prediction error arrive at these synapses via the terminals of dopamine neurons in the substantia nigra (Levey et al. 1993, Schultz 2006, Haber et al. 2000, Haber & Knutson 2010), and multiple types of dopamine receptors in the striatum can modulate the plasticity of corticostriatal synapses accordingo the relative timing of presynaptic vs. postsynaptic action potentials (Shen et al. 2008, Gerfen & Surmeier 2011). However, the nature of specific information stored by these synapses remains poorly understood. In addition, whether the corticostriatal circuit carries appropriate signals related to eligibility traces for chosen actions and other necessary state information at the right time needs to be tested in future studies. Given broad dopaminergic projections to various cortical areas (Lewis et al. 2001), value functions might be updated in many of the same cortical areas encoding the value functions at the time of decision making.

Uncertainty and Learning Rate

The learning rate, which controls the speed of learning, must be adjusted according to uncertainty and volatility of the animal's environment. In natural environments, decision makers face many different types of uncertainty. When the probabilities of different outcomes are known, as when flipping a coin or during economic experiments, uncertainty about outcomes is referred to as risk (Kahneman & Tversky 1979) or expected uncertainty (Yu & Dayan 2005). In contrast, when the exact probabilities are unknown, this is referred to as ambiguity (Ellsberg 1961) or unexpected uncertainty (Yu & Dayan, 2005). Ambiguity or unexpected uncertainty is high in a volatile environment, in which the probabilities of different outcomes expected from a given action change frequently (Behrens et al. 2007), and this requires a large learning rate so that value functions can be modified quickly. By contrast, if the environment is largely known and stable, then the learning rate should be close to 0 so that value functions are not too easily altered by stochastic variability in the environment. Human learners can change their learning rates almost optimally when the rate of changes in reward probabilities for alternative actions is manipulated (Behrens et al. 2007). The level of volatility, and hence the learning rate, is reflected in the activity of the anterior cingulate cortex, suggesting that this region of the brain might be important for adjusting the learning rate according to the stability of the decision maker's environment (Behrens et al. 2007; Figure 4). Similarly, the brain areas known to increase their activity during decision making under ambiguity, such as the lateral prefrontal cortex, orbitofrontal cortex, and amygdala, might also be involved in optimizing the learning rate (Hsu et al. 2005, Huettel et al. 2006). Single-neuron recording studies have also implicated the orbitofrontal cortex in evaluating the amount of uncertainty in choice outcomes (Kepecs et al. 2008, O'Neill & Schultz 2010).

NEURAL SYSTEMS FOR MODEL–FREE VS. MODEL–BASED REINFORCEMENT LEARNING

Model-based Value Functions and Reward Prediction Errors

During model-based reinforcement learning, decision makers utilize their knowledge of the environment to update the estimates of outcomes expected from different actions, even without actual reward or penalty. A wide range of algorithms can be used to implement model-based reinforcement learning. For example, decision makers may learn the configuration and dynamics of their environment separately from the values of outcomes at different locations (Tolman 1948). This enables the animal to re-discover an optimal path of travel quickly whenever the location of a desired item changes. Flexibly combining these two different types of information might rely on the prefrontal cortex (Daw et al. 2005, Pan et al. 2008, Gläscher et al. 2010) and hippocampus (Womelsdorf et al. 2010, Simon & Daw 2011). For example, activity in the human lateral prefrontal cortex increases when unexpected state transitions are observed, suggesting that this area is involved in learning the likelihood of state transitions (Gläscher et al. 2010). The hippocampus might play a role in providing the information about the layout of the environment and other contextual information necessary for updating the value functions. The integration of information about the behavioral context and current task demands encoded in these two areas, especially at the time of decision making, may rely on rhythmic synchronization of neural activity in the theta frequency range (Sirota et al.

2008, Benchenane et al. 2010, Hyman et al. 2010, Womelsdorf et al. 2010).

Whether and to what extent model-free and model-based forms of reinforcement learning are supported by the same brain areas remains an important area of research. Value functions estimated by model-free and model-based algorithms might be updated or represented separately in different brain areas (Daw et al. 2005), but they might be also combined to produce a unique estimate for the outcomes expected from chosen actions. For example, when decision makers are required to combine the information about reward history and social information, reliability of predictions based on these two different types of information is reflected separately in two different regions of the anterior cingulate cortex (Behrens et al. 2008). On the other hand, signals related to reward probability predicted by both types of information were found in the ventromedial prefrontal cortex (Behrens et al. 2008). Similarly, human ventral striatum might represent the chosen values and reward prediction errors regardless of how they are computed (Daw et al. 2011, Simon & Daw 2011). This is also consistent with the finding that reward prediction error signals encoded by the midbrain dopamine neurons, as well as the neurons in the globus pallidus and lateral habenula, are in accordance with both model-free and model-based reinforcement learning (Bromberg-Martin et al. 2010b).

Hypothetical Outcomes and Mental Simulation

From the information that becomes available after completing chosen actions, decision makers can often deduce what alternative outcomes would have been possible from other actions. They can then use this information about hypothetical outcomes to update the action value functions for unchosen actions. In particular, the observed behaviors of other decision makers during social interaction are a rich source of information about such hypothetical outcomes (Camerer & Ho 1999, Camerer 2003, Lee et al. 2005, Lee 2008). Results from lesion and neuroimaging studies have demonstrated that the information about hypothetical or counterfactual outcomes might be processed in the same brain areas that are also involved in evaluating the actual outcomes of chosen actions, such as the prefrontal cortex (Camille et al. 2004, Coricelli et al. 2005, Boorman et al. 2011) and striatum (Lohrenz et al. 2007). The hippocampus might also be involved in simulating the possible outcomes of future actions (Hassabis & Maguire 2007, Schacter et al. 2007, Johnson & Redish 2007, Luhmann et al. 2008). Single-neuron recording studies have also shown that neurons in the dorsal anterior cingulate cortex respond similarly to actual and hypothetical outcomes (Hayden et al. 2009). More recent neurophysiological experiments in the dorsolateral prefrontal cortex and orbitofrontal cortex further revealed that neurons in these areas tend to encode actual and hypothetical outcomes for the same action, suggesting that they might provide an important substrate for updating the action value functions for chosen and unchosen actions simultaneously (Abe & Lee 2011; Figure 5).

CONCLUSIONS

In recent decades, reinforcement learning theory has become a central framework in the newly emerging areas of neuroeconomics and decision neuroscience. This is hardly surprising, because unlike abstract decisions analyzed in economic theories, biological organisms seldom receive complete information about the likelihoods of different outcomes expected from alternative actions. Instead, they face the challenge of learning how to predict the outcomes of their actions by trial and error, which is the essence of reinforcement learning.

The field of reinforcement learning theory has yielded many different algorithms, which provide neurobiologists with exciting opportunities to test whether they can successfully account for the actual behaviors of humans and animals and how different computational elements are implemented in the brain. In some cases, particular theoretical components closely correspond to specific brain structures, as in the case of reward prediction errors and midbrain dopamine neurons. However, in general, a relatively well circumscribed computational step in a given algorithm is often implemented in multiple brain areas, and this relationship might change with the animal's experience and task demands. The challenge that lies ahead is therefore to understand whether and how the signals in different brain areas related to various components of reinforcement learning, such as action value functions and chosen value, make different contributions to the overall behaviors of the animal. An important example discussed in this article is the relationship between model-free and model-based reinforcement learning algorithms. Neural machinery of model-free reinforcement learning might be phylogenetically older, and for simple decision-making problems, it may be more robust. By contrast, for complex decision-making problems, model-based reinforcement learning algorithms can be more efficient, since it can avoid the need to re-learn appropriate stimulus-action associations repeatedly by exploiting the regularities in the environment and the current information about the animal's internal state. Failures in applying appropriate reinforcement learning algorithms can lead to a variety of maladaptive behaviors observed in different mental disorders. Therefore, it would be crucial for future studies to elucidate the mechanisms that allow the brain to coordinate different types of reinforcement learning and their individual elements.

Acknowledgments

We are grateful to Soyoun Kim, Jung Hoon Sul, and Hoseok Kim for their help with the illustrations, and Jeansok Kim, Matthew Kleinman, and Tim Vickery for helpful comments on the manuscript. The research of the authors was supported by the grants from the National Institute of Drug Abuse (DA024855 and DA029330 to D. L.) and the Korea Ministry of Education, Science and Technology (the Brain Research Center of the 21st Century Frontier Research Program, NRF grant 2011-0015618 and the Original Technology Research Program for Brain Science 2011-0019209 to M.W.J.).

LITERATURE CITED

Abe H, Lee D. Distributed coding of actual and hypothetical outcomes in the orbital and dorsolateral prefrontal cortex. Neuron. 2011; 70:731–741. [PubMed: 21609828]

Andersen RA, Essick GK, Siegel RM. Neurons of area 7 activated by both visual stimuli and oculomotor behavior. Exp. Brain Res. 1987; 67:316–322. [PubMed: 3622691]

Balleine BW, Dickinson A. Goal-directed instrumental action: contingency and incentive learning and their cortical substrates. Neuropharmacology. 1998; 37:407–419. [PubMed: 9704982]

Barraclough DJ, Conroy ML, Lee D. Prefrontal cortex and decision making in a mixed-strategy game. Nat. Neurosci. 2004; 7:404–410. [PubMed: 15004564]

Beck JM, Ma WJ, Kiani R, Hanks T, Churchland AK, Roitman J, Shadlen MN, Latham PE, Pouget A. Probabilistic population codes for Bayesian decision making. Neuron. 2008; 60:1142–1152. [PubMed: 19109917]

Belova MA, Paton JJ, Salzman CD. Moment-to-moment tracking of state value in the amygdala. J. Neurosci. 2008; 48:10023–10030. [PubMed: 18829960]

Behrens TE, Woolrich MW, Walton ME, Rushworth MF. Learning the value of information in an uncertain world. Nat. Neurosci. 2007; 10:1214–1221. [PubMed: 17676057]

Behrens TE, Hunt LT, Woolrich MW, Rushworth MF. Associative learning of social value. Nature. 2008; 456:245–249. [PubMed: 19005555]

Benchenane K, Peyrache A, Khamassi M, Tierney PL, Gioanni Y, Battaglia FP, et al. Coherent theta oscillations and reorganization of spike timing in the hippocampal-prefrontal network upon learning. Neuron. 2010; 66:921–936. [PubMed: 20620877]

Boorman ED, Behrens TE, Rushworth MF. Counterfactual choicer and learning in a neural network centered on human lateral frontopolar cortex. PLoS Biol. 2011; 9 e1001093.

Bornstein AM, Daw ND. Multiplicity of control in the basal ganglia: computational roles of striatal subregions. Curr. Opin. Neurobiol. 2011; 21:374–380. [PubMed: 21429734]

Bromberg-Martin ES, Matsumoto M, Hikosaka O. Dopamine in motivational control: rewarding, aversive, and alerting. Neuron. 2010a; 68:815–834. [PubMed: 21144997]

Bromberg-Martin ES, Matsumoto M, Hong S, Hikosaka O. A pallidus-habenula-dopamine pathway signals inferred stimulus values. J. Neurophysiol. 2010b; 104:1068–1076. [PubMed: 20538770]

Bruce CJ, Goldberg ME. Primate frontal eye fields. I. Single neurons discharging before saccades. J.Neurophysiol. 1985; 53:603–635. [PubMed: 3981231]

Cai X, Kim S, Lee D. Heterogeneous coding of temporally discounted values in the dorsal and ventral striatum during intertemporal choice. Neuron. 2011; 69:170–182. [PubMed: 21220107]

Camerer, CF. Behavioral Game Theory: Experiments in Strategic Interaction. Princeton: Princeton Univ. Press; 2003.

Camerer C, Ho TH. Experience-weighted attraction learning in normal form games. Econometrica.1999; 67:827–874.

Camille N, Coricelli G, Sallet J, Pradat-Diehl P, Duhamel JR, Sirigu A. The involvement of the orbitofrontal cortex in the existence of regret. Science. 2004; 304:1167–1170. [PubMed: 15155951]

Churchland MM, Yu BM, Ryu SI, Santhnam G, Shenoy KV. Neural variability in premotor cortex provides a signature of motor preparation. J. Neurosci. 2006; 26:3697–3712. [PubMed: 16597724]

Coe B, Tomihara K, Matsuzawa M, Hikosaka O. Visual and anticipatory bias in three cortical eye fields of the monkey during an adaptive decision-making task. J. Neurosci. 2002; 22:5081–5090. [PubMed: 12077203]

Coricelli G, Critchley HD, Joffily M, O'Doherty JP, Sirigu A, Dolan RJ. Regret and its avoidance: a neuroimaging study of choice behavior. Nat. Neurosci. 2005; 8:1255–1262. [PubMed: 16116457]

Curtis CE, Lee D. Beyond working memory: the role of persistent activity in decision making. Trends Cogn. Sci. 2010; 14:216–222. [PubMed: 20381406]

Daw ND, Gershman SJ, Seymour B, Dayan P, Dolan RJ. Model-based influences on humans' choices and striatal prediction errors. Neuron. 2011; 69:1204–1215. [PubMed: 21435563]

Daw ND, Niv Y, Dayan P. Uncertainty-based competition between prefrontal and dorsolateral striatal systems for behavioral control. Nat. Neurosci. 2005; 8:1704–1711. [PubMed: 16286932]

Dayan P, Niv Y. Reinforcement learning: the good, the bad and the ugly. Curr. Opin. Neurobiol. 2008; 18:185–196. [PubMed: 18708140]

Ding L, Gold JI. Neural correlates of perceptual decision making before, during, and after decision commitment in monkey frontal eye field. Cereb. Cortex. 2011 In press.

Dorris MC, Glimcher PW. Activity in posterior parietal cortex is correlated with the relative subjective desirability of action. Neuron. 2004; 44:365–378. [PubMed: 15473973]

Ellsberg D. Risk, ambiguity, and the Savage axioms. Q. J. Econ. 1961; 61:643–669.

Erev I, Roth AE. Predicting how people play games: reinforcement learning in experimental games with unique, mixed strategy equilibria. Am. Econ. Rev. 1998; 88:848–881.

Fellows LK, Farah MJ. Ventromedial frontal cortex mediates affective shifting in humans: evidence from a reversal learning paradigm. Brain. 2003; 126:1830–1837. [PubMed: 12821528]

Feltovich R. Reinforcement-based vs. belief-based learning models in experimental asymmetric-information games. Econometrica. 2000; 68:605–641.

Funahashi S, Bruce CJ, Goldman-Rakic PS. Mnemonic coding of visual space in the monkey's dorsolateral prefrontal cortex. 1989; 61:331–349.

Furman M, Wang X-J. Similarity effect and optimal control of multiple-choice decision making. Neuron. 2008; 60:1153–1168. [PubMed: 19109918]

Georgopoulos AP, Schwartz AB, Kettner RE. Neural population coding of movement direction. Science. 1986; 233:1416–1419. [PubMed: 3749885]

Gerfen CR, Surmeier DJ. Modulation of striatal projection systems by dopamine. Annu. Rev.Neurosci. 2011; 34:441–466. [PubMed: 21469956]

Gläscher J, Daw N, Dayan P, O'Doherty JP. States versus rewards: dissociable neural prediction error signals underlying model-based and model-free reinforcement learning. Neuron. 2010; 66:585–595. [PubMed: 20510862]

Glimcher PW, Sparks DL. Movement selection in advance of action in the superior colliculus. Nature.1992; 355:542–545. [PubMed: 1741032]

Gnadt JW, Andersen RA. Memory related motor planning activity in posterior parietal cortex of macaque. Exp. Brain Res. 1988; 70:216–220. [PubMed: 3402565]

Haber SN, Fudge JL, McFarland R. Striatonigrostriatal pathways in primates form an ascending spiral from the shell to the dorsolateral striatum. J. Neurosci. 2000; 20:2369–2382. [PubMed: 10704511]

Haber SN, Knutson B. The reward circuit: linking primate anatomy and human imaging. Neuropsychopharmacology. 2010; 35:4–26. [PubMed: 19812543]

Haggard P. Human volition: towards a neuroscience of will. Nat. Rev. Neurosci. 2008; 9:934–946. [PubMed: 19020512]

Hanes DP, Schall JD. Neural control of voluntary movement initiation. Science. 1996; 274:427–430. [PubMed: 8832893]

Hassabis D, Maguire EA. Deconstructing episodic memory with construction. Trends Cogn. Sci. 2007; 11:299–306. [PubMed: 17548229]

Hayden BY, Pearson JM, Platt ML. Fictive reward signals in the anterior cingulate cortex. Science.2009; 324:948–950. [PubMed: 19443783]

Hikosaka O, Nakamura K, Nakahara H. Basal ganglia orient eyes to reward. J. Neurophysiol. 2006; 95:567–584. [PubMed: 16424448]

Hikoaka O, Nakahara H, Rand MK, Sakai K, Lu X, Nakamura K, et al. Parallel neural networks for learning sequential procedures. Trends Neurosci. 1999; 22:464–471. [PubMed: 10481194]

Hong S, Hikosaka O. The globus pallidus sends reward-related signals to the lateral habenula. Neuron.2008; 60:720–729. [PubMed: 19038227]

Hong S, Hikosaka O. Dopamine-mediated learning and switching in cortico-striatal circuit explain behavioral changes in reinforcement learning. Front. Behav. Neurosci. 2011; 5:15. [PubMed: 21472026]

Horwitz GD, Newsome WT. Target selection for saccadic eye movements: prelude activity in the superior colliculus during a direction-discrimination task. J. Neurophysiol. 2001; 86:2543–2558. [PubMed: 11698541]

Hsu M, Bhatt M, Adolphs R, Tranel D, Camerer CF. Neural systems responding to degrees of uncertainty in human decision-making. Science. 2005; 310:1680–1683. [PubMed: 16339445]

Huettel SA, Stowe CJ, Gordon EM, Warner BT, Platt ML. Neural signatures of economic preferences for risk and ambiguity. Neuron. 2006; 49:765–775. [PubMed: 16504951]

Hyman JM, Zilli EA, Paley AM, Hasselmo ME. Working memory performance correlates with prefrontal-hippocampal theta interactions but not with prefrontal neuron firing rates. Front. Integr. Neurosci. 2010; 4:2. [PubMed: 20431726]

Ito M, Doya K. Validation of decision-making models and analysis of decision variables in the rat basal ganglia. J. Neurosci. 2009; 29:9861–9874. [PubMed: 19657038]

Ito M, Doya K. Multiple representations and algorithms for reinforcement learning in the cortico-basal ganglia circuit. Curr. Opin. Neurobiol. 2011; 21:368–373. [PubMed: 21531544]

Iversen SD, Mishkin M. Perseverative interference in monkeys following selective lesions of the inferior prefrontal convexity. Exp. Brain Res. 1970; 11:376–386. [PubMed: 4993199]

Johnson A, Redish AD. Neural ensembles in CA3 transiently encode paths forward of the animal at a decision point. J. Neurosci. 2007; 27:12176–12189. [PubMed: 17989284]

Kahneman D, Tversky A. Prospect theory: an analysis of decision under risk. Econometrica. 1979; 47:263–291.

Kepecs A, Uchida N, Zariwala HA, Mainen ZF. Neural correlates, computation and behavioural impact of decision confidence. Nature. 2008; 455:227–231. [PubMed: 18690210]

Kim H, Sul JH, Huh N, Lee D, Jung MW. Role of striatum in updating values of chosen actions. J. Neurosci. 2009; 29:14701–14712. [PubMed: 19940165]

Kim S, Hwang J, Lee D. Prefrontal coding of temporally discounted values during intertemporal choice. Neuron. 2008; 59:161–172. [PubMed: 18614037]

Kim S, Lee D. Prefrontal cortex and impulsive decision making. Biol. Psychiatry. 2011; 69:1140–1146. [PubMed: 20728878]

Kim Y, Huh N, Lee H, Baeg E, Lee D, Jung MW. Encoding of action history in the rat ventral striatum. J. Neurophysiol. 2007; 98:3548–3556. [PubMed: 17942629]

Knowlton BJ, Mangels JA, Squire LR. A neostriatal habit learning system in humans. Science. 1996; 273:1399–1402. [PubMed: 8703077]

Krajbich I, Armel C, Rangel A. Visual fixations and the computation and comparison of value in simple choice. Nat. Neurosci. 2010; 13:1292–1298. [PubMed: 20835253]

Laming, DRJ. Information Theory of Choice-Reaction Times. London: Academic Press; 1968. Lau B, Glimcher PW. Value representations in the primate striatum during matching behavior.Neuron. 2008; 58:451–463. [PubMed: 18466754]

Lee D. Game theory and neural basis of social decision making. Nat. Neurosci. 2008; 11:404–409. [PubMed: 18368047]

Lee D, Conroy ML, McGreevy BP, Barraclough DJ. Reinforcement learning and decision making in monkeys during a competitive game. Cogn. Brain Res. 2004; 22:45–58.

Lee D, McGreevy BP, Barraclough DJ. Learning and decision making in monkeys during a rock-paper-scissors game. Cogn. Brain Res. 2005; 25:416–430.

Levey AI, Hersch SM, Rye DB, Sunahara RK, Niznik HB, et al. Localization of D1 and D2 dopamine receptors in brain with subtype-specific antibodies. Proc. Natl. Acad. Sci. USA. 1993; 90:8861–8865. [PubMed: 8415621]

Lewis DA, Melchitzky DS, Sesack SR, Whitehead RE, Auh S, Sampson A. Dopamine transporter immunoreactivity in monkey cerebral cortex: regional, laminar, and ultrastructural localization. J. Compar. Neurol. 2001; 432:119–136.

Litt A, Plassmann H, Shiv B, Rangel A. Dissociating valuation and saliency signals during decision making. Cereb. Cortex. 2011; 21:95–102. [PubMed: 20444840]

Lo C-C, Wang X-J. Cortico-basal ganglia circuit mechanism for a decision threshold in reaction time tasks. Nat. Neurosci. 2006; 9:956–963. [PubMed: 16767089]

Lohrenz T, McCabe K, Camerer CF, Montague PR. Neural signature of fictive learning signals in a sequential investment task. Proc. Natl. Acad. Sci. U.S.A. 2007; 104:9493–9498. [PubMed: 17519340]

Louie K, Glimcher PW. Separating value from choice: delay discounting activity in the lateral intrapreital area. J. Neurosci. 2010; 30:5498–5507. [PubMed: 20410103]

Luhmann CC, Chun MM, Yi D-J, Lee D, Wang X-J. Neural dissociation of delay and uncertainty in intertemporal choice. J. Neurosci. 2008; 28:14459–14466. [PubMed: 19118180]

Matsumoto M, Hikosaka O. Lateral habenula as a source of negative reward signals in dopamine neurons. Nature. 2007; 447:1111–1115. [PubMed: 17522629]

Matsumoto M, Matsumoto K, Abe H, Tanaka K. Medial prefrontal cell activity signaling prediction errors of action values. Nat. Neurosci. 2007; 10:647–656. [PubMed: 17450137]

Maunsell JHR. Neuronal representations of cognitive state: reward or attention? Trends Cogn. Sci.2004; 8:261–265. [PubMed: 15165551]

Montague PR, Dayan P, Sejnowski TJ. A framework for mesencephalic dopamine systems based on predictive Hebbian learning. J. Neurosci. 1996; 16:1936–1947. [PubMed: 8774460]

Mookherjee D, Sopher B. Learning behavior in an experimental matching pennies game. Games Econ. Behav. 1994; 7:62–91.

Mookherjee D, Sopher B. Learning and decision costs in experimental constant sum games. Games Econ. Behav. 1997; 19:97–132.

Murray EA, O'Doherty JP, Schoenbaum G. What we know and do not know about the functions of the orbitofrontal cortex after 20 years of cross-species studies. J. Neurosci. 2007; 27:8166–8169. [PubMed: 17670960]

Nachev P, Kennard C, Husain M. Functional role of the supplementary and pre-supplementary motor areas. Nat. Rev. Neurosci. 2008; 9:856–869. [PubMed: 18843271]

Okano K, Tanji J. Neuronal activities in the primate motor fields of the agranular frontal cortex preceding visually triggered and self-paced movement. Exp. Brain Res. 1987; 66:155–166. [PubMed: 3582529]

O'Neill M, Schultz W. Coding of reward risk by orbitofrontal neurons is mostly distinct from coding of reward value. Neuron. 2010; 68:789–800. [PubMed: 21092866]

Oyama K, Hernádi I, Iijima T, Tsutsui K-I. Reward prediction error coding in dorsal striatal neurons. J. Neurosci. 2010; 30:11447–11457. [PubMed: 20739566]

Padoa-Schioppa C. Neurobiology of economic choice: a good-based model. Annu. Rev. Neurosci.2011; 34:333–359. [PubMed: 21456961]

Padoa-Schioppa C, Assad JA. Neurons in the orbitofrontal cortex encode economic value. Nature.2006; 441:223–226. [PubMed: 16633341]

Pan W-X, Schmidt R, Wickens JR, Hyland BI. Dopamine cells respond to predicted events during classical conditioning: evidence for eligibility traces in the reward-learning network. J. Neurosci. 2005; 25:6235–6242. [PubMed: 15987953]

Pan X, Sawa K, Tsuda I, Tsukada M, Sakagami M. Reward prediction based on stimulus categorization in primate lateral prefrontal cortex. Nat. Neurosci. 2008; 11:703–712. [PubMed: 18500338]

Pastor-Bernier A, Cisek P. Neural correlates of biased competition in premotor cortex. J. Neurosci.2011; 31:7083–7088. [PubMed: 21562270]

Platt ML, Glimcher PW. Neural correlates of decision variables in parietal cortex. Nature. 1999; 400:233–238. [PubMed: 10421364]

Premack D, Woodruff G. Does the chimpanzee have a theory of mind? Behav. Brain Sci. 1978; 4:515–526.

Reynolds JNJ, Hyland BI, Wickens JR. A cellular mechanism of reward-related learning. Nature. 2001;

413:67–70. [PubMed: 11544526]

Roesch MR, Singh T, Brown PL, Mullins SE, Schoenbaum G. Ventral striatal neurons encode the value of the chosen action in rats deciding between differently delayed or sized rewards. J. Neurosci. 2009; 29:13365–13376. [PubMed: 19846724]

Roese, NJ.; Olson, JM. What might have been: the social psychology of counterfactual thinking. New York: Psychology Press; 1995.

Roitman JD, Shadlen MN. Response of neurons in the lateral intraparietal area during a combined visual discrimination reaction time task. J. Neurosci. 2002; 22:9475–9489. [PubMed: 12417672]

Samejima K, Ueda Y, Doya K, Kimura M. Representation of action-specific reward values in the striatum. Science. 2005; 310:1337–1340. [PubMed: 16311337]

Schacter DL, Addis DR, Buckner RL. Remembering the past to imagine the future: the prospective brain. Nat. Rev. Neurosci. 2007; 8:657–661. [PubMed: 17700624]

Schall JD. Neuronal activity related to visually guided saccadic eye movements in the supplementary motor area of rhesus monkeys. J. Neurophysiol. 1991; 66:530–558. [PubMed: 1774585]

Schiller PH, Stryker M. Single-unit recording and stimulation in superior colliculus of the alert rhesus monkey. J. Neurophysiol. 1972; 35:915–924. [PubMed: 4631839]

Schlag J, Schlag-Rey M. Evidence for a supplementary eye field. J. Neurophysiol. 1987; 57:179–200. [PubMed: 3559671]

Schoenbaum G, Nugent SL, Saddoris MP, Setlow B. Orbitofrontal lesions in rats impair reversal but not acquisition of go, no-go odor discriminations. Neuroreport. 2002; 13:885–890. [PubMed: 11997707]

Schultz W. Behavioral theories and the neurophysiology of reward. Annu. Rev. Psychol. 2006; 57:87–115. [PubMed: 16318590]

Schultz W, Tremblay L, Hollerman JR. Reward processing in primate orbitofrontal cortex and basal ganglia. Cereb. Cortex. 2000; 10:272–284. [PubMed: 10731222]

Seo H, Barraclough DJ, Lee D. Lateral intraparietal cortex and reinforcement learning during a mixed-strategy game. J. Neurosci. 2009; 29:7278–7289. [PubMed: 19494150]

Seo H, Lee D. Temporal filtering of reward signals in the dorsal anterior cingulate cortex during a mixed-strategy game. J. Neurosci. 2007; 27:8366–8377. [PubMed: 17670983]

Seo H, Lee D. Cortical mechanisms for reinforcement learning in competitive games. Phil. Trans. R. Soc. B. 2008; 363:3845–3857. [PubMed: 18829430]

Seo H, Lee D. Behavioral and neural changes after gains and losses of conditioned reinforcers. J. Neurosci. 2009; 29:3627–3641. [PubMed: 19295166]

Serences JT. Value-based modulations in human visual cortex. Neuron. 2008; 60:1169–1181. [PubMed: 19109919]

Shadlen MN, Newsome WT. Neural basis of a perceptual decision in the parietal cortex of the rhesus monkey. J. Neurophysiol. 2001; 86:1916–1936. [PubMed: 11600651]

Shen W, Flajolet M, Greengard P, Surmeier DJ. Dichotomous dopaminergic control of striatal synaptic plasticity. Science. 2008; 321:848–851. [PubMed: 18687967]

Simon DA, Daw ND. Neural correlates of forward planning in a spatial decision task in humans. J. Neurosci. 2011; 31:5526–5539. [PubMed: 21471389]

Shuler MG, Bear MF. Reward timing in the primary visual cortex. Science. 2006; 311:1606–1609. [PubMed: 16543459]

Sirota A, Montgomery S, Fujisawa S, Isomura Y, Zugaro M, Buzsáki G. Entrainment of neocortical neurons and gamma oscillations by the hippocampal theta rhythm. Neuron. 2008; 60:683–697. [PubMed: 19038224]

Smith PL, Ratcliff R. Psychology and neurobiology of simple decisions. Trends Neurosci. 2004; 27:161–168. [PubMed: 15036882]

Smyrnis N, Taira M, Ashe J, Georgopoulos AP. Motor cortical activity in a memorized delay task.Exp. Brain Res. 1992; 92:139–151. [PubMed: 1486948]

So NY, Stuphorn V. Supplementary eye field encodes option and action value for saccades with variable reward. J. Neurophysiol. 2010; 104:2634–2653. [PubMed: 20739596]

Softky WR, Koch C. The highly irregular firing of cortical cells is inconsistent with temporal integration of random EPSPs. J. Neurosci. 1993; 13:334–350. [PubMed: 8423479]

Soltani A, Lee D, Wang X-J. Neural mechanism for stochastic behaviour during a competitive game. Neural Netw. 2006; 19:1075–1090. [PubMed: 17015181]

Soltani A, Wang X-J. A biophysically based neural model of matching law behavior: melioration by stochastic synapses. J. Neurosci. 2006; 26:3731–3744. [PubMed: 16597727]

Sohn J-W, Lee D. Order-dependent modulation of directional signals in the supplementary and presupplementary motor areas. J. Neurosci. 2007; 27:13655–13666. [PubMed: 18077677]

Soon CS, Brass M, Heinze H-J, Haynes J-D. Unconscious determinants of free decisions in the human brain. Nat. Neurosci. 2008; 11:543–545. [PubMed: 18408715]

Sugrue LP, Corrado GS, Newsome WT. Matching behavior and the representation of value in the parietal cortex. Science. 2004; 304:1782–1787. [PubMed: 15205529]

Sul JH, Kim H, Huh N, Lee D, Jung MW. Distinct roles of rodent orbitofrontal and medial prefrontal cortex in decision making. Neuron. 2010; 66:449–460. [PubMed: 20471357]

Sul JH, Jo S, Lee D, Jung MW. Role of rodent secondary motor cortex in value-based action selection. Nat. Neurosci. 2011 In press.

Sutton, RS.; Barto, AG. Reinforcement Learning: An Introduction. Cambridge, MA: MIT Press; 1998.

Tanji J, Kurata K. Contrasting neuronal activity in supplementary and precentral motor cortex of monkeys. I. Responses to instructions determining motor responses to forthcoming signals of different modalities. J. Neurophysiol. 1985; 53:129–141. [PubMed: 3973654]

Tolman EC. Cognitive maps in rats and men. Psychol. Rev. 1948; 55:189–208. [PubMed: 18870876]

Tremblay L, Schultz W. Relative reward preference in primate orbitofrontal cortex. Nature. 1999;398:704–708. [PubMed: 10227292]

Usher M, McClelland J. On the time course of perceptual choice: the leaky, competing accumulator model. Psychol. Rev. 2001; 108:550–592. [PubMed: 11488378]

van der Meer MAA, Redish AD. Ventral striatum: a critical look at models of learning and evaluation. Curr. Opin. Neurobiol. 2011; 21:387–392. [PubMed: 21420853]

Vickery TJ, Chun MM, Lee D. Ubiquity and specificity of reinforcement signals throughout the human brain. Neuron. 2011. In press.

von Neumann, J. Morgenstern, O. Theory of Games and Economic Behavior. Princeton: Princeton Univ. Press; 1944.

Wallis JD, Kennerley SW. Heterogeneous reward signals in prefrontal cortex. Curr. Opin. Neurobiol.2010; 20:191–198. [PubMed: 20303739]

Wallis JD, Miller EK. Neuronal activity in primate dorsolateral and orbital prefrontal cortex during performance of a reward preference task. Eur. J. Neurosci. 2003; 8:2069–2081. [PubMed: 14622240]

Walton ME, Behrens TE, Buckley MJ, Rudebeck PH, Rushworth MF. Separable learning systems in the macaque brain and the role of orbitofrontal cortex in contingent learning. Neuron. 2010; 65:927–939. [PubMed: 20346766]

Wang X-J. Synaptic reverberation underlying mnemonic persistent activity. Trends Neurosci. 2001; 24:455–463. [PubMed: 11476885]

Wang X-J. Probabilistic decision making by slow reverberation in cortical circuits. Neuron. 2002; 36:955–968. [PubMed: 12467598]

Wang X-J. Decision making in recurrent neuronal circuits. Neuron. 2008; 60:215–234. [PubMed: 18957215]

Weinrich M, Wise SP. The premotor cortex of the monkey. J. Neurosci. 1982; 2:1329–1345. [PubMed: 7119878]

Womelsdorf T, Vinck M, Leung LS, Everling S. Selective theta-synchronization of choice-relevant information subserves goal-directed behavior. Front. Hum. Neurosci. 2010; 4:210. [PubMed: 21119780]

Wunderlich K, Rangel A, O'Doherty JP. Neural computations underlying action-based decision making in the human brain. Proc. Natl. Acad. Sci. U.S.A. 2009; 106:17199–17204. [PubMed: 19805082]

Wurtz RH, Goldberg ME. Activity of superior colliculus in behaving monkey. 3. Cells discharging before eye movements. J. Neurophysiol. 1972; 35:575–586. [PubMed: 4624741]

Yin HH, Knowlton BJ. The role of the basal ganglia in habit formation. Nat. Rev. Neurosci. 2006; 7:464–476. [PubMed: 16715055]

Yu AJ, Dayan P. Uncertainty, neuromodulation, and attention. Neuron. 2005; 46:681–692. [PubMed: 15944135]

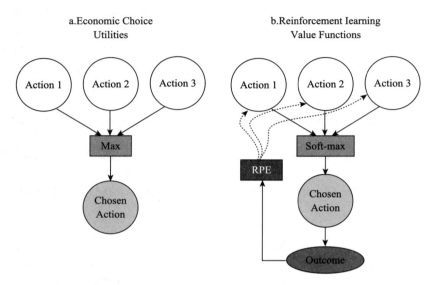

Figure 1　Economic and reinforcement learning theories of decision making

(a) In economic theories, decision making corresponds to selecting an action with the maximum utility. (b) In reinforcement learning, actions are chosen probabilistically (i.e., softmax) on the basis of their value functions. In addition, value functions are updated on the basis of the outcome (reward or penalty) resulting from the action chosen by the animal. RPE, reward prediction error.

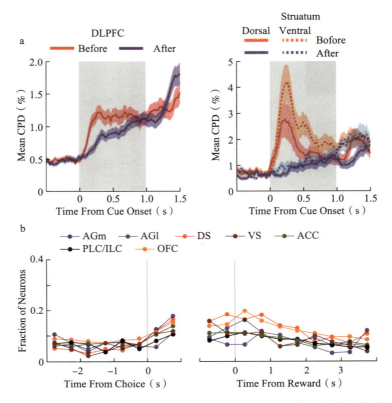

Figure 2   Time course of signals related to different state value functions during decision making

(a) Signals related to the state value functions before (red) and after (blue) decision making in the dorsolateral prefrontal cortex (DLPFC; Kim et al. 2008, Kim & Lee 2011) and striatum (Cai et al. 2011) during an intertemporal choice task. These two state value functions correspond to the average of the action value functions for two options and the chosen value, respectively. During these studies, monkeys chose between a small immediate reward and a large delayed reward, and the magnitude of neural signals related to different value functions were estimated by the coefficient of partial determination (CPD). Lines correspond to the mean CPD for all the neurons recorded in each brain area with the shaded area corresponding to the standard error of the mean. (b) Proportion of neurons carrying chosen value signals in the rodent lateral (AGl) and medial (AGm) agranular cortex, corresponding to the primary and secondary motor cortex, respectively, dorsal (DS) and ventral (VS) striatum, anterior cingulate cortex (ACC), prelimbic (PLC)/infralimbic (ILC) cortex, and orbitofrontal cortex (OFC). During these studies (Kim et al. 2009, Sul et al. 2010, 2011), the rats performed a dynamic foraging task. Large symbols indicate that the proportions are significantly (p<0.05) above the chance level.

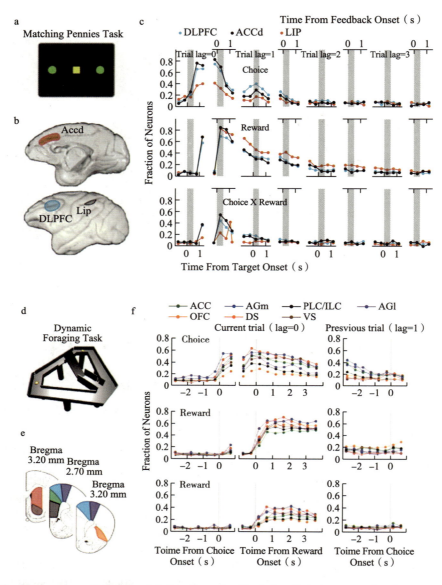

Figure 3 **Time course of signals related to the animal's choice, its outcome, and action-outcome conjunction in multiple brain areas of primates and rodents**

(a) Spatial layout of the choice targets during a matching pennies task used in single-neuron recording experiments in monkeys. (b) Brain regions tested during the studies on monkeys (Barraclough et al. 2004, Seo & Lee 2007, Seo et al. 2009). ACCd, dorsal anterior cingulate cortex; DLPFC, dorsolateral prefrontal cortex; LIP, lateral intraparietal cortex. (c) Fraction of neurons significantly modulating their activity according to the animal's choice (top), its outcome (middle), and choice-outcome conjunction (bottom) during the current (trial lag =0) and 3 previous trials (trial lags =1~3). (d) Modified T-maze used in a rodent dynamic foraging task. (e) Anatomical areas tested in single-neuron recording experiments in rodents (Kim et al. 2009, Sul et al. 2010, 2011). Same abbreviations as in Figure 2b. (f) Fraction of neurons significantly modulating their activity according to the animal's choice (top), its outcome (middle), and choice-outcome conjunction (bottom) during the current (lag =0) and previous trials (lag =1). Large symbols indicate that the proportions are significantly (p<0.05) above the chance level.

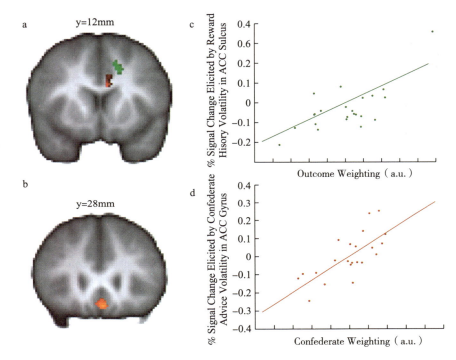

Figure 4   Areas in the human brain involved in updating model-free and model-based value functions (Behrens et al. 2008)

(a) Regions in which the activity is correlated with the volatility in estimating the value functions based on reward history (green) and social information (red). (b) Activity in the ventromedital prefrontal cortex was correlated with the value functions regardless of whether they were estimated from reward history or social information. (c) Subjects more strongly influenced by reward history (ordinate) tended to show greater signal change in the anterior cingulate cortex in association with reward history (abscissa; green region in a). (d) Subjects more strongly influenced by social information (ordinate) showed greater signal changes in the anterior cingulate cortex in association with social information (abnscissa; red region in a).

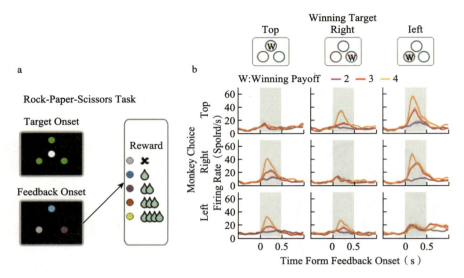

Figure 5   Neuronal activity related to hypothetical outcomes in the primate orbitofrontal cortex

(a) Rock-paper-sciossors task used for single-neuron recording studies in monkeys (Abe &Lee 2011). (b) An example neuron recorded in the orbitofrontal cortex that modulated its activity according to the magnitude of reward that was available from the unchosen winning target (indicated by 'W' in the top panels). The spike density function of this neuron was estimated separately according to the position of the winning target (columns), the position of the target chosen by the animal (rows), and the magnitude of the reward available from the winning target (colors).

Social Control of the Brain

Russell D. Fernald

Biology Department, Stanford University, Stanford, California 94305;
Email: rfernald@standord.edu

Key Words

social behavior, teleost fish, control of the brain, *Astatotilapia burtoni*, immediate early genes, social behavior network

Abstract

In the course of evolution, social behavior has been a strikingly potent selective force in shaping brains to control action. Physiological, cellular, and molecular processes reflect this evolutionary force, particularly in the regulation of reproductive behavior and its neural circuitry. Typically, experimental analysis is directed at how the brain controls behavior, but the brain is also changed by behavior over evolution, during development, and through its ongoing function. Understanding how the brain is influenced by behavior offers unusual experimental challenges. General principles governing the social regulation of the brain are most evident in the control of reproductive behavior. This is most likely because reproduction is arguably the most important event in an animal's life and has been a powerful and essential selective force over evolution. Here I describe the mechanisms through which behavior changes the brain in the service of reproduction using a teleostfish model system.

INTRODUCTION

In many species, social context modulates interactions among animals, which then tune their behavior accordingly. Social interactions clearly influence the brain and circulating hormonal levels, but how does the social environment regulate the physiological, cellular, and molecular processes of the brain? Elucidating this connection is critically important because social interactions, especially those related to reproduction, are essential for the evolutionary success of all species. Social regulation of vertebrate reproduction offers a unique opportunity to understand how behavior influences the brain for two reasons. First, reproductive behaviors are often stereotypic, making them relatively easy to observe and quantify. Second, central control of reproduction is lodged in the brain-pituitary-gonadal (BPG) axis with the hypophysiotropic gonadotropin-releasing hormone (GnRH1) neurons as its final output path. Thus the neuronal circuit that produces the GnRH1 signaling peptide is the key locus for integrating the external and internal factors that control reproduction.

Understanding behavioral regulation of the brain requires an animal model in which (*a*) social interchange is essential for reproductive success; (*b*) animals can be studied in a seminatural context; (*c*) key molecular, cellular, and physiological processes are accessible; and (*d*) behavior and physiology of both individuals and groups of animals can be readily analyzed.

CHOICE OF FISH SPECIES FOR STUDY

In considering experimental systems for studies of the brain that focus on social behavior, it is important to select a system that allows a reasonable replication of the native social and environmental habitat. Although primates are an obvious choice of an experimental animal species that is closely related to humans, fish offer an interesting chance to understand how the brain controls behavior for several reasons. Fish are the largest vertebrate group, with more species than all other vertebrates combined, occupying ecological niches from freezing water to hot soda springs. Moreover, they represent more than 400 million years of vertebrate evolution, and their taxonomic dimensions exceed the distance between frogs and humans (Romer 1959). The variety of sensory modalities required for aquatic life, aside from vision, olfaction, taste, and hearing, include mechanosensory systems (e.g., lateral line), external taste buds, and numerous electroreceptor systems that have led to extensive variation in brain structures. Moreover, it is not an exaggeration to say that essentially every known kind of social system has evolved in fish species from monogamy to harems to sexchanging animals. Given the wide variety of ecological niches exploited, appropriate species comparisons among fish offer the potential for understanding the relative contributions of environmental and social factors to brain evolution. Even more appealing, fish offer the promise of discovering mechanisms through which the brain controls social activities in ecologically meaningful contexts (Bshary et al. 2002).

Studies of animal behavior fall roughly into two groups that have quite different goals (Shettleworth 1993, Kamil 1998). Some investigators seek to identify humanlike skills in other species, a thread derived broadly from a search for general cognitive abilities (Hodos & Campbell 1969). This approach has been termed the anthropocentric perspective and may, in fact, be an indirect descendent of Aristotle's *scala naturae*, a view reinforced by comparative brain research. Another camp treats cognition as strictly biological, best elucidated through observations and exper iments directed at understanding adaptive modifications crafted by evolution to regulate social interactions (Robinson et al. 2008). Such adaptations typically produce cognitive skills that animals need and use in nature to navigate their social worlds. There are several ways to parse these general classes of investigation in other terms, for example, understanding aspects of human behavior using animal models versus understanding animals in meaningful biological contexts for their own sake.

Fish Have Significant Cognitive Abilities

Cognitive skills in fish species offer an opportunity to study basic skills in a tractable animal that can be kept in a seminatural environment where the whole social system can be well mimicked. The following is a brief summary of the range of cognitive capacities that could be tapped in a fish model system.

Numerosity. Counting is an important form of abstraction that led to the development of mathematics. Since the discovery in 1904 that Hans, the famous Russian trotting horse, could not really count or do mathematical calculations (or read either German or musical notation) but rather responded to his owner's cues (Candland 1993), scientists have used more rigorous methods to investigate whether and how animals can keep track of quantities. Can some fish species keep track of amounts, and when might they need to count something? Numerical abilities in a fish species appear to be handy when choosing which group to join in times of danger. "Safety in numbers" predicts that fish should select the larger group when given a choice. In fact, threes-pined sticklebacks (*Gasterosteus aculeatus*), when threatened with a simulated aerial predator, chose to join larger shoals when equidistant from small shoals but made a trade-off between distance to the shoal and its numbers when given that option (Tegeder & Krause 1995). Because the number of fish in larger shoals covaries with several physical attributes of the group (e.g., area, contour, density), sophisticated experiments are required to show which feature of shoal size is actually being discriminated. These types of experiments can be done when the stimuli are presented in succession, rather than in aggregation. In one case, item-by-item presentation showed that the mosquitofish (*Gambusi holbrooki*) can distinguish small (three versus two) and large (eight versus four) shoals independent of other factors (Agrillo et al. 2009). Such numerical skills likely evolved in species in which antipredator benefits of group assemblies have evolved. We can predict that similar capacities will be found in other fish species that aggregate during predation threats.

Recognition of individuals. When an animal encounters another individual, it likely compares sensory information with a template to categorize that individual as conspecific, heterospecific, threatening, or nonthreatening. It may also recognize the individual. Recognition of individuals is a prerequisite for many behavioral interactions and has been demonstrated in a wide range of taxa. In particular, individual recognition is essential for kin recognition and hence required for any kin-selected behavior.

Individuals can be identified using multiple sensory modalities depending on the species and ecological circumstances. For fish, novel sensory systems, including electroreception, pressure reception, and polarization vision, may be important for recognizing individuals.

How many individuals would a fish need to know in some way? Sampling natural populations, Ward et al. (2005) estimated that three-spined sticklebacks living in ~20 m of a channel in a freshwater lake could meet with 900 conspecifics regularly and showed that direct experience and social cues led to relatively quick learning about the categories of individuals. However, the total number of conspecifics that stickleback's remember or whether they recognize individuals was not established (Ward et al. 2005). Bshary et al. (2002) report that some cleaner fish species can probably distinguish ~100 individuals by observing their behavior toward clients.

One of the most important realms for individual recognition is in mating pairs of fish. Noble & Curtis (1939) first described recognition between mated pairs in a cichlid fish with biparental care of the young, *Hemichromis bimaculatus.* Much later, Fricke (1973) performed elegant field experiments on fish (*Amphiprion bicinctus*) that live together among the tentacles of anemones, showing that mated individuals recognized one another on the basis of individual body color patterns rather than mutually recognizing the anemone. Moreover, after arbitrarily pairing a male and female, anemone fish could learn the identity of the new partner in 24h and could also recognize that individual after 10 days of isolation. Given the high rates of predation on *A. bicinctus*, the ability to identify a new partner rapidly would allow animals to continue reproducing despite the loss of a familiar partner. In cichlids living in clear water habitats, individuals used primarily visual cues to recognize other individuals (Noble & Curtis 1939, Fricke 1973, Balshine-Earn & Lotem 1998).

Differences in mating systems and ecology may result in the evolution of different recognition abilities among species. For example, guppies can recognize conspecifics individually as well as distinguish among groups of con specifics on the basis of cues about resource use and habitat (Ward et al. 2009). In contrast, sticklebacks do not recognize individuals in a social context despite prior interactions, though they do have general recognition capacities based on resource use (Ward et al. 2009) that are considered "familiarity" rather than individual recognition. This cognitive skill seems to increase the chances of grouping together and improving foraging (Ward et al. 2005, 2007). More recently, Ward et al. (2009) suggested that sticklebacks may rely on habitat information, specifically odors, to identify particular groups of individuals.

A second important function for individual recognition is to reduce the costs of contesting resources. This has been demonstrated in sea trout (*Salmo trutta*) in which familiar ity with conspecifics enhanced growth (Höjesjö et al. 1998). In a twist on this skill, European minnows (*Phoxinus phoxinus*) recognized and preferred to group together in a shoal with poor competitors, although how they recognized poor competitors is unknown (Metcalfe & Thomson 1995). Evidence also indicates that kin recognition is widespread among fish species, particularly those that school (Quinn & Hara 1986, Havre & FitzGerald 1988, Olsen 1989), and that this skill requires individual recognition.

As in all laboratory experiments, individual recognition may be a consequence of artificially extended interactions among individuals. However, most data cited here were from field experiments and hence provide more convincing examples of individual recognition in fish (see also discussion below on transitive inference data: Assessing Male Fighting Abilities).

Deception

Do fish communicate honestly or can they deceive? Deception is a fundamental issue in animal communication (Maynard Smith & Harper 2003), so can fish deceive? Several authors beginning with Byrne & Whiten (1988) have distinguished between functional and intentional deception. Functional deception is widespread and includes many examples that do not require cognitive skill (e.g., mimicry, crypsis, although see Chittka & Osorio 2007), whereas intentional deception implies behavior based on intentional states (e.g., beliefs, desires). As cogently discussed by Shettleworth (1998), translating anthropocentric concepts into predictions and experiments that are testable is a serious challenge. Suitable experiments have been described for chimpanzees (Hare et al. 2000, Hare & Tomasello 2001), scrub jays (Dally et al. 2006), and ravens (Bugnyar & Heinrich 2006) with a test to demonstrate an ultimate fitness benefit of deception to the deceiver and the cost to the deceived (Hauser 1997).

Perhaps the cleanest example of deception has been described for the cleaner wrasse, *Labroides dimidiatus*. This marine cleaner fish removes ectoparasites from visiting reef fish clients. This relationship is mutual because the client gets cleaned of ectoparasites and the cleaner gets a meal, the parasite, delivered. But a problem exists: Cleaner fish prefer the client's tasty layer of mucus to its ectoparasites (Grutter & Bshary 2003). Because cleaner fish service up to 2,000 clients every day (Grutter 1997) and many of these encounters happen in the presence of observing bystanders, including future clients, does the presence of bystanders alter the cleaner's behavior? Pinto et al. (2011) used two species of client fish to ask whether being watched matters. Cleaners were tested on clients, and the introduction of a bystander led to an immediate increase in cooperation by the cleaner fish: The cleaners spent more time removing ectoparasites than eating mucus when being watched. This brief discussion of the social skills of teleost fish reveals why they are such useful model systems for understanding the neural bases of social behavior.

SOCIAL CONTROL OF THE BRAIN: WHY STUDY THE CICHLID FISH, ASTATOTILAPIA BURTONI?

Astatotilapia (Haplochromis) burtoni, the African cichlid fish model system we developed, has numerous social skills and, perhaps most importantly, social interactions related to the fact that social dominance tightly controls reproductive physiology. By manipulating the social system we can essentially turn on or off an animal's reproductive competence, mimicking natural changes to identify key regulatory processes. This fish model system offers several important advantages for understanding how social behavior changes the brain: (*a*) The social system of this fish can be easily replicated in the laboratory; (*b*) male status is signaled by obvious rapid color changes, making it easy to detect and quantify; (*c*) GnRH1 neurons are directly regulated by male social status and hence are causally related to behavior; (*d*) *A. burtoni* offers easy access to the brain, allowing sampling of cells and molecules of interest; and (*e*) the *A. burtoni* genome has been sequenced, enabling a class of experiments not previously possible (http://www.genome.gov/11007951). Taken together, these attributes make *A. burtoni* uniquely useful for studying social regulation of reproduction. Using this system, we have manipulated the social situation to produce phenotypic change in a variety of ways and have measured relevant molecular, neural, and hormonal systems.

    A. burtoni males exist as one of two socially controlled, reversible phenotypes: reproductively competent dominant (D) males and reproductively incompetent nondominant (ND) males (see **Figure 1**).

    D males display bright coloration, aggressively defend territories, and court females, whereas ND males display dull gray coloration, mimic females, and limit their behavior to schooling and fleeing. The major differences between these two phenotypic states of male *A. burtoni* can be summarized as follows.

Social signals regulate GnRH1 cell size, peptide level, GnRH1, and gonadotropin-releasing hormone (GnRH) mRNA receptor levels. When a D male is moved into a social system with larger (>5% in length) D males, it abruptly loses its color (<1min) and joins other ND males and females in a school. Its GnRH1 containing neurons in the preoptic area (POA) shrink to one eighth their volume and produce less GnRH1 mRNA and peptide, causing hypogonadism and loss of reproductive competence (~2 weeks) (Davis & Fernald 1990, Francis et al. 1993, White et al. 2002). Similarly, androgen, estrogen, and GnRH receptor mRNA expression levels depend on social status (Au et al. 2006, Burmeister et al. 2007, Harbott et al. 2007) as do electrical properties of the GnRH1 neurons (Greenwood & Fernald 2004). Conversely, when an ND male is moved into a social system in which it is larger than other males, these changes are reversed. The male quickly (<20s) assumes a bright territorial coloration, engages in aggressive encounters, and acquires a territory. We know that ascending animals rapidly (~20min) activate the molecular processes related to the subsequent neural transformations (Burmeister et al. 2005).

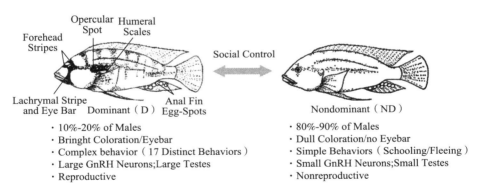

Figure 1
Socially regulated differences between dominant (D) and nondominant (ND) males. Dominant males are brightly colored yellow or blue, express complex behaviors, are reproductively competent, and comprise a small fraction of the population, whereas ND males have dull coloration, express only a few behaviors, and are reproductively incompetent.

HOW DOES SOCIAL INFORMATION INFLUENCE BEHAVIOR AND THE BRAIN?

Social living requires sophisticated cognitive abilities because successful social individuals must collect information to guide their future behavior. We have tested the abilities of *A. burtoni* to understand what they know and how they know it about their social world, and more importantly, how this knowledge affects their brain and subsequent behavior. Knowledge of how conspecifics act in key social situations would be a useful guide to predict future behavior. Consequently, in social settings, individuals gather and use information about others' behavior (Brown & Laland 2003) to regulate their own behavior. But how do animals collect social information (McGregor 2005, Brown et al. 2006), and what do they learn as a result? Here we describe several experiments in which we tested individuals' responses after social observations.

Female Mate Choice

People often assume that females choose males displaying the most exaggerated sexual traits—whether behavioral, morphological, or material, such as food and shelter. However, other factors may also contribute importantly to female mate-choice decisions. A wide range of subtle and complex external factors has been shown to influence female mate choice, suggesting sophisticated integration of cues by females. Less well understood are the physiological substrates that are likely also crucial for successful female reproductive choice. Choice of a mate by a female is very important and is dictated by a variety of factors. For example, genetic and epigenetic factors, circulating hormones, and learned behavior can contribute to a female's final mate choice (see review by Argiolas 1999). In addition to their role in solicitation behavior, hormones are intimately involved in establishing a preference for conspecifics of the opposite sex during ontogeny (e.g., Adkins-Regan 1998).

Although manipulating hormone levels experimentally can be informative, we took advantage of the naturally fluctuating levels of hormones in the female reproductive cycle to discover whether and how decisions to affiliate with males of different reproductive quality change as a function of the female's stage in the reproductive cycle. We showed in *A. burtoni* that gravid females preferentially associated with D males, whereas nongravid females showed no preference and that preference did not depend on male size (Clement et al. 2005). These data suggest that females use a hierarchy of internal and external cues in deciding on a mate.

Females who choose an inappropriate mate may pay a high cost in lost reproductive opportunity. But which cues should they use to select a mate that might be a successful reproductive partner? In general, females should be choosier about prospective mates than should males because bad mating decisions typically result in higher costs for females than for males (Trivers 1972). Consequently, studies measuring female assessment of male characteristics have produced conceptual, theoretical, and empirical hypotheses suggesting key factors that may mediate female mate choice (for reviews, see Ryan 1980, Andersson & Simmons 2006). Perhaps unsurprisingly, some evidence indicates that females may use information about male–male social interactions in their matechoice decisions (Otter et al. 1999, Doutrelant & McGregor 2000, Mennill et al. 2002, Earley & Dugatkin 2005), but little is known about how the brain responds to this kind of information. Specifically, what are the consequences for mate choice after collectingkey information? Male–male social interactions may affect female mate choice because information about a potential mate triggers changes in female reproductive physiology.

Using immediate early gene (IEG) expression as a proxy for brain activity, we asked whether social information about a preferred male influenced neural activity in females (Desjardins et al. 2010). After a gravid female, *A. burtoni* chose between two socially and physically equivalent males, we staged a fight between these males. Her preferred male either won or lost. We then measured IEG expression levels in several brain nuclei including those in the vertebrate social behavior network (SBN), a collection of brain nuclei known to be important for social behavior (Newman 1999, Goodson 2005). When the female saw her preferred male win a fight, SBN brain nuclei associated with reproduction were activated, but when she saw her preferred male lose a fight, the lateral septum, a nucleus associated with anxiety, was activated instead. Thus social information alone, independent of actual social interactions, activated specific brain regions that differ significantly depending on what the female sees. These effects are seen only in gravid females, consistent with our earlier data showing that hormones are important for female mate choices.

These experiments, assessing the role of matechoice information in the brain using a paradigm of successive presentations of mate information, suggest a method for identifying the neural consequences of social information on animals using IEG activation (Desjardins et al. 2010). IEGs are the earliest genomic responses to stimuli and require no prior activation by any other gene (Clayton 2000). This response, however, is the tip of the genetic activation iceberg because the total number of genes that comprise one neuron's inducible genomic response has been estimated from tens to hundreds (Nedivi et al. 1993); likewise, the set of rapidly inducible genes in any particular cell in the brain may be still larger (Miczek 1977), meaning that the IEG expression measured here is likely only a tiny fraction of the total gene expression. Nonetheless, this glimpse of the genetic response to social information shows not only that females attend to the information received from watching males interact but also that such information has dramatic effects on their brains in key nuclei rather than producing widespread, general arousal, generating the genetic substrate for subsequent behavioral responses.

Assessing Male Fighting Abilities

A second example of social information being used by *A. burtoni* is the process of drawing inferences through observation. Because males in this species need to be dominant and defend a territory to reproduce, individuals are innately driven to acquire a territory through fighting with an incumbent male. Males engage invigorous aggressive fighting bouts that determine their access to a territory and subsequent mating opportunities. However, in a colony containing ~60 D males, it would take a large number of pairwise fights for a male to figureout which dominant individual is vulnerable for a territorial takeover. Plus, the cost to an individual male to engage in repeated conflicts would be substantial. The process of transitive inference (TI) could shorten this process significantly if the animals could infer their chances of winning a fight in advance. TI involves using known relationships to deduce unknown ones. For example, using A > B and B > C to infer A > C allows an individual to acquire hierarchical information essential to logical reasoning. First described as a developmental milestone in children (Piaget 1928), TI has since been reported in nonhuman primates (Gillian 1981, McGonigle & Chalmers 1977, Rapp et al. 1996), rats (Davis 1992, Roberts & Phelps 1994), and birds (Bond et al. 2003, Steirn et al. 1995, von Fersen et al. 1991). Still, how animals acquire and represent transitive relationships and why such abilities might have evolved remain unknown.

We have shown that *A. burtoni* males can draw inferences about a hierarchy implied by pairwise fights between rival males, demonstrating TI. These fish learned the implied hierarchy vicariously as bystanders watching fights between rivals (Grosenick et al. 2007) and can use TI when trained on socially relevant stimuli. Note that they can make such inferences using indirect information alone. The key to this experiment was to show bystanders staged fights between matched animals to assure the outcomes (see Grosenick et al. 2007 for details). Testing the animals' choice robustly demonstrated TI in both the home tank and a novel tank.

As noted above, social interactions require knowledge of the environment and status of others that can be acquired indirectly by observing others' behavior. When being observed, animals can also alter their signals on the basis of who is watching. We measured how male *A. burtoni* behave when being watched in two different contexts. In the first, we showed that aggressive and courtship behaviors displayed by subordinate males depend critically on whether D males can see them; in the second, we manipulated who was watching aggressive interactions and showed that D males will change their behavior depending on audience composition. In both cases, when a more dominant individual is out of view and the audience consists of more subordinate individuals, those males signal key social information to females by displaying courtship and dominant behaviors. In contrast, when a D male is present, males cease both aggression and courtship. These data suggest that males are keenly aware of their social environment and modulate their aggressive and courtship behaviors strategically for reproductive and social advantage (Desjardins et al. 2012).

FROM SOCIAL INFORMATION TO CHANGES IN THE BRAIN

The brief description of the variety of social interactions found in fish described above offers just a glimpse of the potential mechanisms through which animals sample information from their social environments. But how does that social information change their behavior and ultimately their brain? More specifically, how does information change circuits and cells to control reproduction, and how does it ultimately reach the organs responsible for it? Although investigators have suggested potential genomic substrates for these processes in several systems (Robinson et al. 2008), the actual circuits and other parts of the responsible nervous system remain unknown.

To approach this problem, we exploit the extreme, reversible phenotypic switch between *A. burtoni* males from D to ND to understand the effects of social environment on reproduction. By combining manipulations of the social milieu with direct intervention and/or measurement of relevant neural and hormonal systems, we have discovered many socially regulated changes that provide insight into the subtle interplay among the factors responsible for causal links to reproductive function. Here I focus on the signaling pathway that begins in the brain with GnRH1 release to the pituitary and ends by controlling reproductive competence and reproduction itself. The macroscopic consequences of a change in phenotype on the reproductive axis is shown in **Figure 2**.

We measured the dendritic extent using laser confocal microscopy and found that the total dendrite length of GnRH1 neurons in D males (n = 12, average total length per cell = 838 µm) is dramatically greater than the total dendrite length of GnRH neurons in NDs [n = 8, average total length per cell = 459 µm (Scanlon et al. 2003)] (**Figure 3**).

A Sholl (Sholl 1953) analysis of these data, which measures the dendrites in concentric circles, shows significant trends for the overall difference in dendrite arbor size between D and ND males and statistically significant differences in dendritic lengths between 30 and 80 µm. As can be seen in the graph (**Figure 4**), the difference in length emerges ~30 µm from the cell body, suggesting increased branching or extension of the dendrite ends rather than an addition of primary dendrites from the soma.

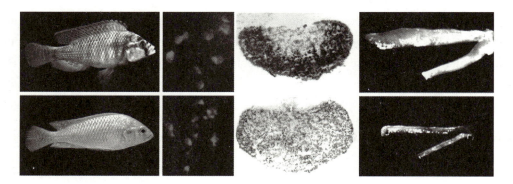

Figure 2

Social regulation of the hypothalamicpituitarygonadal axis in *A. burtoni*. Phenotypic characters of reproductively active dominant (D) males (*top row*) and socially suppressed ND males (*bottom row*) are shown. D males have larger GnRH1 neurons (*red*; immunohistochemical staining) in the preoptic area of the brain (Davis & Fernald 1990, White et al. 2002), higher GnRHR1 levels (*black*, GnRHR1 in situ hybridization; *purple*, cresyl violet counterstain) in the pituitary gland (Au et al. 2006, Flanagan et al. 2007, Maruska et al. 2011), and larger testes (Fraley & Fernald 1982, Davis & Fernald 1990, Maruska & Fernald 2011) compared with subordinate males (modified from Maruska & Fernald 2011). Comparing the dendritic morphology of GnRH1 neurons between D and ND male *A. burtoni* using confocal images provides preliminary evidence that features of the dendritic arbor morphology depend on reproductive state.

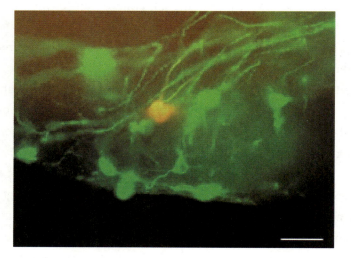

Figure 3

Individual neurons in the POA were filled with neurobiotin using a microelectrode and immunostained with antibodies to GnRH (*green*) to confirm their identity. Yellow cell is filled with neurobiotin and colabeled with GnRH antibody (A. K. Greenwood and R. D. Fernald, unpublished).

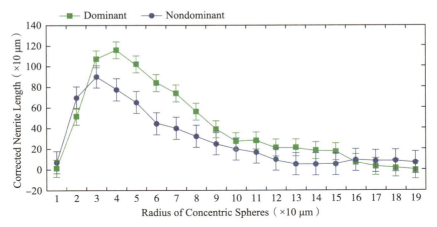

Figure 4

Neurite length measured within each shell of 10 μm concentric spheres (Sholl analysis). GnRH1 neurons from D males have a greater dendrite length beyond30 μm from the soma (see text for discussion): D males (*n* = 12; averagelength = 838 μm), ND males (*n* = 8; average length = 459 μm). From M.D. Scanlon, A.K. Greenwood & R.D. Fernald, unpublished observations.

In some cases, dye from an injected neuron passed into a second uninjected neuron (e.g., unlabeled cell), suggesting that GnRH cells may be coupled, possibly by electrical synapses. Coupling by gap junctions is perhaps the most widely studied mechanism of direct electrical communication and exists in multiple regions of the mammalian brain (see Bennett & Zukin 2004 for review). Direct electrical communication between neurons can give rise to a wide repertoire of dynamic network outputs, and the particular network pattern that does arise will depend on intrinsic cellular properties that affect firing frequencies and heterogeneity across the population and at the location of the communication (e.g., soma, proximal or distal dendrite; Roberts et al. 2006).

The role of direct coupling in GnRH neurons is unknown but could be important in the production of rhythmic GnRH1 output. Fluorescence recovery after photobleaching (FRAP) (Matesic et al. 1996) and neurobiotin labeling (Hu et al. 1999) in immortalized GnRH1 neurons (GT17 cell line) indicated direct connections between 20% and 75% of cultured cells. In GT17 cells, gap junctions also contribute to synchronized pulses (VazquezMartinez et al. 2001), and the same group (Bose et al. 2010) recently reported the necessity of connexin 43 (Cx43) for synchrony. Moreover, gap junction protein levels appear to be regulated by at least one intracellular messenger important in GnRH secretion, cyclic adenosine monophosphate (AMP) (Matesic et al. 1996). In the rat, hypothalamic GnRH neurons express mRNA for one of the more common connexin proteins (connexin 32), but gap junctions between GnRH neurons have not been demonstrated directly in this system (Hosny & Jennes 1998).

Evidence increasingly indicates a role for gap junctions in several systems related to reproduction. For example, gap junctions have been shown to be important in the pituitary of teleost fish (Levavi-Sivan et al. 2005), and they are important in the rat spinal column motor nuclei, where they are regulated by testosterone (Coleman & Sengelaub 2002).

Most studies of electrical communication in GnRH1 neurons have focused on communication at the soma. However, two of the above studies found *in vitro* evidence for the potential involvement of neurites, some of which could be rudimentary dendrites. First, Matesic et al. (1996) reported that pharmacological agents that increase cAMP also increase connectivity, an effect possibly mediated by dendrites because these treatments increased both the number of neurites and the number of connexin 26 positive neurites. Second, Hu et al. (1999) demonstrated that interactions among neurites could account for at least some of the direct coupling observed in GT17 cells, but dendritic coupling of hypothalamic GnRH1 neurons remains to be studied systematically. Nonetheless, emerging evidence in mammalian GnRH neurons is consistent with our findings, namely that GnRH neurons may interact extensively at the dendrite level. Thus, the social cues that drive dendritic remodeling may result in the formation of a dendritemediated network of GnRH1 neurons. Modification of dendritic structure by social status is another of the socially regulated parameters in the GnRH1 neurons.

As noted above, in all vertebrates, both development and differences in reproductive state are controlled by the GnRH1 neurons in the basal forebrain (e.g., Gore 2002). In *A. burtoni* these neurons are located in the anterior parvocellular preoptic nucleus (aPPn), which is the most anterior part of the preoptic area in teleosts, a conserved vertebrate brain region (Wullimann & Mueller 2004). Other GnRH peptides exist in all vertebrates (White et al. 1998) as in *A. burtoni.* Specifically, GnRH2 and GnRH3 (White et al. 1995) are expressed in the midbrain tegmentum and the forebrain terminal nerve ganglion, respectively, but neither is found in the pituitary (Powell et al. 1995) nor do they exhibit socially induced neural plasticity in soma size or gene expression (Davis & Fernald 1990, White et al. 2002).

Social modulation of reproductive function is widespread among vertebrates with dominance status as a particularly salient factor. Examples include the suppression of ovulation (e.g., Abbott & Hearn 1978; Rood 1980; Abbott et al. 1978, 1998), control of maturation in social mammals (e.g., Vandenbergh 1973, Lombardi & Vandenbergh 1976, Bediz & Whitsett 1979, Faulkes et al. 1990), delay of first breeding in birds (e.g., Selander 1965, Wiley 1974), control of maturation in fish and other vertebrates (e.g., Sohn 1977, Fraley & Fernald1982, McKenzie et al. 1983, Leitz 1987), changes in stress levels influencing reproduction (Fox et al. 1997, Abbott et al. 2003), and even sex change (e.g., Robertson 1972, Fricke & Fricke 1977, Cole & Robertson 1988). Thus the mechanistic discoveries from *A. burtoni* may be instructive for understanding the neural control of this process in many other species.

In vertebrates, numerous studies demonstrate activation of the reproductive axis caused by different sensory systems [e.g., olfactory (Gore et al. 2000, Rekwot et al. 2001, Murata et al. 2011), auditory (Bentley et al. 2000, Burmeister & Wilczynski 2005, Maney et al. 2007), tactile (Pfaus & Heeb 1997, Wersinger & Baum 1997), and visual (Castro et al. 2009)]. Changes in the reproductive axis are measured as changes in the number, size, or axonal densities of GnRH1-immunoreactive neurons, alterations in neuronal firing patterns, surges in circulating leutinizing hormone (LH) or steroid levels, increased testicular activity, or increases in sexual arousal and behavior. In addition to sensory channel-specific signals, contextual social interactions with multimodal sensory information such as courtship, mating, exposure to the opposite sex, parental care, and opportunities to rise in social rank are also known to influence GnRH neurons and the hypothalamic pituitary gonadal (HPG) axis in many vertebrates (Wu et al. 1992; Dellovade & Rissman 1994; Wersinger & Baum 1997; Rissman et al. 1997; Bakker et al. 2001; Scaggiante et al. 2004, 2006; Burmeister et al. 2005; Cameron et al. 2008; Lake et al. 2008; Mantei et al. 2008; Stevenson et al. 2008). However, despite numerous studies that demonstrate links between important social signals and activation of the reproductive axis, very few experiments examine how these social sensory signals cause changes, either directly or indirectly, within GnRH1 at the neuronal or genomic level.

One method for approaching this problem has been to use IEGs such as *egr-1, c-fos, jun*, and *arc* to identify activated neurons within reproductive and neuroendocrine circuits (Pfaus & Heeb 1997, Clayton 2000), but neuronal activation is not always associated with IEG induction. IEGs are typically not expressed in chronically activated neurons, and IEG induction is often not related to challenge-induced neuropeptide expression (Farivar et al. 2004, Hoffman & Lyo 2002, Kovacs 2008, Pfaus & Heeb 1997). Nevertheless, socially relevant reproductive stimuli are known to induce IEG expression within GnRH1 neurons across vertebrates from fishes (Burmeister & Fernald 2005) to mammals (Pfaus et al. 1994, Meredith & Fewell 2001, Gelez & Babre-Nys 2006). Burmeister et al. (2005) showed that in *A. burtoni* the perception of a social opportunity by a subordinate male who then ascends to become a D male produces a rapid (20–30 min) induction of the IEG *egr-1* (a transcription factor–encoding gene; also called *zenk, zif*-268, *ngfi-a, krox-24, tis8*) in the preoptic area and in GnRH1 neurons (Burmeister & Fernald 2005). This molecular response results from the recognition of a social opportunity because it is not elicited in males who are already dominant. Recent studies in *A. burtoni* also suggest that visual cues alone are not sufficient to fully suppress the reproductive axis of subordinate males and that other senses such as olfaction are likely involved (Chen & Fernald 2011, Maruska & Fernald 2012).

SOCIAL REGULATION OF GENE EXPRESSION IN THE PITUITARY: GNRH RECEPTORS AND GONADOTROPIN HORMONES

GnRH1 from the brain travels to the gonadotrope producing cells in the anterior pituitary gland directly via neuronal projections in fish species in contrast with the specialized vascular system in other vertebrates. Once in the pituitary, GnRH1 binds to its cognate receptors on secretory cells that release LH and follicle-stimulating hormone (FSH), which stimulate steroid production and gamete development in the gonads (testes or ovaries). Multiple forms of GnRH receptors (i.e., types I, II, III) are found in mammals (Millar 2005), amphibians (Wang et al. 2001), and fishes (Robinson et al. 2001, Lethimonier et al. 2004, Moncaut et al. 2005, Flanagan et al. 2007), and they often show differential distributions, expression patterns (e.g., across season, reproductive stage, or dominance status), and varying responses to regulation by steroids, GnRH, and monoamines, all of which suggest functional specializations (Cowley et al. 1998, Levavi-Sivan et al. 2004, Au et al. 2006, Chen & Fernald 2006, Lin et al. 2010). Although we have considerable information on the signal transduction pathways and how different neurohormones and steroids influence gonadotropin synthesis and release (Bliss et al. 2010, Thackray et al. 2010), little is known about how social information modulates gonadotrope output at the pituitary.

In male *A. burtoni*, pituitary mRNA levels of *GnRHR1*, but not *GnRHR2*, are socially regulated such that stable D males have higher levels compared with subordinate males, and the increase during the social transition appears to occur more slowly (days) than do changes in mRNA levels of other genes, which occur within minutes to hours (Au et al. 2006, Maruska et al. 2011). However, pituitary mRNA levels of the IEG *egr-1* and of the β-subunits of LH and FSH are increased at just 30 min after social ascent, suggesting that GnRH1 release has quickly activated the pituitary gland (Maruska et al. 2011).

SOCIAL REGULATION OF GENE EXPRESSION IN THE TESTES: SPERMATOGENESIS AND STEROID PRODUCTION

Although many studies have shown how social information including mating opportunities, female presence or attractiveness, and social status can influence testicular function in terms of sperm quality (e.g., velocity, motility, number) from fishes to humans (Kilgallon & Simmons 2005, Cornwallis & Birkhead 2007, Gasparini et al. 2009, Ramm & Stockley 2009, Maruska & Fernald 2011), less is known about how social cues induce molecular changes in the testes. In *A. burtoni*, however, perception of social opportunity triggers genomic changes in mRNA levels on both rapid (minutes to hours: FSHR, androgen receptors, corticosteroid receptors) and slower [days: leutinizing hormone releasing hormone (LHR), aromatase, estrogen receptors] time scales (Maruska & Fernald 2011). During the subordinate to D male social transition, the morphological and structural changes in testicular cell composition and relative testes size take several days, whereas many molecular changes in the testes are detected more quickly (Maruska & Fernald 2011). This rapid genomic response in the most distal component of the HPG axis highlights the sensitivity and plasticity of the entire reproductive system to social information. Furthermore, the quick genomic changes in the testes raise the possibility that there may be additional and parallel signaling pathways that perhaps bypass the traditional linear cascade from brain GnRH1 release to pituitary LH/FSH release to a testicular gonadotropin receptor activation scheme.

CONCLUSIONS

Clearly, social information has a profound influence on the function of the reproductive axis in all vertebrates; however, far less is known about how this social information influences the HPG axis at the cellular and molecular levels (e.g., changes in gene expression), and some critical questions about the links between social behaviors, reproductive axis function, and the genome remain unanswered. For example, which signal pathways link reception/perception of a constellation of social cues that produce changes in gene expression along the HPG axis? It seems clear that the next step is to use the many advances in proteomics, transcriptomics, microtranscriptomics, and epigenomics, combined with comparative systems approaches including singlecell analyses, optogenetics, and transgenic methods. Determining the regulatory roles of epigenetic and small RNAs (e.g., microRNAs) in mediating socially induced changes along the reproductive axis is also an exciting area of future work (Robinson et al. 2005, 2008; Huang et al. 2011; Rajender et al. 2011) that should provide insights into our understanding of the mechanisms governing social and seasonal reproductive plasticity across taxa. The cichlid fish *A. burtoni*, with its complex and experimentally manipulable social system, the wealth of background knowledge on the social control of HPG axis function, and the recently available genomic resources will all become a valuable vertebrate model system for studying how the social environment influences genomic plasticity and function of the reproductive axis.

DISCLOSURE STATEMENT

The author is not aware of any affiliations, memberships, funding, or financial holdings that might be perceived as affecting the objectivity of this review.

ACKNOWLEDGMENTS

My thanks go to the members of my laboratory for their contributions to this work. R. D. F. was supported by NIH NS 034950, MH087930, and NSF IOS 0923588.

LITERATURE CITED

Abbott DH, Hearn JP. 1978. Physical, hormonal and behavioral aspects of sexual development in marmoset monkeys, *Callthrix jaccus. J. Reprod. Fertil.* 53:155–66

Abbott DH, Keverne EB, Bercovitch FB, Shively CA, Mendoza SP, et al. 2003. Are subordinates always stressed? A comparative analysis of rank differences in cortisol levels among primates. *Horm. Behav.* 43:67–82

Abbott DH, Saltzman W, SchultzDarken NJ, Tannenbaum PL. 1998. Adaptations to subordinate status in female marmoset monkeys. *Comp. Biochem. Physiol. C* 119:261–74

Adkins-Regan E. 1998. Hormonal mechanisms of mate choice. *Am. Zool.* 38:166–78

Agrillo C, Dadda M, Serena G, Bisazza A. 2009. Use of number by fish. *PLoS One* 4:e 4786 Andersson M, Simmons LW. 2006. Sexual selection and mate choice. *Trends Ecol. Evol.* 21:296–302

Argiolas A. 1999. Neuropeptides and sexual behavior. *Neurosci. Biobehav. Rev.* 23:1127–42

Au TM, Greenwood AK, Fernald RD. 2006. Differential social regulation of two pituitary gonadotropin-releasing hormone receptors. *Behav. Brain Res.* 170:342–46

Bakker J, Kelliher KR, Baum MJ. 2001. Mating induces gonadotropin-releasing hormone neuronal activation in anosmic female ferrets. *Biol. Reprod.* 64:1100–5

Balshine-Earn S, Lotem A. 1998. Individual recognition in a cooperatively breeding cichilid: evidence from video playback experiments. *Behaviour* 135:369–86

Bediz GM, Whitsett JM. 1979. Social inhibition of sexual maturation in male prairie mice. *J. Comp. Physiol. Psychol.* 93:493–500

Bennett MV, Zukin RS. 2004. Electrical coupling and neuronal synchronization in the mammalian brain. *Neuron* 41:495–511

Bentley GE, Wingfield JC, Morton ML, Ball GF. 2000. Stimulatory effects on the reproductive axis in female songbirds by conspecific and heterospecific male song. *Horm. Behav.* 37:179–89

Bliss SP, Navratil AM, Xie J, Roberson MS. 2010. GnRH signaling, the gonadotrope and endocrine control of fertility. *Front. Neuroendocrinol.* 31:322–40

Bond AB, Kamil AC, Balda RP. 2003. Social complexity and transitive inference in corvids. *Anim. Behav.* 65:479–87

Bose SK, Leclerc GM, VazquezMartinez R, Boockfor FR. 2010. Administration of connexin43 siRNA abolishes secretory pulse synchronization in GnRH clonal cell populations. *Mol. Cell. Endocrinol.* 314:75–83

Brown C, Laland K. 2003. Social learning in fishes: a review. *Fish Fish.* 4:280–88

Brown C, Laland K, Krause J, eds. 2006. *Fish Cognition and Behavior*. Oxford, UK: Blackwell

Byrne RW, Whiten A, eds. 1988. *Machiavellian intelligence: social complexity and the evolution of intellect in monkeys, apes, and humans*. Oxford, UK: Oxford Univ. Press

Bshary R, Wickler W, Fricke H. 2002. Fish cognition: a primate's eye view. *Anim. Cogn.* 5:1–13

Bugnyar T, Heinrich B. 2006. Pilfering ravens, Corvus corax, adjust their behaviour to social context and identity of competitors. *Anim. Cogn.* 9:369–76

Burmeister SS, Fernald RD. 2005. Evolutionary conservation of the egr1 immediateearly gene response in a teleost. *J. Comp. Neurol.* 481:220–32

Burmeister SS, Jarvis ED, Fernald RD. 2005. Rapid behavioral and genomic responses to social opportunity. *PLoS Biol.* 3:e 363

Burmeister SS, Kailasanath V, Fernald RD. 2007. Social dominance regulates androgen and estrogen receptor gene expression. *Horm. Behav.* 51:164–70

Burmeister SS, Wilczynski W. 2005. Social signals regulate gonadotropin-releasing hormone neurons in the green treefrog. *Brain Behav. Evol.* 65:26–32

Cameron N, Del Corpo A, Diorio J, McAllister K, Sharma S, Meaney MJ. 2008. Maternal programming of sexual behavior and hypothalamicpituitarygonadal function in the female rat. *PLoS One* 3:e 2210

Candland EJ. 1993. *Feral Children and Clever Animals*. New York: Oxford Univ. Press

Castro AL, Gonc¸alvesdeFreitas E, Volpato GL, Oliveira C. 2009. Visual communication stimulates reproduction in Nile tilapia, *Oreochromis niloticus* (L.). *Braz. J. Med. Biol. Res.* 42:368–74

Chen CC, Fernald RD. 2006. Distributions of two gonadotropin-releasing hormone receptor types in a cichlid fish suggest functional specialization. *J. Comp. Neurol.* 495:314–23

Chen CC, Fernald RD. 2011. Visual information alone changes behavior and physiology during social inter actions in a cichlid fish (*Astatotilapia burtoni*). *PLoS One* 6:e20313

Chittka LA, Osorio D. 2007. Cognitive dimensions of predator responses to imperfect mimicry. *PLoS Biol.* 5:e339

Clayton DF. 2000. The genomic action potential. Neurobiol. *Learn. Mem.* 74:185–216

Clement TS, Parikh V, Schrumpf M, Fernald RD. 2005. Behavioral coping strategies in a cichlid fish: the role of social status and acute stress response in direct and displaced aggression. *Horm. Behav.* 47:336–42

Cole KS, Robertson DR. 1988. Protogyny in the caribbean reef goby, *Coryphopterus personatus:* gonadontogeny and social influences on sexchange. *Bull. Mar. Sci.* 42:317–33

Coleman AM, Sengelaub DR. 2002. Patterns of dye coupling in lumbar motor nuclei of the rat. *J. Comp. Neurol.* 454:34–41

Cornwallis CK, Birkhead TR. 2007. Changes in sperm quality and numbers in response to experimental manipulation of male social status and female attractiveness. *Am. Nat.* 170:758–70

Cowley MA, Rao A, Wright PJ, Illing N, Millar RP, Clarke IJ. 1998. Evidence for differential regulation of multiple transcripts of the gonadotropin releasing hormone receptor in the ovine pituitary gland; effect of estrogen. *Mol. Cell. Endocrinol.* 146:141–49

Dally JM, Emery NJ, Clayton NS. 2006. Foodcaching western scrubjays keep track of who was watching when. *Science* 312:1662–65

Davis H. 1992. Transitive inference in rats (*Rattus norvegicus*). *J. Comp. Psychol.* 106:342–49

Davis MR, Fernald RD. 1990. Social control of neuronal soma size. *J. Neurobiol.* 21:1180–88

Dellovade TL, Rissman EF. 1994. Gonadotropin-releasing hormoneimmunoreactive cell numbers change in response to social interactions. *Endocrinology* 134:2189–97

Desjardins JK, Klausner JQ, Fernald RD. 2010. Female genomic response to mate information. *Proc. Natl. Acad. Sci. USA* 107:21176–80

Desjardins JK, Hofmann HA, Fernald RD. 2012. Social context influences aggressive and courtship behavior in a cichlid fish. *PloS ONE.* In press

Doutrelant C, McGregor PK. 2000. Eavesdropping and mate choice in female fighting fish. *Behavior* 137:1655–69

Earley RL, Dugatkin LA. 2005. Three poeciliid pillars: fighting, mating and networking. See McGregor 2005, pp. 84–113

Farivar R, Zangenehpour S, Chaudhuri A. 2004. Cellularresolution activity mapping of the brain using immediateearly gene expression. *Front. Biosci.* 9:104–9

Faulkes CG, Abbott DH, Jarvis JUM. 1990. Social suppression of ovarian cyclicity in captive and wild colonies of naked molerats, *Heterocephalus glaber. J. Reprod. Fertil. Suppl.* 88:559–68

Flanagan CA, Chen CC, Coetsee M, Mamputha S, Whitlock KE, et al. 2007. Expression, structure, function, and evolution of gonadotropin-releasing hormone (GnRH) receptors GnRHR1SHS and GnRHR2PEY in the teleost, *Astatotilapia burtoni. Endocrinology* 148:5060–71

Fox HE, White SA, Kao MH, Fernald RD. 1997. Stress and dominance in a social fish. *J. Neurosci.* 17:6463–69

Fraley NB, Fernald RD. 1982. Social control of developmental rate in the African cichlid, *Haplochromis burtoni. Z. Tierpsychol.* 60:66–82

Francis RC, Soma K, Fernald RD. 1993. Social regulation of the brainpituitarygonadal axis. *Proc. Natl. Acad. Sci.* 90:7794–98

Fricke HW. 1973. Individual partner recognition in fish: field studies on *Amphiprion bicinctus. Naturwis senschaften* 60:204–5

Fricke HW, Fricke S. 1977. Monogamy and sex change by aggressive dominance in coral reef fish. *Nature* 266:830–32

Gasparini C, Peretti AV, Pilastro A. 2009. Female presence influences sperm velocity in the guppy. *Biol. Lett.* 5:792–94

Gelez H, FabreNys C. 2006. Neural pathways involved in the endocrine response of anestrous ewes to the male or its odor. *Neuroscience* 140:791–800

Gillian DJ. 1981. Reasoning in the chimpanzee: II. Transitive inference. *J. Exp. Psychol. Anim. Behav. Process.* 7:87–108

Goodson JL. 2005. The vertebrate social behavior network: evolutionary themes and variations. *Horm. Behav.* 48:11–22

Gore AC. 2002. Gonadotropin-releasing hormone (GnRH) neurons: gene expression and neuroanatomical studies. *Prog. Brain Res.* 141:193–208

Gore AC, Wersinger SR, Rissman EF. 2000. Effects of female pheromones on gonadotropin-releasing hormone gene expression and luteinizing hormone release in male wildtype and oestrogen receptoralpha knockout mice. *J. Neuroendocrinol.* 12:1200–4

Greenwood AK, Fernald RD. 2004. Social regulation of the electrical properties of gonadotropin-releasing hormone neurons in a cichlid fish (*Astatotilapia burtoni*). *Biol. Reprod.* 71:909–18

Grosenick L, Clement TS, Fernald RD. 2007. Fish can infer social rank by observation alone. *Nature* 445:429–32

Grutter AS. 1997. Spatiotemporal variation and feeding selectivity in the diet of the cleaner fish *Labroides dimidiatus. Copeia* 1997:345–55

Grutter AS, Bshary R. 2003. Cleaner wrasse prefer client mucus: support for partner control mechanisms in cleaning interactions. *Proc. R. Soc. Lond. B* 270(Suppl. 2):242–44

Harbott LK, Burmeister SS, White RB, Vagell M, Fernald RD. 2007. Androgen receptors in a cichlid fish, *Astatotilapia burtoni:* structure, localization, and expression levels. *J. Comp. Neurol.* 504:57–73

Hare B, Call J, Agnetta B, Tomasello M. 2000. Chimpanzees know what conspecifics do and do not see. *Anim.Behav.* 59:771–85

Hare B, Tomasello M. 2001. Do chimpanzees know what conspecifics know? *Anim. Behav.* 61:139–51

Hauser MD. 1997. Minding the behaviour of deception. In *Machiavellian Intelligence II: Extensions and Evaluations,* ed. A Whiten, RW Byrne, pp. 112–43. Cambridge, UK: Cambridge Univ. Press

Havre N, Fitzgerald GJ. 1988. Shoaling and kin recognition in the threespined stickleback (*Gasterosteus aculeatus*). *Biol. Behav.* 13:190–201

Hodos W, Campbell CBG. 1969. The scala naturae: why there is no theory in comparative psychology. *Psychol.Rev.* 76:337–50

Hoffman GE, Lyo D. 2002. Anatomical markers of activity in neuroendocrine systems: Are we all 'fosed out'? *J. Neuroendocrinol.* 14:259–68

Höjesjö J, Johnsson JI, Petersson E, Järvi T. 1998. The importance of being familiar: individual recognition and social behavior in sea trout (*Salmo trutta*). *Behav. Ecol.* 9:445–51

Hosny S, Jennes L. 1998. Identification of gap junctional connexin32 mRNA and protein in gonadotropin releasing hormone neurons of the female rat. *Neuroendocrinology* 67:101–8

Hu L, Olson AJ, Weiner RI, Goldsmith PC. 1999. Connexin 26 expression and extensive gap junctional coupling in cultures of GT17 cells secreting gonadotropin-releasing hormone. *Neuroendocrinology* 70:221–27

Huang Y, Shen XJ, Zou Q, Wang SP, Tang SM, Zhang GZ. 2011. Biological functions of microRNAs: a review. *J. Physiol. Biochem.* 67:129–39

Kamil AC. 1998. On the proper definition of cognitive ethology. In *Animal Cognition in Nature,* ed. RP Balda, IM Pepperberg, AC Kamil, pp. 1–28. San Diego, CA: Academic

Kilgallon SJ, Simmons LW. 2005. Image content influences men's semen quality. *Biol. Lett.* 1:253–55

Kovacs KJ. 2008. Measurement of immediateearly gene activation—cfos and beyond. *J. Neuroendocrinol.* 20:665–72

Lake JI, Lange HS, O'Brien S, Sanford SE, Maney DL. 2008. Activity of the hypothalamic-pituitary-gonadal axis differs between behavioral phenotypes in female whitethroated sparrows (*Zonotrichia albicollis*). *Gen. Comp. Endocrinol.* 156:426–33

Leitz T. 1987. Social control of testicular steroidogenic capacities in the Siamese fighting fish *Betta splendens.J. Exp. Zool.* 244:473–78

Lethimonier C, Madigou T, MuñozCueto JA, Lareyre JJ, Kah O. 2004. Evolutionary aspects of GnRHs, GnRH neuronal systems and GnRH receptors in teleost fish. *Gen. Comp. Endocrinol.* 135:1–16

Levavi-Sivan B, Bloch CL, Gutnick MJ, Fleidervish IA. 2005. Electrotonic coupling in the anterior pituitary of a teleost fish. *Endocrinology* 146:1048–52

Levavi-Sivan B, Safarian H, Rosenfeld H, Elizur A, Avitan A. 2004. Regulation of gonadotropin-releasing hormone (GnRH)receptor gene expression in tilapia: effect of GnRH and dopamine. *Biol. Reprod.* 70:1545–51

Lin CJ, Wu GC, Lee MF, Lau EL, Dufour S, Chang CF. 2010. Regulation of two forms of gonadotropin releasing hormone receptor gene expression in the protandrous black porgy fish, *Acanthopagrus schlegeli. Mol. Cell Endocrinol.* 323:137–46

Lombardi JR, Vandenbergh JG. 1976. Pheromonally induced sexual maturation in females: regulation by the social environment of the male. *Science* 196:545–46

Maney DL, Goode CT, Lake JI, Lange HS, O'Brien S. 2007. Rapid neuroendocrine responses to auditory courtship signals. *Endocrinology* 148:5614–23

Mantei KE, Ramakrishnan S, Sharp PJ, Buntin JD. 2008. Courtship interactions stimulate rapid changes in GnRH synthesis in male ring doves. *Horm. Behav.* 54:669–75

Maruska KP Fernald RD. 2011. Social regulation of gene expression in the hypothalamic-pituitary-gonadalaxis. *Physiology* 26:412–23

Maruska KP, Fernald RD. 2011. Plasticity of the reproductive axis caused by social status change in an African cichlid fish: II. Testicular gene expression and spermatogenesis. *Endocrinology* 152:291–302

Maruska KP, Fernald RD. 2012. Contextual chemosensory signaling in an African cichlid fish. *J. Exp. Biol.* 215:68–74

Maruska KP, Levavi-Sivan B, Biran J, Fernald RD. 2011. Plasticity of the reproductive axis caused by social status change in an African cichlid fish: I. Pituitary gonadotropins. *Endocrinology* 152:281–90

Matesic DF, Hayashi T, Trosko JE, Germak JA. 1996. Upregulation of gap junctional intercellular communication in immortalized gonadotropin-releasing hormone neurons by stimulation of the cyclic AMP pathway. *Neuroendocrinology* 64:286–97

Maynard Smith J, Harper D. 2003. *Animal Signals.* Oxford, UK: Oxford Univ. Press

McGonigle BO, Chalmers M. 1977. Are monkeys logical? *Nature* 267:694–96

McGregor PK. 2005. *Animal Communication Networks.* Cambridge, UK: Cambridge Univ. Press

McKenzie WDJ, Crews D, Kallman KD, Policansky D, Sohn JJ, et al. 1983. Age, weight, and the genetics of sexual maturation in the platyfish, *Xiphophorus maclatus. Copeia* 1983:770–73

Mennill DJ, Ratcliffe LM, Boag PT. 2002. Female eavesdropping on male song contests in songbirds. Science296:873

Meredith M, Fewell G. 2001. Vomeronasal organ: electrical stimulation activates Fos in mating pathways and in GnRH neurons. *Brain Res.* 922:87–94

Metcalfe NB, Thomson BC. 1995. Fish recognize and prefer to shoal with poor competitors. *Proc. R. Soc. Lond. B* 259:207–10

Miczek KA. 1977. Effects of Ldopa, damphetamine and cocaine on intruderevoked aggression in rats and mice. *Prog. NeuroPharmacol.* 1:271–77

Millar RP. 2005. GnRHs and GnRH receptors. *Anim. Reprod. Sci.* 88:5–28

Moncaut N, Somoza G, Power DM, Canario AV. 2005. Five gonadotrophin-releasing hormone receptors in a teleost fish: isolation, tissue distribution and phylogenetic relationships. *J. Mol. Endocrinol.* 34:767–79

Murata K, Wakabayashi Y, Sakamoto K, Tanaka T, Takeuchi Y, et al. 2011. Effects of brief exposure of male pheromone on multipleunit activity at close proximity to kisspeptin neurons in the goat arcuate nucleus. *J. Reprod. Dev.* 57:197–202

Nedivi E, Hevroni D, Naot D, Israeli D, Citri Y. 1993. Numerous candidate plasticityrelated genes revealed by differential cDNA cloning. *Nature* 363:718–22

Newman SW. 1999. The medial extended amygdala in male reproductive behavior. A node in the mammalian social behavior network. *Ann. N. Y. Acad. Sci.* 877:242–57

Noble GK, Curtis B. 1939. The social behavior of the jewel fish, *Hemichromis bimaculatus* (Gill). *Am. Mus. Nat. Hist.* 76:1–46

Olsen KH. 1989. Sibling recognition in juvenile Arctic charr, *Salvelinus alpinus. J. Fish Biol.* 34:571–81

Otter K, McGregor PK, Terry AMR, Burford FRL, Peake TM, Dabelsteen T. 1999. Do female great tits (*Parus major*) assess males by eavesdropping? A field study using interactive song playback. *Proc. Biol. Sci.* 266:1305–9

Pfaus JG, Heeb MM. 1997. Implications of immediateearly gene induction in the brain following sexual stimulation of female and male rodents. *Brain Res. Bull.* 44:397–407

Pfaus JG, Jakob A, Kleopoulos SP, Gibbs RB, Pfaff DW. 1994. Sexual stimulation induces Fos immunore activity within GnRH neurons of the female rat preoptic area: interaction with steroid hormones. *Neuroendocrinology* 60:283–90

Piaget J. 1928. *Judgement and Reasoning in the Child.* London: Routledge & Kegan Paul

Pinto A, Oates J, Grutter AS, Bshary R. 2011. Cleaner wrasses (*Labroides dimidiatus*) are more cooperative in the presence of an audience. *Curr. Biol.* 21:1140–44

Powell JF, Fischer WH, Park M, Craig AG, Rivier JE, et al. 1995. Primary structure of solitary form of gonadotropin-releasing hormone (GnRH) in cichlid pituitary; three forms of GnRH in brain of cichlid and pumpkinseed fish. *Regul. Pept.* 57:43–53

Quinn TP, Hara TJ. 1986. Sibling recognition and olfactory sensitivity in juvenile coho salmon (Oncorhynchus kisutch). *Can. J. Zool.* 64:921–25

Rajender S, Avery K, Agarwal A. 2011. Epigenetics, spermatogenesis and male infertility. *Mutat. Res.* 727:62–71

Ramm SA, Stockley P. 2009. Adaptive plasticity of mammalian sperm production in response to social experience. *Proc. Biol. Sci.* 276:745–51

Rapp PR, Kansky MT, Eichenbaum H. 1996. Learning and memory for hierarchical relationships in the monkey: effects of aging. *Behav. Neurosci.* 110:887–97

Rekwot PI, Ogwu D, Oyedipe EO, Sekoni VO. 2001. The role of pheromones and biostimulation in animal reproduction. *Anim. Reprod. Sci.* 65:157–70

Rissman EF, Li X, King JA, Millar RP. 1997. Behavioral regulation of gonadotropin-releasing hormone production. *Brain Res. Bull.* 44:459–64

Roberts CB, Best JA, Suter KJ. 2006. Dendritic processing of excitatory synaptic input in hypothalamic gonadotropin releasing-hormone neurons. *Endocrinology* 147:1545–55

Roberts WA, Phelps MT. 1994. Transitive inference in rats—a test of the spatial coding hypothesis. *Psychol. Sci.* 5:368–74

Robertson DR. 1972. Social control of sex reversal in a coralreef fish. *Science* 177:1007–9

Robinson GE, Fernald RD, Clayton DF. 2008. Genes and social behavior. *Science* 322:896–900

Robinson GE, Grozinger CM, Whitfield CW. 2005. Sociogenomics: social life in molecular terms. *Nat. Rev. Genet.* 6:257–70

Robison RR, White RB, Illing N, Troskie BE, Morley M, et al. 2001. Gonadotropin-releasing hormone receptor in the teleost *Haplochromis burtoni:* structure, location, and function. *Endocrinology* 142:1737–43

Romer AS. 1959. *The Vertebrate Story.* Chicago, IL: Univ. Chicago Press

Rood JP. 1980. Mating relationships and breeding suppression in the dwarf mongoose. *Anim. Behav.* 23:143–50

Ryan MJ. 1980. Female mate choice in a neotropical frog. *Science* 209:523–25

Scaggiante M, Grober MS, Lorenzi V, Rasotto MB. 2004. Changes along the male reproductive axis in response to social context in a gonochoristic gobiid, *Zosterisessor ophiocephalus* (Teleostei, Gobiidae), with alternative mating tactics. *Horm. Behav.* 46:607–17

Scaggiante M, Grober MS, Lorenzi V, Rasotto MB. 2006. Variability of GnRH secretion in two goby species with socially controlled alternative male mating tactics. *Horm. Behav.* 50:107–17

Scanlon MD, Greenwood AK, Fernald RD. 2003. Dendritic plasticity in gonadotropin-releasing hormone neurons following changes in reproductive status. *Soc. Neurosci. Abstr.* No. 828.20

Selander RK. 1965. On mating systems and sexual selection. *Am. Nat.* 99:129–41

Shettleworth SJ. 1993. Varieties of learning and memory in animals. *J. Exp. Psychol. Anim. Behav. Process* 19:5–14

Shettleworth SJ. 1998. *Cognition, Evolution and Behaviour.* New York: Oxford Univ. Press

Sholl DA. 1953. Dendritic organization in the neurons of the visual and motor cortices of the cat. *J. Anat.* 87:387–406

Sohn JL. 1977. Socially induced inhibition of genetically determined maturation in the platyfish, *Xiphophorus macalatus. Science* 195:199–201

Steirn JN, Weaver JE, Zentall TR. 1995. Transitive inference in pigeons: simplified procedures and a test of value transfer theory. *Anim. Learn. Behav.* 23:76–82

Stevenson TJ, Bentley GE, Ubuka T, Arckens L, Hampson E, MacDougall-Shackleton SA. 2008. Effects of social cues on GnRHI, GnRHII, and reproductive physiology in female house sparrows (*Passer domesticus*). *Gen. Comp. Endocrinol.* 156:385–94

Tegeder RW, Krause J. 1995. Density dependence and numerosity in fright stimulated aggregation behaviour of shoaling fish. *Philos. Trans. R. Soc. Lond. B* 350:381–90

Thackray VG, Mellon PL, Coss D. 2010. Hormones in synergy: regulation of the pituitary gonadotropin genes. *Mol. Cell. Endocrinol.* 314:192–203

Trivers RL. 1972. Parental investment and sexual selection. In Sexual Selection and the *Descent of Man,* ed. B Campbell, pp. 136–79. Chicago: Aldine

Vandenbergh JG. 1973. Acceleration and inhibition of puberty in female mice by pheromones. *J. Reprod. Fertil. Suppl.* 19:411–19

Vazquez-Martinez R, Shorte SL, Boockfor FR, Frawley LS. 2001. Synchronized exocytotic bursts from gonadotropin-releasing hormone-expressing cells: dual control by intrinsic cellular pulsatility and gap junctional communication. *Endocrinology* 142:2095–101

von Fersen L, Wynne CDL, Delius JD, Staddon JER. 1991. Transitive inference formation in pigeons. *J. Exp. Psychol. Anim. Behav. Process* 17:334–41

Wang L, Bogerd J, Choi HS, Seong JY, Soh JM, et al. 2001. Three distinct types of GnRH receptor charac terized in the bullfrog. *Proc. Natl. Acad. Sci. USA* 98:361–66

Ward AJW, Holbrook RI, Krause K, Hart PJB. 2005. Social recognition in sticklebacks: the role of direct experience and habitat cues. *Behav. Ecol. Sociobiol.* 57:575–83

Ward AJW, Webster MM, Hart PJB. 2007. Social recognition in wild fish populations. *Proc. R. Soc.B* 274:1071–77

Ward AJW, Webster MM, Magurran AE, Currie S, Krause J. 2009. Species and population differences in social recognition between fishes: a role for ecology. *Behav. Ecol.* 20:511–16

Wersinger SR, Baum MJ. 1997. Sexually dimorphic processing of somatosensory and chemosensory inputs to forebrain luteinizing hormone-releasing hormone neurons in mated ferrets. *Endocrinology* 138:1121–29

White RB, Eisen JA, Kasten TL, Fernald RD. 1998. Second gene for gonadotropin releasing hormone in humans. *Proc. Natl. Acad. Sci.* 95:305–9

White SA, Kasten TL, Bond CT, Adelman JP, Fernald RD. 1995. Three gonadotropin-releasing hormone genes in one organism suggest novel roles for an ancient peptide. *Proc. Natl. Acad. Sci. USA* 92:8363–67

White SA, Nguyen T, Fernald RD. 2002. Social regulation of gonadotropin-releasing hormone. *J. Exp. Biol.* 205:2567–81

Wiley RH. 1974. Effects of delayed reproduction on survival, fecundity, and the rate of population increase. *Am. Nat.* 108:705–9

Wu TJ, Segal AZ, Miller GM, Gibson MJ, Silverman AJ. 1992. FOS expression in gonadotropin-releasing hormone neurons: enhancement by steroid treatment and mating. *Endocrinology* 131:2045–50

Wullimann MF, Mueller T. 2004. Teleostean and mammalian forebrains contrasted: evidence from genes to behavior. *J. Comp. Neurol.* 475:143–62

Cortical Control of Arm Movements: A Dynamical Systems Perspective

Krishna V. Shenoy,[1,2] *Maneesh Sahani,*[1,3] *and Mark M. Churchland*[4]

[1] Departments of Electrical Engineering, [2] Bioengineering, and Neurobiology, Bio-X and Neurosciences Programs, Stanford Institute for Neuro-Innovation and Translational Neuroscience, Stanford University, Stanford, California 94305; email: shenoy@stanford.edu

[3] Gatsby Computational Neuroscience Unit, University College London, London WC1N 3AR, United Kingdom; email: maneesh@gatsby.ucl.ac.uk

[4] Department of Neuroscience, Grossman Center for the Statistics of Mind, David Mahoney Center for Brain and Behavior Research, Kavli Institute for Brain Science, Columbia University Medical Center, New York, NY 10032; email: mc3502@columbia.edu

Key Words

premotor cortex, primary motor cortex, neural control of movement, dimensionality reduction

Abstract

Our ability to move is central to everyday life. Investigating the neural control of movement in general, and the cortical control of volitional arm movements in particular, has been a major research focus in recent decades. Studies have involved primarily either attempts to account for single-neuron responses in terms of tuning for movement parameters or attempts to decode movement parameters from populations of tuned neurons. Even though this focus on encoding and decoding has led to many seminal advances, it has not produced an agreed-upon conceptual framework. Interest in understanding the underlying neural dynamics has recently increased, leading to questions such as how does the current population response determine the future population response, and to what purpose? We review how a dynamical systems perspective may help us understand why neural activity evolves the way it does, how neural activity relates to movement parameters, and how a unified conceptual framework may result.

INTRODUCTION

It is difficult to appreciate just how central movement is to everyday life until the ability to move is lost owing to neurological injury or disease. Moving is how we interact and communicate with the world. We move our legs and feet to walk, we move our arms and hands to manipulate the objects around us, and we move our tongues and vocal cords to speak. Movement is therefore also central to self-image and psychological well-being. Decades of research have explored the neural basis of movement preparation, generation, and control. In particular, a substantial body of knowledge about the cortical control of arm movements in rhesus macaques has grown from Evarts' pioneering research (e.g., Evarts 1964, 1968; Georgopoulos et al. 1982, 1986; Kalaska 2009; Tanji & Evarts 1976; Weinrich & Wise 1982; Wise 1985). This knowledge recently helped investigators to design cortically controlled neural prosthetic systems aimed at restoring motor function to paralyzed patients (for recent reviews, see, e.g., Green & Kalaska 2011, Hatsopoulos & Donoghue 2009).

Extensive as these discoveries have been, and encouraging as these medical applications are, our understanding of the neural control of movement remains incomplete. Indeed, there is remarkably little agreement regarding even the basic response properties of the motor cortex, including PMd and M1 (e.g., Churchland et al. 2010a; Churchland & Shenoy 2007b; Fetz 1992; Graziano 2009, 2011a; Hatsopoulos 2005; Mussa-Ivaldi 1988; Reimer & Hatsopoulos 2009; Scott 2000, 2008; Scott & Kalaska 1995; Todorov 2000 and associated articles). This lack of agreement contrasts starkly with, say, the primary visual cortex, where basic response properties have been largely agreed upon for decades. To understand the motor cortex is thus a major challenge, as well as an essential step toward designing more capable, accurate, and robust neural prostheses (e.g., Gilja et al. 2011, 2012; Shenoy et al. 2011).

M1: primary motor cortex

PMd: premotor cortex, dorsal aspect

Much of the controversy over motor cortex responses has hinged on the question of whether the cortical activity codes (or represents) muscle action on the one hand or higher-level movement parameters such as effector velocity on the other. **Figure 1** illustrates the dichotomy. Cortical activity passes, via the spinal cord, to the muscles, which contract to move the arm; but the temporal patterns of muscle activity and hand movement differ. Which signal is found in the cortex? Does the firing of cortical cells drive muscle contraction with little intervening translation, so that cortical activity resembles muscle activity; or does it encode the intended movement end point or path, to be transformed by the spinal cord into commands that contract the muscles? Studies correlating neural activity with electromyographic (EMG) muscle activity or with movement kinematics (factors such as velocity and position) have proven equivocal; investigators have seen both patterns (for a recent review, see Kalaska 2009). Just as critically, the activity of most neurons is poorly explained by either pattern (e.g., Churchland & Shenoy 2007b, Graziano 2011b, Scott 2008). Thus, the controversy has continued.

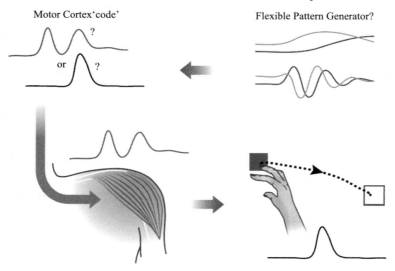

Figure 1
Schematic illustrating the focus of the representational perspective and of the dynamical systems perspective. The traditional perspective has concentrated on the representation or code employed by the motor cortex. For example, does the motor cortex (*upper left panel*) code muscle activity (*red trace*) or reach velocity (*black trace*)? Thus, the traditional perspective attempts to determine the output or controlled parameters of the motor cortex. The dynamical systems perspective focuses less on the output itself and more on how that output is created (*upper right panel*). It attempts to isolate the basic patterns (*blue*) from which the final output might be built. It further attempts to understand the dynamics that produced that set of patterns and the role of preparatory activity in creating the right set of patterns for a particular movement. The red trace indicates the activity of the deltoid versus time during a rightward reach (e.g., Churchland et al. 2012). The black trace is the hand velocity for that same reach; the black trace between the beginning and ending reach targets is the hand path. The light and dark blue traces (*upper right*) illustrate a potential dynamical basis set from which the red trace might be built.

In fact, determining the 'code' or 'representation' in motor cortex is but half the challenge. Whatever the cortical output, its temporal pattern must be generated by the circuitry of the cortex and reciprocally connected subcortical structures. Where is this flexible pattern generator—which can produce the wide

Dynamical system:
a physical system whose future state is a function of its current state, its input, and possibly some noise

variety of motor commands necessary to drive our large repertoire of movements—to be found? Is it up-stream of M1, handing down a 'motor program' to be executed there (Miles & Evarts 1979), or is the pattern at least partly generated in M1 itself? Questions such as these suggest a different way to study the motor cortex, shifting the focus from the meaning of the output to the nature of the dynamical system that creates the required, precisely patterned, command (e.g., Graziano 2011b). A core prediction of this perspective is that activity in the motor system reflects a mix of signals: Some will be outputs to drive the spinal cord and muscles, but many will be internal processes that help to compose the outputs but are themselves only poorly described in terms of the movement. They may, for instance, reflect a much larger basis set of patterns from which the eventual commands are built (see **Figure 1**). Some of these internal signals may well correlate coarsely with movement parameters: For example, one of the blue traces in **Figure 1** resembles hand position, whereas another resembles a filtered version of hand velocity. But such coincidental correlations may not generalize across different tasks and need not constitute a representation of the movement parameters that is actively used by the brain (e.g., Churchland & Shenoy 2007b, Fetz 1992, Todorov 2000). Indeed, the dynamical systems perspective predicts that the evolution of neural activity should be best captured not in terms of movement parameter evolution, but in terms of the dynamical rules by which the current state causes the next state.

The dynamical systems perspective is not new to motor neuroscience. Brown, a student of Sherrington, argued in 1914 that internal pattern generators are at least as important as feedforward reflex arcs (Brown 1914, Yuste et al. 2005). Since then, the approach has shaped our understanding of central pattern generators (e.g., Grillner 2006, Kopell & Ermentrout 2002) and the brain stem circuitry that guides eye movements (e.g., Lisberger & Sejnowski 1992, Skavenski & Robinson 1973). In studying the motor cortex, Fetz (1992) argued for the dynamical systems perspective 20 years ago in an article entitled, "Are Movement Parameters Recognizably Coded in the Activity of Single Neurons?" He noted that

> over the last three decades this formula [recording single neurons in behaving animals] has generated numerous papers illustrating neurons whose activity appears to code (i.e., to covary with) various movement parameters or representations of higher-order sensorimotor functions.... the search for neural correlates of motor parameters may actually distract us from recognizing the operation of radically different neural mechanisms of sensorimotor control. (p. 77)

The same point has been reiterated recently by Cisek 2006b, who summarizes that "the role of the motor system is to produce movement, not to describe it" (p. 2843). The dynamical systems perspective is also reflected in recent attempts to understand motor cortex as it relates to optimal feedback control (e.g., Scott 2004, Todorov & Jordan 2002). Indeed, the dynamical systems perspective may be experiencing a renaissance in neuroscience as a whole (e.g., Briggman et al. 2005, 2006; Broome et al. 2006; Mazor & Laurent 2005; Rabinovich et al. 2008; Stopfer et al. 2003; Yu et al. 2006), largely as the result of the widening adoption of multichannel recording techniques (e.g., Churchland et al. 2007, Harvey et al. 2012, Maynard et al. 1999), machine-learning based algorithms for estimating the population state from those recordings (e.g., Yu et al. 2009), and the computational resources necessary for data analysis and the exploration of plausible models. Just as importantly, there are growing bodies of neural data that are difficult to interpret from a purely representational framework but may be more approachable when dynamical systems concepts are brought to bear (e.g., Ganguli et al. 2008, Machens et al. 2010).

In this review we focus on one such body of literature, that from the field of motor control. We focus less on the role of dynamics in the context of sensory feedback (e.g., Scott 2004) and more on the internal neural dynamics that occur during movement preparation and the subsequent dynamics that translate preparatory activity into movement activity.

A DYNAMICAL SYSTEMS PERSPECTIVE OF MOTOR CONTROL

An Alternative to the Representational View

In principle, the representational and dynamical perspectives are compatible: The first seeks to determine the parameters controlled by cortical output, whereas the latter seeks to determine how that output is generated. However, in practice, adoption of the representational perspective has led to attempts to explain most neural activity in terms of tuning for movement parameters. That is, studies have sought to describe the firing (r) of each neuron (n) in the motor cortex as a function of various parameters ($param_i$) of an upcoming or concurrent movement:

$$r_n(t) = f_n(param_1(t), param_2(t),...).$$
1.

Admittedly, the available range of parameters is extensive, so such models may be adjusted to exhibit considerable richness (e.g., Fu et al. 1995, Hatsopoulos & Amit 2012, Pearce & Moran 2012, Reimer & Hatsopoulos 2009, Wang et al. 2010). Possible covariates include the intended target location, the kinematics of the hand or of the joints, the activity of individual muscles or synergistic groups, the activity of proprioceptors, the predicted end point error, and many others. These parameters may also be filtered, allowing for varying time lags, differentiation, or integration of the corresponding time-dependent signals. The common theme, however, is that neuronal activity should be understood in terms of such representational functions.

By contrast, the dynamical systems perspective stresses the view that the nervous system is a machine that must generate a pattern of activity appropriate to drive the desired movement. That is, the cortical activity [a time-varying vector $\mathbf{r}(t)$], when mapped to muscle activity [a time-varying vector $\mathbf{m}(t)$] by downstream circuitry,

$$\mathbf{m}(t) = G\,[\mathbf{r}(t)], \tag{2.}$$

must produce forces that move the body in a way that achieves the organism's goals. The mapping $G[\ ]$ captures the action of all the circuits that lie between the cortex and the muscles, which may themselves implement sophisticated controllers. The dimension of $\mathbf{m}(t)$, set approximately by the number of independent muscle groups or synergies, is much lower than that of $\mathbf{r}(t)$, the number of different neurons in the motor cortex. Thus it is unlikely that $G[\ ]$ will be invertible. That is, knowledge of the final output alone (e.g., desired muscle activity or kinematics) may be insufficient to determine fully the pattern of neural activity that generated the output. This view thus accords with the observation that the apparent tuning of many neurons changes idiosyncratically with time (Churchland & Shenoy 2007b), with arm starting location (Caminiti et al. 1991), with posture (Kakei et al. 1999, Scott & Kalaska 1995), and with movement speed (Churchland & Shenoy 2007b). More broadly, it may help to understand why, despite many well-designed experiments, the issue of representation in the motor cortex has remained unresolved (e.g., Reimer & Hatsopoulos 2009, Scott 2008). In this view, a confusion of representation is not unexpected: the functions (f_n) of Equation 1 may not exist for any proposed set of movement parameters (Churchland et al. 2010a).

By moving the activity $\mathbf{r}(t)$ to the right-hand side of the equation, the dynamical systems perspective brings into focus the system that must generate that firing pattern (Graziano 2011b). Mathematically, population activity evolves with a derivative $\dot{\mathbf{r}}$, scaled by time constant τ, that is determined by the local circuitry of the motor cortex acting on its current activity through a function $h()$ and by inputs that arrive from other areas, $\mathbf{u}(t)$:

$$\tau\,\dot{\mathbf{r}}\,(t) = h(\mathbf{r}(t)) + \mathbf{u}(t). \tag{3.}$$

With the appropriate input, this dynamical system causes the population activity to trace a path in time that maps through G to generate the correct movement. As this occurs, the neurons in the population may exhibit a variety of response patterns. Some patterns will directly influence the output of G, but others will reflect the act of pattern generation itself. A central aim of the dynamical systems approach is to understand these responses and thus to understand how the dynamics of a neural population produce the temporal patterns needed to drive movement.

Thus, the representational and dynamical perspectives often suggest very different forms of experiment and analysis. If seeking a representation, one asks which parameters are represented by neural activity, in which reference frames, and how these representations are transformed from one reference frame to another (e.g., Andersen & Buneo 2002, Batista et al. 2007, McGuire & Sabes 2011, Mullette Gillman et al. 2009, Pesaran et al. 2006). Equation 1 suggests that the pattern of neural activity in time, and across different movements, should resemble that of the encoded parameters. Thus by designing experiments in which movement variables (e.g., muscle activity) vary systematically, one searches for neural firing patterns that vary in the same way. Conversely, a failure to find neural activity that covaries systematically with muscle activity would be taken to falsify the hypothesis that the cortex is concerned with control of muscles (e.g., Hatsopoulos 2005) or at the very least to suggest a 'messiness' of representation.

From the dynamical standpoint, the possibility that some cortical activity patterns are only indirectly related to the movement and reflect instead internal states of the dynamical process dictates a different approach. At least three practical possibilities present themselves. First, one might seek evidence of this internal state-space via direct visualization (Yu et al. 2009) or by testing the prediction that the population response is more complex than expected given the final output (Churchland & Shenoy 2007b). Second, one might attempt to trace the causality of the

State-space trajectory: evolution of network activity in a space where each axis captures one neuron's response, or a factor shared among neurons

dynamical system. One might ask how the population's premovement state is determined (e.g., Churchland et al. 2006c), how this state influences the subsequent neural activity (e.g., Churchland et al. 2010a, 2012), and how variability in this state influences both neural activity and the movement (e.g., Afshar et al. 2011, Churchland et al. 2006a). Finally, and perhaps most challenging, one might seek to characterize the function $h(\ )$ of Equation 3 by mapping out attractor states (whether fixed-points, limit cycles, or more complex), probing the system's transient behavior (e.g., Buesing et al. 2012, Macke et al. 2011, Sussillo et al. 2013, Sussillo & Abbott 2009, Yu et al. 2006) and studying the effects of perturbations applied to the neural activity (e.g., Churchland & Shenoy 2007a, Diester et al. 2011, Gerits et al. 2012, O'Shea et al. 2013). These different investigations are reviewed in greater detail below.

The Population Dynamical State

A dynamical description of cortical function is inherently a description of activity at the population level. This notion is evident in Equations 2 and 3 above, neither of which can easily be separated into single-neuron components. Unfortunately, obtaining direct empirical access to the relevant scale of population activity is challenging. The full dynamical system is an extensively connected recurrent network of millions of neurons, coupled through input and feedback signals with much of the rest of the brain. The best current measurement technology can record either individual activity of no more than hundreds of neurons (using silicon or microwire arrays or calcium imaging) or aggregate signals that pool over thousands or more neurons at a time (local field potentials, fluorescence changes in voltage-sensitive dyes, or hemodynamic responses). Neither recording scale would seem suited to describing in detail the activity of the whole population. The unreliability of neuronal spiking introduces further challenges (e.g., Faisal et al. 2008, Manwani et al. 2002). Activity cannot be time-averaged on a scale longer than the dynamical time constant of Equation 3 without distorting the resulting dynamics. Similarly, one cannot average over repeated movements to construct a peri-stimulus-time histogram (PSTH) without suppressing intrinsic trial-to-trial variability, which is often of interest (e.g., Afshar et al. 2011, Yu et al. 2009).

These challenges can be addressed using at least two approaches. The first approach avoids the attempt to visualize or describe the dynamical process directly. Instead, hypotheses derived from the dynamical systems viewpoint are tested by assessing related predictions. One example is the prediction that trial-to-trial variance should fall as movement preparation brings the activity of the cortex to a suitable initial point from which appropriate movement activity can be generated (Churchland et al. 2006c, 2010b).

The second approach uses statistical methods to infer the population state from the available data and to examine how that state changes with time. Neurons within a single cortical population do not act alone; instead the coordinated firing of all the neurons presumably guides the evolution of activity within the population and the evolution of its outputs.

This coordination may be intuitively most clear in the context of representation. Activity in a sensory population that encodes the features of a stimulus will covary as those features change. If the number of features is fewer than the number of neurons in the population, then population-level activity must be confined to a space the dimension of which is lower than the number of neurons. Even if the stimulus were unknown, the relevant aspects of the population activity may still be read out by looking for this lower-dimensional coordination. The same idea applies when the low-dimensional structure derives from the population dynamics rather than from a stimulus representation.

There are at least two reasons to think that the essential dimensions of the dynamical state will be few and will be distributed across many, if not all, of the neurons within a local area. First, the tight recurrent connectivity of the network will naturally tend to spread activity between cells. Second, and more subtly, the need for the network to be robust against the very unreliability that hampers experimental observation favors redundant activation patterns. The vector $\mathbf{r}$ in Equations 2 and 3 spans many neurons, and we assume that independent noise in the activity of those neurons (or, indeed, injury to some of them) has only minimal impact on the output of the map G. Thus both G and the function h that determines the dynamics are likely to pool responses from many neurons, compressing the high-dimensional activity into a smaller set of meaningful degrees of freedom and thereby rejecting noise. These meaningful degrees of freedom may be viewed as defining a restricted space of lower dimension that is embedded within the space of all possible activity patterns. Because only the projection of the activity into this lower-dimensional space matters both to the dynamics of the area and to the influence it exerts on the muscles, the meaningful outputs of h must also be confined to this space. Thus, the projection of $\mathbf{r}$ into this lower-dimensional space defines a population dynamical state (see Definition of Terms).

DEFINITION OF TERMS

Population dynamical state: a set of coordinates, often represented as a vector, describing the instantaneous configuration of a dynamical system and that is sufficient to determine the future evolution of that system and its response to inputs. The population dynamical state of a neuronal network might be the vector of instantaneous firing of all its cells or may incorporate aspects of the neurons' biophysical states. It may also be a lower-dimensional projection of this network-wide description. See Dimensionality Reduction.

Dimensionality reduction: in this context, a technique for mapping the responses of many neurons onto a small number of variables that capture the basic patterns present in those responses. For example, the first variable/dimension might capture the response of a large proportion of neurons that all have very similar responses.

The dimensions explored by the population dynamical state may depend on the type of

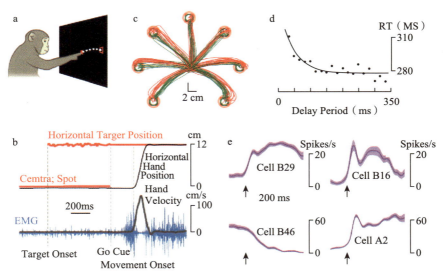

Figure 2

Overview of experimental paradigm, behavioral measurements, muscle measurements, and neural measurements. (*a*) Illustration of the instructed-delay task. Monkeys sit in a primate chair ~25 cm from a fronto-parallel display. A trial begins by fixating (eye) and touching(hand) a central target (*red filled square*) and holding for a few hundred milliseconds. A peripheral target (*red open square*) then appears, cuing the animal about where a movement must ultimately be made. After a randomized delay period (e.g., 0–1 s) a go cue is given (e.g., extinction of central fixation and touch targets) signaling that an arm movement to the peripheral target may begin. (*b*) Sample hand measurements and electromyographic (EMG) recordings for the same trial as in panel *a.* *Top*: Horizontal hand (*black*) and target (*red*) positions are plotted. For this experiment, the target jittered on first appearing and stabilized at the go cue. *Bottom:* Hand velocity superimposed on the voltage recorded from the medial deltoid. (*c*) Sample reach trajectories and end points in a center-out two-instructed-speed version of the instructed-delay task. Red and green traces/symbols correspond to instructed-fast and instructed-slow conditions. (*d*) Mean reaction time (RT) plotted versus delay-period duration. The line shows an exponential fit. (*e*) Examples of typical delay-period firing-rate responses in PMd. Mean ± Standard Error firing rates for four sample neurons are shown. Figure adapted from Churchland et al. (2006c).

movements being performed. Over the full repertoire of movements (e.g., Foster et al. 2012, Gilja et al. 2010, Szuts et al. 2011), the range of dimensions might number in the thousands or more. However, in limited experimental settings with well-controlled motor outputs, the state may be confined to many fewer degrees of freedom. If so, then it should be possible to access the population dynamical state by means of dimensionality reduction techniques applied to the recorded data (e.g., Yu et al. 2009). These methods trace out the trajectory followed by the dynamical system, often on a single trial. Such trajectories make it possible to observe the dynamics more directly—indeed, in some cases the dimensionality reduction itself depends on forming a simultaneous estimate of the dynamical equations—and also to ask qualitative questions about the nature of the dynamics. For example, does the population state observed before the arm moves relate sensibly to the state trajectory traced out during the subsequent movement?

CORTICAL ACTIVITY DURING MOVEMENT PREPARATION

Studies indicate that voluntary movements are prepared before they are executed (Day et al. 1989, Ghez et al. 1997, Keele 1968, Kutas & Donchin 1974, Riehle & Requin 1989, Rosenbaum 1980, Wise 1985). To build intuition, consitder the sudden, rapid, and accurate movement needed to swat a fly. An immediate, unpremeditated attack could miss, allowing the fly to escape. Conversely, a short preparatory delay may permit the accuracy and velocity of movement to be improved, increasing the chances of success. In the laboratory, movement preparation has been studied by instructing a similar, but experimentally controlled, delay prior to a rapid, accurate movement (e.g., Mountcastle et al. 1975). **Figure 2*a–c*** illustrates the experimental design and task timing, along with sample hand position and EMG measurements.

Evidence that subjects use this instructed delay period to prepare a movement comes in part from the observation that reaction time (RT) is shorter on trials with a delay (e.g., Churchland et al. 2006c, Ghez et al. 1989, Riehle et al. 1997, Riehle & Requin 1989, Rosenbaum 1980). **Figure 2*d*** illustrates how RT first decreases and then plateaus with delay duration, suggesting that a time-consuming preparatory process has been given a head start during the initial ~200 ms of delay (e.g., Crammond & Kalaska 2000, Riehle & Requin 1989, Rosenbaum 1980).

Further evidence for movement preparation comes from neural recordings. Neurons in many cortical areas, including the parietal reach region (e.g., Snyder et al. 1997), PMd (e.g., Weinrich & Wise 1982), and M1 (e.g., Tanji & Evarts 1976), systematically modulate their activity during the delay. Thus, these motorrelated areas appear to be engaged in computation prior to the movement (Crammond & Kalaska 2000). **Figure 2*e*** shows four PMd neurons that exemplify the range of delay-period firing patterns: Some neurons' firing rates increase, some decrease, and some stabilize after an initial transient, whereas others vary throughout. This variety of neural responses contrasts with the simple monotonic decline of behavioral RT (**Figure 2*d***) and complicates efforts to understand the role of this activity.

Preparatory Activity as a Subthreshold Representation

Early proposals extended the representational view with the suggestion that preparatory neural activity represents the desired movement at a subthreshold level with the same tuning as that used during movement but with lower overall firing rates (e.g., Tanji & Evarts 1976). This lower-intensity activity is thought not to evoke movement by itself, but instead to reduce the time taken to achieve the correct suprathreshold firing pattern, thus shortening RT. This hypothesis has been assumed by many models of reach generation (e.g., Bastian et al. 1998, Cisek 2006a, Erlhagen & Schöner 2002, Schöner 2004) and agrees with our understanding of the saccadic system (e.g., Hanes & Schall 1996).

Many studies, particularly those exploring summary measures such as the population vector, have indeed reported consistently tuned neural activity before and during movement (e.g., Bastian et al. 1998, 2003; Cisek 2006a; Erlhagen et al. 1999; Georgopoulos et al. 1989; Requin et al. 1988; Riehle & Requin 1989).

Reaction time (RT): the time from the go cue until the start of the movement

However, some studies at the single-neuron level have come to the opposite conclusion: that preparatory and movement tuning are often dissimilar (e.g., Wise et al. 1986) and nearly uncorrelated on average (e.g., Churchland et al. 2010a, Crammond & Kalaska 2000, Kaufman et al. 2010, 2013). Attempts to verify a threshold mechanism have also proven inconclusive. Higher premovement firing rates are not consistently associated with shorter RTs (e.g., Bastian et al. 2003, Churchland et al. 2006c). Furthermore, responses of cortical inhibitory interneurons seem inconsistent (Kaufman et al. 2010) with the common hypothesis that subthreshold preparatory activity is released from inhibition to initiate the movement (Bullock & Grossberg 1988, Sawaguchi et al. 1996).

Preparatory Activity as the Initial Dynamical State

The dynamical systems view suggests a different purpose for preparatory activity. Equations 2 and 3 describe the evolution of neural activity and its translation to muscle activity and thus to movement. The population state trajectory, and thus the movement produced, will clearly depend on the dynamics by which the population state evolves, captured by the function $h(\ )$. It may also be affected by descending input or feedback [$\mathbf{u}(t)$] and by any sources of noise (e.g., van Beers et al. 2004). Finally, and crucially for our current purposes, the trajectory will depend on the population state $\mathbf{r}(t_0)$ at the time (t_0) that movement-related activity begins to be generated. Thus, all else remaining equal, different initial states will lead to different movements. This suggests that one role of preparation is to bring the population dynamical state to an initial value from which accurate movement-related activity will follow efficiently.

In general, more than one initial population dynamical state may lead to a movement that is sufficiently accurate: for example, a reach adequate to earn a reward. Assuming smoothness in the dynamics, and in the mapping to muscle activity and thus to kinematics, we might expect preparation for each movement to be associated with a compact subregion of the space of all possible population states (**Figure 3a**). State-space trajectories [$\mathbf{r}(t)$ for $t > t_0$] originating from different points in this subregion may vary; however, for the reach to be successful, such variation must (*a*) be confined to dimensions that are discounted by the mapping to muscles, (*b*) perturb the movement by too little to affect the desired outcome, or (*c*) be contained by compensatory changes in the external input provided by other areas, including corrective feedback signals.

Thus, the representational and dynamical perspectives both suggest that different movements should require different preparatory activity. Indeed, premovement firing is found to vary with every movement parameter studied so far (Cisek 2006b), including direction (e.g., Kurata 1989, Wise 1985), distance (e.g., Crammond & Kalaska 2000), speed (Churchland et al. 2006b), and curvature (Hocherman & Wise 1991); even apparently random variability in the preparatory state correlates with variability in the subsequent movement (Churchland et al. 2006a). However, short of a rapid de- and re-coding of activity between preparation and movement, the representational view predicts that preparatory and movement tuning should be congruent, which contrasts with the single-unit data as reviewed above. If the link between pre- and peri-movement activity were simply dynamical, on the other hand, then there would be no reason to necessarily expect such congruence.

Two recent studies have extended further support for the dynamical view. First, Churchland et al. (2010a) showed that although preparatory activity does indeed covary with movement parameters such as direction, distance, speed, and curvature, it is more closely related to the pattern of cortical neural activity during the movement—as would be expected if the premovement population state led directly to the subsequent trajectory of movement-period neural activity, and only indirectly to the movement. Second, Kaufman et al. (2011) observed that preparatory states associated with different reaches were arranged along dimensions orthogonal to the dimensions of activity that correlate with changes in muscle force during movement. This result is consistent with a view in which preparatory activity does not itself engage changes in muscle output through the mapping G[ ] but nonetheless leads to movement control signals that do. In a representational picture, where prepatory and movement activity are similarly tuned, such orthogonality would be unexpected.

The Dynamics of Preparation

The end point of motor preparation is hypothesized to be an initial population dynamical state, from which the movement-period neural activity evolves to generate the desired movement. How is the correct initial state achieved between the times when the subject first sees the target and subsequently is told to move? Clearly the dynamics of movement preparation cannot be the same as the dynamics of movement activity. During movement preparation, the dynamical system must bring the population state toward the optimal preparatory region (as in **Figure 3a**) not away from it. Is it possible to detect signatures of this convergent dynamical process?

Activity in the experimental premovement period starts from a baseline condition, in which the only behavioral constraints are that the eyes remain fixated and the hand remains still (**Figure 2a**). There is little to prevent motor cortical activity in this state varying substantially across trials. During preparation, the activity then approaches the preparatory state, while avoiding the premature generation of movement. Again, because the intervening states do not themselves engage muscles, they may well be less constrained than those traversed during the movement's active phase. (See schematic trials 1 and 2 in **Figure 3a**.) The final preparatory state, however, is constrained by the need to generate the correct movement. Thus, we might expect that as preparation progresses, the relative variability across different trials should fall. Such a decrease has indeed been identified in the Fano factor of individual neurons in both the premotor cortex and motor cortex (Churchland et al. 2006c, MandelblatCerf et al. 2009, Rickert et al. 2009). The decline in variability is also apparent in the population dynamical state directly visualized via dimensionality reduction (Churchland et al. 2010b), as shown in **Figure 3b**. As predicted, the reduction in variance comes primarily from convergence in the low-dimensional population dynamical state rather than in the spiking noise of each cell (Churchland et al. 2010b). Finally, variability is only partially reduced when incomplete information about the target is provided (Rickert et al. 2009). These findings support the hypothesis that motor preparation requires network activity to converge to a relatively tight set of population dynamical states. As an aside, a similar decline in neural variability is present in a variety of different cortical areas whenever a relevant stimulus is presented (Churchland et al. 2010b). These findings suggest that many different cortical computations may involve attractor-like dynamics. Nonetheless, the significance of such computation must depend on the function of the area. In visual cortex, the decline in variability may reflect the formation of a more consistent representation of the visual stimulus. In the motor areas considered here, the evidence (discussed below) indicates that the decline in variability reflects convergence to a preparatory state that has motor consequences.

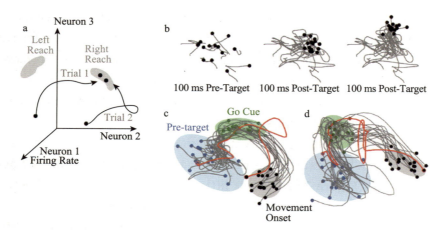

Figure 3

Schematic illustration of the optimal subspace hypothesis and single-trial neural trajectories computed using Gaussian process factor analysis (GPFA). (*a*) The configuration of firing rates is represented in a state-space, with the firing rate of each neuron contributing an axis, only three of which are drawn. For each possible movement, we hypothesize that there exists a subspace/subregion of states that are optimal in the sense that they will produce the desired result, with a minimal reaction time, when the movement is triggered. Different movements will have different optimal subspaces (*shaded areas*). The goal of motor preparation would be to optimize the configuration of firing rates so that it lies within the optimal subspace for the desired movement. For different trials (*arrows*), this process may take place at different rates, along different paths, and from different starting points. Figure from Churchland et al. (2006c). (*b*) Projections of PMd activity into a two-dimensional state-space. Each black point represents the location of neural activity on one trial. Gray traces show trajectories from 200 ms before target onset until the indicated time. The stimulus was a reach target(135°, 60 mm distant), with no reach allowed until a subsequent go cue; 15 (of 47) randomly selected trials are shown. (*c*) Trajectories were plotted until movement onset. Blue dots indicate 100 ms before stimulus (reach target) onset. No reach was allowed until after the go cue (*green dots*), 400–900 ms later. Activity between the blue and green dots thus relates to movement planning. Movement onset (*black dots*) was ~300 ms after the go cue. For display, 18 randomly selected trials are plotted, plus one hand-selected outlier trial (red, trial ID 211). Covariance ellipses were computed across all 47 trials. This is a two-dimensional projection of a ten-dimensional latent space. In the full space, the black ellipse is far from the edge of the blue ellipse. This projection was chosen to preserve accurate relative sizes (on a per-dimension basis) of the true ten-dimensional volumes of the ellipsoids. Data are from the G20040123 dataset. (*d*) Data are presented as in panel *c*, with the same target location, but for data from another day's data set (G20040122; *red* outlier trial: ID 793). Figure panels *b–d* adapted from Churchland et al. (2010b).

What are the consequences if the convergence of the preparatory state is not complete at the time of the go cue? Instructed-delay experiments in which accuracy was emphasized have gathered some data to address this question. First, Churchland et al. (2006c) found that neural variability was lower among trials with short RT, in which motor preparation was likely to have been complete at the time of the go cue, than among trials with longer RT, in which motor preparation may have been incomplete or not quite accurate. This result is consistent with lower variability indicating greater preparatory accuracy (i.e., closer to the optimal preparatory state). Second, when subthreshold electrical microstimulation disrupted the preparatory state, RT was increased (Churchland & Shenoy 2007a). This effect could reflect additional time taken to recover the appropriate preparatory state (see also Ames et al. 2012 for another possibility). The effect was specific (Churchland & Shenoy 2007a). First, RT was more strongly affected when the microstimulation targeted the premotor cortex (where preparatory activity is more common) rather than the motor cortex (where preparatory activity is less common). Second, effects were seen only when the preparatory state was disrupted around the time of the go cue; disruption of the preparatory state before it was needed had little impact on RT. Third, the impact on reach RT was much greater than the impact on saccadic RT, consistent with the role of the premotor cortex in preparing reaches rather than saccades and inconsistent with the possibility that microstimulation simply distracted the animal. Finally, O'Shea et al. (2013) recently found that optical stimulation of optogenetically transfected PMd neurons during the preparatory period similarly increases RT.

THE TRANSITION TO MOVEMENT

By itself, the idea that motor cortical activity represents or codes movement parameters (Equation 1) does not constrain the relationship between preparatory and movement activity. Nonetheless, this transition has frequently been thought to depend on the strengthening of a representation until it crosses a firing-rate threshold, by analogy to the oculomotor system. However, direct evidence for such a threshold has been lacking in the case of reaches (Bastian et al. 2003, Churchland et al. 2006c). By contrast, under the dynamical systems perspective (Equations 2 and 3), the transition to movement is a transition between two different types of network dynamics, most likely to be mediated by a change in the external input term $\mathbf{u}(t)$ of Equation 3. Preparatory dynamics, which brings the population to a suitable state of readiness, then gives way to the dynamics that generate movement. As the movement is triggered, the population dynamical state departs from the prepared initial state and follows a trajectory through state-space. It is that state-space trajectory—determined by the initial state, the neural dynamics, and any feedback—that drives the movement.

The transition from preparatory to movement dynamics may be directly observable. Petreska et al. (2011) used an unsupervised machine-learning technique to study changes in dynamics within multielectrode neuronal data gathered while animals performed instructed-delay reaches. They observed stereotyped dynamical transitions occurring at times shortly after target presentation as well as between the go cue and the beginning of movement. The timing of the dynamical transition that followed the go cue was correlated with the timing of subsequent movement onset. Indeed, this transition—the identity and timing of which were determined by the neural data alone—predicted trial-to-trial variation in RT much better than did alternatives based on a threshold applied to overall firing or to the length of the population vector, even when that threshold was chosen with direct reference to the behavior.

Afshar et al. (2011) addressed a related issue. They reasoned that natural variability might occasionally displace the population dynamical state from the average point of preparation toward the direction in which that state will need to evolve when the movement is to be initiated. Any such variability could actually be beneficial to the initiation of movement and might reduce RT. Indeed, Afshar et al. found that the displacement of the dynamical state at the time of the go cue in the direction defined by the movement-period activity was negatively correlated with RT. Furthermore, RT was even lower if the preparatory state happened to be moving in that direction at the time of the go cue. Thus, although previous studies (Churchland et al. 2006c, Rickert et al. 2009) stressed the importance of an accurate and consistent preparatory state (RTs being lower on average when neural activity is near that state), displacement from the preparatory state can, in fact, result in a lower RT when the displacement is in the direction that is to be traversed during movement.

CORTICAL ACTIVITY DURING MOVEMENT

The dynamical systems perspective focuses on the population dynamical state and its evolution. Testing specific hypotheses therefore often requires direct visualization of that state. The traces in **Figure 3c,d** illustrate the trajectories of the population dynamical state on 19 trials from just before target onset to the moment when movement begins. After target onset, the dynamical state approaches a preparatory region and its variability falls. Then, following the go cue, the neural state moves rapidly away from the preparatory state in a curved trajectory. Some trial-to-trial variability is evident even after the go cue. In particular, for two outlier trials, the neural state wanders before falling back on track. On these trials, the monkey hesitated for an abnormally long few hundred milliseconds before beginning to reach. These observations underscore the ability of dimensionality reduction methods, when applied to data collected from multielectrode arrays, to reveal single-trial (and potentially rare) phenomena that would normally have been lost to averaging or discarded owing to abnormal behavior. That said, for this highly practiced task, such trials were rare (about 0.1%). On the vast majority of trials, the population state evolved along a stereotyped curved trajectory. How can we characterize that trajectory: its shape, its time evolution, and the principles that give rise to it?

This relates to Fetz's original question, "Are movement parameters recognizably coded in the activity of single neurons?" If they are, then the neural state-space trajectory should reflect the trajectory of the represented parameters. For example, consider the model in which neural activity is cosine-tuned for reach velocity (Moran & Schwartz 1999). This relationship can be written in matrix form as $\mathbf{r}(t - \tau) = \mathbf{M}\mathbf{v}(t)$, where $\mathbf{r}(t)$ is an $n \times 1$ vector describing the firing rate of each neuron, $\mathbf{v}(t)$ is the three-dimensional reach velocity vector at time t, τ is the lag by which neural activity leads movements, and $\mathbf{M}$ is an $n \times 3$ matrix describing (in each row) each neuron's preferred direction. Under this model, the population state would be three dimensional, with those dimensions capturing the neural representation of velocity. The population vector is a dimensionality reduction method made specifically for just such a situation.

However, a simple velocity-tuned model is inadequate to fully capture the richness of the neural responses. **Figure 4** illustrates the PSTH responses of two typical neurons recorded from the motor cortex of two monkeys. Monkey B performed a standard center-out reaching task, with two distances and two instructed speeds. Monkey J performed a similar task, but in it some reaches were required to curve around a barrier.

Four features of the neural responses are relevant to the controversy over what is being coded. First, the same neurons exhibit both preparatory- and movement-period activity, yet tuning during the preparatory period often differs from that during the movement period (e.g., cell 12 prefers up-left during the preparation, but down-right by movement onset) (Churchland et al. 2010a). Second, the movement-period responses are complex and multiphasic (Churchland et al. 2010a, 2012; Churchland & Shenoy 2007b). Third, the responses of different neurons are heterogeneous, even in the same animal and the same local region of the cortex (Churchland et al. 2010a, Churchland & Shenoy 2007b, Fetz 1992). In dynamical systems terms, the neural responses occupy a relatively high-dimensional space, on the order of 15–30 dimensions (Churchland & Shenoy 2007b). Thus, if neurons are to represent movement parameters, there must be many such parameters (e.g., Pearce & Moran 2012). Finally, neural firing fluctuates over 400–800ms, even when the reaches themselves are quite brief (e.g., the reaches for Monkey B lasted ~150–300ms). Thus if movement parameters are represented directly, there must be some unexpected temporal multiplexing (e.g., Fu et al. 1995).

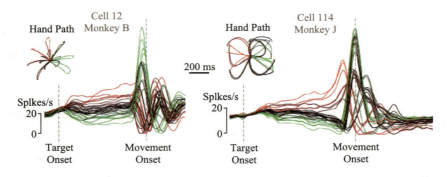

Figure 4

Peri-stimulus-time histogram (PSTHs) and arm-movement kinematics. PSTH from two sample neurons (*left and right panels*), including hand paths for sample reaches (*insets*). Monkey B performed a standard center-out reaching task, with two distances and two instructed speeds. Monkey J performed a more complex version of this task where some reaches were required to curve around a barrier. Traces are colored red to green on the basis of the relative level of preparatory activity for each condition. Insets show hand trajectories and are color-coded to reveal the directional nature of preparatory tuning. Note that what was preferred during the preparatory period was typically not preferred during the movement period; e.g., the first neuron shows a preference for left and up during the preparatory period but a preference for right and down by movement onset. Figure adapted from Churchland et al. (2010a).

In large part because the neural responses are complex, our field does not yet agree on the relationship between movement-period neural responses and movement itself; the nature of G in Equation 2 remains mostly unresolved (although, see, e.g., Fetz et al. 2000). Yet some recent progress has been made in characterizing the nature of h in Equation 3 and in describing the dynamics that generate movement-period neural responses. First, the collective activity of motor cortical neurons is better described by a model in which activity is driven by a low-dimensional dynamical model, relative to a model in which firing coordination emerges from direct connections between recorded cells (Macke et al. 2011).

Second, the dynamics at play during movement appear to have some simple aspects. In many lower-dimensionality projections, one of which is shown in **Figure 5**, the neural trajectory simply rotates with a phase and amplitude set by the preparatory state (Churchland et al. 2012). The rotational trajectories of the neural state resemble what is seen during rhythmic movement, even though the reaches were not overtly rhythmic. These trajectories suggest that the role of motor cortex may be most naturally thought of in the context of pattern generation. Consistent with this idea, EMG activity was well fit by the sum of two rhythmic components that were fixed in frequency but varied in phase and amplitude across the different movements (Churchland et al. 2012).

Monkey N-array

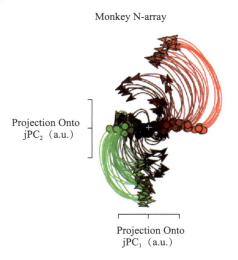

Projection Onto
jPC$_2$ (a.u.)

Projection Onto
jPC$_1$ (a.u.)

Figure 5

Projections of the neural population response, produced by applying jPCA to the first six principal components of the data. Two-dimensional projection using 218 single-and multiunit isolations, 108 conditions employing straight and curved reaches, from the N-array data set. Each trace is the average trajectory for one of 108 conditions. Trial-averaged neural trajectories are colored red to green on the basis of the level of preparatory activity for that projection. Each trace (one condition) plots the first 200 ms of movement-related activity away from the preparatory state (circles). Figure adapted from Churchland et al. (2012).

It should be stressed that the neural dynamics have aspects that are not well fit by a simple linear model (Churchland et al. 2012, Sussillo et al. 2013). However, a simple linear time-invariant system accounted for a large proportion of the variance (48.5% of variance explained over the nine data sets tested) (Elsayed et al. 2013). Furthermore, the linear component was almost entirely normal and rotational (linear systems constrained to be normal and rotational performed 93.2% and 91.3% as well as did an unconstrained linear system) (Elsayed et al. 2013).

Although purely rotational dynamics are only an approximation to the true nonlinear dynamics, the observation that neural pattern generation involves rotations of the neural state illustrates two key points of the dynamical systems perspective. First, the goal of the preparatory state is not to act as an overt representation, but rather to set the amplitude and phase of the subsequent rotations. Second, those state-space rotations produce, in the temporal domain, brief sinusoidal oscillations that form a natural basis set for building more complex patterns (e.g., the blue traces in **Figure 1** can be linearly combined to fit the red EMG trace quite well). Thus, the neural dynamics and the resulting patterns can be understood in simple terms, even though they do not constitute an overt representation of movement parameters.

The observed similarity between the average trajectories during rhythmic and nonrhythmic movement might come about because the nervous system has redeployed old principles for a new purpose. Alternatively, rotations are a common dynamical motif and produce a natural basis set (Rokni & Sompolinsky 2012). Thus we see more than one potential explanation for the key features of the observed dynamics. These key features are that dynamics are similar across different reach types, have a strong rotational component, and have their phase and amplitude determined by the neural state achieved during movement preparation. A number of dynamical models, possibly including control-theory style models, may be able to account for these features (e.g., Scott 2004, Todorov & Jordan 2002). That said, the data are inconsistent with many classes of dynamics (Churchland et al. 2012). Because the observed dynamics are similar across reach types, they are not consistent with a system that is dominated by reach-specific inputs (e.g., a dynamical system that converts velocity commands to muscle activity). The rotational patterns are also not consistent with rise-to-threshold or burst-generator models. This discrepancy is important because many previous intuitions regarding single-neuron activity derived from such models. Most centrally, these intuitions included the expectation that the preparatory- and movement-period preferred direction should be closely related. In contrast, the empirical preferred direction is in a constant state of flux, a natural consequence of the underlying rotations (Churchland et al. 2010a, 2012). This observation illustrates how the complex responses of individual neurons can often hide simple underlying structure—structure that is readily interpretable from a state-space, dynamical systems perspective.

DISCUSSION

The past 50 years have witnessed remarkable progress in our understanding of the cortical control of arm movements. Many fundamental principles have been discovered, and this basic scientific knowledge has led to rapid advances, including early clinical trials of cortically controlled neural prostheses for paralyzed participants (Collinger et al. 2013; Hochberg et al. 2006, 2012). Despite this considerable progress, it remains arguable whether an adequate conceptual framework has yet been identified, around which experimental, computational, and theoretical research can be oriented. Indeed without an adequate conceptual framework, it is unclear how a unified and comprehensive understanding for cortical motor control could be assembled or even recognized. This sentiment, initially expressed by Fetz in his 1992 article, appears to be of at least as great a concern today as it was 20 years ago. The following excerpts from recent articles serve as examples.

> A shift in how to examine the motor system occurred in the 1980s from a problem of control back to a problem of what variables were coded in the activity of neurons.... [P]erhaps it is time to re-evaluate what we are learning about M1 function from continuing to ask what coordinate frames or neural representations can be found in M1. Perhaps it is time to stop pursuing the penultimate goal of identifying the coordinate frame(s) represented in the discharge patterns of M1 and again move back to the question of control (Scott 2008, p. 1220).

> Neurophysiological experiments have revealed neural correlates of many arm movement parameters, ranging from the spatial kinematics of hand path trajectories to muscle activation patterns. However, there is still no broad consensus on the role of the motor cortex in the control of voluntary movement. The answer to that question will depend as much on further theoretical insights into the computational architecture of the motor system as on the design of the definitive neurophysiological experiment (Kalaska 2009, p. 172).

> An epic, twenty-year battle was fought over the cortical representation of movement. Do motor cortex neurons represent the direction of the hand during reaching, or do they represent other features of movement such as joint rotation or muscle output? As vigorous as this debate may have been, it still did not address the nature of the network within the motor cortex. Indeed, it tended to emphasize the properties of individual neurons rather than network properties....The battles over the cortical representation of movement never satisfactorily addressed those questions (Graziano 2011b, p. 388).

It appears that the field has reached a point where a new way of conceptualizing cortical motor control is needed. We have reviewed here a relatively new dynamical systems framework that appears to have several of the desired attributes. As such, the dynamical systems framework may help deepen our understanding of the neural control of movement. It may help do so by (*a*)making relatively few assumptions (e.g., being agnostic to tuning curves, specifically their lack of invariance and generalization);(*b*) observing and documenting the dominant, and perhaps essential, features of neural state-space trajectories; (*c*)providing single-trial neural correlates of behavior that offer insights beyond those available using average relationships alone; (*d*) generating hypotheses that can be answered in the quantitative terms of population dynamical states and evolution rules (equations of motion), without the need to ascribe representational meaning to the detailed response of single neurons; and (*e*) being entirely open to the nature of the dynamics uncovered (e.g., ranging from pattern generators through sophisticated feedback controllers).

The dynamical systems framework is not single-neuron nihilistic: It does not ignore or attempt to average away the complex features of single-neuron responses. Indeed, we hope that by capturing the underlying dynamics it will become possible to explain the many seemingly odd aspects of individual-neuron responses. The dynamical systems perspective also provides a clear road map and goal, which is to quantify the dynamical systems instantiated by neural circuits. This mathematical quantification comes in two inter-related parts: the state-space neural trajectories (a focus of this review article) and the form and meaning of the evolution rule or equations of motion (a focus of ongoing research; e.g., Abeles et al. 1995, Churchland et al. 2012, Petreska et al. 2011, Rabinovich et al. 2008, Seidemann et al. 1996, Smith et al. 2004, Vaadia et al. 1995, Yu et al. 2006). It appears possible that three primary dynamical systems underlie reaching arm movements: one to prepare the neural state in an appropriate manner; a second system to use this computationally optimized starting point to generate movement activity, muscle contraction, and thus movement itself; and a third system that uses feedback for control. Much future research is certainly needed to explore this possibility, to extend and relate it to numerous behaviors beyond the instructed-delay point-to-point reaching task, and to see whether the dynamical systems perspective can ultimately help provide a more comprehensive understanding of cortical motor control.

FUTURE ISSUES

The predictions, experiments, and analyses described above stem from a dynamical systems perspective, and to some degree their confirmation argues for that perspective. Yet many central questions remain largely unaddressed. What is the nature of the relevant dynamics (*h* in Equation 3), and why are they what they are? Do they relate to the dynamics of movement-generating circuits in simpler organisms? Do they reflect sophisticated mechanisms of online control and feedback (e.g., Scott 2004, Todorov & Jordan 2002)? How and why do dynamics change as a function of input from other brain areas (e.g., resting versus planning versus moving)? What is the nature of the circuitry, both local and feedback, that produces these dynamics? What is the mapping between the population dynamical state and muscular activity (*G* in Equation 2)? Answering such questions will likely depend on progress in three domains of research: first, the increased ability to resolve dynamical structure in neural data; second, the increased ability to perturb the population dynamical state while observing dynamical structure; and third, the increased ability to relate state-space trajectories to externally measurable parameters.

For example, it is becoming possible to employ optogenetic techniques to briefly increase (or decrease) the firing rate of excitatory or inhibitory neurons in the cortex of the rhesus monkey (e.g., Diester et al. 2011, O'Shea et al. 2013). This can be accomplished at various times relative to withholding, preparing, or generating arm movements while simultaneously observing the resulting perturbation and recovery of the population dynamical state. This class of pump-probe experiment should enable more quantitative measurement of the neural dynamics in operation during various phases of the behavioral task and should help illuminate the nature and operation of the neural circuitry underlying these neural dynamics.

SUMMARY POINTS

1. Movement preparation has long been thought to be a critical step in generating movement. Recent work supports this idea and argues that achieving the correct preparatory state is important for producing the desired movement.
2. Measurements of the preparatory state predict both reaction time and trial-to-trial movement variability. Disruption of the preparatory state increases reaction time.
3. Preparatory activity is not a subthreshold version of movement activity but instead appears to serve as an initial state for dynamics that engage shortly before movement onset.
4. The onset of these dynamics is tied to movement onset rather than to the go cue and is predictive of trial-by-trial reaction time.
5. Neural responses during the movement appear complex but have at least some simple aspects: Dynamics can be approximated by a linear differential equation in which the same dynamics govern many reaching movements.
6. Because dynamics are similar across conditions, the pattern of movement-related activity is determined largely by the preparatory state.
7. The best linear approximation (to the true nonlinear dynamics) is dominated by rotational dynamics. Preparatory activity sets the amplitude and phase of the movement-period rotations.
8. The resulting firing rate patterns form a natural basis set for building more complex patterns such as muscle activity.

DISCLOSURE STATEMENT

The authors are not aware of any affiliations, memberships, funding, or financial holdings that might be perceived as affecting the objectivity of this review.

ACKNOWLEDGMENTS

We apologize in advance to all the investigators whose research could not be appropriately cited owing to space limitations. This work was supported by Burroughs Wellcome Fund Career Awards in the Biomedical Sciences (to K.V.S. and M.M.C.), DARPA REPAIR N66001-10-C-2010 (to K.V.S. and M.S.), NSF-NIH CRCNS NINDS R01NS054283 (to K.V.S. and M.S.), an NIH Director's Pioneer Award 1DP1OD006409 (to K.V.S.), the Gatsby Charitable Foundation (to M.S.), Searle Scholars Program (to M.M.C.), and NIH Director's New Innovator Award DP2NS083037 (to M.M.C.).

LITERATURE CITED

Abeles M, Bergman H, Gat I, Meilijson I, Seidemann E, et al. 1995. Cortical activity flips among quasistationary states. *Proc. Natl. Acad. Sci. USA* 92:8616–20

Afshar A, Santhanam G, Yu BM, Ryu SI, Sahani M, Shenoy KV. 2011. Single-trial neural correlates of arm movement preparation. *Neuron* 71:555–64

Ames KC, Ryu SI, Shenoy KV. 2012. Neural dynamics of reaching following incomplete or incorrect planning. *Front. Neurosci. Conf.: Comput. Syst. Neurosci. (COSYNE)*. Abstr. T–5

Andersen RA, Buneo CA. 2002. Intentional maps in posterior parietal cortex. *Annu. Rev. Neurosci.* 25:189–220

Bastian A, Riehle A, Erlhagen W, Schöner G. 1998. Prior information preshapes the population representation of movement direction in motor cortex. *Neuroreport* 9:315–19

Bastian A, Schöner G, Riehle A. 2003. Preshaping and continuous evolution of motor cortical representations during movement preparation. *Eur. J. Neurosci.* 18:2047–58

Batista AP, Santhanam G, Yu BM, Ryu SI, Afshar A, Shenoy KV. 2007. Reference frames for reach planning in macaque dorsal premotor cortex. *J. Neurophysiol.* 98:966–83

Briggman KL, Abarbanel HD, Kristan WB Jr. 2005. Optical imaging of neuronal populations during decision-making. *Science* 307:896–901

Briggman KL, Abarbanel HD, Kristan WB Jr. 2006. From crawling to cognition: analyzing the dynamical interactions among populations of neurons. *Curr. Opin. Neurobiol.* 16:135–44

Broome BM, Jayaraman V, Laurent G. 2006. Encoding and decoding of overlapping odor sequences. *Neuron* 51:467–82

Brown TG. 1914. On the nature of the fundamental activity of the nervous centres; together with an analysis of the conditioning of rhythmic activity in progression, and a theory of the evolution of function in the nervous system. *J. Physiol.* 48:18–46

Buesing L, Macke JH, Sahani M. 2012. Learning stable, regularised latent models of neural population dynamics. *Network* 23:24–47

Bullock D, Grossberg S. 1988. Neural dynamics of planned arm movements: emergent invariants and speed-accuracy properties during trajectory formation. *Psychol. Rev.* 95:49–90

Caminiti R, Johnson PB, Galli C, Ferraina S, Burnod Y. 1991. Making arm movements within different parts of space: the premotor and motor cortical representation of a coordinate system for reaching to visual targets. *J. Neurosci.* 11:1182–97

Churchland MM, Afshar A, Shenoy KV. 2006a. A central source of movement variability. Neuron 52:1085–96

Churchland MM, Cunningham JP, Kaufman MT, Foster JD, Nuyujukian P, et al. 2012. Neural population dynamics during reaching. *Nature* 487:51–56

Churchland MM, Cunningham JP, Kaufman MT, Ryu SI, Shenoy KV. 2010a. Cortical preparatory activity: representation of movement or first cog in a dynamical machine? *Neuron* 68:387–400

Churchland MM, Santhanam G, Shenoy KV. 2006b. Preparatory activity in premotor and motor cortex reflects the speed of the upcoming reach. *J. Neurophysiol.* 96:3130–46

Churchland MM, Shenoy KV. 2007a. Delay of movement caused by disruption of cortical preparatory activity. *J. Neurophysiol.* 97:348–59

Churchland MM, Shenoy KV. 2007b. Temporal complexity and heterogeneity of single-neuron activity in premotor and motor cortex. *J. Neurophysiol.* 97:4235–57

Churchland MM, Yu BM, Cunningham JP, Sugrue LP, Cohen MR, et al. 2010b. Stimulus onset quenches neural variability: a widespread cortical phenomenon. *Nat. Neurosci.* 13:369–78

Churchland MM, Yu BM, Ryu SI, Santhanam G, Shenoy KV. 2006c. Neural variability in premotor cortex provides a signature of motor preparation. *J. Neurosci.* 26:3697–712

Churchland MM, Yu BM, Sahani M, Shenoy KV. 2007. Techniques for extracting single-trial activity patterns from large-scale neural recordings. *Curr. Opin. Neurobiol.* 17:609–18

Cisek P. 2006a. Integrated neural processes for defining potential actions and deciding between them: a computational model. *J. Neurosci.* 26:9761–70

Cisek P. 2006b. Preparing for speed. Focus on "Preparatory activity in premotor and motor cortex reflects the speed of the upcoming reach." *J. Neurophysiol.* 96:2842–43

Collinger JL, Wodlinger B, Downey JE, Wang W, Tyler-Kabara EC, et al. 2013. High-performance neuroprosthetic control by an individual with tetraplegia. *Lancet* 381:557–64

Crammond DJ, Kalaska JF. 2000. Prior information in motor and premotor cortex: activity during the delay period and effect on pre-movement activity. *J. Neurophysiol.* 84:986–1005

Day BL, Rothwell JC, Thompson PD, Maertens de Noordhout A, Nakashima K, et al. 1989. Delay in the execution of voluntary movement by electrical or magnetic brain stimulation in intact man. Evidence for the storage of motor programs in the brain. *Brain* 112(Pt. 3):649–63

Diester I, Kaufman MT, Mogri M, Pashaie R, Goo W, et al. 2011. An optogenetic toolbox designed for primates. *Nat. Neurosci.* 14:387–97

Elsayed G, Kaufman MT, Ryu SI, Shenoy KV, Churchland MM, Cunningham JP. 2013. Characterization of dynamical activity in motor cortex. *Front. Neurosci. Conf.: Comput. Syst. Neurosci.* (*COSYNE*). Abstr. III-61

Erlhagen W, Bastian A, Jancke D, Riehle A, Schöner G. 1999. The distribution of neuronal population activation (DPA) as a tool to study interaction and integration in cortical representations. *J. Neurosci. Methods* 94:53–66.

Erlhagen W, Schöner G. 2002. Dynamic field theory of movement preparation. *Psychol. Rev.* 109:545–72

Evarts EV. 1964. Temporal patterns of discharge of pyramidal tract neurons during sleep and waking in the monkey. *J. Neurophysiol.* 27:152–71

Evarts EV. 1968. Relation of pyramidal tract activity to force exerted during voluntary movement. *J. Neurophysiol.* 31:14–27

Faisal AA, Selen LPJ, Wolpert DM. 2008. Noise in the nervous system. *Nat. Rev. Neurosci.* 9:292–303

Fetz EE. 1992. Are movement parameters recognizably coded in the activity of single neurons? *Behav. Brain Sci.* 15:679–90

Fetz EE, Perlmutter SI, Prut Y. 2000. Functions of mammalian spinal interneurons during movement. *Curr. Opin. Neurobiol.* 10:699–707

Foster JD, Nuyujukian P, Freifeld O, Ryu SI, Black MJ, Shenoy KV. 2012. A framework for relating neural activity to freely moving behavior. *Conf. Proc. IEEE Eng. Med. Biol. Soc.* 2012:2736–39

Fu QG, Flament D, Coltz JD, Ebner T. 1995. Temporal encoding of movement kinematics in the discharge of primate primary motor and premotor neurons. *J. Neurophysiol.* 73:836–54

Ganguli S, Bisley JW, Roitman JD, Shadlen MN, Goldberg ME, Miller KD. 2008. One-dimensional dynamics of attention and decision making in LIP. *Neuron* 58:15–25

Georgopoulos AP, Crutcher MD, Schwartz AB. 1989. Cognitive spatial-motor processes 3. Motor cortical prediction of movement direction during an instructed delay period. *Exp. Brain Res.* 75:183–94

Georgopoulos AP, Kalaska JF, Caminiti R, Massey JT. 1982. On the relations between the direction of two-dimensional arm movements and cell discharge in primate motor cortex. *J. Neurosci.* 2:1527–37

Georgopoulos AP, Schwartz AB, Kettner RE. 1986. Neuronal population coding of movement direction. *Science* 233:1416–19

Gerits A, Farivar R, Rosen BR, Wald LL, Boyden ES, Vanduffel W. 2012. Optogenetically induced behavioral and functional network changes in primates. *Curr. Biol.* 22:1722–26

Ghez C, Favilla M, Ghilardi MF, Gordon J, Bermejo R, Pullman S. 1997. Discrete and continuous planning of hand movements and isometric force trajectories. *Exp. Brain Res.* 115:217–33

Ghez C, Hening W, Favilla M. 1989. Gradual specification of response amplitude in human tracking performance. *Brain Behav. Evol.* 33:69–74

Gilja V, Chestek CA, Diester I, Henderson JM, Deisseroth K, Shenoy KV. 2011. Challenges and opportunities for next-generation intracortically based neural prostheses. *IEEE Trans. Biomed. Eng.* 58:1891–99

Gilja V, Chestek CA, Nuyujukian P, Foster J, Shenoy KV. 2010. Autonomous head-mounted electrophysiology systems for freely behaving primates. *Curr. Opin. Neurobiol.* 20:676–86

Gilja V, Nuyujukian P, Chestek C, Cunningham J, Yu B, et al. 2012. A high-performance neural prosthesis enabled by control algorithm design. *Nat. Neurosci.* 15:1752–57

Graziano MSA. 2008. *The Intelligent Movement Machine: An Ethological Perspective on the Primate Motor System,* Vol. 224. Oxford/New York: Oxford Univ. Press

Graziano MSA. 2011a. Cables vs. networks: old and new views on the function of motor cortex. *J. Physiol.* 589:2439

Graziano MSA. 2011b. New insights into motor cortex. *Neuron* 71:387–88

Green AM, Kalaska JF. 2011. Learning to move machines with the mind. *Trends Neurosci.* 34:61–75

Grillner S. 2006. Biological pattern generation: the cellular and computational logic of networks in motion. *Neuron* 52:751–66

Hanes DP, Schall JD. 1996. Neural control of voluntary movement initiation. *Science* 274:427–30

Harvey CD, Coen P, Tank DW. 2012. Choice-specific sequences in parietal cortex during a virtual-navigation decision task. *Nature* 484:62–68

Hatsopoulos NG. 2005. Encoding in the motor cortex: Was Evarts right after all? Focus on "motor cortex neural correlates of output kinematics and kinetics during isometric-force and arm-reaching tasks." *J. Neurophysiol.* 94:2261–62

Hatsopoulos NG, Amit Y. 2012. Synthesizing complex movement fragment representations from motor cortical ensembles. *J. Physiol. Paris* 106:112–19

Hatsopoulos NG, Donoghue JP. 2009. The science of neural interface systems. *Annu. Rev. Neurosci.* 32:249–66

Hochberg LR, Bacher D, Jarosiewicz B, Masse NY, Simeral JD, et al. 2012. Reach and grasp by people with tetraplegia using a neurally controlled robotic arm. *Nature* 485:372–75

Hochberg LR, Serruya MD, Friehs GM, Mukand JA, Saleh M, et al. 2006. Neuronal ensemble control of prosthetic devices by a human with tetraplegia. *Nature* 442:164–71

Hocherman S, Wise SP. 1991. Effects of hand movement path on motor cortical activity in awake, behaving rhesus monkeys. *Exp. Brain Res.* 83:285–302

Kakei S, Hoffman DS, Strick PL. 1999. Muscle and movement representations in the primary motor cortex. *Science* 285:2136–39

Kalaska JF. 2009. From intention to action: motor cortex and the control of reaching movements. *Adv. Exp. Med. Biol.* 629:139–78

Kaufman MT, Churchland MM, Santhanam G, Yu BM, Afshar A, et al. 2010. Roles of monkey premotor neuron classes in movement preparation and execution. *J. Neurophysiol.* 104:799–810

Kaufman MT, Churchland MM, Shenoy KV. 2011. Cortical preparatory activity avoids causing movement by remaining in a muscle-neutral space. *Front. Neurosci. Conf.: Comput. Syst. Neurosci. (COSYNE)*. Abstr. II-61

Keele SW. 1968. Movement control in skilled motor performance. *Psychol. Bull.* 70:387–403

Kopell N, Ermentrout GB. 2002. Mechanisms of phase-locking and frequency control in pairs of coupled neural oscillators. In *Handbook on Dynamical Systems,* Vol. 2: *Toward Applications,* ed. B Fiedler, pp. 3–54. Philadelphia: Elsevier

Kurata K. 1989. Distribution of neurons with set-and movement-related activity before hand and foot movements in the premotor cortex of rhesus monkeys. *Exp. Brain Res.* 77:245–56

Kutas M, Donchin E. 1974. Studies of squeezing: handedness, responding hand, response force, and asymmetry of readiness potential. *Science* 186:545–48

Lisberger SG, Sejnowski TJ. 1992. Motor learning in a recurrent network model based on the vestibulo ocular reflex. *Nature* 360:159–61

Machens CK, Romo R, Brody CD. 2010. Functional, but not anatomical, separation of "what" and "when" in prefrontal cortex. *J. Neurosci.* 30:350–60

Macke JH, Bü sing L, Cunningham JP, Yu BM, Shenoy KV, and Sahani M. 2011. Empirical models of spiking in neural populations. See Shawe-Taylor et al. 2011, pp. 1350–58

Mandelblat-Cerf Y, Paz R, Vaadia E. 2009. Trial-to-trial variability of single cells in motor cortices is dynamically modified during visuomotor adaptation. *J. Neurosci.* 29:15053–62

Manwani A, Steinmetz PN, Koch C. 2002. The impact of spike timing variability on the signal-encoding performance of neural spiking models. *Neural Comput.* 14:347–67

Maynard EM, Hatsopoulos NG, Ojakangas CL, Acuna BD, Sanes JN, et al. 1999. Neuronal interactions improve cortical population coding of movement direction. *J. Neurosci.* 19:8083–93

Mazor O, Laurent G. 2005. Transient dynamics versus fixed points in odor representations by locust antennal lobe projection neurons. *Neuron* 48:661–73

McGuire LM, Sabes PN. 2011. Heterogeneous representations in the superior parietal lobule are common across reaches to visual and proprioceptive targets. *J. Neurosci.* 31:6661–73

Miles FA, Evarts EV. 1979. Concepts of motor organization. *Annu. Rev. Psychol.* 30:327–62

Moran DW, Schwartz AB. 1999. Motor cortical representation of speed and direction during reaching. *J. Neurophysiol.* 82:2676–92

Mountcastle VB, Lynch JC, Georgopoulos A, Sakata H, Acuna C. 1975. Posterior parietal association cortex of the monkey: command functions for operations within extrapersonal space. *J. Neurophysiol.* 38:871–908

Mullette-Gillman OA, Cohen YE, Groh JM. 2009. Motor-related signals in the intraparietal cortex encode locations in a hybrid, rather than eye-centered reference frame. *Cereb. Cortex* 19:1761–75

Mussa-Ivaldi FA. 1988. Do neurons in the motor cortex encode movement direction? An alternative hypothesis. *Neurosci. Lett.* 91:106–11

O'Shea D, Goo W, Kalanithi P, Diester I, Ramakrishnan C, et al. 2013. Neural dynamics following optogenetic disruption of motor preparation. *Front. Neurosci. Conf.: Comput. Syst. Neurosci. (COSYNE).* Abstr. III-60

Pearce TM, Moran DW. 2012. Strategy-dependent encoding of planned arm movements in the dorsal premotor cortex. *Science* 337:984–88

Pesaran B, Nelson MJ, Andersen RA. 2006. Dorsal premotor neurons encode the relative position of the hand, eye, and goal during reach planning. *Neuron* 51:125–34

Petreska B, Yu BM, Cunningham JP, Santhanam G, Ryu SI, et al. 2011. Dynamical segmentation of single trials from population neural data. See Shawe-Taylor et al. 2011, pp. 756–64

Rabinovich M, Huerta R, Laurent G. 2008. Transient dynamics for neural processing. *Science* 321:48–50

Reimer J, Hatsopoulos NG. 2009. The problem of parametric neural coding in the motor system. *Adv. Exp. Med. Biol.* 629:243–59

Requin J, Riehle A, Seal J. 1988. Neuronal activity and information processing in motor control: from stages to continuous flow. *Biol. Psychol.* 26:179–98

Rickert J, Riehle A, Aertsen A, Rotter S, Nawrot MP. 2009. Dynamic encoding of movement direction in motor cortical neurons. *J. Neurosci.* 29:13870–82

Riehle A, Grün S, Diesmann M, Aertsen A. 1997. Spike synchronization and rate modulation differentially involved in motor cortical function. *Science* 278:1950–53

Riehle A, Requin J. 1989. Monkey primary motor and premotor cortex: single-cell activity related to prior information about direction and extent of an intended movement. *J. Neurophysiol.* 61:534–49

Rokni U, Sompolinsky H. 2012. How the brain generates movement. *Neural Comput.* 24:289–331

Rosenbaum DA. 1980. Human movement initiation: specification of arm, direction, and extent. *J. Exp. Psychol. Gen.* 109:444–74

Sawaguchi T, Yamane I, Kubota K. 1996. Application of the GABA antagonist bicuculline to the premotor cortex reduces the ability to withhold reaching movements by well-trained monkeys in visually guided reaching task. *J. Neurophysiol.* 75:2150–56

Schöner G. 2004. Dynamical systems approaches to understanding the generation of movement by the nervous system. In *Progress in Motor Control*, Vol. 3, ed. ML Latash, MF Levin. Champaign, IL: Hum. Kinet.

Scott SH. 2000. Population vectors and motor cortex: neural coding or epiphenomenon? *Nat. Neurosci.* 3:307–8

Scott SH.2004. Optimal feedback control and the neural basis of volitional motor control. *Nat. Rev. Neurosci.* 5:532–46

Scott SH. 2008. Inconvenient truths about neural processing in primary motor cortex. *J. Physiol.* 586:1217–24

Scott SH, Kalaska JF. 1995. Changes in motor cortex activity during reaching movements with similar hand paths but different arm postures. *J. Neurophysiol.* 73:2563–67

Seidemann E, Meilijson I, Abeles M, Bergman H, Vaadia E.1996. Simultaneously recorded single units in the frontal cortex go through sequences of discrete and stable states in monkeys performing a delayed localization task. *J. Neurosci.* 16:752–68

Shawe-Taylor J, Zemel RS, Bartlett P, Pereira F, Weinberger KQ, eds. 2011. *Advances in Neural Information Processing Systems (NIPS)*, Vol. 24. Red Hook, NY: Curran

Shenoy KV, Kaufman MT, Sahani M, Churchland MM. 2011. A dynamical systems view of motor preparation: implications for neural prosthetic system design. *Prog. Brain Res.* 192:33–58

Skavenski AA, Robinson DA. 1973. Role of abducens neurons in vestibuloocular reflex. *J. Neurophysiol.* 36:724–38

Smith AC, Frank LM, Wirth S, Yanike M, Hu D, et al. 2004. Dynamic analysis of learning in behavioral experiments. *J. Neurosci.* 24:447–61

Snyder LH, Batista AP, Andersen RA. 1997. Coding of intention in the posterior parietal cortex. *Nature* 386:167–70

Stopfer M, Jayaraman V, Laurent G. 2003. Intensity versus identity coding in an olfactory system. *Neuron* 39:991–1004

Sussillo D, Abbott LF. 2009. Generating coherent patterns of activity from chaotic neural networks. Neuron 63:544–57

Sussillo D, Churchland MM, Kaufman MT, Shenoy KV. 2013. A recurrent neural network that produces EMG from rhythmic dynamics. *Front. Neurosci. Conf.: Comput. Syst. Neurosci. (COSYNE)*. Abstr. III-67

Szuts TA, Fadeyev V, Kachiguine S, Sher A, Grivich MV, et al. 2011. A wireless multi-channel neural amplifier for freely moving animals. *Nat. Neurosci.* 14:263–69

Tanji J, Evarts EV. 1976. Anticipatory activity of motor cortex neurons in relation to direction of an intended movement. *J. Neurophysiol.* 39:1062–68

Todorov E. 2000. Direct cortical control of muscle activation in voluntary arm movements: a model. *Nat. Neurosci.* 3:391–98

Todorov E, Jordan MI. 2002. Optimal feedback control as a theory of motor coordination. *Nat. Neurosci.* 5:1226–35

Vaadia E, Haalman I, Abeles M, Bergman H, Prut Y, et al. 1995. Dynamics of neuronal interactions in monkey cortex in relation to behavioural events. *Nature* 373:515–18

van Beers RJ, Haggard P, Wolpert DM. 2004. The role of execution noise in movement variability. *J. Neurophysiol.* 91:1050–63

Wang W, Chan SS, Heldman DA, Moran DW. 2010. Motor cortical representation of hand translation and rotation during reaching. *J. Neurosci.* 30:958–62

Weinrich M, Wise SP. 1982. The premotor cortex of the monkey. *J. Neurosci.* 2:1329–45

Wise SP. 1985. The primate premotor cortex: past, present, and preparatory. *Annu. Rev. Neurosci.* 8:1–19

Wise SP, Weinrich M, Mauritz KH. 1986. Movement-related activity in the premotor cortex of rhesus macaques. *Prog. Brain Res.* 64:117–31

Yu BM, Afshar A, Santhanam G, Ryu SI, Shenoy KV, Sahani M. 2006. Extracting dynamical structure embedded in neural activity. In *Advances in Neural Information Processing Systems (NIPS), Vol.* 18, ed. Y Weiss, B Scholkopf, J Platt, pp. 1545–52. Cambridge, MA: MIT Press

Yu BM, Cunningham JP, Santhanam G, Ryu SI, Shenoy KV, Sahani M. 2009. Gaussian-process factor analysis for low-dimensional single-trial analysis of neural population activity. *J. Neurophysiol.* 102:614–35

Yuste R, MacLean JN, Smith J, Lansner A. 2005. The cortex as a central pattern generator. *Nat. Rev. Neurosci.* 6:477–83

Emotion and Decision Making: Multiple Modulatory Neural Circuits

Elizabeth A. Phelps,[1,2,3] *Karolina M. Lempert,*[1] *and Peter Sokol-Hessner*[1,2]

[1] Department of Psychology, [2] Center for Neural Science, New York University, New York, NY 10003; [3]Nathan Kline Institute, Orangeburg, New York, NY 10963; email: liz.phelps@nyu.edu, karolina.lempert@gmail.com, psh234@nyu.edu

Key Words

striatum, orbitofrontal cortex, amygdala, insular cortex, mood, stress

Abstract

Although the prevalent view of emotion and decision making is derived from the notion that there are dual systems of emotion and reason, a modulatory relationship more accurately reflects the current research in affective neuroscience and neuroeconomics. Studies show two potential mechanisms for affect's modulation of the computation of subjective value and decisions. Incidental affective states may carry over to the assessment of subjective value and the decision, and emotional reactions to the choice may be incorporated into the value calculation. In addition, this modulatory relationship is reciprocal: Changing emotion can change choices. This research suggests that the neural mechanisms mediating the relation between affect and choice vary depending on which affective component is engaged and which decision variables are assessed. We suggest that a detailed and nuanced understanding of emotion and decision making requires characterizing the multiple modulatory neural circuits underlying the different means by which emotion and affect can influence choices.

BEYOND DUAL SYSTEMS IN THE MIND AND BRAIN: A MODULATORY ROLE FOR EMOTION IN DECISION MAKING

The prevalent view of the role of emotion in decision making in economics, psychology, and, more recently, neuroscience is the dual systems approach. In economics, choices have been characterized as relying on either System 1 or System 2, with emotion as one of the factors contributing to the more automatic, less deliberative system 1 (Kahneman 2011). In psychology, the terms "hot" versus "cool" have been used to describe decisions driven by affect or not (e.g., Figner et al. 2009; see sidebar, Dual-Process Theories). In neuroscience, brain-imaging research has been used to argue that the human mind is "vulcanized" such that our highly developed prefrontal cortex can be used to overcome the emotional or limbic responses that may sway us to perform irrationally (Cohen 2005). These modern dual-system accounts of the relation between emotion and decision making have a long history. The idea that opposing forces of emotion and reason compete in the human mind is prevalent in Western thought, highlighted by a range of scholars including philosophers such as Plato and Kant and the father of psychoanalysis, Sigmund Freud (Peters 1970). The intuitive nature of this distinction is also apparent in the everyday language used when reflecting on decisions as being made with the heart or the head.

The notion that there are distinct systems for emotion and cognition was also apparent in early theories of brain anatomy. Building on earlier work by Paul Broca and James Papez, Paul Maclean introduced the term "limbic system" in 1952 to describe the phylogenetically older brain regions that lined the inner border of the cortex that he proposed were responsible for basic emotional responses (Lambert 2003) (see **Figure 1a**). The limbic system quickly became known as the emotional center of the brain, with the neocortex underlying higher cognitive functions, including reason. This early theory was highly influential in its time; however, as basic research into neuroanatomy and structure-function relationships progressed over the past several decades, the limbic system concept did not hold up. For example, a region of the neocortex, the orbitofrontal cortex, is important in emotion (Damasio 2005), and the hippocampus, a key component of the limbic system, is critical for the basic cognitive function of memory (Squire 2004). Some researchers have tried to modify the limbic system concept to more accurately reflect the emotion/cognition division (e.g., Cohen 2005, Rolls 2013) (see **Figure 1b**). However, as our understanding of both the complexity of emotion and its underlying neural systems expands, there is clearly no clean delineation between brain regions underlying emotion and cognition. There is no clear evidence for a unified system that drives emotion. Thus affective neuroscientists and neuroanatomists have suggested that the limbic system concept is no longer useful and should be abandoned to facilitate the development of a more complete and detailed understanding of the representation of emotion in the brain (see LeDoux 2000 for a discussion).

DUAL-PROCESS THEORIES

Dual-process theories are a dominant class of theories of human decision making that argue for the existence of two separate, opposing decision systems. Choice results from competition between these two systems: One is generally emotional, fast, automatic, "hot," and/or subconscious, whereas the other is cognitive, slow, deliberative, "cool," and/or explicit. Some specific theories using this structure include the following.

- System 1/System 2: Emotion is considered part of System 1, which "operates automatically and quickly, with little or no effort and no sense of voluntary control. System 2 allocates attention to the effortful mental activities that demand it, including complex computations. The operations of System 2 are often associated with the subjective experience of agency, choice, and concentration" (Kahneman 2011, pp. 20–21).

- Hot/cool: In this theory, positing a hot emotional system and a cool cognitive system, "risk taking [is] the result of a competition between two neural systems.. .Affective processing is spontaneous and automatic, operates by principles of similarity and contiguity, and influences behavior by affective impulses. ... The cognitive-control system ... is the neural basis of deliberative processing, which is effortful, controlled, and operates according to formal rules of logic ... [and] it is the neural basis of inhibitory control, a mechanism that can block affective impulses and therefore enables deliberative decision making even in affect-charged situations" (Figner et al. 2009, p. 710).

Given that affective neuroscientists now generally view the limbic system concept as obsolete, perhaps it is also time to revisit the usefulness of dual system models to characterize the relation between emotion and decision making. Without a clear instantiation of an emotion system in the brain, it is difficult to conceive of a psychological model that relies on such a system. The importance of neural instantiations for psychological theories has become increasingly apparent as the discipline of cognitive neuroscience evolves. When examining other cognitive functions, such as memory, attention, and perception, a more fine-grained analysis of specific brain circuitries underlying the relation between factors indicative of emotion and those of cognition has emerged. This research suggests a modulatory role for emotion's influence on cognition, and vice versa (see Phelps 2006 for a review). Translating this modulatory view to the study of decision making suggests that affective processes may influence a primary factor underlying choice behavior: the computation of subjective value.

In this review, we explore a range of means by which affective factors may influence choices and highlight investigations of the neural circuitry mediating emotion's modulation of decision making. One challenge in approaching this literature is the recognition that emotion is not a unitary construct, but rather a compilation of component affective processes. Although the precise nature of these component processes is a topic of theoretical debate that goes beyond the scope of this review (e.g.,

Ekman & Davidson 1994, Scherer 2000, Barrett 2006), a few general distinctions have emerged that will aid in characterizing this literature. The term affect is generally used as the overarching term to describe this collection of component processes, whereas the term emotion refers to a discrete reaction to an internal or external event that can yield a range of synchronized responses, including physiological responses (e.g., flight or fight), facial and/or bodily expressions, subjective feelings, and action tendencies, such as approach or avoid. Although these reactions are synchronized in response to an event, they may not all be present and their intensity can vary independently. The discrete nature of these emotional reactions is their distinct quality relative to other affective components, although how long they last, from relatively transient to much longer, can vary depending on the nature of the eliciting event and intensity. For the purposes of this review, we differentiate a discrete emotional reaction from a stress response, which is characterized by specific physiological and neurohormonal changes that disrupt homeostasis resulting from a real, imagined, or implied threat (Ulrich-Lai & Herman 2009). The impact of these neurohormonal changes lasts beyond the stressor itself (Dickerson & Kemeny 2004) and induces a relatively lasting affective state. Another lasting affective state is mood, which is defined predominantly by subjective feelings that are not necessarily linked to a specific event. Like emotions, moods can elicit action tendencies. Although these affective processes provide a basis for our characterization of the existing literature on the neural basis of affect and decision making, they do not capture the range of affective experience that may be relevant to understanding decision processes more broadly (see Scherer 2000, 2005 for a more in-depth discussion of component process models of emotion).

One could argue that choice itself is indicative of an affective response because it signals an evaluation of preference, motivation, or subjective value assigned to the choice options. The view that value and emotion are inherently intertwined is more common among psychologists and neuroscientists (e.g., Rolls & Grabenhorst 2008) than economists (e.g., Kahneman 2011), but for this review we focus on evidence that independent affective components modulate the assessment of subjective value and the decision. With this aim in mind, we limit our discussion largely to studies that have explicitly measured and/or manipulated a factor commonly linked to emotion or affect. We do not include studies in which emotion is inferred from choices or blood-oxygenation-level-dependent (BOLD) signal because there is limited evidence of unique BOLD patterns linked to specific affective components (Phelps 2009), a problem commonly known as reverse inference (Poldrack 2006). Given that our primary interest is to characterize the current literature examining the neuroscience of emotion's modulation of decision making, we discuss the growing psychological literature on affect and decision making (Lerner et al. 2015) only if there is also some link to the underlying neural circuitry.

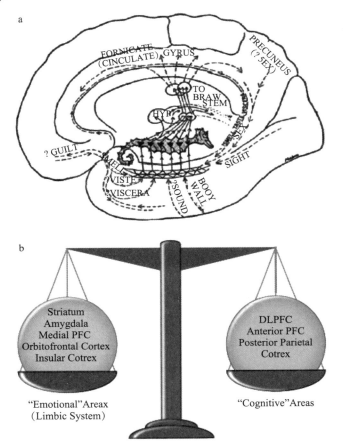

Figure 1

(*a*) The limbic system, centered on the hippocampus, as conceptualized by MacLean (1949). The limbic system concept, as an integrated brain circuit for emotion, has not been supported by more recent neuroanatomical evidence and investigations of brain function (LeDoux 2000). Panel reproduced with permission from Lippincott Williams & Wilkins. (*b*) Dual-process theories of emotion and decision making suggest that choices reflect the outcome of a competition between systems. In this framework, emotional or limbic areas (updated to include other regions; Cohen 2005) are associated with automatic, often irrational choices, whereas cognitive areas are implicated in more deliberative, rational decision making. DLPFC, dorsolateral prefrontal cortex; PFC, prefrontal cortex.

Two broad categories of research explore the modulation of decision making by emotion or affect. The first explores how a decision is altered when it occurs during a specific affective state. In this class of studies, the affective state is incidental to the choice itself but nevertheless modulates the decision. The second class of studies examines how the emotional reaction elicited by the choice itself is incorporated into the computation of subjective value. In the final section, we examine how processes that alter emotion can change choices, highlighting the reciprocal, modulatory relationship between emotion and decision making.

AFFECT CARRYOVER: HOW INCIDENTAL AFFECT MODULATES DECISIONS

One means by which emotion can influence choices is through incidental affect. Incidental affect is a baseline affective state that is unrelated to the decision itself. Studies investigating incidental affect trigger an affective state prior to the decision-making task and evaluate its impact on choices. Below we describe two incidental affective states that have been shown to influence decisions.

Stress

Although stress is a term that is widely used to mean many different things, one clear neurobiological indication of a stress reaction is activation of the hypothalamic-pituitary-adrenal (HPA) axis. Stress reactions are also accompanied by sympathetic nervous system arousal, which can be more transient, but HPA axis activation results in a cascade of neuroendocrine changes, most notably glucocorticoid release, that can have a relatively lasting impact. These neurohormonal changes influence function in several brain regions implicated in decision making (see Arnsten 2009, Ulrich-Lai & Herman 2009 for reviews) (see **Figure 2a**).

Several studies examining stress and decision making highlight the impact of stress on the prefrontal cortex (PFC). Even relatively mild stress can impair performance on PFC-dependent tasks, such as working memory, owing to the negative impact of catecholamines and glucocorticoids on PFC function (Arnsten 2009). The impact of stress on other brain regions varies. For example, although mild stress can enhance hippocampal function, more intense and/or prolonged stress impairs the hippocampus (McEwen 2007). In contrast, performance on striatal-dependent tasks is often enhanced with stress (Packard & Goodman 2012), and dopaminergic neurons in the ventral tegmental area and the striatum show transient and lasting stress-specific responses (Ungless et al. 2010). In addition, amygdala function is generally enhanced with stress, and the amygdala modulates some stress effects on the hippocampus, striatum, and PFC (e.g., Roozendaal et al. 2009). Given the uneven effects of stress on the neural circuits that mediate decision making, the impact of stress may vary depending on the intensity and duration of the stressor, as well as on the specific variables assessed in the decision-making task.

The PFC is proposed to play a role in goal-directed decisions, whereas the striatum is generally linked to choices based on habits (Balleine & O'Doherty 2010). To explore this trade-off between PFC- and striatal-mediated choices under stress, Dias-Ferreira and colleagues (2009) examined how chronic stress affected later performance on a devaluation task in rodents. Devaluation tasks assess whether choices are habitual or directed toward a reinforcement goal by altering the value of the reward. If reducing the value of the reward changes behavior, the task is said to be goal directed, whereas if devaluing the reward does not alter behavior, the task is habitual. After training rats to press a lever to receive a food reward, the rats were fed the food to satiety, thus devaluing subsequent reward presentations. Rats who had not been previously stressed reduced their response rate, reflecting the devalued reward outcome. In contrast, stressed rats failed to modify their behavior following devaluation, consistent with habitual responding. Dias-Ferreira et al. (2009) found that chronic, restraint stress resulted in neuronal atrophy of the medial PFC and dorsal medial striatum, a circuit known to be involved in goal-directed actions. They also observed neuronal hypertrophy of the dorsal lateral striatum, a region linked to habit learning.

In humans, it is not possible to experimentally induce chronic stress and observe its long-term consequences, but several techniques have been used to induce acute, mild stress that reliably results in HPA-axis activation (see Dickerson & Kemeny 2004). These acute, mild stressors have been shown to impair performance on PFC-dependent tasks and reduce PFC BOLD responses (e.g., Qin et al. 2009, Raio et al. 2013). Using a devaluation paradigm similar to that described above, Schwabe & Wolf (2009) found that acute stress yields a similar shift from goal-directed to habitual choices in humans. In a follow-up series of studies, Schwabe and colleagues (2010, 2011) administered drugs targeting glucocorticoid and noradrenergic activity to explore the neurohormonal changes that might underlie this stress-induced shift to habitual actions. They observed that administering both hydrocortisone and an $\alpha 2$-adrenoceptor antagonist (yohimbine), which increases noradrenaline levels in the brain, resulted in the shift to habitual actions typically observed with stress, but neither drug alone was sufficient to do so (Schwabe et al. 2010). Conversely, if stressed participants were administered a beta-adrenergic antagonist (propranolol), they failed to demonstrate the typical shift to habitual actions, despite intact increased cortisol with stress (Schwabe et al. 2011). This observation suggests that both glucocorticoids and noradrenaline are necessary neurohormonal components underlying the shift from goal-directed to habitual actions with stress.

These findings are consistent with research in nonhuman animals showing that elevated noradrenaline levels during stress alter executive control and PFC function, and glucocorticoids play a role in the exaggeration and persistence of this effect. The impact of these neurohormonal changes on the PFC is mediated through the amygdala (Arnsten 2009).

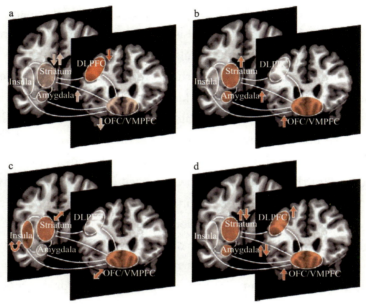

Figure 2

Potential candidates for multiple modulatory neural circuits involved in affect and decision making. (*a*) Decision making under stress. Even mild stress impairs the dorsolateral prefrontal cortex (DLPFC), leading to decreased goal-directed behavior and increased habitual behavior (Schwabe & Wolf 2009, Otto et al. 2013). Stress also impairs the orbitofrontal cortex (OFC)/ventromedial prefrontal cortex (VMPFC) and enhances amygdala function, whereas different subdivisions of the striatum may show increased or decreased reactivity with stress (Arnsten 2009, Roozendaal et al. 2009). (*b*) Emotion contributes to valuation. The amygdala influences the computation of subjective value in the striatum and the OFC/ VMPFC and modulates learning from reinforcement (Li et al. 2011, Sokol-Hessner et al. 2012, Rudebeck et al. 2013). (*c*) The relationship between subjective value and BOLD activity. A meta-analysis of fMRI studies (Bartra et al. 2013) shows a linear relationship between subjective value and activity in the OFC/ VMPFC and the ventral striatum, whereas the relationship between insula activity and subjective value is U-shaped, suggesting that the insula may contribute to value computation in situations of high arousal or salience. (*d*) Cognitive emotion regulation. The influence of emotion on choice can be altered using cognitive emotion regulation techniques mediated by the DLPFC and VMPFC. Emotional reactions can be increased or decreased with these techniques (Ochsner & Gross 2005), leading to corresponding changes in the amygdala and striatum (Delgado et al. 2008a,b; Sokol-Hessner et al. 2013).

This balance between PFC and striatal contributions to decision making can also be observed in tasks that tap into model-based and model-free reinforcement learning

(Daw et al. 2005). In model-free learning, one learns which choice is beneficial through previous experience with its reinforcing consequences, whereas model-based learning requires a model of the environment that allows one to engage in a series of choices that maximize reward. Theoretical models (Daw et al. 2005) and functional magnetic resonance imaging (fMRI) studies (Gläscher et al. 2010) suggest that although both model-based and model-free decisions engage striatal-based reinforcement learning mechanisms, model-based choices also depend on interactions with the lateral PFC. Using a decision-making task that yields different patterns of choices depending on whether one is using a model-free or model-based strategy, Otto and colleagues (2013) found that stress attenuated model-based, but not model-free, contributions to choice behavior. Relatively high baseline working-memory capacity had a protective effect, attenuating the deleterious effect of stress on choices.

Several studies have examined the impact of stress on tasks of risky decision making, although the nature of these findings varies depending on a number of factors. Porcelli & Delgado (2009) found that stress exacerbated the "reflection effect," which is the tendency to be risk seeking when choosing between possible losses and risk averse when choosing between potential gains. However, other risky decision-making tasks have reported that participants are more risk seeking overall under stress (Starcke et al. 2008) or more risk seeking in the loss domain only (Pabst et al. 2013). In addition, the impact of stress on risky decisions may depend on the level of risk (von Helversen & Reiskamp 2013) and may interact with gender (e.g., Preston et al. 2007, Lighthall et al. 2009). As this series of studies indicates, there are likely several decision and individual difference variables that will need to be disentangled to determine how stress may influence different aspects of risky decisions.

The only brain-imaging study on stress and risky decision making to date (Lighthall et al. 2012) used a task in which participants earn points for inflating virtual balloons but must "cash out" before the balloon explodes or risk losing their points. Consistent with earlier findings (Lighthall et al. 2009), this task demonstrated a gender interaction: Males were more risk seeking, and females less risk seeking, following the stressor. There was also an interaction in BOLD responses: Males in the stress condition showed greater activation in the insula and putamen while making decisions, whereas females showed the opposite pattern. Although the precise roles of these regions in this task is unclear, the insula has been implicated in signaling aversive outcomes and weighing differences in expected value in risky decision making (Clark et al. 2008), whereas the putamen is known to play a role in habitual behavior (Balleine & O'Doherty 2010).

Limited research in other decision-making domains has examined the impact of stress. Studies of intertemporal choice have found that stress exaggerates the tendency to discount future rewards in favor of smaller immediate rewards (Kimura et al. 2013) or that this effect depends on the level of perceived stress (Lempert et al. 2012). Studies of moral decision making find that stress decreases the likelihood of making utilitarian judgments in personal moral decisions (i.e., inflicting harm to maximize good consequences; Youssef et al. 2012) and correlates with egocentric moral decisions (Starcke et al. 2011). Finally, stress results in more prosocial decisions (i.e., more trust and less punishment) but less generosity as well (von Dawans et al. 2012, Vinkers et al. 2013). Most of these studies hypothesize that their findings could be attributed to the impact of stress on executive control and PFC function, although direct evidence of diminished PFC involvement due to stress in these tasks is lacking.

As this emerging research on stress and decision making indicates, a range of processes are tapped in decision-making tasks, and stress has broad, and uneven, effects on brain function. In addition, there are significant individual differences in response to stress and differential effects of chronic, acute, mild, or severe stress. In spite of these caveats, the extensive literature characterizing the impact of stress on brain function can be leveraged to understand one of the means by which an affective state might influence choices. To the extent that we can identify specific neural circuits linked to specific decision variables, such as in goal-directed versus habitual actions and model-based versus model-free decisions, we can start to characterize the distinct impact of stress.

Mood

Although stress responses are often accompanied by negative feelings, mood states are characterized primarily by subjective feelings with little concordant psychophysiological or neurohormonal changes (Scherer 2005). Research on the neural basis of moods is limited by two significant constraints. First, moods are relatively lasting states, which prevents quickly switching from one mood to another—a necessary requirement to detect within-subject differential BOLD responses, which are optimal for fMRI studies. Second, given that the primary measure of a mood state is via subjective report, it is challenging to assess moods in nonhuman animals. However, substantial psychological evidence indicates that moods affect decisions and provides some hints of the neural changes that may mediate these effects.

The influence of mood on the neural systems of decision making has been explored in a social decision-making task, the ultimatum game. In this game, there are two players: a proposer who is given a sum of money to divide with a responder, who can choose whether to accept or reject the proposer's offer. If the responder rejects the offer, both players receive nothing. In theory, the responder should accept any offer because the alternative is nothing at all. However, previous research has shown that offers around 20% of the total sum are rejected approximately half the time, presumably to punish the proposer for an unfair offer (Thaler 1988).

Studies inducing mood states prior to the ultimatum game show that participants in the role of the responder were more likely to reject unfair offers when they were in a sad (Harlé & Sanfey 2007) or disgust mood (Moretti & di Pellegrino 2010). BOLD responses during the ultimatum game were examined in two groups of participants in the role of the responder. One group underwent a sad mood induction procedure prior to scanning, and the other underwent a control, neutral mood induction task (Harlé et al. 2012). As expected, the sad mood group rejected more offers. During the presentation of unfair offers, investigators noted significantly more BOLD activation in the bilateral insula, the ventral striatum, and the anterior cingulate in the sad group relative to the control group. During the presentation of fair offers, there were no group differences in the insula; in the ventral striatum, however, the control group showed greater BOLD reward activity relative to the sad group. BOLD responses in the insula mediated the relationship between self-reported sadness and the tendency to reject unfair offers. In the context of this study, the authors suggested that the insula supports the integration of a negative mood state into the decision process. The findings in the ventral striatum are interpreted as reflecting reduced reward sensitivity when sad. This network of regions is proposed to underlie the infusion of sadness into the choice (Harlé et al. 2012).

Psychological research suggests that the infusion of mood into the computation of subjective value results from the carryover of the general action tendencies elicited by mood states onto the decision process. This proposed carryover effect is known as an appraisal tendency (Lerner et al. 2004). For example, Lerner and colleagues (2004) induced a sad, disgust, or neutral mood and explored its impact on the endowment effect—the phenomenon in which the price one is willing to accept to sell an owned item is greater than the price one would pay to buy the same item. They found that a sad mood reversed the endowment effect (i.e., higher buy prices than sell prices), whereas a disgust mood led to a reduction in both buy and sell prices. It was suggested that sadness is an indication that the current situation is unfavorable, which enhances the appraisal of the subjective value of choice options that change the situation. Disgust, however, is linked to a tendency to move away from or expel what is disgusting, which carries over to a tendency to reduce the subjective value of all items.

Numerous studies have shown that moods also influence risky choices. For example, sad moods can increase preferences toward high-risk options, whereas anxious moods bias preferences toward low-risk options (Raghunathan & Pham 1999). Consistent with this concept, fear results in less risk seeking and anger results in more risk seeking (Lerner & Keltner 2001). Finally, positive moods can exaggerate the tendency to overweigh losses relative to gains (i.e., loss aversion) in risky gambles (Isen et al. 1988), and some of these effects of mood on risky decisions may vary by gender (Fessler et al. 2004). Studies investigating the neural systems of risky decision making have highlighted the roles of the orbitofrontal cortex (OFC) in risk-prediction errors, or in updating assessments of risk (e.g., O'Neill & Schultz 2013), and the insular cortex in the representation of risk (e.g., Knutson & Bossaerts 2007). Both of these regions have also been implicated in the representation of mood states (Lane et al. 1997, Damasio et al. 2000). Although neural evidence has yet to indicate how mood states shift the neural representation of risk assessment, this overlap in the neural circuitry mediating mood and risk provides a starting point for investigations on this topic.

Incidental Affect: Summary and Other Factors

A range of incidental affective states may bias decisions. We have highlighted two such states, stress and mood, which have different effects on choices. Stress results in changes in brain function in several regions that have been implicated in different aspects of the decision process, most notably impaired function of the PFC. To date, we know relatively more concerning stress effects on the brain than we do about how to distinguish different decision variables that engage unique neural circuits. Investigations of mood and decision making, however, are limited by the sparse literature on the neural basis of moods. The best hypothesis at this point is that moods somehow shift neural processing in regions that are involved in the assessment of subjective value, such as the OFC. Our relatively poor understanding of the neural basis of moods is exacerbated by the fact that animal models are intrinsically limited for studying phenomena characterized by subjective states. However, given the extensive psychological research on this topic, and the prevalence of mood states in everyday life, unpacking the neural mechanisms of moods and decisions is critical if we ever hope to achieve a relatively nuanced and rich understanding of human decision making.

The carryover effect of incidental affect on decisions has also been linked to other factors not discussed above because they lacked either the measurement or manipulation of affect or evidence for the underlying neural mechanisms. For instance, studies of Pavlovian-to-instrumental transfer demonstrate that actions occurring in the presence of affective Pavlovian cues are modified consistent with the motivational valence of these cues (i.e., performed with more or less vigor; see Huys et al. 2011). Both the striatum and amgydala are highlighted as regions important in integrating the affective value of the Pavlovian cue with the value of the instrumental action (e.g., Corbit & Balleine 2005, Corbit et al. 2007). A similar line of psychological research, known as affective priming, examines how the presence of an emotional cue, such as an angry or happy face, shifts subsequent choices (Winkielman et al. 2005). These studies suggest that the emotional reaction to the cue carries over to the decision. Although emotional reactions are event driven and discrete and moods are lasting subjective states that do not require an eliciting event, the mechanism by which Pavlovian-to-instrumental transfer and affective priming are proposed to alter choices is similar to the notion of appraisal tendency discussed earlier. That is, emotional reactions, like moods, produce action tendencies that bleed over to the appraisal of the subjective value of concurrent choice options.

Finally, an understanding of the influence of incidental affect on choice behavior would be incomplete without considering how individual variability in baseline affective tendencies may alter decisions. Individuals' affect dispositions vary (Scherer 2005). For instance, some people are generally more anxious and others are more cheery. These traits with an affective flavor can influence choices, much like transient mood states. Just as anxious or fear mood states result in less risk taking, higher trait anxiety is also linked to less risky decisions, perhaps because anxiety results in more negative appraisals of subjective value (see Hartley & Phelps 2012 for a review). Of course, more extreme negative affect dispositions, such as trait anxiety, are linked to psychopathologies that have profound functional consequences, including maladaptive decisions. Accordingly, patients with anxiety disorders are more risk averse than are healthy individuals (Giorgetta et al. 2012) and are also more likely to punish in social decision-making tasks (Grecucci et al. 2013a). Given the clear link between maladaptive decisions and psychopathology, Sharp et al. (2012) proposed that decision science is an important tool to aid in our characterization of a range of psychological disorders.

EMOTION AS VALUE: HOW EMOTIONAL REACTIONS TO THE CHOICE MODULATE DECISIONS

Theories concerning the function of emotions universally highlight the role of emotions in driving actions (e.g., Frijda 2007). A classic example is the fight-or-flight response first characterized by Cannon (1915), in which a potentially threatening event (such as a predator) alters the physiological state to facilitate adaptive action (i.e., quickly escaping). Unlike the influence of incidental stress in biasing concurrent but unrelated choices, described above, in the predator example it is the choice options that evoke the emotional reaction, which in turn drives the choice. In this case, the emotional reaction is an important component of the value computation. Although this example may seem extreme, since we rarely encounter threats to our survival in our everyday lives, the principle applies in more subtle ways in our daily choices. That is, our emotional reactions to choice options or outcomes contribute to the determination of subjective value.

Emotional reactions vary widely in both intensity and quality. The neural circuits mediating the influence of emotion on choices may vary with these emotional qualities, and our understanding of the neural representation of different kinds of emotional reactions is still relatively limited. However, for emotions related to threat, extensive, cross-species research into the underlying neural circuits has been carried out. This literature provides a starting point to explore emotion's modulation of choice (see Phelps & LeDoux 2005 for a review).

The amygdala is a central component of this circuitry and is known to have a critical role in associating aversive, threatening events with neutral cues (i.e., Pavlovian fear conditioning). One of the amygdala's subregions, the lateral nucleus, is the site of synaptic plasticity linking neutral cues and aversive events. The lateral nucleus projects to both the central and basal nuclei. The central nucleus sends signals to the hypothalamus and brain stem nuclei, which mediate the physiological threat response, whereas the basal nucleus projects to the striatum. The striatum helps integrate motivation with action values, and in the presence of a conditioned threat cue, the basal nucleus input is critical for avoidance actions (LeDoux & Gorman 2001). This circuitry, with independent pathways mediating physiological threat reactions and avoidance actions, may also play a broader role in different decision contexts.

In addition to the amygdala's influence on the striatum, the amygdala has reciprocal connections with the OFC, and lesion studies in nonhuman primates have demonstrated that this connectivity with the amygdala contributes to the representation of value in the OFC (Rudebeck et al. 2013). Finally, the amygdala, the OFC, and the striatum share connectivity with the insula, a region also commonly linked to emotion's influence on decisions (e.g., Naqvi et al. 2007).

To determine how and when emotion influences the value computation, it is necessary to both measure and/or manipulate the emotional response and specify the decision variables. How emotion may influence a choice depends not only on the qualities of the emotional reaction, but also on the characteristics of the choice. Below we review studies exploring the neural systems mediating emotion's modulation of subjective value for different types of decision tasks (see **Figure 2*b*,*c***).

Risky Decisions

Risky decisions involve comparing choice options with varying probabilities of losses or gains. One of the first studies that assessed a specific emotion variable in humans and linked it to brain function used a risky decision-making paradigm known as the Iowa gambling task (IGT). The IGT presents participants with four decks of cards: two yielding small gains and losses (safe decks) and two yielding larger gains, but also occasional large losses (risky decks). Participants are asked to select cards sequentially from the different decks to win or lose money. Preferentially choosing from the safe decks results in a more favorable long-term outcome, and healthy control participants learn this through trial and error. In contrast, patients with OFC or amygdala lesions fail to shift their preference to the safe decks over time. Bechara and colleagues (1997, 1999) measured the skin conductance response (SCR), an indication of autonomic nervous system arousal, during choices and found that control participants developed an anticipatory SCR prior to selecting from the risky decks, whereas OFC- and amygdala-lesioned patients did not. The authors proposed that the anticipatory arousal response is a bodily (somatic) signal that steers participants away from less profitable, risky choices, an idea they refer to as the somatic marker hypothesis. Several studies over the years have challenged the primary assumption of the somatic marker hypothesis (Maia & McClelland 2004, Fellows & Farah 2005), a challenge further supported by evidence suggesting that autonomic responses and avoidance actions are driven by separate neural circuits (LeDoux & Gorman 2001). In spite of these caveats, this study was the first to clearly identify some of the neural circuitry mediating the integration of emotion in risky decisions.

Risky decision-making tasks vary widely, and several decision factors that influence choices may come into play. Two decision variables that are often confounded are risk sensitivity and loss aversion. Loss aversion is the tendency to weigh losses more than gains when considering the choice options. Someone who is highly loss averse may also appear to be risk averse, even if she or he is generally risk seeking in choices with minimal potential loss. Using a gambling task that enabled independent assessment of risk sensitivity and loss aversion, Sokol-Hessner and colleagues (2009) found that higher relative SCRs to losses versus gains were linked to greater loss aversion. No relationship was noted between arousal and risk sensitivity. Similarly, greater BOLD signal in the amygdala to losses relative to gains also correlated with loss aversion, but this response was unrelated to risk sensitivity (Sokol-Hessner et al. 2012). Consistent with these imaging results, patients with amygdala lesions show reduced loss aversion overall

(DeMartino et al. 2010), and administering a beta-adrenergic blocker (propranolol), which has previously been shown to diminish the amygdala's modulation of memory (Phelps 2006), also reduces loss aversion but does not affect risk sensitivity (Sokol-Hessner et al. 2013). This series of studies provides strong evidence that the amygdala plays a critical role in mediating aversion to losses but that it is not linked to risk tendencies. Given that most risky decision tasks do not independently model loss aversion and risk sensitivity, some observed effects of emotion in risky decision-making tasks may be due to loss aversion and not to risk attitudes per se.

Both the IGT and the task used by Sokol-Hessner and colleagues (2009) engage neural and physiological systems that serve to identify potential negative outcomes, prompting avoidance actions. However, other risky decision tasks have relatively few losses, as illustrated in pay-to-play games similar to slot machines, in which the only loss that occurs is when participants pay to play and do not win. Studies with these tasks have shown that emotional responses, including pleasantness ratings, SCR, and cardiovascular measures, to wins and near misses predict gambling propensity, including probable pathological gambling (Lole et al. 2011, Clark et al. 2012). These findings have been linked to increased BOLD activity in the striatum and insula during near misses (Clark et al. 2008, Chase & Clark 2010). In this decision-making context, emotions may drive people to take risky choices and not avoid them.

Another potentially important factor in risky decision-making tasks is whether the risks are known and static or unknown and changing. In dynamic and uncertain environments, the decision maker must learn the risk involved in different choices, and this risk may change over time. In dynamic, risky decision-making tasks, autonomic arousal, assessed via pupil dilation, was associated with more uncertain, exploratory decisions (Jepma & Niewenhuis 2011) and surprising outcomes (Preuschoff et al. 2011). Arousal, as well as amygdala BOLD responses, has been linked to associability (Li et al. 2011), a learning signal related to the unexpected or surprising nature of the cue, which serves to gate updating values from prediction errors coded in the striatum (see also Roesch et al. 2012). These studies suggest that in dynamic choice contexts the emotional response may be a component of ongoing predictions and evaluation necessary for learning.

Risky decision-making tasks vary widely on several dimensions, including the content of the choice (e.g., gains versus losses) and its context (e.g., static or dynamic). As the above studies demonstrate, such factors matter in part because they may shift the modulatory role of emotion: from avoiding bad outcomes to seeking favorable ones, to weighing and incorporating new information in changing environments. Future research will need to dissociate these possible roles and influences of emotion by carefully identifying the decision variables at play, the shared and separate neural circuitry used, and the underlying computations driving choices.

Social Decisions

In our everyday lives, the stimuli most likely to elicit emotional responses are other people. Social decision-making tasks investigate how choices are influenced by social context. For most of these tasks, the shift in decisions is simply due to the presence of another person, even if that person is anonymous. This observation is apparent in the ultimatum game described above in which responder participants routinely reject potential profit to punish the proposer for unfair offers. Not surprisingly, such rejections of financial gain are not observed if the proposer is a computer (van't Wout et al. 2006).

Studies examining the neural basis of rejection in the ultimatum game report an increased BOLD signal in the insula during unfair offers that is correlated with rejection rate (Sanfey et al. 2003). Arousal, assessed via SCR, was also increased during unfair offers and correlated with rejection rates—a pattern not observed when playing against a computer (van't Wout et al. 2006). In this case, the subjective value of the unfair offers was modulated by the social context of the choice. Increased physiological arousal has also been correlated with choice behavior in a social, moral decision-making task, and patients with damage to the ventromedial PFC (VMPFC), including the OFC, show both reduced physiological arousal and diminished impact of the social context on decisions (Moretto et al. 2010). The insula and OFC are two regions linked to emotion's influence on risky decisions (see above), but in these tasks the emotional reaction is driven by the interpersonal nature of the decision, as opposed to other decision variables.

Although the simple presence of another person can evoke an emotional reaction that may influence choices, who that person is may also matter. The influence of individual characteristics in social decision making has been investigated in cross-race interactions. The impact of race group on decisions was examined using the trust game in which a participant must decide whether to invest money with a partner. Trust decisions correlated with nonconscious, negative evaluative race attitudes for Black versus White, such that participants with stronger negative implicit attitudes invested less with Black compared with White partners (Stanley et al. 2011). BOLD responses during this task showed greater amygdala activation for Black versus White partners, scaled for the size of the investment, whereas striatum activation reflected the race-based discrepancy in trust decisions (Stanley et al. 2012). These findings are consistent with a model in which the amygdala codes race-related evaluative information and the striatum integrates this information with the action value.

The impact of social factors on decisions has been demonstrated with many different decision tasks, and there have been numerous investigations of the neural circuitry and neurochemistry mediating these effects (see Rilling & Sanfey 2011, Kubota et al. 2012 for reviews). However, relatively few studies have examined whether the impact of social context on choice is related to emotional responses. Given the emotional salience of other people, it is possible that emotion mediates the influence of many social factors on decisions. Only by assessing emotional reactions during these decision tasks can we start to delineate the impact of emotion evoked by the social situation from other factors that are uniquely social.

Intertemporal Choice

Intertemporal choice tasks measure preferences between options available at different points in time. In general, people tend to prefer immediate rewards to rewards received after a delay, even when the delayed reward is larger. This phenomenon, known as temporal discounting, has been linked to many maladaptive behaviors, including poor retirement savings, obesity, and drug addiction.

Investigations of the neural systems mediating intertemporal choice have reported conflicting results. One study reported greater BOLD responses in the OFC and the striatum during choices with an immediate reward option and greater BOLD signal in the DLPFC related to choosing the delay option (McClure et al. 2004). This BOLD pattern was interpreted as supporting a theory proposed in economics (Laibson 1997) suggesting that immediate rewards engender a greater emotional response, as reflected in the striatum and OFC BOLD responses, whereas choosing the delayed reward requires cognitive control of this emotional impulse, thus engaging the DLPFC. Consistent with this proposed inhibitory role for the DLPFC, Figner and colleagues (2010) found that disrupting DLPFC function through transcranial magnetic stimulation resulted in greater temporal discounting; however, in contrast to this proposed model, so did lesions of the OFC (Sellitto et al. 2010). Another study found that BOLD signal in the VMPFC and striatum correlated with subjective value of both immediate and delayed rewards (Kable & Glimcher 2010), consistent with the known roles for these regions in the representation and updating of value (see Bartra et al. 2013 for review). The investigators suggested that increased BOLD responses to immediate reward options observed in the earlier study were due to the fact that immediate rewards generally had a greater subjective value than did delayed rewards. However, none of these studies assessed emotional responses.

To determine if emotion plays a role in temporal discounting, Lempert and colleagues (2013) measured arousal, as assessed with pupil dilation, during an intertemporal choice task. Surprisingly, emotional arousal did not reliably correlate with the subjective value of either immediate or delayed rewards, but rather this relationship varied depending on the structure of the choice set. Greater arousal responses were observed when rewards were better than expected, regardless of whether those rewards were immediate or delayed. These findings conflict with the model proposed by McClure and colleagues (2004), which suggests that it is the emotional response to the immediate choice that drives discounting, and more closely align with the study by Kable & Glimcher (2007), which proposed a unified neural representation of subjective value of immediate and delayed rewards; both may be influenced by emotion depending on the task environment.

Further support for the notion that both immediate and delayed rewards elicit emotional responses that influence choices comes from studies investigating how altering the emotional salience of the delayed reward increases patience. For example, manipulating the mental representation of a future reward to make it more concrete can change its emotional intensity and the choice. Benoit and colleagues (2011) gave participants a typical intertemporal choice task but asked them to imagine specific ways they could spend the delayed reward in the future. This manipulation increased subjective ratings of vividness and emotional intensity of the future reward and resulted in less temporal discounting. This effect was associated with increased coupling between the VMPFC and the hippocampus. A study using a similar task replicated these behavioral results and found, consistent with Kable & Glimcher (2007), that subjective value for delayed rewards correlated with BOLD signal in the striatum and the OFC, whereas activation of the dorsal anterior cingulate, and its connectivity with the hippocampus and the amygdala, mediated the change in discount rate (Peters & Buchel 2010). These same neural circuits are known to be involved in the future projection of personal events and their modulation by emotion (Sharot et al. 2007).

Studies of intertemporal choice provide a compelling example of the influence of affective neuroscience on decision science: The predominant theory used to explain the tendency to discount future rewards in economics relies on dual systems, one impulsive (emotion) and one that controls these impulses (cognitive control; Laibson 1997). As the discussion above indicates, to the extent that emotion plays a role in this behavior, emotion's contribution varies depending on the choice environment and the task structure. This variability in the role of emotion provides an opportunity for investigators to manipulate task parameters that alter emotion to influence the tendency of subjects to discount future rewards, a topic we discuss in more detail in the next section.

Emotion as Value: Summary and Related Phenomena

In addition to the decision tasks described above, emotion is also thought to contribute to subjective value computation in drug addiction, although the measurement and quantification of emotion and the understanding of the underlying neural mechanisms lag behind theory. For example, intense cue-driven motivation, termed craving, is central to addiction theory (Skinner & Aubin 2010). Cravings are a major factor that contribute to relapse, but their source is complex; studies have variously connected them to the insula (Naqvi et al. 2007), the striatum (Kober et al. 2010), and the PFC (Rose et al. 2011). Nevertheless, understanding the systems that induce these motivational desires will ultimately lead to significant advances in the treatment of such disorders of choice.

A primary function of emotion is to provide a signal to the organism that a stimulus or event may be relevant for present or future survival or well-being (e.g., Frijda 2007). Thus it is not surprising that emotional reactions modulate a range of cognitive functions, such as memory, attention, and perception (Phelps 2006). It is also not surprising that part of the calculation of the value of decision options should include the nature of the emotional response elicited by those options or potential outcomes. How this occurs, however, varies depending on the decision variables assessed and the specific emotional reaction. As our review of this literature indicates, there is likely a collection of neural circuits underlying emotion's modulation of the value calculation.

Across studies of the neural basis of decision making, the OFC/VMPFC and the striatum are cited as necessary for the coding of subjective value; the striatum is specifically linked to updating values from reinforcement (prediction errors) via dopaminergic projections from the ventral tegmental area (e.g., Bartra et al. 2013). The studies outlined above examining the impact of emotion also implicate, in addition to these regions, the insula and amygdala. The insula is a large region linked to numerous functions relevant to decision making, including the anticipation of pain (Ploghaus et al. 1999) and monetary loss (Knutson & Bossaerts 2007), as well as the representation of disgust (Phillips et al. 1997) and physiological arousal (Critchley et al. 2000). A recent meta-analysis of fMRI studies examining the coding of subjective value found, not surprisingly, that the OFC/VMPFC and the ventral striatum emerged as two regions with BOLD responses that positively correlate with subjective value. In contrast, the insula, along with some other striatal regions and the dorsomedial PFC, showed greater BOLD responses for both more positive or negative subjective value. Bartra et al. (2013) suggest that the insula may integrate emotional salience or arousal linked to the decision variables into the value computation, regardless of its valence. As mentioned above, connectivity between the amygdala and the ventral striatum is critical for enabling avoidance behavior to acquired threats (LeDoux & Gorman 2001), and the amygdala contributes to value coding in the OFC (Rudebeck et al. 2013). The amygdala may play a role in avoidance across a range of decision tasks (e.g., Stanley et al. 2012, Sokol-Hessner et al. 2013), as well as in modulating learning from both positive and negative reinforcement more broadly (e.g., Roesch et al. 2012; Murray & Rudebeck 2013) (see **Figure 2*b,c***). The limited research to date on the integration of emotion into value computation is starting to yield a network of regions, but our understanding of precisely how these regions interact in more complex human decision-making tasks is still relatively unclear.

CHANGING AFFECT, CHANGING CHOICES

Clinical interventions for a range of psychopathologies are focused on changing affect. Outside the clinic, the ability to regulate the appropriateness of emotional responses to circumstances is a major component of healthy, adaptive social behavior and well-being. Although we often describe emotions as reactions to environmental stimuli that are beyond our control, affective scientists have long recognized the fluidity of our emotional lives and our ability to alter or determine our emotions. A major focus of basic research in affective neuroscience over the past decade has been to understand how emotions can be modified and how we can utilize this flexibility of emotion to develop more effective clinical interventions or more satisfying and healthy lives (Hartley & Phelps 2010, Davidson & Begley 2012).

To the extent that affect and emotions are incorporated into the assessment of subjective value, changing emotions should also change choices. Although several techniques have been used to change emotion in the laboratory across species (see Hartley & Phelps 2010 for a review), and all these techniques are presumed to influence later choices, only a few have been implemented directly during decision-making tasks to assess how choices are altered. Below we review the research examining one such technique and highlight some other potential mechanisms for future investigation (see **Figure 2*d***).

Cognitive Emotion Regulation

The common wisdom that one can see the glass as half full or half empty captures the essence of cognitive emotion regulation. Our emotional reactions are determined, in part, by how we appraise or interpret the circumstance or event (Scherer 2005). Although some individuals may have a general tendency to see the world in a positive or negative light, the ability to shift emotion through changing one's interpretation of an event, known as reappraisal, can also be taught and consciously applied. In a typical reappraisal task, the participant is asked to think about the stimulus differently to reduce its negative emotional consequences.

Many studies have investigated the neural systems that mediate the cognitive regulation of negative emotions as assessed through subjective reports or physiological responses (see Ochsner & Gross 2005). They typically report increased DLPFC BOLD responses during regulation versus attend conditions accompanied by decreased amygdala activation. The DLPFC is proposed to implement the executive control needed to actively reinterpret the stimulus during reappraisal, whereas the amygdala is involved in the expression of the emotional response. There is relatively sparse direct connectivity between the DLPFC and amygdala, so it is unlikely that the DLPFC directly influences amygdala function but rather does so through more ventral PFC regions. The VMPFC is known to have reciprocal connections with the amygdala that inhibit emotional reactions following extinction learning in Pavlovian fear-conditioning tasks, and it is proposed to mediate the influence of the DLPFC on the amygdala (Delgado et al. 2008b); however, other studies have suggested that the ventrolateral PFC (VLPFC) plays this role (Buhle et al. 2014). This DLPFC-VMPFC/VLPFC-amygdala circuitry is thought to underlie the cognitive control of diminishing negative emotional reactions, but it may also play a role in increasing negative affect depending on the reappraisal strategy (Otto et al. 2014). Emotion regulation strategies can also be employed to reduce arousal associated with anticipated monetary reward. These strategies engage overlapping regions of the DLPFC and VMPFC and yield decreased BOLD reward responses in the striatum (Delgado et al. 2008a). A similar circuitry has been implicated in the cognitive control of cravings (Kober et al. 2010).

In a risky decision-making task, a reappraisal strategy altered both arousal and choices. As described earlier, Sokol-Hessner and colleagues (2009) found that the relative SCR response to losses relative to gains correlated selectively with loss aversion but was unrelated to risk sensitivity. A similar pattern was observed for amygdala BOLD signal (Sokol-Hessner et al. 2013). In a variation of this task, participants were instructed to reappraise the significance of the choice by thinking of it as one of many, or to "think like a trader" building a portfolio. Using this strategy reduced the SCR to losses, and this reduction was correlated with diminished loss aversion, with no effect on risk sensitivity (Sokol-Hessner et al. 2009). Mirroring the SCR results, reduced amygdala BOLD responses to losses during regulation also correlated with a reduction in loss aversion. In contrast, baseline BOLD responses in the DLPFC, VMPFC, and striatum increased with regulation (Sokol-Hessner et al. 2013). These findings suggest that using a reappraisal strategy to change emotion and choices engages the same neural circuitry that is observed in more typical emotion regulation tasks. A similar reappraisal strategy that either emphasized or de-emphasized the importance of each individual choice was found to both increase and decrease subjective value in a risky decision task (Braunstein et al. 2014). In addition, in an intertemporal choice study described above, reframing the interpretation of a future reward resulted in more patience (Benoit et al. 2011).

Emotion regulation strategies have also been used to change the tendency to punish in the ultimatum game. Van't Wout and colleagues (2010) asked participants to play the ultimatum game while utilizing a cognitive emotion reappraisal strategy. In the responder role, participants who reappraised the motivations of the proposer in suggesting an unfair offer were less likely to reject it. This cognitive emotion regulation manipulation carried over to future choices. When the participants were subsequently put in the proposer role, they were less likely to propose unfair offers. In a follow-up fMRI study, participants in the responder role were asked to imagine either negative intentions of the proposer or positive intentions. Relative to a baseline condition, these reappraisal strategies resulted in rejecting more or fewer unfair offers, respectively, and subjective emotional responses varied as well. Consistent with previous research, activation of the insula predicted the rejection of unfair offers (Sanfey et al. 2003, Harlé et al. 2012), and the regulation strategies resulted in both increased insula BOLD responses with the negative intention strategy and decreased BOLD signal with the positive intention strategy. As expected, given the general emotion regulation circuitry outlined above, the DLPFC showed increased activation during both reappraisal conditions relative to baseline (Grecucci et al. 2013b).

As these studies indicate, cognitive emotion regulation techniques are flexible strategies that can rapidly change emotional reactions. The reappraisal strategies described above were adapted to the specific decision situation but had the same effect of altering the emotional response and modulating decisions, and they engaged typical cognitive emotion regulation regions. This confluence of evidence provides strong support for the notion that emotion is a critical component of the assessment of subjective value in these tasks because changing emotion also changed the choice.

Changing Affect: Summary and Other Potential Techniques

The notion that changing affect alters decisions was also demonstrated in the incidental affect manipulations described above. In those studies, inducing stress or a mood in the laboratory changed choices. In contrast, cognitive emotion regulation techniques alter choices by changing the appraisal of the decision variables that elicit emotional reactions. These techniques are powerful because they are flexible, can alter emotion in different ways, and can be quickly acquired and utilized without changing the situation. However, they require an effortful application of the strategy, which may not always be ideal. With practice, these strategies may become more automatic and less deliberate. Consistent with this, novice stock traders have demonstrated more physiological arousal to volatility in the stock market, a finding attributed to loss aversion, as compared with more senior stock traders, who show less arousal and better choices (Lo & Repin 2002), perhaps resulting from a broader perspective on market volatility gained from experience.

Although the flexibility of cognitive emotion regulation techniques can be an advantage, there are also some potential disadvantages. Cognitive emotion regulation strategies are less successful in stressful situations (Raio et al. 2013), perhaps owing to their dependence on the DLPFC. In addition, when emotional reactions consistently result in maladaptive choices, it may be useful to have a technique that leads to a more lasting change. Affective neuroscience has identified a few such strategies, but their impact on emotional reactions linked to decision making has not yet been widely investigated. For example, extinction training has been used to reduce acquired affective responses by repeatedly presenting the cue without the associated reinforcement. Although this technique can be effective, it leaves the original association intact; thus the unwanted affective response may return. To induce a more lasting change in learned associations underlying emotional responses and instrumental actions, researchers have recently investigated techniques that change the original associative, affective memory by altering its re-storage after retrieval or reconsolidation. Our understanding of reconsolidation mechanisms and how to target these processes in humans is still in its infancy, but this technique may lead to exciting advances in reducing the impact of maladaptive emotional reactions on choices (see Hartley & Phelps 2010 for a review).

Finally, an interesting twist in investigations examining the relation between emotion and decision making is that choices themselves can alter emotions. For example, animals given the opportunity to learn to avoid shocks show a lasting benefit, exhibiting diminished fear responses and faster and more robust extinction in subsequent tasks in which they do not have control over the shock reinforcer (Maier & Watkins 2010, Hartley et al. 2014). This research suggests that this persistent impact of choice on future threat reactions results from alterations in the brain stem–prefrontal–amygdala circuitry underlying the generation and control of learned threat associations (Maier & Watkins 2010). In humans, the opportunity for choice enhances subjective affective ratings of choice options and concurrently increases BOLD reward responses in the striatum (Leotti & Delgado 2011, 2014). Psychological theories have emphasized the importance of perception of control over one's environment on well-being (e.g., Bandura et al. 2003), as well as the impact of choices on preferences (e.g., Festinger 1957). Studies examining the neural basis of the impact of choice on emotional reactions and preferences (e.g., Sharot et al. 2009) are starting to provide a neurobiological framework for these psychological findings.

To the extent that affect and emotion influence choices, changing affective responses will alter our decisions. The emerging research on techniques to change affect shows that a range of mechanisms to modify emotions can be differentially applied in different decision contexts. Some are flexible and rapid, such as cognitive emotion regulation, and others are more lasting, such as targeting reconsolidation. In addition, choices themselves can change affect, which in theory should change subsequent choices. If we can discover and characterize more effective means to alter emotion, we should be able to harness these techniques to help optimize decisions.

MOVING PAST DUAL SYSTEMS TO MULTIPLE MODULATORY NEURAL CIRCUITS

In this overview of the current neuroscience literature exploring the relationship between affect and decision making, we have attempted to identify the neural circuits that mediate this interaction. What is emerging is clearly incompatible with the notion of two systems. Rather the literature suggests that there are multiple neural circuits underlying the modulation of decision making by emotion or affect. As our breakdown of affective components and decision tasks demonstrates, the specific neural circuits involved vary depending on which affective component is engaged and which decision variables are assessed. Thus, we suggest an alternative approach to understand the relation between emotion and decision making, which entails characterizing and identifying the multiple neural circuits underlying the different means by which emotion and affect influence decisions (see also Sanfey & Chang 2008).

Of course, this multiple modulatory neural circuits approach is not nearly as parsimonious as the dual systems account, and one could argue that dual systems is simply a rough and useful heuristic to characterize the role of emotion in decision making. Although referring to the heart and head may be useful in thinking about some types of decisions outside the laboratory, the reference to dual systems as the primary psychological and neural theory for understanding the relationship between emotion and decisions is still relatively common in the scientific literature (e.g., Cohen 2005, Greene 2007, Figner et al. 2009, Reyna & Brainerd 2011, Paxton et al. 2012). Much as some researchers have suggested that scientists abandon the limbic system concept because its continued use impedes progress in understanding the detailed and complex neural basis of affect (LeDoux 2000), we argue that the repeated reference to dual systems of emotion and reason in research on decision making potentially limits scientific advances by discouraging investigations that capture the detailed and nuanced relationships between unique aspects of affect and choices. Furthermore, it suggests to the layperson and scientists in other disciplines that the intuitive and historical notion of competing forces of emotion and reason is based on scientific fact. Perpetuation of this idea may dampen enthusiasm for efforts to further explore the complex interactions of affect and decision making and may result in the development of potentially misguided or nonoptimal techniques to inhibit emotion in order to promote rational decision making.

Despite its complexity, our proposed conceptualization of the relationship between affect and decision making begins to capture the subtleties involved in understanding their interaction. Both affect and decision making are general terms that describe a collection of factors and processes, only some of which are explored above. Investigating affect and emotion is challenging by itself, both because manipulating and measuring affect in the laboratory is difficult and because there is debate about how best to characterize affective variables. Differentiating the collection of unique variables that influence choice in any given situation is also challenging for decision science. In spite of these caveats, initial attempts to measure or manipulate affective components and to relate them to specific aspects of decision tasks have yielded exciting advances. As the disciplines of affective neuroscience and neuroeconomics advance, we can build on this progress to further characterize the multiple neural circuits that mediate the modulatory relationship between emotion and decision making.

DISCLOSURE STATEMENT

The authors are not aware of any affiliations, memberships, funding, or financial holdings that might be perceived as affecting the objectivity of this review.

ACKNOWLEDGMENTS

The authors thank Catherine Stevenson, Jackie Reitzes, and Catherine Hartley for assistance with manuscript preparation and Sandra Lackovic for assistance with figure preparation. This work was partially funded by a grant from the National Institutes of Health (AG039283) to E.A.P.

LITERATURE CITED

Arnsten AF. 2009. Stress signaling pathways that impair prefrontal cortex structure and function. *Nat. Rev. Neurosci.* 10(6):410–22

Balleine BW, O'Doherty JP. 2010. Human and rodent homologies in action control: cortico-striatal determinants of goal-directed and habitual action. *Neuropsychopharmacology* 35(1):48–69

Bandura A, Caprara GV, Barbaranelli C, Gerbino M, Pastorelli C. 2003. Role of affective self-regulatory efficacy in diverse spheres of psychosocial functioning. *Child Dev.* 74(3):769–82

Barrett LF. 2006. Are emotions natural kinds? *Perspect. Psychol. Sci.* 1(1):28–58

Bartra O, McGuire JT, Kable JW. 2013. The valuation system: a coordinate-based meta-analysis of BOLD fMRI experiments examining neural correlates of subjective value. *NeuroImage* 76:412–27

Bechara A, Damasio H, Damasio AR, Lee GP. 1999. Different contributions of the human amygdala and ventromedial prefrontal cortex to decision-making. *J. Neurosci.* 19(13):5473–81

Bechara A, Damasio H, Tranel D, Damasio AR. 1997. Deciding advantageously before knowing the advantageous strategy. *Science* 275:1293–95

Benoit RG, Gilbert SJ, Burgess PW. 2011. A neural mechanism mediating the impact of episodic prospection on farsighted decisions. *J. Neurosci.* 31:6771–79

Braunstein ML, Herrera SJ, Delgado MR. 2014. Reappraisal and expected value modulate risk taking. *Cogn. Emot.* 28(1):172–81

Buhle JT, Silvers JA, Wager TD, ONyemekwu C, Kober H, et al. 2014. Cognitive reappraisal of emotion: A meta-analysis of human neuroimaging studies. *Cereb. Cortex.* In press

Cannon WB. 1915. *Bodily Changes in Pain, Hunger, Fear and Rage: An Account of Recent Researches into the Function of Emotional Excitement.* New York: D. Appleton

Chase HW, Clark L. 2010. Gambling severity predicts midbrain response to near-miss outcomes. *J. Neurosci.* 30(18):6180–87

Clark L, Bechara A, Damasio H, Aitken MR, Sahakian BJ, Robbins TW. 2008. Differential effects of insular and ventromedial prefrontal cortex lesions on risky decision-making. *Brain* 131:1311–22

Clark L, Crooks B, Clarke K, Aitken MR, Dunn BD. 2012. Physiological responses to near-miss outcomes and personal control during simulated gambling. *J. Gambl. Stud.* 28(1):123–37

Cohen JD. 2005. The vulcanization of the human brain: a neural perspective on interactions between cognition and emotion. *J. Econ. Perspect.* 19:3–24

Corbit LH, Balleine BW. 2005. Double dissociation of basolateral and central amygdala lesions on the general and outcome-specific forms of Pavlovian-instrumental transfer. *J. Neurosci.* 25(4):962–70

Corbit LH, Janak PH, Balleine BW. 2007. General outcome-specific forms of Pavlovian-instrumental transfer: the effect of shifts in motivational state and inactivation of the ventral tegmental area. *Eur. J. Neurosci.* 26:3141–49

Critchley HD, Elliot R, Mathias CJ, Dolan RJ. 2000. Neural activity relating to generation and representation of galvanic skin conductance responses: a functional magnetic resonance imaging study. *J. Neurosci.* 20(8):3033–40

Damasio AR. 2005. *Descartes'Error: Emotion, Reason and the Human Brain.* New York: Penguin

Damasio AR, Grabowski TJ, Bechara A, Damasio H, Ponto LL, et al. 2000. Subcortical and cortical brain activity during the feeling of self-generated emotions. *Nat. Neurosci.* 3(10):1049–56

Davidson RJ, Begley S. 2012. *The Emotional Life of Your Brain: How Its Unique Patterns Affect the Way You Think, Feel, and Live—and How You Can Change Them.* New York: Hudson Street

Daw ND, Niv Y, Dayan P. 2005. Uncertainty-based competition between prefrontal and dorsolateral striatal systems for behavioral control. *Nat. Neurosci.* 8(27):1704–11

De Martino B, Camerer CF, Adolphs R. 2010. Amygdala damage eliminates monetary loss aversion. *Proc. Natl. Acad. Sci. USA* 107(8):3788–92

Delgado MR, Gillis MM, Phelps EA. 2008a. Regulating the expectation of reward via cognitive strategies. *Nat. Neurosci.* 11:880–81

Delgado MR, Nearing KI, LeDoux JE, Phelps EA. 2008b. Neural circuitry underlying the regulation of conditioned fear and its relation to extinction. *Neuron* 59(5):829–38

Dias-Ferreira E, Sousa JC, Melo I, Morgado P, Mesquita AR, et al. 2009. Chronic stress causes frontostriatal reorganization and affects decision-making. *Science* 325:621–25

Dickerson SS, Kemeny ME. 2004. Acute stressors and cortisol responses: a theoretical integration and synthesis of laboratory research. *Psychol. Bull.* 130:355–91

Ekman P, Davidson RJ. 1994. *The Nature of Emotion: Fundamental Questions.* New York: Oxford Univ. Press

Fellows LK, Farah MJ. 2005. Different underlying impairments in decision-making following ventromedial and dorsolateral frontal lobe damage in humans. *Cereb. Cortex* 15:58–63

Fessler DMT, Pillsworth EG, Flamson TJ. 2004. Angry men and disgusted women: an evolutionary approach to the influence of emotion on risk-taking. *Organ. Behav. Hum. Decis. Process.* 95:107–23

Festinger L. 1957. *A Theory of Cognitive Dissonance.* Stanford, CA: Stanford Univ. Press

Figner B, Knoch D, Johnson EJ, Krosch AR, Lisanby SH, et al. 2010. Lateral prefrontal cortex and self-control in intertemporal choice. *Nat. Neurosci.* 13:538–39

Figner B, Mackinley RJ, Wilkening F, Weber EU. 2009. Affective and deliberative processes in risky choice: age differences in risk taking in the Columbia Card Task. *J. Exp. Psychol. Learn. Mem. Cogn.* 35(3):709–30

Frijda NH. 2007. *The Laws of Emotion.* Mahwah, NJ: Lawrence Erlbaum

Giorgetta C, Grecucci A, Zuanon S, Perini L, Balestrieri M, et al. 2012. Reduced risk-taking behavior as a trait feature of anxiety. *Emotion* 12(6):1373–83

Gläscher J, Daw N, Dayan P, O'Doherty JP. 2010. States versus rewards: dissociable neural prediction error signals underlying model-based and model-free reinforcement learning. *Neuron* 66(4):585–95

Greene JD. 2007. Why are VMPFC patients more utilitarian? A dual-process theory of moral judgment explains. *Trends Cogn. Sci.* 11(8):322–23

Grecucci A, Giorgetta C, Brambilla P, Zuanon S, Perini L, et al. 2013a. Anxious ultimatums: how anxiety disorders affect socioeconomic behaviour. *Cogn. Emot.* 27(2):230–44

Grecucci A, Giorgetta C, Van't Wout M, Bonini N, Sanfey AG. 2013b. Reappraising the ultimatum: an fMRI study of emotion regulation and decision making. *Cereb. Cortex* 23(2):399–410

Harlé K, Sanfey AG. 2007. Incidental sadness biases social economic decisions in the ultimatum game. *Emotion* 7:876–81

Harlé KM, Chang LJ, van't Wout M, Sanfey AG. 2012. The neural mechanisms of affect infusion in social economic decision-making: a mediating role of the anterior insula. *NeuroImage* 61:32–40

Hartley CA, Gorun A, Reddan MC, Ramirez F, Phelps EA. 2014. Stressor controllability modulates fear extinction in humans. *Neurobiol. Learn. Mem.* In press

Hartley CA, Phelps EA. 2010. Changing fear: the neurocircuitry of emotion regulation. *Neuropsychopharma- cology* 35:136–46

Hartley CA, Phelps EA. 2012. Anxiety and decision-making. *Biol. Psychiatry* 72:113–18

Huys QJM, Cools R, Gölzer M, Friedel E, Heinz A, et al. 2011. Disentangling the roles of approach, activation and valence in instrumental and Pavlovian responding. *PLoS Comput. Biol.* 7(4):e1002028

Isen AM, Nygren TE, Ashby FG. 1988. Influence of positive affect on the subjective utility of gains and losses:It is just not worth the risk. *J. Personal. Soc. Psychol.* 55:710–17

Jepma K, Niewenhuis S. 2011. Pupil diameter predicts changes in the exploration-exploitation tradeoff: evidence for the adaptive gain theory. *J. Cogn. Neurosci.* 23:1587–96

Kable JW, Glimcher PW. 2007. The neural correlates of subjective value during intertemporal choice. *Nat. Neurosci.* 10(12):1625–33

Kable JW, Glimcher PW. 2010. An "as soon as possible" effect in human intertemporal decision making: behavioral evidence and neural mechanisms. *J. Neurophysiol.* 103(5):2513–31

Kahneman D. 2011. *Thinking, Fast and Slow.* New York: Farrar, Strauss, and Giroux

Kimura K, Izawa S, Sugaya N, Ogawa N, Yamada KC, et al. 2013. The biological effects of acute psychosocial stress on delay discounting. *Psychoneuroendocrinology* 38(10):2300–8

Knutson B, Bossaerts P. 2007. Neutral antecedents of financial decisions. *J. Neurosci.* 27(31):8174–77

Kober H, Mende-Siedlecki P, Kross EF, Weber J, Mischel W, et al. 2010. Prefrontal-striatal pathway underlies cognitive regulation of craving. *Proc. Natl. Acad. Sci. USA* 107(33):14811–16

Kubota JT, Banaji MR, Phelps EA. 2012. The neuroscience of race. *Nat. Neurosci.* 15:940–48

Laibson D. 1997. Golden eggs and hyberbolic discounting. *Q. J. Econ.* 112(2):443–78

Lambert KG. 2003. The life and career of Paul Maclean: a journey toward neurobiological and social harmony. *Physiol. Behav.* 79(3):343–49

Lane RD, Reiman EM, Ahern GL, Schwartz GE, Davidson RJ. 1997. Neuroanatomical correlates of happiness, sadness and disgust. *Am. J. Psychiatry* 154(7):926–33

LeDoux JE. 2000. Emotion circuits in the brain. *Annu. Rev. Neurosci.* 23:155–84

LeDoux JE, Gorman JM. 2001. A call to action: overcoming anxiety through active coping. *Am. J. Psychiatry* 158:1953–55

Lempert KM, Glimcher PW, Phelps EA. 2013. *Reference-dependence in intertemporal choice.* Poster presented at Annu. Meet. Soc. Neuroecon., 12th, Lausanne, Switz.

Lempert KM, Porcelli AJ, Delgado MR, Tricomi E. 2012. Individual differences in delay discounting under acute stress: the role of trait perceived stress. *Front. Psychol.* 3:251

Leotti LA, Delgado MR. 2011. The inherent reward of choice. *Psychol. Sci.* 22:1310–18

Leotti LA, Delgado MR. 2014. The value of exercising control over monetary gains and losses. *Psychol. Sci.* 25:596–604

Lerner JS, Keltner D. 2001. Fear, anger, and risk. *J. Pers. Soc. Psychol.* 81:146–59

Lerner JS, Li Y, Valdesolo P, Kassam K. 2015. Emotion and decision making. *Annu. Rev. Psychol.* 66:In press

Lerner JS, Small DA, Loewenstein G. 2004. Heart strings and purse strings: carryover effects of emotions on economic decisions. *Psychol. Sci.* 15:337–41

Li J, Schiller D, Schoenbaum G, Phelps EA, Daw ND. 2011. Differential roles of human striatum and amygdala in associative learning. *Nat. Neurosci.* 14:1250–52

Lighthall NR, Mather M, Gorlick MA. 2009. Acute stress increases sex differences in risk seeking in the balloon analogue risk task. *PLoS ONE* 4:e6002

Lighthall NR, Sakaki M, Vasunilashorn S, Nga L, Somayajula S, et al. 2012. Gender differences in reward-related decision processing under stress. *Soc. Cogn. Affect. Neurosci.* 7:476–84

Lo AW, Repin DV. 2002. The psychophysiology of real-time financial risk processing. *J. Cogn. Neurosci.* 14(3):323–39

Lole L, Gonsalvez CJ, Blaszczynski A, Clarke AR. 2012. Electrodermal activity reliably captures physiological differences between wins and losses during gambling on electronic machines. *Psychophysiology* 49(2):154–63

Maclean PD. 1949. Psychosomatic disease and the visceral brain; recent developments bearing on the Papez theory of emotion. *Psychosom. Med.* 11(6):338–53

Maia TV, McClelland JL. 2004. A reexamination of the evidence for the somatic marker hypothesis: what participants really know in the Iowa gambling task. *Proc. Natl. Acad. Sci. USA* 101:16075–80

Maier SF, Watkins LR. 2010. Role of the medial prefrontal cortex in coping and resilience. *Brain Res.* 1355:52–60

McClure SM, Laibson DI, Loewenstein G, Cohen JD. 2004. Separate neural systems value immediate and delayed monetary rewards. *Science* 306(5695):503–7

McEwen BS. 2007. Physiology and neurobiology of stress and adaptation: central role of the brain. *Physiol. Rev.* 87(3):873–904

Moretti L, di Pellegrino G. 2010. Disgust selectively modulates reciprocal fairness in economic interactions. *Emotion* 10:169–80

Moretto G, Làdavas E, Mattioli F, di Pellegrino G. 2010. A psychophysiological investigation of moral judgment after ventromedial prefrontal damage. *J. Cogn. Neurosci.* 22:1888–99

Murray EA, Rudebeck. 2013. The drive to strive: goal generation based on current needs. *Front. Neurosci.* 7:112

Naqvi NH, Rudrauf D, Damasio H, Bechara A. 2007. Damage to the insula disrupts addiction to cigarette smoking. *Science* 315(5811):531–34

Ochsner KN, Gross JJ. 2005. The cognitive control of emotion. *Trends Cogn. Sci.* 9:242–49

O'Neill M, Schultz W. 2013. Risk prediction error in orbitofrontal neurons. *J. Neurosci.* 33(40):15810–14

Otto AR, Raio CM, Chiang A, Phelps EA, Daw NA. 2013. Working-memory capacity protects model-based learning from stress. *Proc. Natl. Acad. Sci. USA* 110(52):20941–46

Otto B, Misra S, Prasad A, McRae K. 2014. Functional overlap of top-down emotion regulation and generation: an fMRI study identifying common neural substrates between cognitive reappraisal and cognitively generated emotions. *Cogn. Affect Behav. Neurosci.* In press

Pabst S, Brand M, Wolf OT. 2013. Stress effects on framed decisions: There are differences for gains and losses. *Front. Behav. Neurosci.* 7:142

Packard MG, Goodman J. 2012. Emotional arousal and multiple memory systems in the mammalian brain. *Front. Behav. Neurosci.* 6:14

Paxton JM, Ungar L, Greene JD. 2012. Reflection and reasoning in moral judgment. *Cogn. Sci.* 36(1):163–77

Peters J, Büchel C. 2010. Episodic future thinking reduces reward delay discounting through an enhancement of prefrontal-mediotemporal interactions. *Neuron* 66:138–48

Peters RS. 1970. Reason and passion. *R. Inst. Philos. Lect.* 4:132–53

Phelps EA. 2006. Emotion and cognition: insights from studies of the human amygdala. *Annu. Rev. Psychol.* 57:27–53

Phelps EA. 2009. The study of emotion in neuroeconomics. In *Neuroeconomics: Decision Making and the Brain*, ed. PW Glimcher, C Camerer, E Fehr, RA Poldrack, pp. 233–50. London: Elsevier

Phelps EA, LeDoux JE. 2005. Contributions of the amygdala to emotion processing: from animal models to human behavior. *Neuron* 48(2):175–87

Phillips ML, Young AW, Senior C, Brammer M, Andrew C, et al. 1997. A specific neural substrate for perceiving facial expressions of disgust. *Nature* 389(6650):495–98

Ploghaus A, Tracey I, Gati JS, Clare S, Menon RS, et al. 1999. Dissociating pain from its anticipation in the human brain. *Science* 284(5422):1979–81

Poldrack RA. 2006. Can cognitive processes be inferred from neuroimaging data? *Trends Cogn. Sci.* 10:59–63

Porcelli AJ, Delgado MR. 2009. Acute stress modulates risk taking in financial decision making. *Psychol. Sci.* 20:278–83

Preston SD, Buchanan TW, Stansfield RB, Bechara A. 2007. Effects of anticipatory stress on decision making in a gambling task. *Behav. Neurosci.* 121:257–63

Preuschoff K, 't Hart BM, Einhäuser W. 2011. Pupil dilation signals surprise: evidence for noradrenaline's role in decision making. *Front. Neurosci.* 5:115

Qin S, Hermans EJ, van Marle HJ, Luo J, Fernández G. 2009. Acute psychological stress reduces working memory-related activity in the dorsolateral prefrontal cortex. *Biol. Psychiatry* 66(1):25–32

Raghunathan R, Pham MT. 1999. All negative moods are not equal: motivational influences of anxiety and sadness on decision-making. *Organ. Behav. Hum. Decis. Process.* 79:56–77

Raio CM, Orederu TA, Palazzolo L, Shurick AA, Phelps EA. 2013. Cognitive emotion regulation fails stress test. *Proc. Natl. Acad. Sci. USA* 110(37):15139–44

Reyna VF, Brainerd CJ. 2011. Dual processes in decision making and developmental neuroscience: a fuzzy-trace model. *Dev. Rev.* 31:180–206

Rilling JK, Sanfey AG. 2011. The neuroscience of social decision-making. *Annu. Rev. Psychol.* 62:23–48

Roesch MR, Esber GR, Li J, Daw ND, Schoenbaum G. 2012. Surprise! Neural correlates of Pearce-Hall and Rescorla-Wagner coexist within the brain. *Eur. J. Neurosci.* 35:1190–200

Rolls ET. 2013. Limbic systems for emotion and for memory, but no single limbic system. *Cortex.* In press Rolls Rolls ET,Grabenhorst F. 2008. The orbitofrontal cortex and beyond: from affect to decision-making. *Prog. Neurobiol.* 86(3):216–44

Roozendaal B, McEwen BS, Chattarji S. 2009. Stress, memory and the amygdala. *Nat. Rev. Neurosci.* 10:423–33

Rose JE, McClernon FJ, Froeliger B, Behm FM, Preud'homme X, Krystal AD. 2011. Repetitive transcranial magnetic stimulation of the superior frontal gyrus modulates craving for cigarettes. *Biol. Psychiatry* 70(8):794–99

Rudebeck PH, Mitz AR, Chacko RV, Murray EA. 2013. Effects of amygdala lesions on reward-value coding in orbital and medial prefrontal cortex. *Neuron* 80(6):1519–31

Sanfey AG, Chang LJ. 2008. Multiple systems in decision making. *Ann. N.Y. Acad. Sci.* 1128:53–62

Sanfey AG, Rilling JK, Aronson JA, Nystrom LE, Cohen JD. 2003. The neural basis of economic decision-making in the ultimatum game. *Science* 300:1755–58

Scherer KR. 2000. Psychological models of emotion. In *The Neuropsychology of Emotion,* ed. J Borod, pp. 137–62. Oxford, UK: Oxford Univ. Press

Scherer KR. 2005. What are emotions? And how can they be measured? *Soc. Sci. Inf.* 44:695–729

Schwabe L, Höffken O, Tegenthoff M, Wolf OT. 2011. Preventing the stress-induced shift from goal-directed to habit action with a β-adrenergic antagonist. *J. Neurosci.* 31(47):17317–25

Schwabe L, Tegenthoff M, Höffken O, Wolf OT. 2010. Concurrent glucocorticoid and noradrenergic activity shifts instrumental behavior from goal-directed to habitual control. *J. Neurosci.* 30(24):8190–96

Schwabe L, Wolf OT. 2009. Stress prompts habit behavior in humans. *J. Neurosci.* 39(22):7191–98

Sellitto M, Ciaramelli E, di Pellegrino G. 2010. Myopic discounting of future rewards after medial orbitofrontal damage in humans. *J. Neurosci.* 8:16429–36

Sharot T, De Martino B, Dolan RJ. 2009. How choice reveals and shapes expected hedonic outcome. *J. Neurosci.* 29(12):3760–65

Sharot T, Riccardi AM, Raio CM, Phelps EA. 2007. Neural mechanisms mediating optimism bias. *Nature* 450:102–5

Sharp C, Monterosso J, Montague PR. 2012. Neuroeconomics: a bridge for translational research. *Biol. Psychiatry* 72:87–92

Skinner MD, Aubin H-J. 2010. Craving's place in addiction theory: contributions of the major models. *Neurosci. Biobehav. Rev.* 34(4):606–23

Sokol-Hessner P, Camerer CF, Phelps EA. 2012. Emotion regulation reduces loss aversion and decreases amygdala responses to losses. *Soc. Cogn. Affect. Neurosci.* 8(3):341–50

Sokol-Hessner P, Hsu M, Curley NG, Delgado MR, Camerer CF, Phelps EA. 2009. Thinking like a trader selectively reduces individuals' loss aversion. *Proc. Natl. Acad. Sci. USA* 106:5035–40

Sokol-Hessner P, Lackovic SF, Tobe RH, Leventhal BL, Phelps EA. 2013. *The effect of propranolol on loss aversion and decision-making.* Poster presented at Soc. Neurosci. Annu. Meet., 43rd, San Diego, CA. Program no. 99.15

Squire LR. 2004. Memory systems of the brain: a brief history and current perspective. *Neurobiol. Learn. Mem.* 82(3):171–77

Stanley DA, Sokol-Hessner P, Banaji MR, Phelps EA. 2011. Implicit race attitudes predict trustworthiness judgments and economic trust decisions. *Proc. Natl. Acad. Sci. USA* 108(19):7710–15

Stanley DA, Sokol-Hessner P, Fareri DS, Perino MT, Delgado MR, et al. 2012. Race and reputation: Perceived racial group trustworthiness influences the neural correlates of trust decisions. *Philos. Trans. R.*

Soc. B. 367(1589):744–53

Starcke K, Polzer C, Wolf OT, Brand M. 2011. Does stress alter everyday moral decision-making? *Psychoneuroendocrinology* 36:210–19

Starcke K, Wolf OT, Markowitsch HJ, Brand M. 2008. Anticipatory stress influences decision making under explicit risk conditions. *Behav. Neurosci.* 122(6):1352–60

Thaler RH. 1988. Anomalies: the ultimatum game. *J. Econ. Perspect.* 2:195–206

Ulrich-Lai YM, Herman JP. 2009. Neural regulation of endocrine and autonomic stress responses. *Nat. Rev.Neurosci.* 10:397–409

Ungless MA, Argilli E, Bonci A. 2010. Effects of stress and aversion on dopamine neurons: implications for addiction. *Neurosci. Biobehav. Rev.* 35:151–56

van't Wout M, Chang LJ, Sanfey AG. 2010. The influence of emotion regulation on social interactive decision-making. *Emotion* 10:815–21

van't Wout M, Kahn R, Sanfey AG, Aleman A. 2006. Affective state and decision-making in the ultimatum game. *Exp. Brain Res.* 169:564–68

Vinkers CH, Zorn JV, Cornelisse S, Koot S, Houtepen LC, et al. 2013. Time-dependent changes in altruistic punishment following stress. *Psychoneuroendocrinology* 38(9):1467–75

von Dawans B, Fischbacher U, Kirschbaum C, Fehr E, Heinrichs M. 2012. The social dimension of stress reactivity: acute stress increases prosocial behavior in humans. *Psychol. Sci.* 23(6):651–60

von Helversen B, Rieskamp J. 2013. Does the influence of stress on financial risk taking depend on the riskiness of the decision? *Proc. Natl. Acad. Sci. USA* 35:1546–51

Winkielman P, Berridge KC, Wilbarger JL. 2005. Unconscious affective reactions to masked happy versus angry faces influence consumption behavior and judgments of value. *Personal. Soc. Psychol. Bull.* 31:121–35

Youssef FF, Dookeeram K, Basdeo V, Francis E, Doman M, et al. 2012. Stress alters personal moral decision making. *Psychoneuroendocrinology* 37:491–98

Embodied Cognition and Mirror Neurons: A Critical Assessment

Alfonso Caramazza,[1,2] *Stefano Anzellotti,*[1,2] *Lukas Strnad,*[1] *and Angelika Lingnau*[2,3]

[1]Department of Psychology, Harvard University, Cambridge, Massachusetts 02138;
email: caramazz@fas.harvard.edu
[2]Center for Mind/Brain Sciences, University of Trento, 38100, Mattarello, Italy
[3]Department of Psychological and Cognitive Sciences, University of Trento, 38068, Rovereto, Trento, Italy

Key Words

concepts, embodied cognition, mirror neurons, simulation, action understanding

Abstract

According to embodied cognition theories, higher cognitive abilities depend on the reenactment of sensory and motor representations. In the first part of this review, we critically analyze the central claims of embodied theories and argue that the existing behavioral and neuroimaging data do not allow investigators to discriminate between embodied cognition and classical cognitive accounts, which assume that conceptual representations are amodal and symbolic. In the second part, we review the main claims and the core electrophysiological findings typically cited in support of the mirror neuron theory of action understanding, one of the most influential examples of embodied cognition theories. In the final part, we analyze the claim that mirror neurons subserve action understanding by mapping visual representations of observed actions on motor representations, trying to clarify in what sense the representations carried by these neurons can be claimed motor.

INTRODUCTION

Over the past 25 years, numerous theories have been proposed that emphasize the role of perceptual and motor processes for higher cognitive abilities such as language comprehension and action understanding. According to these theories, which we broadly group under the term embodied cognition theories, higher cognitive abilities are achieved in large part or entirely through the reenactment of processes used primarily for sensory input processing or for action execution. This review aims to critically evaluate the central tenets of embodied cognition theories by considering some of the most significant examples of such theories and the evidence supporting them.

Although there are many flavors and varieties of embodied cognition theories, the vast majority of them agree on at least two claims (e.g., Barsalou 2008, Gallese & Lakoff 2005). First, they all converge on the claim that semantic knowledge is carried by sensorimotor representations: The neural systems that are causally involved in forming and retrieving semantic knowledge are the same systems necessary for perceiving different sensory modalities or for producing actions. In line with this claim, studies have proposed that retrieving semantic knowledge of perceptual properties such as the colors of objects critically depends on the neural systems implicated in the perception of these properties (in this example, color perception; Simmons et al. 2007) and that understanding another person's actions requires the contribution from one's own motor system (Rizzolatti & Sinigaglia 2010). Second, most embodied cognition theories emphasize the importance of simulation in conceptual processing (Jeannerod 2001, Zwaan & Taylor 2006). On this account, retrieving semantic knowledge requires neural systems that are involved in perception or action execution and also requires that they perform the same processes utilized during perception or action execution. Semantic processing amounts to a reenactment of stored modality-specific representations in the relevant sensorimotor cortices. For instance, all the semantic knowledge we have about chairs could be exhaustively described as a collection of interacting modality-specific records of what a chair looks like, of the action of sitting, of the somatosensory experiences associated with sitting in a chair, etc. (Barsalou 2008). In this review, we focus on the evidence that investigators have used to support the validity of these two claims.

In both the behavioral and neuroimaging literatures, the arguments offered in support of embodiment are based on numerous interesting findings. However, investigators do not agree on whether the findings actually provide support to the central claims of embodied cognition (Barsalou 2008, Fischer & Zwaan 2008, Glenberg & Kaschak 2002, Kiefer & Pulvermuller 2012) or whether they are orthogonal to such claims, that is, consistent with classical, nonembodied theories of cognition (Caramazza et al. 1990, Chatterjee 2010, Csibra 2008, Jacob & Jeannerod 2005, Mahon & Caramazza 2008). Here we refer to the latter theories as cognitive theories. In the first part of this review, we attempt to clarify the nature of the controversy by presenting the neuroimaging and behavioral results that have been cited in support of embodied theories of semantic knowledge, and we discuss the alternative, cognitive interpretations of these results from the literature. Even though our discussion in this part centers on semantic knowledge, we emphasize that the issues at the core of the controversy are analogous to those in other contexts.

The remainder of the review is then devoted to a critical evaluation of perhaps the most influential embodied theory of cognition: the mirror neuron theory. Mirror neurons were originally discovered in the premotor cortex of the macaque monkey and are characterized by responses produced not only when the animal performs an action but also when it observes a similar action (di Pellegrino et al. 1992, Gallese et al. 1996). For instance, the same neuron would fire at an increased rate both when the monkey grasps an object with its hand and when it passively observes the object being grasped by the hand of the experimenter. This intriguing property inspired a theory that postulates a causal involvement of such motor neurons in action understanding (mirror neuron theory). We review the theory and point to the aspects of the theory that remain debated. In particular, we discuss whether there is sufficient evidence to show that mirror neurons play a causal role in action understanding, and we evaluate the direct-matching hypothesis (one of the central tenets of the mirror neuron theory) in light of the available evidence. Finally, we analyze the central claim that mirror neurons subserve action understanding through the reenactment of motor representations, trying to clarify in what sense the representations carried by mirror neurons can be claimed motor. [For a critical assessment of embodied theories of decision making, see Freedman & Assad (2011)].

EMBODIED COGNITION THEORIES: THE NATURE OF THE CONTROVERSY

Embodiment and Semantic Knowledge: Neuroimaging Studies

A central claim of embodied cognition states that semantic knowledge is represented in sensorimotor systems (Buccino etal. 2005; Goldberg etal. 2006; Pulvermuller etal. 2000, 2005; Simmons et al. 2005, 2007). Numerous studies looked for an overlap between brain areas involved in sensorimotor processes and those involved in the retrieval of semantic knowledge (Hauk et al. 2004, Postle et al. 2008, Simmons et al. 2007; for a recent meta-analysis, see Watson et al. 2013). If semantic knowledge were represented in sensorimotor areas, those areas should be active during sensorimotor processing as well as during retrieval of semantic knowledge. This prediction has been tested for color knowledge and action word comprehension.

The Case of Color Knowledge

Simmons et al. (2007), using fMRI, found evidence for an overlap between areas involved in color perception and those involved in the retrieval of color knowledge. The authors first individuated areas involved in color perception by contrasting activity during a color-discrimination task with activity during a task discriminating between hues of gray. Then, within these areas, they tested whether the activity during retrieval of color knowledge (e.g., TAXI = yellow) was greater than that during a control task requiring subjects to evaluate whether a particular motor property was associated with an object (e.g., HAIR = combed). The authors found a greater signal for the color knowledge task than for the control task in a left fusiform area demonstrated to be more active during color perception than during discrimination of hues of gray. They concluded that this result supported embodied theories of color knowledge. However, their conclusion was too strong. An overlap between brain areas active in two different tasks does not imply an overlap between the neural mechanisms involved in performing those tasks. Several neural populations coexist in a single brain area, and the results cannot rule out that the observed overlap derives from the activity of two different, nonoverlapping networks of neurons (see also Dinstein et al. 2008).Furthermore, cognitive theories of conceptual representation also predict some overlap between activity noted during color perception and that shown during the retrieval of color knowledge. According to cognitive theories, some representations of color do not depend on the specific modality through which the information is accessed, that is, those that are activated when one hears a color word, when one thinks of a color, and when one sees a color. Therefore, these representations would also be active during both color perception and retrieval of color knowledge.

The findings reported by Simmons et al. (2007) may actually be problematic for embodied theories of color knowledge. The double dissociation between color discrimination and color knowledge in cases of brain damage has been well documented (Miceli et al. 2001), ruling out the strong embodied view that would reduce color knowledge entirely to reactivations of mechanisms used primarily for color discrimination. The overlap detected by Simmons et al. (2007) does not occur in areas involved in relatively early stages of color processing (lingual gyrus), as determined by lesion overlap analyses of deficits in color perception (Bouvier & Engel 2006). Instead, and in accord with cognitive theories, the overlap occurs more anteriorly (left fusiform gyrus), in areas that, when damaged, do not affect color perception but impair retrieval of object color knowledge (Miceli et al. 2001)

The Case of Action Words

Another prominent embodied theory in the contemporary literature concerns the understanding of the meaning of action words, which is assumed to depend on the reenactment of motor processes involved in performing those actions (Pulvermuller 2005). The results of many studies have been interpreted to support this theory (Aziz-Zadeh et al. 2006, Boulenger et al. 2009, Pulvermuller et al. 2006, Tettamanti et al. 2005). Here, we discuss a typical example.

In an fMRI study, Hauk et al. (2004) investigated participants' brain activity during passive reading of hand, foot, and mouth action words (e.g., pick, kick, lick) and during the performance of actions with the corresponding body parts. They found that passive reading of action words activates premotor and frontal areas in a somatotopic manner. However, the study did not reporta direct analysis of the overlap between the activity in the word and the motor localizer conditions. Although the activity during the two different conditions seems to partially overlap for some effectors, many of the areas of activity during the two conditions are markedly different; therefore, it is difficult to assess whether and to what extent the activations in the two tasks overlap. Nonetheless, the authors interpreted the results as support for the embodied view of action word processing (for similar studies and conclusions, see Aziz-Zadeh et al. 2006, Boulenger et al. 2009, Pulvermuller et al. 2006, Tettamanti et al. 2005).

However, somatotopic activity during action word processing is not by itself evidence supporting an embodied theory of action processing. As we have discussed in the case of color, overlap is predicted by embodied theories, but it is also predicted by nonembodied theories of cognition. Therefore, even in the presence of an overlap, we must ask where the area of overlap is located. Postle and colleagues (2008) investigated the overlap between areas involved in action execution and areas involved in understanding action words more rigorously, and they failed to find reliable somatotopic recruitment of the primary or premotor cortex during the processing of action words (for similarly problematic results, see de Zubicaray et al. 2013, Kemmerer et al.2008,Kemmerer & Gonzalez-Castillo 2010, Lorey et al. 2013). A recent meta-analysis of fMRI studies on action concepts found no support for the idea that the activation of premotor and motor regions plays a significant role in processing action concepts (Watson et al. 2013; see also Bedny & Caramazza 2011).

In a series of studies, Pulvermuller and colleagues tried to support the embodied view of action word understanding, arguing that when participants read action words their motor cortex is activated rapidly (within 200 ms) and somatotopically (Hauk & Pulvermuller 2004; Pulvermuller et al. 2000, 2005). However, these findings do not address whether such fast and somatotopic activation plays a causal role in semantic processing or is merely the consequence of semantic processing in other, nonmotor areas (Mahon & Caramazza 2008). Studies that used transcranial magnetic stimulation (TMS) over the precentral motor cortex to assess its causal role in semantic processing of action words (Buccino et al. 2004, D'Ausilio et al. 2009, Gerfo et al. 2008, Mottonen & Watkins 2009, Papeo et al. 2009, Willems et al. 2011) have produced inconsistent results (for a review, see Papeo et al. 2013). In contrast, a recent study that used repetitive TMS to interfere with processing in the left posterior middle temporal gyrus (lpMTG), an area known to represent action verb semantics (Peelen et al. 2012), eliminated the action–nonaction verb distinction in the precentral motor cortex (Papeo et al. 2014). This result suggests that activity in the precentral motor cortex during action word comprehension is driven by semantic processing in lpMTG.

In sum, as in the case of color knowledge, the overlap between sensorimotor mechanisms and semantic knowledge does not seem to occur within areas involved in low-level sensorimotor processing, and the activity in precentral motor areas is driven by semantic processing outside the motor system.

Embodiment and Semantic Knowledge: Behavioral Findings

A wide range of behavioral evidence has been produced supporting the claims of embodied cognition theories. All of these studies follow the same general pattern in that they demonstrate various interactions between semantic knowledge and sensorimotor processes. Such interactions are then interpreted as evidence that sensorimotor processes or simulation plays a central role in mediating semantic knowledge.

The relevant evidence comes from various domains. For instance,Hansenetal.(2006) observed that one can sometimes perceive achromatic objects as having a color and that the perceived color is systematically related to the canonical color of the object (e.g., yellow for a banana). These data have been taken to suggest that observers automatically simulate the canonical color of an object as they categorize it. The effects of objects' perceptual properties can become apparent even in tasks that involve a substantial amount of semantic processing. Stanfield & Zwaan (2001) and Zwaan et al. (2002) have shown that in a picture-naming paradigm subjects name an object more quickly if it is preceded by text that implies perceptual properties that match those in the object's depiction. For example, subjects name a picture of an eagle with outstretched wings faster compared with a picture of an eagle with folded wings if the former is preceded by the sentence, "The ranger saw the eagle in the sky." Embodied cognition theory proposes that in order to understand the sentence, subjects simulate the perceptual processes implied by its meaning and are therefore faster at naming a perceptually congruent picture.

Studies have shown similar facilitation effects with other experimental paradigms in the perceptual domain (e.g., Borghi et al. 2004, Bosbach et al. 2005, Meteyard et al. 2008, Solomon & Barsalou 2004) but also in the motor domain(Gentilucci & Gangitano 1998, Glen-berg & Kaschak 2002). For example, participants are faster to respond to target words (e.g., "typewriter") following prime words referring to objects that, if manipulated in a typical way, require a similar motor response (e.g., "piano") (Myung et al. 2006; but see Postle et al. 2013). And Rueschemeyer et al. (2010) found that prior planning of motor actions facilitates processing of words denoting objects typically associated with such actions. These results, as explained by embodied cognition theories, are taken to suggest that at least some aspects of semantic knowledge about words and objects are stored in the form of motor representations.

Even though such behavioral evidence is of great interest in its own right, it plays only a very limited role in assessing the two central claims of embodied cognition theories. In particular, the interactions between semantic knowledge and sensorimotor processes do not address whether sensorimotor processes are, in fact, necessary for mediating conceptual representations. Cognitive, nonembodied accounts of semantic knowledge also predict such interactions; however, on these accounts, sensorimotor processes are triggered by retrieving semantic knowledge through association. For example, when one hears the sentence "The ranger saw the eagle in the sky," it is perfectly plausible that one retrieves a visual representation of a flying eagle and is therefore primed to name the picture of a flying eagle faster than a picture of a standing eagle. However, in this case, the activation of sensorimotor representations is a consequence of retrieving semantic knowledge rather than an integral part of it (Chatterjee 2010, Mahon & Caramazza 2008). The mere fact that sensorimotor processes interact with retrieval of semantic knowledge provides no clue about the direction of the causal link between the two. Thus, extant behavioral data do not allow one to discriminate between embodied and cognitive accounts.

THE MIRROR NEURON THEORY OF ACTION UNDERSTANDING

The mirror neuron theory has been immensely influential both as the most complete instantiation of an embodied cognition theory in one particular domain, action understanding, and as the foundation for embodied cognition theories in many other domains, such as language and social cognition. Below, we review the theory's main claims; the core evidence cited as supporting those claims, which stem from monkey physiology and human studies; and some of the problematic issues that some researchers have raised.

Main Claims of the Mirror Neuron Theory

Since its original formulation, several different versions of the mirror neuron theory have been put forth. The basic claim, which has not changed substantially across its various versions, concerns the overlap of neural mechanisms mediating action understanding and action production. It is most clearly expressed in a review by Rizzolatti et al. (2001). These authors maintain that "we understand actions when we map the visual representation of the observed action onto our motor representation of the same action. According to this view, an action is understood when its observation causes the motor system of the observer to resonate. So, when we observe a hand grasping an apple, the same population of neurons that control the execution of grasping movements becomes active in the observer's motor areas. By this approach, the motor knowledge of the observer is used to understand the observed action" (Rizzolatti et al. 2001, p. 661).

The neurons that are active, for instance, both when an individual grasps an apple and when the individual observes someone else grasp an apple are, by definition, mirror neurons. The theory in effect asserts that these neurons constitute a key mechanism shared by action production and action understanding. More specifically, it suggests that populations of mirror neurons are causally involved in mediating both these functions.

The presence of a mechanism that is recruited during both production and understanding does not distinguish the mirror neuron theory from classical cognitive theories according to which central, abstract representations are involved in both comprehension and production of actions and language. However, the mirror neuron theory makes at least three strong claims about the character of the shared mechanism of action production and action understanding, which distinguish it from cognitive, nonembodied theories.

First, the core, novel claim of the mirror neuron theory concerns the motor nature of the representations carried by mirror neurons. The idea that the motor system is involved not only in movement generation but also in understanding actions and intentions is radically different from classical theories for which these processes require the involvement of abstract (or amodal/symbolic) representations.

Second, action understanding mediated by the mirror neuron mechanism is assumed to be direct in the sense that it can be achieved without needing "inferential processing" or other "high-level mental processes" (Rizzolatti & Sinigaglia 2010). The mapping of sensory inputs onto corresponding representations of actions within the motor system is thus postulated to be largely automatic. It presumably does not account for factors such as prior beliefs, specifics of the situation, or the context in which an observed action is carried out because all these likely require the "high-level mental processes" that direct matching between an observed action and a motor representation circumvents. Instead, the matching depends on a "natural response" of the mirror system to the visual input (Rizzolatti & Sinigaglia 2010; but see Cook et al. 2014).

Finally, action understanding involves simulation of the observed actions in the motor system of the observer: Whenever an individual observes an action, his or her understanding is mediated by the same population of premotor neurons that also control his or her own execution of that action. The relevant action is effectively reenacted within the observer's premotor cortex (Rizzolatti & Sinigaglia 2010).

Basic Properties of Macaque Mirror Neurons

Early on after the discovery of mirror neurons, many studies focused on characterizing their basic response properties. At least three important findings emerged from these investigations.

Mirror neurons are activated only by particular kinds of actions. Mirror neurons fire only when the monkey is presented with a natural, transitive action that targets a simultaneously presented object (di Pellegrinoetal. 1992), for instance, when the experimenter grasps a piece of food in front of the monkey. Mirror neurons would not fire when the experimenter only moves his hand toward the food but does not grasp it. Furthermore, they would not fire during the presentation of the food alone or when the experimenter performs a grasping movement in absence of an object. Thus, mirror neuron activity during visual observation appears to be triggered by object-directed actions (but see Kraskov et al. 2009).

Mirror neurons have different degrees of congruency. The actions that cause the mirror neuron to fire during both motor production and action observation tend to be congruent (Gallese et al. 1996). For instance, a mirror neuron that is active when the monkey grasps an object with a precision grip is likely to be activated when the monkey observes the same or a similar action. However, the degree of congruency varies considerably across different mirror neurons. Some mirror neurons exhibit a strict relationship between the performed and observed actions that activate them, such as grasping with a specific type of grip, whereas others fire even when the relationship between the observed and performed action is very loose, such as neurons that respond when an action is performed but are activated by the sight of multiple different actions. Finally, for some mirror neurons there is no clear relationship between observed and performed actions.

Mirror neurons are sensitive to the goal of an action. The observations of congruence between the observed and the executed actions triggering some of the mirror neurons led investigators to propose that mirror neuron activity correlates with action understanding(di Pellegrino et al. 1992, Gallese et al. 1996, Rizzolatti et al. 1996). One of the most influential studies cited in support of such claims is an experiment by Umilta et al. (2001). In the experiment, two monkeys viewed hand actions performed by an experimenter such as grasping, holding, or placing in two conditions. In one of the conditions, the monkeys observed the actions from start to finish without interruption. In the other condition, the monkeys could only observe the initial stage of the action, but the final stage, during which the hand interacted with the object, was occluded. The researchers found that some neurons in area F5, which showed mirror properties when the monkeys observed the entire hand action sequence, also responded when the final stage of the action was occluded. The authors interpreted this result to mean that, on the basis of the observed part of the action sequence, the monkey understood the action being performed; thus, its understanding was reflected in the activity of the mirror neurons. Because the monkey typically understands the action before it is completed, the firing of these mirror neurons is sustained even if the final part of the action is occluded. Therefore, mirror neuron activity correlates with action understanding.

A study by Fogassi et al. (2005) provided another piece of evidence supporting the correlation between mirror neuron activity and action understanding. The authors recorded neurons in the convexity of the inferior parietal lobule (IPL) of a monkey that responded selectively to reaching actions with extremely similar motor profiles: either reaching for a piece of food and placing it in its mouth or reaching for a piece of food and placing it in a container affixed close to its head. Some of these neurons retained their selectivity for one specific type of action when the monkey was passively viewing the experimenter's actions. Fogassi et al. (2005) take these findings to indicate that the mirror neurons in question selectively encode goals of motor acts and thus facilitate action understanding. They also suggest that intentions are understood by activating one of several possible motor chains (e.g., grasp-to-place versus grasp-to-eat).

The experiments by Umiltà and colleagues (2001) and by Fogassi and colleagues (2005) provide clear cases of mirror neuron activity being sensitive to fairly subtle distinctions between different kinds of observed actions. However, whether these neurons actively contribute to action understanding or whether their activity is only correlated with it is not directly addressed by these experiments. That is, they do not rule out the possibility that mirror neuron activity results from processes that occur in other parts of the brain that mediate action understanding. When a monkey observes an action whose final part is occluded (as in Umilta et al. 2001), assuming the monkey correctly infers the kind of action being performed, at least two accounts of the mirror neuron activity pattern are equally plausible. On the one hand, the mirror neurons could be actively contributing to the categorization of the observed action. On the other hand, the action could be categorized outside the motor system, and a corresponding nonmotor representation of the action (e.g., crack a nut to get food) could be retrieved (Mahon & Caramazza 2008).

Properties of the Human Mirror System

In the human brain, studies show that the inferior limb of the precentral sulcus/ posterior part of the inferior frontal gyrus, the inferior parietal lobe, and the superior temporal sulcus, and recently also the supplementary motor cortex, the primary somatosensory cortex, and visual area MT are recruited during both observation and imitation/execution of actions (Chong et al. 2008, Dinstein et al. 2007, Grezes et al. 2003, Iacoboni et al. 1999, Kilner et al. 2009, Press et al. 2012; forarecent meta-analysis, see Caspers et al. 2010). Using TMS, many studies have demonstrated that action observation leads to highly effector-specific and even muscle-specific modulations of corticospinal excitability (e.g., Cattaneo et al. 2009, Fadiga et al. 1995, Maeda et al. 2002, Urgesi et al. 2010).

Using multi-voxel pattern analysis (MVPA), Oosterhof et al. (2010) observed above-chance classification of actions across modalities in the left postcentral gyrus and the left anterior parietal cortex. Using a similar approach, Oosterhof et al. (2012) found that the parietal and occipitotemporal cortices contained cross-modal action-specific representations irrespective of the viewpoint of the observed action. By contrast, the ventral premotor cortex contained action-specific representations across modalities for the first- but not the third-person perspective (Caggiano et al. 2011, Maeda et al. 2002). These studies show that high-level representations of actions are not restricted to early sensorimotor areas (but see Cattaneo et al. 2010). Despite various methodological advances, the types of content represented in the various regions of the human mirror system and whether these contents are specifically motor or more abstract remain unclear (Dinstein et al. 2008, Hickok 2009, Oosterhof et al. 2013).

Direct Matching and Simulation

The proposal that conceptual understanding is achieved through sensorimotor simulation is integral to embodied theories of cognition, and in the context of the mirror neuron theory of action understanding, it is intimately linked with the notion of direct matching. On this theory, direct matching is a mechanism through which sensory inputs associated with actions of other individuals are mapped unmediatedly, without involving "higher-level mental processes" such as "inferential processing," onto motor representations in the observer's brain (Rizzolatti & Sinigaglia 2010).

Data from numerous experiments have been interpreted to support the claims of direct matching in action understanding (see especially Fogassi et al. 2005, Gallese et al. 1996, Kohler et al. 2002, Rizzolatti et al. 1996, Umiltaetal. 2001). In one of the most widely cited studies, Kohlerand colleagues (2002) report finding mirror neurons in the macaque monkey that become active both when the animal visually observes an action and when it hears a sound that is associated with that action. For example, in one experimental condition, some neurons became active when the monkey cracked a peanut, when visually observing the experimenter crack a peanut, and when hearing the sound of the action alone. The authors interpret these observations as evidence supporting the direct matching hypothesis.

Such an interpretation raises an important question about the nature of the link between incoming sensory representations and the subsequently retrieved motor representations. Mapping the sound of an action onto the motor program corresponding to it requires relatively rich prior knowledge about the action. In the case of visual observation, one could establish a correspondence between the low-level visual inputs and motor representations in the premotor cortex. The information about an action contained in the visual signal allows one to determine which effectors were used and what their position and speed were, among other properties of the action. In contrast, the auditory signal alone provides much less information to establish a correspondence with a motor representation of an action; for example, the sound does not carry information about the effector involved in the action. The triggering of mirror neurons by action sounds represents a learned association that could, in principle, be established between an arbitrary sound and an arbitrary motor representation. The fact that the motor representation of the correct action has been retrieved in the motor system even though the sensory signal alone does not contain sufficient information to determine which motor action was performed implies that the action has already been categorized by the time the motor system is activated. It is not obvious how to reconcile the data about auditory triggering of mirror neurons with the direct matching hypothesis.

ARE MIRROR NEURONS MOTOR?

The centrality of motor representations in action understanding—the claim that mirror neurons are essentially motor—is the defining characteristic of the mirror neuron theory. However, the sense in which mirror neurons can be considered motor, what evidence supports such a claim, and its implications for embodied theories of action understanding are not clear.

Mirror neurons can be considered motor in several ways. First, and most straightforward, is that these neurons fire during active movements, and their responses are selective, responding during certain movements and not others (di Pellegrino et al. 1992, Rizzolatti et al. 1996). However, by definition, mirror neurons are also activated during action observation. Therefore, in this sense, the representations carried by mirror neurons are also visual, and one cannot conclude that actions are understood by reenacting motor representations without also concluding that actions are executed by reenacting visual representations. Thus, the visuomotor character of these neurons does not favor choosing one modality over the other.

A second sense in which the representations carried by mirror neurons are motor is that these neurons were found in areas of the brain that are historically considered motor. Rizzolatti & Sinigaglia (2010) seem to argue that the motor function of mirror neurons depends on their anatomical location. Neurons in area F5, where mirror neurons were found originally (di Pellegrino et al. 1992, Gallese et al. 1996, Rizzolatti et al. 1996), respond during action execution (Kurata & Tanji 1986, Rizzolatti et al. 1981). Thus, mirror neurons could then be considered motor. However, mirror neurons, by definition, also respond to visual stimuli. Area F5 also contains mirror-like neurons that do not fire during action execution and fire only during action observation (di Pellegrino et al. 1992; Gallese et al. 1996, 2002). Accepting the assumption above would lead to the conclusion that even these mirror-like neurons are motor, despite that they are not activated at all during action execution.

In what sense, then, could mirror neuron representations be motor in a way that justifies an embodied theory of action understanding? The motor modality may be predominant over the visual modality in mirror neuron representations in the sense that the informational content of the representations carried by mirror neurons specifies details that are particularly relevant for motor execution (e.g., which muscles are used to perform an action) but not for visual processing (e.g., where they are presented in the visual field). In this case, mirror neurons may be considered predominantly motor in the sense that they carry details specific to the motor modality, but they do not carry other details specific to the visual modality. This determination would allow investigators to interpret mirror neuron activation in terms of reenacting specific motor programs.

A study by Umilta et al. (2008) is relevant to this issue. These authors investigated the response properties of neurons in areas F5 and F1 of the premotor cortex of monkeys after they were trained to use normal pliers, which require a squeeze action to hold an object, and inverse pliers, which require a squeeze action to release an object. The response pattern of most of the F5 and some of the F1 neurons when monkeys grasped with pliers was extremely similar to the response these neurons exhibited when monkeys grasped objects with reverse pliers. These data thus shed light on how specific the representations mediated by the F5 and a portion of the F1 neurons actually are. The data suggest that neuron activity reflects abstract action properties, such as outcome, rather than just the sequence of motor programs that need to be executed in order to obtain that outcome.

The neurons studied by Umilta et al. (2008) were not mirror neurons. However, the authors hypothesize that because mirror neurons are found in the same brain regions as those studied in the experiment, one would expect at least some mirror neurons to exhibit the same degree of generalization across different motor actions with the same overall goal. Consistent with this view, Gallese and colleagues (1996) report finding mirror neurons that fire during observation of grasping performed by a monkey either with the hand or with the mouth, a clear indication that mirror neurons represent abstract action goals as opposed to specific motor contents.

Thus far we have adopted a simplified distinction between low-level motor representations and higher-level (abstract, cognitive) representations previously employed in the mirror neuron literature (Rizzolatti & Sinigaglia 2010). Although this distinction can be helpful as a first approximation, it remains unclear on the basis of which criteria the boundary should be drawn. The empirical findings indicate that a richer view is required to appropriately describe the wealth of evidence available in the literature. Action observation and understanding seem to be the outcomes of numerous processing stages at different levels, from early visual areas to the superior temporal sulcus (STS) to the mirror system in the inferior parietal lobe and F5, etc. In a recent study, Mukamel et al. (2010) reported that neurons in the human medial temporal lobe, including the hippocampus and the amygdala, fired both during the execution and the observation of similar actions. These findings indicate that representations active during action execution and action observation are also present outside the regions historically considered motor. The human medial temporal lobe is a highly multimodal brain area known to contain neurons that carry high-level representations of objects that generalize beyond specific views (Kreiman et al. 2000, Quiroga et al. 2005), supporting the hypothesis that these cells store the meaning of a stimulus.

SUMMARY AND CONCLUSIONS

We have provided an overview of the most important empirical results concerning embodied cognition theories and have presented a partial assessment of them, as well. Research motivated by embodied accounts of cognition led to the discovery of many phenomena supporting the close interaction between conceptual processing and sensorimotor representations.

In the context of embodied cognition theories, this body of extraordinarily interesting empirical data has been used by some investigators to argue that conceptual knowledge is mediated primarily by sensorimotor representations and that sensorimotor simulation is an essential part of conceptual processing. We have shown that these claims are unwarranted for two main reasons. First, a substantial part of the evidence cited in support of embodied cognition theories concerns phenomena for which the predictions of embodied and cognitive theories coincide. Therefore, such evidence does not discriminate between embodied and cognitive accounts. In fact, every cognitive theory assumes that perception and action, comprehension and production are bridged through shared, abstract conceptual representations. Cognitive theories would suffer from a strange duality of the mind if there were no possibility for an exchange among perception, action, and conceptual processing. Second, in the field of action understanding, studies on mirror neurons have shown that areas that were thought to carry relatively low-level representations contain neurons that show surprisingly high levels of abstraction (Caggiano et al. 2011, 2012; Ferrari et al. 2005; Gallese et al. 1996; Umilta et al. 2001, 2008) that, we argue, cannot plausibly be considered motor. At the same time, single-cell recordings in humans individuated neurons located outside the so-called motor system that represent actions with perhaps even greater abstraction (Mukamel et al. 2010). These results suggest that conceptual processing relies on high-level, nonsensorimotor, abstract representations.

DISCLOSURE STATEMENT

The authors are not aware of any affiliations, memberships, funding, or financial

holdings that might be perceived as affecting the objectivity of this review.

LITERATURE CITED

Aziz-Zadeh L, Wilson SM, Rizzolatti G, Iacoboni M. 2006. Congruent embodied representations for visually presented actions and linguistic phrases describing actions. *Curr. Biol.* 16:1818–23

Barsalou LW. 2008. Grounded cognition. *Annu. Rev. Psychol.* 59:617–45

Bedny M, Caramazza A. 2011. Perception, action, and word meanings in the human brain: the case from action verbs. *Ann. N. Y. Acad. Sci.* 1224:81–95

Borghi AM, Glenberg AM, Kaschak MP. 2004. Putting words in perspective. *Mem. Cogn.* 32:863–73

Bosbach S, Prinz W, Kerzel D.2005. Is direction position? Position- and direction-based correspondence effects in tasks with moving stimuli. *Q. J. Exp. Psychol. A* 58:467–506

Boulenger V, Hauk O, Pulvermuller F. 2009. Grasping ideas with the motor system: semantic somatotopy in idiom comprehension. *Cereb. Cortex* 19:1905–14

Bouvier SE, Engel SA. 2006. Behavioral deficits and cortical damage loci in cerebral achromatopsia. *Cereb. Cortex* 16:183–91

Buccino G, Riggio L, Melli G, Binkofski F, Gallese V, Rizzolatti G. 2005. Listening to action-related sentences modulates the activity of the motor system: a combined TMS and behavioral study. *Brain Res. Cogn. Brain Res.* 24:355–63

Buccino G, Vogt S, Ritzl A, Fink GR, Zilles K, et al. 2004. Neural circuits underlying imitation learning of hand actions: an event related fMRI study. *Neuron* 42:323–34

Caggiano V, Fogassi L, Rizzolatti G, Casile A, Giese MA, Thier P. 2012. Mirror neurons encode the subjective value of an observed action. *Proc. Natl. Acad. Sci. USA* 109:11848–53

Caggiano V, Fogassi L, Rizzolatti G, Pomper JK, Thier P, et al. 2011. View-based encoding of actions in mirror neurons of area F5 in macaque premotor cortex. *Curr. Biol.* 21:144–48

Caramazza A, Hillis AE, Rapp BC, Romani C. 1990. The Multiple Semantics Hypothesis: multiple confusions? *Cogn. Neuropsychol.* 7:161–89

Caspers S, Zilles K, Laird AR, Eickhoff SB. 2010. ALE meta-analysis of action observation and imitation in the human brain. *NeuroImage* 50:1148–67

Cattaneo L, Caruana F, Jezzini A, Rizzolatti G. 2009. Representation of goal and movements without overt motor behavior in the human motor cortex: a transcranial magnetic stimulation study. *J. Neurosci.* 29:11134–38

Cattaneo L, Sandrini M, Schwarzbach J. 2010. State-dependent TMS reveals a hierarchical representation of observed acts in the temporal, parietal, and premotor cortices. *Cereb. Cortex* 20:2252–58

Chatterjee A. 2010. Disembodying cognition. *Lang. Cogn.* 2:79–116

Chong TT, Cunnington R, Williams MA, Kanwisher N, Mattingley JB. 2008. fMRI adaptation reveals mirror neurons in human inferior parietal cortex. *Curr. Biol.* 18:1576–80

Cook R, Bird G, Catmur C, Press C, Heyes C. 2014. Mirror neurons: from origin to function. *Behav. Brain Sci.* In press

Csibra G.2008.Action mirroring and action understanding: an alternative account. In *Sensorimotor Foundations of Higher Cognition: Attention and Performance XXII,* ed. P Haggard, Y Rossetti, M Kawato, pp. 435–59. New York: Oxford Univ. Press

D'Ausilio A, Pulvermuller F, Salmas P, Bufalari I, Begliomini C, Fadiga L. 2009. The motor somatotopy of speech perception. *Curr. Biol.* 19:381–85

de Zubicaray G, Arciuli J, McMahon K. 2013. Putting an "end" to the motor cortex representations of action words. *J. Cogn. Neurosci.* 25:1957–74

di Pellegrino G, Fadiga L, Fogassi L, Gallese V, Rizzolatti G. 1992. Understanding motor events: a neurophysiological study. *Exp. Brain Res.* 91:176–80

Dinstein I, Hasson U, Rubin N, Heeger DJ. 2007. Brain areas selective for both observed and executed movements. *J. Neurophysiol.* 98:1415–27

Dinstein I, Thomas C, Behrmann M, Heeger DJ. 2008. A mirror up to nature. *Curr. Biol.* 18:R13–18

Fadiga L, Fogassi L, Pavesi G, Rizzolatti G. 1995. Motor facilitation during action observation: a magnetic stimulation study. *J. Neurophysiol.* 73:2608–11

Ferrari PF, Rozzi S, Fogassi L. 2005. Mirror neurons responding to observation of actions made with tools in monkey ventral premotor cortex. *J. Cogn. Neurosci.* 17:212–26

Fischer MH, Zwaan RA. 2008. Embodied language: a review of the role of the motor system in language comprehension. *Q. J. Exp. Psychol.* 61:825–50

Fogassi L, Ferrari PF, Gesierich B, Rozzi S, Chersi F, Rizzolatti G. 2005. Parietal lobe: from action organization to intention understanding. *Science* 308:662–67

Freedman DJ, Assad JA. 2011. A proposed common neural mechanism for categorization and perceptual decisions. *Nat. Neurosci.* 14:143–46

Gallese V, Fadiga L, Fogassi L, Rizzolatti G. 1996. Action recognition in the premotor cortex. *Brain* 119(Pt. 2):593–609

Gallese V, Fadiga L, Fogassi L, Rizzolatti G. 2002. Action representation and the inferior parietal lobule. In *Common Mechanisms in Perception and Action: Attention and Performance,* ed. W Prinz, B Hommel, pp. 334–55. Oxford, UK: Oxford Univ. Press

Gallese V, Lakoff G. 2005. The Brain's concepts: the role of the sensory-motor system in conceptual knowledge. *Cogn. Neuropsychol.* 22:455–79

Gentilucci M, Gangitano M. 1998. Influence of automatic word reading on motor control. *Eur. J. Neurosci.* 10:752–56

Gerfo EL, Oliveri M, Torriero S, Salerno S, Koch G, Caltagirone C. 2008. The influence of rTMS over prefrontal and motor areas in a morphological task: grammatical vs. semantic effects. *Neuropsychologia* 46:764–70

Glenberg AM, Kaschak MP. 2002. Grounding language in action. *Psychon. Bull. Rev.* 9:558–65

Goldberg RF, Perfetti CA, Schneider W. 2006. Perceptual knowledge retrieval activates sensory brain regions. *J. Neurosci.* 26:4917–21

Grezes J, Armony JL, Rowe J, Passingham RE. 2003. Activations related to "mirror" and "canonical" neurons in the human brain: an fMRI study. *NeuroImage* 18:928–37

Hansen T, Olkkonen M, Walter S, Gegenfurtner KR. 2006. Memory modulates color appearance. *Nat. Neurosci.* 9:1367–68

Hauk O, Johnsrude I, Pulvermüller F. 2004. Somatotopic representation of action words in human motor and premotor cortex. *Neuron* 41:301–7

Hauk O, Pulvermuller F. 2004. Neurophysiological distinction of action words in the fronto-central cortex. *Hum. Brain Mapp.* 21:191–201

Hickok G. 2009. Eight problems for the mirror neuron theory of action understanding in monkeys and humans. *J. Cogn. Neurosci.* 21:1229–43

Iacoboni M, Woods RP, Brass M, Bekkering H, Mazziotta JC, Rizzolatti G. 1999. Cortical mechanisms of human imitation. *Science* 286:2526–28

Jacob P, Jeannerod M. 2005. The motor theory of social cognition: a critique. *Trends Cogn. Sci.* 9:21–25

Jeannerod M. 2001. Neural simulation of action: a unifying mechanism for motor cognition. *NeuroImage* 14:S103–9

Kemmerer D, Castillo JG, Talavage T, Patterson S, Wiley C. 2008. Neuroanatomical distribution of five semantic components of verbs: evidence from fMRI. *Brain Lang.* 107:16–43

Kemmerer D, Gonzalez-Castillo J.2010.The Two-Level Theory of verb meaning: an approach to integrating the semantics of action with the mirror neuron system. *Brain Lang.* 112:54–76

Kiefer M, Pulvermuller F. 2012. Conceptual representations in mind and brain: theoretical developments, current evidence and future directions. *Cortex* 48:805–25

Kilner JM, Neal A, Weiskopf N, Friston KJ, Frith CD. 2009. Evidence of mirror neurons in human inferior frontal gyrus. *J. Neurosci.* 29:10153–59

Kohler E, Keysers C, Umiltà MA, Fogassi L, Gallese V, Rizzolatti G. 2002. Hearing sounds, understanding actions: action representation in mirror neurons. *Science* 297:846–48

Kraskov A, Dancause N, Quallo MM, Shepherd S, Lemon RN. 2009. Corticospinal neurons in macaque ventral premotor cortex with mirror properties: a potential mechanism for action suppression? *Neuron* 64:922–30

Kreiman G, Koch C, Fried I. 2000. Category-specific visual responses of single neurons in the human medial temporal lobe. *Nat. Neurosci.* 3:946–53

Kurata K, Tanji J. 1986. Premotor cortex neurons in macaques: activity before distal and proximal forelimb movements. *J. Neurosci.* 6:403–11

Lorey B, Naumann T, Pilgramm S, Petermann C, Bischoff M, et al. 2013. How equivalent are the action execution, imagery, and observation of intransitive movements? Revisiting the concept of somatotopy during action simulation. *Brain Cogn.* 81:139–50

Maeda F, Kleiner-Fisman G, Pascual-Leone A. 2002. Motor facilitation while observing hand actions: specificity of the effect and role of observer's orientation. *J. Neurophysiol.* 87:1329–35

Mahon BZ, Caramazza A. 2008. A critical look at the embodied cognition hypothesis and a new proposal for grounding conceptual content. *J. Physiol. Paris* 102:59–70

Meteyard L, Zokaei N, Bahrami B, Vigliocco G. 2008. Visual motion interferes with lexical decision on motion words. *Curr. Biol.* 18:R732–33

Miceli G, Fouch E, Capasso R, Shelton JR, Tomaiuolo F, Caramazza A. 2001. The dissociation of color from form and function knowledge. *Nat. Neurosci.* 4:662–67

Mottonen R, Watkins KE. 2009. Motor representations of articulators contribute to categorical perception of speech sounds. *J. Neurosci.* 29:9819–25

Mukamel R, Ekstrom AD, Kaplan J, Iacoboni M, Fried I. 2010. Single-neuron responses in humans during execution and observation of actions. *Curr. Biol.* 20:750–56

Myung JY, Blumstein SE, Sedivy JC. 2006. Playing on the typewriter, typing on the piano: manipulation knowledge of objects. *Cognition* 98:223–43

Oosterhof NN, Tipper SP, Downing PE. 2012. Viewpoint (in) dependence of action representations: an MVPA study. *J. Cogn. Neurosci.* 24:975–89

Oosterhof NN, Tipper SP, Downing PE. 2013. Crossmodal and action-specific: neuroimaging the human mirror neuron system. *Trends Cogn. Sci.* 17:311–18

Oosterhof NN, Wiggett AJ, Diedrichsen J, Tipper SP, Downing PE. 2010. Surface-based information mapping reveals crossmodal vision-action representations in human parietal and occipitotemporal cortex. *J. Neurophysiol.* 104:1077–89

Papeo L, Lingnau A, Agosta S, Pascual-Leone A, Battelli L, Caramazza A. 2014. The origin of word-related motor activity. *Cereb. Cortex.* In press. doi: 10.1093/cercor/bht423

Papeo L, Pascual-Leone A, Caramazza A. 2013. Disrupting the brain to validate hypotheses on the neurobiology of language. *Front. Hum. Neurosci.* 7:148

Papeo L, Vallesi A, Isaja A, Rumiati RI. 2009. Effects of TMS on different stages of motor and non-motor verb processing in the primary motor cortex. *PLoS ONE* 4:e4508

Peelen M, Romagno D, Caramazza A. 2012. Independent representations of verbs and actions in left temporal cortex. *J. Cogn. Neurosci.* 24:2096–107

Postle N, Ashton R, McFarland K, de Zubicaray GI. 2013. No specific role for the manual motor system in processing the meanings of words related to the hand. *Front. Hum. Neurosci.* 7:11

Postle N, McMahon KL, Ashton R, Meredith M, deZubicaray GI. 2008. Action word meaning representations in cytoarchitectonically defined primary and premotor cortices. *NeuroImage* 43:634–44

Press C, Weiskopf N, Kilner JM. 2012. Dissociable roles of human inferior frontal gyrus during action execution and observation. *NeuroImage* 60:1671–77

Pulvermuller F. 2005. Brain mechanisms linking language and action. *Nat. Rev. Neurosci.* 6:576–82

Pulvermuller F, Harle M, Hummel F. 2000. Neurophysiological distinction of verb categories. *NeuroReport* 11:2789–93

Pulvermuller F, Huss M, Kherif F, Moscoso del Prado Martin F, Hauk O, Shtyrov Y. 2006. Motor cortex maps articulatory features of speech sounds. *Proc. Natl. Acad. Sci. USA* 103:7865–70

Pulvermuller F, Shtyrov Y, Ilmoniemi R. 2005. Brain signatures of meaning access in action word recognition. *J. Cogn. Neurosci.* 17:884–92

Quiroga RQ, Reddy L, Kreiman G, Koch C, Fried I. 2005. Invariant visual representation by single neurons in the human brain. *Nature* 435:1102–7

Rizzolatti G, Fadiga L, Gallese V, Fogassi L. 1996. Premotor cortex and the recognition of motor actions. *Brain Res. Cogn. Brain Res.* 3:131–41

Rizzolatti G, Fogassi L, Gallese V. 2001. Neurophysiological mechanisms underlying the understanding and imitation of action. *Nat. Rev. Neurosci.* 2:661–70

Rizzolatti G, Scandolara C, Matelli M, Gentilucci M. 1981. Afferent properties of periarcuate neurons in macaque monkeys. I. Somatosensory responses. *Behav. Brain Res.* 2:125–46

Rizzolatti G, Sinigaglia C. 2010. The functional role of the parieto-frontal mirror circuit: interpretations and misinterpretations. *Nat. Rev. Neurosci.* 11:264–74

Rueschemeyer SA, Lindemann O, van Rooij D, van Dam W, Bekkering H. 2010. Effects of intentional motor actions on embodied language processing. *Exp. Psychol.* 57:260–66

Simmons WK, Martin A, Barsalou LW. 2005. Pictures of appetizing foods activate gustatory cortices for taste and reward. *Cereb. Cortex* 15:1602–8

Simmons WK, Ramjee V, Beauchamp MS, McRae K, Martin A, Barsalou LW. 2007. A common neural substrate for perceiving and knowing about color. *Neuropsychologia* 45:2802–10

Solomon KO, Barsalou LW. 2004. Perceptual simulation in property verification. *Mem. Cogn.* 32:244–59

Stanfield RA, Zwaan RA. 2001. The effect of implied orientation derived from verbal context on picture recognition. *Psychol. Sci.* 12:153–56

Tettamanti M, Buccino G, Saccuman MC, Gallese V, Danna M, et al. 2005. Listening to action-related sentences activates fronto-parietal motor circuits. *J. Cogn. Neurosci.* 17:273–81

Umilta MA, Escola L, Intskirveli I, Grammont F, Rochat M, et al. 2008. When pliers become fingers in the monkey motor system. *Proc. Natl. Acad. Sci. USA* 105:2209–13

Umilta MA, Kohler E, Gallese V, Fogassi L, Fadiga L, et al. 2001. I know what you are doing: a neurophysiological study. *Neuron* 31:155–65

Urgesi C, Maieron M, Avenanti A, Tidoni E, Fabbro F, Aglioti SM. 2010. Simulating the future of actions in the human corticospinal system. *Cereb. Cortex* 20:2511–21

Watson CE, Cardillo ER, Ianni GR, Chatterjee A. 2013. Action concepts in the brain: an activation likelihood estimation meta-analysis. *J. Cogn. Neurosci.* 25:1191–205

Willems RM, Labruna L, D'Esposito M, Ivry R, Casasanto D. 2011. A functional role for the motor system in language understanding: evidence from theta-burst transcranial magnetic stimulation. *Psychol. Sci.* 22:849–54

Zwaan RA, Stanfield RA, Yaxley RH. 2002. Language comprehenders mentally represent the shapes of objects. *Psychol. Sci.* 13:168–71

Zwaan RA, Taylor LJ. 2006. Seeing, acting, understanding: motor resonance in language comprehension. *J. Exp. Psychol. Gen.* 135:1–11

Establishing Wiring Specificity in Visual System Circuits: From the Retina to the Brain

Chi Zhang,[1] Alex L. Kolodkin,[2] Rachel O. Wong,[1] and Rebecca E. James[2]

[1] Department of Biological Structure, University of Washington, Seattle, Washington 98195; email: zhangc29@uw.edu, wongr2@uw.edu

[2] Solomon H. Snyder Department of Neuroscience, Howard Hughes Medical Institute, Johns Hopkins University School of Medicine, Baltimore, Maryland 21205; email: kolodkin@jhmi.edu, rjames20@jhmi.edu

Key Words

retina, connectivity, retinorecipient targeting, lamination, guidance

Abstract

The retina is a tremendously complex image processor, containing numerous cell types that form microcircuits encoding different aspects of the visual scene. Each microcircuit exhibits a distinct pattern of synaptic connectivity. The developmental mechanisms responsible for this patterning are just beginning to be revealed. Furthermore, signals processed by different retinal circuits are relayed to specific, often distinct, brain regions. Thus, much work has focused on understanding the mechanisms that wire retinal axonal projections to their appropriate central targets. Here, we highlight recently discovered cellular and molecular mechanisms that together shape stereo-typic wiring patterns along the visual pathway, from within the retina to the brain. Although some mechanisms are common across circuits, others play unconventional and circuit-specific roles. Indeed, the highly organized connectivity of the visual system has greatly facilitated the discovery of novel mechanisms that establish precise synaptic connections within the nervous system.

INTRODUCTION

The well-characterized and stereotypic arrangements of cells and their connectivity in the vertebrate retina enable the discovery of diverse cellular and molecular mechanisms that elaborate precise neuronal circuitry during development. These include mechanisms essential for shaping the synapses between photoreceptors (PRs) and their downstream targets and for orchestrating the complex convergence of inputs onto retinal ganglion cell (RGC) output neurons. RGCs also connect with specific targets in the brain, and their axons engage a variety of cues while pathfinding during development to complete their wiring diagram. Here, we reflect upon recent studies—focusing on those performed in the mouse—that have capitalized on genetic manipulations to unravel key cellular and molecular mechanisms that pattern connectivity within the retina and between the eye and the brain.

LAMINAR WIRING OF THE RETINA

The retina is composed of five types of neurons—PRs, horizontal cells (HCs), bipolar cells (BCs), amacrine cells (ACs), and RGCs—three types of glia, including Müller glia, astrocytes, and microglia, and the endothelial cells that contribute to blood vessels. In the mature retina, neurons and their processes are distributed in a highly ordered, laminated fashion (**Figure 1a**). PRs reside in the outer nuclear layer (ONL) and are separated from BC and HC cell bodies by the outer plexiform layer (OPL), where PRs synapse onto BC and HC neurites. In the OPL, light stimuli are converted into electrochemical signals by modulation of tonic glutamate release from the PRs. Rod and cone PRs together enable the retina to detect visual stimuli across a wide luminance range. Rods are responsible for nighttime (scotopic) vision, given their ability to detect single photons. Cones are less sensitive than rods and operate primarily in daytime (photopic) vision. Both rods and cones are active at an intermediate (mesopic) light level, such as twilight. HCs provide lateral feedback inhibition to regulate neurotransmitter release from PRs (Perlman et al. 2012, Thoreson & Mangel 2012). BC and HC cell bodies reside in the outer region of the inner nuclear layer (INL), and a diverse array of ACs align primarily near the inner region of the INL, adjacent to the inner plexiform layer (IPL). In the IPL, glutamatergic BCs signal to RGCs, and ACs modulate this excitatory transmission via GABAergic or glycinergic inhibition. The cell bodies of RGCs, and several subtypes of displaced AC, reside in the ganglion cell layer (GCL), the innermost nuclear layer.

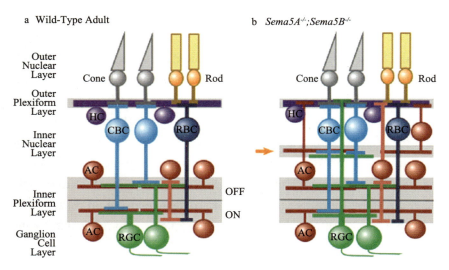

Figure 1

Lamination of neurons in the retina. (*a*) Basic organization of the vertebrate retina. The major neuron types—rod and cone photoreceptors, horizontal cells (HCs), rod (RBCs) and cone (CBCs) bipolar cells, amacrine cells (ACs), and retinal ganglion cells (RGCs)—and their connections are distributed into laminae. The inner and outer nuclear layers and the ganglion cell layer form the cell body layers, and the inner and outer plexiform layers make up the synaptic layers. (*b*) In *semaphorin 5A*; *semaphorin 5B* (*Sema5A⁻/⁻; Sema5B⁻/⁻*) double mutants, inner retinal neurons misproject into the outer plexiform layer and form an ectopic plexiform layer that splits the inner nuclear layer in two (*orange arrow*).

MECHANISMS DIRECTING RETINAL LAMINATION AND WIRING

The synaptic plexiform layers develop in concert with cell body lamination, and IPL formation generally precedes OPL formation (Chalupa 1998). The IPL consists of five major laminae (S1–S5, from the outer to the inner IPL) and additional sublaminae. The processes of cells that depolarize in response to luminance decrements laminate within S1 and S2, the OFF laminae, whereas those that depolarize in response to luminance increments wire within S3–S5, the ON laminae. Much work has focused on elucidating the cellular and molecular mechanisms that set the location of the OPL and IPL and on understanding the mechanisms that organize the lamination of axonsand dendrites of inner retinal neurons into the ON and OFF IPL layers. Although axonal and dendritic lamination are often shaped during development by pruning neurites at inappropriate locations (Hoon et al. 2014), for some retinal neurons, neurite elaboration is constrained within specific IPL stratification depths from the onset of growth (Kim et al. 2010, Morgan et al. 2006). Thus, distinct cellular mechanisms coordinate to create the laminar arrangement of pre- and postsynaptic processes in the IPL. Additionally, despite a temporal sequence in the generation of the major retinal cell classes, no single cell type has been found to be essential for synaptic lamination (Günhan et al. 2002, Günhan-Agar et al. 2000, Keeley et al. 2013, Randlett et al. 2013).

Although a complete picture of the cellular interactions involved in retinal synaptic lamination is still evolving, relatively recent work identifies key molecular factors that contribute to lamination. In the absence of the repulsive transmembrane guidance cues semaphorin 5A (Sema5A) and Sema5B or their receptors plexin A1 (PlexA1) and PlexA3, many retinal neurons that normally restrict their neurites to the IPL extend processes into the OPL, sometimes as far as the PR layer; they can even form an ectopic plexiform layer that splits the INL in two (**Figure 1*b***) (Matsuoka et al. 2011a). Although Sema5A and Sema5B define a broad signaling program that delineates the OPL from the IPL, other molecular cues serve to constrain the processes of specific retinal neuron subtypes. For example, the transmembrane repellent Sema6A and its receptor PlexA4 are required to prevent mistargeting of dopaminergic AC neurites, as well as the dendrites of their postsynaptic target, M1 intrinsically photosensitive RGCs (ipRGCs), away from their proper OFF S1 lamina into the IPL ON layers (Matsuoka et al. 2011b). Sema6A is further required for segregation of ON and OFF starburst ACs (SACs) that are critical components of direction-selective circuits (discussed below). In the outer retina, Sema6A and PlexA4 confine HC neurites within the OPL (Matsuoka et al. 2012).

In the chick, four related immunoglobulin superfamily (IgSf) molecules—Down's syndrome cell adhesion molecule (Dscam), DscamL, sidekick 1 (Sdk1), and Sdk2—comprise a homophilic adhesion code that wires together subsets of RGCs and retinal interneurons within their appropriate IPL laminae (Yamagata & Sanes 2008). Yet in the mouse, Dscam and DscamL apparently function to mediate retinal neurite self-avoidance, as opposed to adhesive laminar targeting (Fuerst et al. 2009). One recent study, however, shows that homophilic interactions mediated by Sdk2 are required for connectivity between, but not lamination of, W3B-RGCs and their presynaptic partners, VG3-ACs (Krishnaswamy et al. 2015). Another set of IgSf molecules, the contactins (CNTNs), are differentially expressed in the chick INL and GCL (Yamagata & Sanes 2012), suggesting they may similarly mediate wiring of IPL circuitry. Indeed, CNTN2 restricts the neurites belonging to CNTN2[*] neurons to their proper IPL laminae. Whether CNTNs and Sdks con tribute to laminar targeting in the mammalian IPL remains to be determined. Interestingly, the cadherin family of cell adhesion molecules also wires together subtypes of BCs with their post synaptic targets in the mouse IPL (discussed below). Furthermore, rod PRs express synaptic cell adhesion molecule-1 (SynCAM1), and HCs express netrinG ligand-2 (NGL2), both of which are required to confine HC processes to the OPL and also for proper rod output (Ribic et al. 2014, Soto et al. 2013). All these observations underscore the critical role that selective adhesion plays in assembling vertebrate retinal plexiform layer development.

Why laminate? The picture we have painted so far is a fairly simplistic view of retinal synaptic organization. When considering that precise synaptic connections form between rod and cone PRs, HCs, 13 types of BCs, approximately 40 different ACs, and approximately 30 RGC subtypes, it is easy to see how convoluted wiring parallel visual channels could become. This intricacy emphasizes the need for developmental strategies that reduce this complexity, to ensure that pre- and postsynaptic partners can find each other and wire together. Lamination maximizes the potential for correct synaptic contacts by bringing pre-and postsynaptic partners in close apposition. Lamination may also prevent wiring of inappropriate synaptic partners, as exemplified by some ON RGCs miswiring with OFF BCs they would not normally connect to in the absence of their correct BC partners (Okawa et al. 2014). Lamination is also a means to reduce the number of signaling molecules required to match synaptic partners. For example, employing redundant Sema5A and Sema5B signaling through PlexA1 and PlexA3 to establish an INL nogo zone decreases the demand for retinal neuron subtypespecific pathways to define IPL versus OPL synaptic layers.

Finally, although lamination simplifies the wiring process, it is not absolutely necessary for retinal neurons to contact their appropriate synaptic partners. As mentioned above, in *Sema6A* and *PlexA4* mutant mice, mistargeted M1 ipRGC and dopaminergic AC neurites in the IPL still find each other and appear connected, albeit at an ectopic location (Matsuoka et al. 2012). Conversely, processes of many retinal neurons are apposed within a given lamina, or even sublamina, in the IPL, but synapses are not necessarily formed at contact sites. These observations support the notion that synaptogenesis between the appropriate partners is largely governed by molecular matching, and although lamination facilitates this process, it may not always be required.

MECHANISMS THAT ASSEMBLE RETINAL CIRCUITS WITH DEFINED FUNCTIONS

Rod Circuits and Pathways

Even though rods are commonly associated with scotopic vision, signals from rods actually propagate through several parallel pathways that convey information at different light levels (**Figure 2**). In the primary rod pathway, the terminals of rods (spherules) make glutamatergic synapses with the dendrites of rod BCs (RBCs), which express the metabotropic glutamate receptor mGluR6 (Nakajima et al. 1993). Rods also synapse onto HCs, which provide feedback inhibition back onto the rods. The rod synapse is composed of two HC axon terminals and one RBC dendritic ending, each of which is apposed to the synaptic ribbon to form a synaptic triad (**Figure 2**). Rods also transmit at higher scotopic to mesopic levels via the cone pathway in at least two ways (**Figure 2**; see also Völgyi et al. 2004). First, rods form gap junctions with cones (Bloomfield & Völgyi 2009). The connexin composition of rod-cone gap junctions remains unclear, but recent studies indicate that it is unlikely to include connexin-36, the connexin responsible for transmission along the primary rod pathway (Bolte et al. 2016, Cowan et al. 2016). Second, rods can make direct chemical synapses with cone BCs (CBCs) (**Figure 2**). Histological and electrophysiological studies now show that rods synapse with type 7 (T7) ON CBCs in mice (Tsukamoto et al. 2007) and several types of OFF CBCs in many mammalian species, including mice and primates (Pang et al. 2012, Tsukamoto & Omi 2014a). Researchers have also identified novel roles for rods in cone pathway visual processing via interactions with HCs. When rods are not saturated, these PRs con tribute to center-surround color opponency in the mouse retina motion-detecting J-RGC, likely by influencing HC feedback onto cones (Joesch & Meister 2016). Moreover, at photopic light levels, cone drive onto HCs could relieve saturation of the rods, leading to rod signals that then contribute to the inhibitory surrounds of inner retina neurons (Szikra et al. 2014).

Unlike CBCs, RBCs do not contact RGCs directly but instead direct their output onto two types of AC: the bistratified glycinergic AII AC and the large-field GABAergic A17 AC (**Figure 2**) (Kolb & Famiglietti 1974). A single RBC ribbon is apposed to the processes of one AII and one A17 AC, forming a synaptic dyad (**Figure 2**) (Kolb 1979). AII ACs relay visual signals from RBCs to both ON and OFF pathways (**Figure 2**) (Demb & Singer 2015). Glutamate release from the RBCs is regulated largely by inhibition from the A17 AC. In the ON layer, AII ACs make connexin-36–containing gap junctions with ON CBCs to convey sign-conserving signals (Hartveit & Veruki 2012). In the OFF layer, AII ACs form glycinergic synapses onto OFF CBCs and RGCs, mediated by glycinergic receptors containing the α1 subunit, to convey sign-inverting signals from the AII AC lobular appendages (Ivanova et al. 2006, Zhang et al. 2014). Neighboring AII ACs are widely electrically coupled to amplify these visual signals (Hartveit & Veruki 2012). Although synaptic convergence across the rod pathway differs among species, these basic synaptic connectivity patterns are preserved (Anderson et al. 2011, Marc et al. 2014, Tsukamoto & Omi 2013).

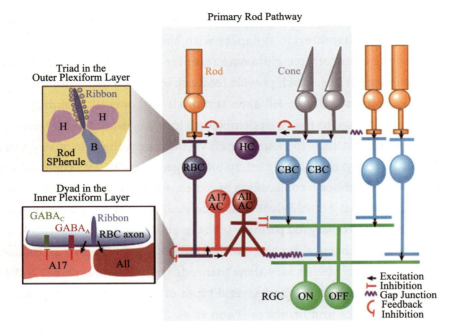

Figure 2

The rod pathway. In the outer retina, horizontal cells (HCs) modulate transmission from rods onto the rod bipolar cells (RBCs). Each ribbon synapse in the rod spherule is apposed to two HC axons (H) and one RBC dendrite (B) to form a triad (*upper left box*). In the inner retina, RBCs contact AII amacrine cells (ACs) that provide glycinergic inhibition onto OFF BCs and excite ON BCs via gap-junctional coupling (*middle gray panel*). RBCs receive reciprocal inhibition from A17 ACs that is mediated by GABA$_A$ and GABA$_C$ receptors at different locations (*lower left box*). Each ribbon synapse in the RBC axon is apposed to one AII and one A17 process to form a dyad (*lower left box*). Rod signals also traverse the secondary and tertiary rod pathways, where rods are electrically coupled to cones via gap junctions, or make direct synapses with ON and OFF BCs (*far right*). Other abbreviations: CBC, cone bipolar cell; RGC, retinal ganglion cell.

Development of Rod Pathway Circuits

In recent years, molecular and genetic approaches have revealed mechanisms that underlie the specificity with which synaptic connections within the rod and cone pathways are achieved during development. Both activity-dependent and-independent mechanisms are implicated. Whereas some mechanisms are common to both the rod and cone pathways [e.g., the essential interaction between RIBEYE and Bassoon for anchoring ribbons at synaptic release sites (tom Dieck et al. 2005)], others influence the development of connectivity within these pathways differentially. We highlight here recent progress in our understanding of the mechanisms that regulate synaptic development in the rod pathway.

Development of rod connectivity with rod bipolar cells. Several molecules found at the rod synapse (**Figure 3*a,b***) play major roles in establishing or maintaining its triad architecture and function. A key event that reflects maturation of the rod triad is the invagination of RBC dendrites and HC axonal tips at locations opposite the synaptic ribbon. Several molecules have been implicated in this process. For example, interaction between pikachurin in the synaptic cleft and dystroglycan (DG) on rod axon terminals is necessary for the invagination of RBC dendrites into rod spherules (Kanagawa et al. 2010, Omori et al. 2012, Sato et al. 2008). RBC dendritic invagination also requires the presence of extracellular leucine-rich repeat and fibronectin type III domain containing 1 (ELFN1) (Cao et al. 2015), an adhesion molecule expressed by rods that binds mGluR6, the voltage-gated calcium channel Ca$_v$1.4 (Zabouri & Haverkamp 2013), and the planar cell polarity transmembrane receptor Frizzled3 (Shen et al. 2016). In mice lacking Sema6A or PlexA4, rod spherules form synapses but do not exhibit the triad arrangement because only one HC process, instead of two, invaginates the spherule (Matsuoka et al. 2012). However, synaptic triads do form, but with smaller ribbons, in other mice harboring mutations in adhesion or guidance molecule genes, including *NGL-2* and *SynCAM-1* mutants (Ribic et al. 2014, Soto et al. 2013).

Localization of mGluR6 to RBC dendritic tips also marks the emergence of a functional rodRBC synapse. mGluR6 fails to accumulate at RBC dendritic tips in the absence of ELFN1 or Ca$_v$1.4 (Cao et al. 2015). The involvement of a volta-gegated calcium channel suggests the possibility that calcium entry subsequent to depolarization is necessary for expression of ELFN1 at rod spherules, which in turn regulates mGluR6 localization at this synapse. Although disrupting vesicular transmitter release from PRs by expression of the tetanus toxin (TeNT) light chain results in a loss of mGluR6 localization, both ELFN1 and mGluR6 localize properly in the *Trpm1* knockout, in which RBC depolarization is diminished (Cao et al. 2015). Thus, even though presynaptic processes associated with transmitter release are involved in regulating postsynaptic mGluR6 localization at RBC dendritic tips, RBC depolarization does not appear to be critical. Moreover, although dark rearing diminishes both mGluR6 expression on CBCs and cone-mediated responses significantly, it does not affect the development of rod-RBC connectivity (Dunn et al. 2013). Loss of mGluR6 (Tsukamoto & Omi 2014b) or other critical elements of the mGluR6-mediated signaling cascade, such as Gαo or Gβ3 (Tummala et al. 2016), can perturb rod triad synapse assembly. However, this occurs by disrupting presynaptic matrix–associated protein expression (e.g., pikachurin and DG) (**Figure 3*a***).

New insight into the molecular interactions that control the recruitment of ELFN1 at the rod synapse was gained recently by examination of knockout mice that lack the auxiliary calcium channel subunit α2δ4 (Wang et al. 2017). In the absence of α2δ4, ELFN1 fails to localize to the rod synapse, and Ca$_v$1.4 expression is diminished. Rescue experiments further revealed that α2δ4 alone is insufficient and that both α2δ4 and Ca$_v$1.4 are needed together to ensure the proper arrangement of the rod-RBC transsynaptic complex, resulting in the alignment of the presynaptic transmitter release machinery and postsynaptic mGluR6 receptors. Interestingly, although α2δ4 is present in both rods and cones, only rod synaptogenesis is drastically perturbed in its absence. This may be because—despite the major role played by ELFN1-mGluR6 binding in synaptic partner matching—ELFN1 is present on rods but not on cones (**Figure 3c**, ①). However, ON CBCs also express mGluR6. What prevents these bipolar cell types from connecting with-rods? One possibility is that ELFN1 binding to mGluR6, although important, is not essential for rod-RBC matching because RBC dendrites invaginate rod spherules in the mGluR6 knockout (Tsukamoto & Omi 2014b). In the rodfull retina, where only a few cones remain, CBCs still prefer to contact cones (Keeley & Reese 2010). This raises the possibility that rods, through a mechanism independent of ELFN1-mGluR6 interactions, may prevent CBCs from contacting them (**Figure 3c**, ②). Another factor that might limit CBCs from contacting rods is that rod synaptogenesis occurs later than cone synaptogenesis (Cao et al. 2015). However, some CBCs continue to make synapses during the period of peak rod synaptogenesis (Dunn & Wong 2012); thus, developmental timing may play a limited role. Conversely, why do RBCs avoid most cones in wildtype animals, when they can make connections with cones in the absence of rods (Strettoi et al. 2004)? One possibility is that the presence of rods discourages RBCs from contacting cones (**Figure 3c**, ③). In addition, adhesion molecules or other interactions at cone-CBC contacts may secure this connection strongly (**Figure 3c**, ④), and select repulsive interactions may also play a role in segregating CBC and RBC contacts with cone and rod PRs.

Connectivity between rods and HCs requires not only matching of partner types but also targeting of a specific HC cellular compartment to the rod terminals. Rods generally contact the axon-like terminal of HCs, whereas cones contact HC dendrites (Kolb 1974). The molecular mechanisms that dictate this unique pattern of connectivity are unknown. It is evident, however, that HC axons persist in the absence of rods, albeit they are smaller in size, and that they become larger in rodfull or coneless retinas (Raven et al. 2007). What remains to be determined is whether or not HC axons make bona fide synapses with cones in the absence of rods, and if so, how closely these synapses resemble normal cone triads. Because HCs differentiate and contact PRs before BCs, genetic tools have recently been used to remove HCs prior to BC differentiation. In the adult *Lim1* mutant retina, HC processes are largely absent due primarily to a failure of many HCs to migrate to the outer retina (Keeley et al. 2013). In *Lim1* mutants, ectopic

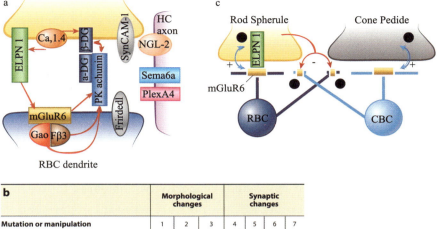

b	Mutation or manipulation	Morphological changes			Synaptic changes			
		1	2	3	4	5	6	7
Mutation or KO of molecules	Elfn1 −/−							
	Fzd3 f/−							
	Pikachurin −/−							
	Dag1 CKO (DG)							
	Ngl-2 −/−							
	Cadm1 −/− (SynCAM-1)							
	PlexA4 −/−							
	Sema6A −/−							
HC absence	Lim1 CKO							
Disrupted neurotransmission	Gnao 1 −/−							
	Cacna1f −/− [Cav1.4(α1F)]							
	Grm6 −/− (mGluR6)							
	Pcdh21-TeNT							
	Dark rearing*							
	Slc32a1 −/− (VIAAT)*							

Phenotype

1. Spherule retraction
2. RBC dendrite sprouting
3. HC process sprouting
4. Ectopic synapse
5. Disrupted rod triad arrangement
6. Decreased mGluR6 at RBC dendrites
7. Change in morphology/ number of rod ribbons

- Abnormal
- Unaffected
- Not assessed

Figure 3

Mechanisms regulating synaptic connectivity within the rod pathway. (*a*) Molecules at rod synapses in the outer plexiform layer whose absence disrupts connectivity in this pathway. Loss of some molecules alters expression or function of other molecules (red arrows). (*b*) Morphological and synaptic changes at synapses associated with the rod pathway in the outer retina in various mouse mutants and models. Red indicates the abnormal phenotype was observed, and yellow indicates an unaffected phenotype. Numbers indicate the phenotype described. Asterisks indicate manipulations that do not affect the rod pathway but disrupt the cone pathway. (*c*) Possible mechanisms involved in rod–rod bipolar cell (RBC) and cone–cone bipolar cell (CBC) specific wiring: ① interaction between ELFN1 and mGluR6 (*yellow rectangle*) that promotes rod-RBC connectivity, unknown negative regulation by the rod that inhibits ② rod-CBC or ③ cone-RBC connection, and ④ strong interaction between the cone and CBC that facilitates their connection. Genes (proteins): *Cacna1f* [CaV1.4(α1F) subunit]; *Cadm1* (SynCAM-1); *Dag1* (dystroglycan, DG); *Fzd3* (Frizzled3); *Gnao1* (G protein αo, Gαo); *Grm6* (mGluR6); *Ngl2* (netrin-G ligand-2, NGL-2); *Slc32a1* (vesicular inhibitory amino acid transporter, VIAAT). Other abbreviations: CKO, conditional knockout; ELFN1, extracellular leucine-rich repeat and fibronectin type III domain containing 1; Gαo, G protein αo; Gβ3, G protein β3; HC, horizontal cell; KO, knockout; *Lim1*, LIM homeobox protein 1 transcription factor; mGluR6, metabotropic glutamate receptor 6; Pcdh21-TeNT, specific expression of tetanus toxin (TeNT) in photoreceptors upon crossing Pcdh21-cre (protocadherin 21) mice with ROSA-TeNT mice; PlexA4, plexin A4; Sema6A, semaphorin 6A. See main text for references.

synapses between RBCs and retracted rod spherules emerge, but how these aberrations evolve is unknown. Interestingly, cones and CBCs retain their normal lamination in the absence of HCs in *Lim1* mutants, suggesting a differential role for HCs in the maintenance of PR-BC synaptic organization in the adult rod and cone pathways. Whether this differential effect is related to the compartment-specific connections of HCs to rods versus cones requires further investigation. For example, expression of specific molecules, or even neurotransmission, within HC axons could be critical for maintaining the position of the rod spherule in the OPL, whereas HC dendrites are not necessary for lamination of the cone circuits.

Development of rod bipolar cell output connectivity. Compared to rod input, the molecular mechanisms governing RBC outputs are less well understood. Several observations implicate reelin, a large secreted extracellular matrix glycoprotein, in the development of RBC→AII connectivity (Rice et al. 2001, Trotter et al. 2011). Loss of reelin, or its apolipoprotein E receptor 2 results in ectopically located contacts between RBCs and AII ACs; however, the structure and function of these synapses are unclear. Specifically, defining whether or not these ectopic contacts between RBC axons and AII processes in S1 of the IPL involve the AII lobular output appendages would help establish a role for reelin in matching the appropriate RBC cellular compartments to AII ACs. Furthermore, because the processes of the other RBC postsynaptic partner, the A17 AC, course through IPL S1, it would also be interesting to ascertain whether or not these ACs contribute to the ectopically located contacts.

Like the outer retina, glutamatergic transmission is not essential for the development and maintenance of synapses within the rod pathway in the inner retina. Expression of TeNT in RBCs does not prevent the formation of RBC→AII/A17 synapses (Kerschensteiner et al. 2009, Schubert et al. 2013). However, synapses that include TeNT-expressing RBCs have multiple ribbons rather than a single ribbon at each contact site, suggesting an attempt to increase glutamate release at the silenced synapse. Although glutamatergic transmission is not critical for RBC→AC synaptogenesis, this form of transmission regulates gap-junctional coupling in AII ACs. In the rabbit retina, connexin-36 expression in AII ACs is upregulated under visual deprivation or glutamatergic transmission blockade (Tu & Chiao 2016).

Researchers have known for some time that inhibition from A17 ACs onto RBCs is mediated by both ionotropic GABAA and GABAC receptors (Chávez et al. 2006, Fletcher et al. 1998). GABA receptor expression on RBC axons is regulated by GABAergic neurotransmission in a receptor type–specific fashion: GABA release is needed to maintain GABA$_A$α1 and GABA$_C$ receptors, but not GABA$_A$α3 receptors, on

RBC axons. An immuno–electron microscopy study now demonstrates that GABAA receptors are localized close to the ribbon, whereas GABAC receptors are more distally localized (**Figure 2**) (Grimes et al. 2015). The specific locations of these two GABA receptor types appear to correlate with distinct functional roles for these receptors in RBC transmission. $GABA_A$-mediated inhibition near the transmitter release site (ribbon) largely regulates tonic glutamate release from RBCs in darkness. In contrast, the more distally located but higher affinity $GABA_C$ receptors can readily detect light-evoked GABA release from A17 ACs, driven by the RBCs, which shapes the temporal profile of the RBC output (Eggers & Lukasiewicz 2011, Eggers et al. 2007). How this specialized GABAergic feedback synapse is assembled precisely during development remains to be unraveled.

Overall, both activity-dependent and independent mechanisms clearly play unconventional and complex roles in regulating synapse development and maintenance in the rod pathway. It is important to realize, however, that on occasion, molecules associated with neurotransmission, such as $Ca_v1.4$, play key roles through interactions that are independent of activity (**Figure 3b**). Moreover, neurotransmission can influence synaptic development of RBCs in a cell compartment–specific manner, but these mechanisms have not yet been explored. For example, in retina-specific knockouts of the vesicular inhibitory amino acid transporter, $GABA_A$ receptors on the axons are lost, whereas those on the dendrites are unaffected (Hoon et al. 2015). Finally, much remains to be discovered about the molecular and cellular interactions that coordinate the development of the input and output synapses of the RBC.

Alpha ON-Sustained Retinal Ganglion Cell Circuitry

Alpha, or A-type, RGCs in the mouse retina are characterized by their relatively large somata and dendritic fields. The three main types of alpha RGCs are classified by differences in dendritic lamination and temporal responses to light: sustained ON (AON-S), transient OFF (AOFF-T), and sustained OFF (AOFF-S). Although the anatomy and physiology of these cell types have been studied extensively, the connectivity underlying their unique response properties is less well understood. Of all the alpha cells, the structure and function of AON-S RGCs has been studied the most extensively (**Figure 4a**). Recent work uncovered a nasal-temporal gradient in AON-S RGC density across the retina, with the peak density located in a dorsal-temporal region (**Figure 4b**) (Bleckert et al. 2014). The AON-S RGC dendritic field area is inversely related to soma density, following a common architectural theme in the retina (Wässle et al. 1981). The relatively higher density and smaller dendritic fields of AON-S RGCs in temporal retina are consistent with the requirements for higher spatial sampling frequency (acuity) in the region of central vision within the zone of binocular overlap. This topographic specialization was unexpected for the mouse retina because overall RGC density appears quite uniform across the retina.

Transgenic labeling of retinal BCs helped reveal the microcircuitry of inputs onto AON-S RGC dendrites. Excitatory input is provided primarily by two types of CBCs. Type 6 (T6) CBCs provide the dominant excitatory input (approximately 70%), whereas T7 CBCs contribute a minor (<5%) proportion (**Figure 4c**) (Morgan et al. 2011, Okawa et al. 2014, Schwartz et al. 2012). A structure-function study that mapped T6 CBC connectivity with AON-S RGCs revealed a fine-scale heterogeneity in the spatial receptive field of AON-S RGCs that could be attributed to differences in the number of synapses individual T6 BC axons make with the RGC dendritic arbor (Schwartz et al. 2012). Inhibition onto AON-S RGCs is mediated by GABA$_A$ (Bleckert et al. 2014, Greferath et al. 1995) and glycine α1 receptors on the RGC dendrites (Majumdar et al. 2007). Recent studies identified some AC types that are presynaptic to the AON-S RGCs. GABAergic wide-field ACs with axon-like processes may generate the suppressive surround of AON-S RGCs (Farrow et al. 2013). A8 ACs make glycinergic synaptic contacts with the AON-S RGCs (Lee et al. 2015), but how they or other glycinergic ACs regulate AON-S function is not yet clear. Given the increasing knowledge of the types of synaptic inputs converging onto AON-S RGCs, this RGC type will likely remain the subject of intense investigation, with the aim to better understand how diverse excitatory and inhibitory inputs act in concert to shape the output of an RGC type common to mammals.

Development of alpha ON-sustained retinal ganglion cell inputs. The stereotypic connectivity of AON-S RGCs (Figure 4c) affords many opportunities to investigate the mechanisms that circuits engage not only to select synaptic partners but also to distribute connections differentially among connected partners, such that some partners receive more connections than others. T6 CBCs, the dominant input, increase connectivity with the AON-S RGC gradually during maturation. In contrast, although T7 CBCs, which provide minor input, establish contact at the same time as T6 CBCs, T7 connections do not increase over developmental time (**Figure 4c**). Thus, differential connectivity with the two CBC types is established by preferential synaptogenesis with one partner type, rather than by selective elimination of synapses with the less preferred partner. Synapse elimination, however, does play a role in shaping BC to AON-S RGC connectivity, as the contacts RBCs make early in development are later eliminated (**Figure 4c**) (Morgan et al. 2011).

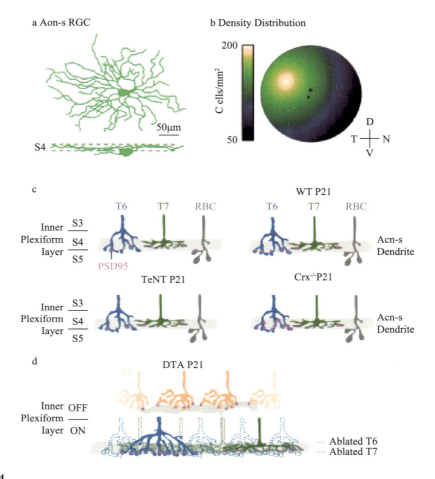

Figure 4

Alpha sustained ON retinal ganglion cell (Aon-s RGC) pathway. (*a*) Dendritic morphology and ON stratification pattern of a representative Aon-s RGC. (*b*) Density distribution of Aon-s RGCs exhibits a peak in the dorsotemporal retina. D, V, T, and N indicate dorsal, ventral, temporal, and nasal, respectively. (*c*)Development and activity-dependent regulation of connectivity between bipolar cells (BCs) and Aon-s RGCs. Postnatal day 21 (P21) Aon-s RGC dendrites receive input from type 6 (T6) and T7 BCs but also receive input transiently from rod BCs (RBCs) at P9. Loss of transmission from ON cone BCs (CBCs) due to expression of tetanus toxin (TeNT) reduces synapses from the T6, but not T7, BCs. In cone-rod homeobox (*Crx*)$^{-/-}$ mice, BCs are hyperactive, and T6 but not T7 synapses are doubled compared to wildtype (WT). (*d*) Activity-independent regulation of connectivity between BCs and Aon-s RGCs. When subsets of ON BCs are ablated by expression of diphtheria toxin (DTA), surviving T6 or T7 BCs expand their axonal territories and increase synaptogenesis with Aon-s RGCs. Moreover, AONS RGCs have ectopic dendrites in the OFF layer that make functional synapses with T2 BCs. Panel *b* adapted with permission from Bleckert et al. (2014). Other abbreviation: PSD95, postsynaptic density protein 95.

Preferential synaptogenesis with the favored T6 CBCs is normally constrained by homotypic interactions between these BC types. T6 CBCs that survive after genetic ablation of many ON CBCs expand their axonal territory and increase synaptic contact with AON-S RGC dendrites (**Figure 4*d***) (Okawa et al. 2014). These studies show that in the absence of T6 CBCs, synaptogenesis with T7 CBCs exceeds that expected for a simple increase in T7 CBC axonal size. Thus, the presence of the major, favored input type may normally suppress synaptogenesis with the unfavored input type. As yet, the exact cellular and molecular interactions that result in preferential synaptogenesis with one partner and not another have not been identified. Furthermore, in the absence of T6 CBCs, the AON-S RGC makes ectopic synapses with OFF T2 BCs, suggesting that contact with the dominant input also restricts the AON-S RGC's ability to wire only with ON CBCs (**Figure 4*d***). This observation underscores the importance of laminar cues that restrict access to inappropriate partner types, without which erroneous connections can form.

Glutamatergic neurotransmission from BCs during development does not influence the stereotypic ratio of synapses between the AON-S RGC and the two CBC types. However, transmission clearly controls the number of synapses an individual T6 CBC makes with the RGC. Reduction of glutamate release from the T6 CBC leads to fewer synapses, whereas increased CBC activity causes increased contact (**Figure 4*c***) (Kerschensteiner et al. 2009, Morgan et al. 2011, Okawa et al. 2014, Soto et al. 2012). Surprisingly, the number of synapses from T7 CBCs onto AON-S RGCs is not dependent on synaptic transmission. This differential control by transmission among the two input types raises the question of whether one input type's (T7) connectivity is completely independent of activity, or whether synapses with the major partner are always more susceptible to alterations in activity levels.

Direction-Selective Circuitry

One of the best-studied circuits in the vertebrate retina is that which encodes the direction of moving objects in the visual scene (Barlow & Levick 1965). Direction-selective ganglion cells (DSGCs) fire robust action potentials in response to objects moving in their preferred direction, and they fire fewer, if any, action potentials in response to motion in the opposite or null direction (**Figure 5*a***). Direction-selective (DS) circuits detect object motion across a broad range of illumination, including mesopic light conditions in which rods are active, indicating that both rods and cones transmit visual information about object motion (Vaney & Taylor 2002). DSGCs can respond to light onset (ON, or oDSGCs), and also to both light onset and light offset (ON-OFF, or ooDSGCs). ooDSGCs respond to objects moving at a broad range of speeds, and they detect motion in the upward, downward, nasal, or temporal directions (Vaney et al. 2012). oDSGCs, by contrast, respond to bright objects moving at slower velocities (Sanes

& Masland 2015), and they are thought to respond primarily to motion in the upward, downward, and nasal directions.

Work on direction selectivity initially led to the conclusion that the ability of a DSGC to respond preferentially to motion in a particular direction derived from both DS BC excitatory inputs and DS inhibitory inputs from SACs onto these RGCs (Fried et al. 2002, 2005). Recently, excitatory inputs onto DSGCs were found not to exhibit direction selectivity (Park et al. 2014, Yonehara et al. 2013). Instead, the basis for DS firing of DSGCs comes from unique functional properties of SACs and their asymmetric wiring with DSGCs (Briggman et al. 2011, Morrie & Feller 2015). Null-side SACs provide GABAergic input onto DSGCs that prevents their firing in response to null-directed motion (**Figure 5a,b**). SACs consist of two mirror populations across the IPL (**Figure 5c**). OFF SAC cell bodies reside in the INL and project their neurites into the S2 OFF layer of the IPL; in contrast, ON SAC cell bodies reside in the GCL and project their neurites into the S4 ON layer of the IPL (Vaney 1990, Vaney et al. 2012). Several characteristics of SACs are unique and serve to define this cell type. SAC dendritic arbors are strikingly radially symmetric, with approximately five primary dendrites confined to a single IPL lamina branching out in a starburst pattern (**Figure 5b**). Like other ACs, SACs lack axons and instead release neurotransmitter—both GABA and acetylcholine, in the case of SACs—from terminal varicosities that are restricted to the distal third of their dendritic arbor (Famiglietti 1991, Taylor & Smith 2012, Vaney et al. 2012).

SACs are themselves DS, releasing neurotransmitter in response to centrifugal motion of stimuli moving from their soma to their dendritic tips (Euler et al. 2002). The source (or sources) of SAC DS neurotransmitter release is a topic that currently receives much attention. Excited SACs release GABA onto neighboring SACs, in addition to DSGCs, fine-tuning their own DS responses via lateral inhibition (Vaney et al. 2012). The contribution of excitatory inputs to the DS computation within SACs has been controversial; however, much of the controversy can be attributed to species-specific differences in SAC connectivity. Researchers initially thought presynaptic inputs onto SACs in the rabbit retina occurred along the full length of the SAC dendritic arbor (Famiglietti 1991). In the mouse, one study attributed DS SAC responses to the varied release kinetics of differentially distributed transient versus sustained BC inputs (Kim et al. 2014). However, another group found that SAC excitatory input kinetics are independent of dendritic location (Stincic et al. 2016). Most recently, mouse BC excitatory inputs onto SACs were found to be restricted to the proximal 70% of the SAC dendritic arbor (Ding et al. 2016, Vlasits et al. 2016). This restriction is predicted to establish DS SAC neurotransmitter release, whereas lateral inhibition enhances DS SAC output (Vlasits et al. 2016).

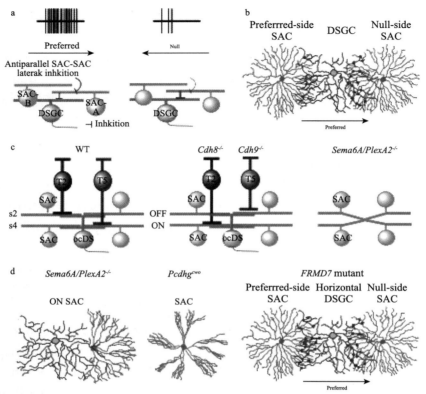

Figure 5

Mechanisms underlying organization and development of direction-selective (DS) circuits in the retina.
(*a*) Direction-selective ganglion cells (DSGCs) fire robust action potentials in response to motion in their preferred direction. Antiparallel lateral inhibition from starburst amacrine cell B (SAC-B) onto SAC-A decreases SAC-A's inhibition of the DSGC in response to motion in the DSGC's preferred direction. When motion originates from the DSGC's null side, lateral inhibition of SAC-A does not occur (*shaded curved arrow*), and SAC-A strongly inhibits the DSGC (*wide red arrow*), diminishing its output (*top*). (*b*) En face view of asymmetric inhibitory inputs from null-side SACs onto DSGCs, which underlie the DS computation. (*c*) Schematics illustrating ON and OFF laminar connectivity of SACs and type 2 and type 5 (T2 and T5) bipolar cells (BCs) with ON-OFF DSGCs (ooDSGCs) in wild-type (WT) and mutant retinas. In *cadherin 8* (*Cdh8*) and *Cdh9* mutants, T2 BC axons misproject into ON layers, and T5 axons misproject into OFF layers, respectively. In *semaphorin 6A* (*Sema6A*) and in *plexin A2* (*PlexA2*) mutants, developmental OFF and ON SAC neurite repulsion is compromised, and crossovers occur between the S2 and S4 inner plexiform layers. (*d*) En face view of SAC phenotypes in various mutant backgrounds. In *Sema6A* and in *PlexA2* mutants, portions of ON SAC dendritic arbors are missing, and aberrant self-avoidance leads to crossover events in the distal third of the dendritic arbor. In *protocadherin gamma* (*Pcdhg*) retinal conditional knockouts (*Pcdhg^{rCKO}*), SAC dendritic self-avoidance fails, resulting in fused dendritic branches. In *FERM domain–containing protein 7* (*FRMD7*) mutants, horizontal motion-preferring DSGCs are no longer asymmetrically inhibited by SACs, as shown in panel b.

Wiring bipolar cells into direction-selective circuitry. Even though BC input onto SACs or DSGCs is not in itself DS, altering the connectivity between BCs and ooDSGCs disrupts direction selectivity. Investigators have recently made progress in understanding how two types of CBCs, type 2 (T2) and T5 BCs, target their postsynaptic partners: ventrally preferring Hb9-GFP+ ooDSGCs, and OFF (T2) or ON (T5) SACs (Duan et al. 2014). In this study, the transmembrane protein cadherin 8 (Cdh8) was found to be selectively expressed in T2, whereas cadherin 9 (Cdh9) was expressed in T5. In the absence of Cdh8, T2 axon terminals are no longer restricted to the OFF layers of the IPL, but instead they misproject into the ON layers; similarly, in the absence of Cdh9, T5 axons misproject into the OFF layers of the IPL (**Figure 5c**). Cadherins are classic homophilic cell adhesion molecules. Yet in this study, the authors found that Cdh8 and Cdh9 appear to interact with downstream binding partners in a heterophilic manner. Sparse misexpression of Cdh8 in T5 or of Cdh9 in T2, in each case in a null genetic background for either *Cdh8* or *Cdh9*, was sufficient to redirect these BCs to mistarget into the OFF or ON layers, respectively, resulting in attenuation of DS responses in Hb9-GFP⁺ ooDSGCs (Duan et al. 2014). Interestingly, a subset of SACs and ooDSGCs express Cdh6 (Kay et al. 2011), raising the possibility that Cdh6 could also serve as a target recognition molecule.

The skewed distribution of excitatory inputs in the SAC dendritic arbor may be a prominent contributor to the SAC DS computation, which is in turn required for shaping the direction selectivity of output responses from DSGCs. This restricted distribution of excitatory BC inputs onto SACs suggests additional mechanisms that specify connectivity, beyond the level of wiring individual neurons, and raises several important issues related to how subcellular connectivity arises. For example, what molecular cues direct presynaptic BC growth cones to the proximal two-thirds of the SAC dendritic arbor while avoiding the distal third? Furthermore, what mechanisms direct the specialization of presynaptic SAC varicosities, confining the SAC presynaptic release machinery to the distal third of the SAC dendritic arbor?

Starburst amacrine cell circuit elaboration. Recent studies emphasize the importance of SAC morphology in determining DS tuning of DSGCs. First, a semaphoring-plexin signaling axis was found to regulate multiple aspects of SAC circuit elaboration. The repulsive cue Sema6A is expressed by ON SACs, whereas its receptor PlexA2 is expressed by both ON and OFF SACs (Sun et al. 2013). Both are required for ON and OFF SAC neurite segregation into discrete laminae of the developing IPL. Additionally, both are required in ON SACs specifically for the acquisition of radial dendritic morphology: In *Sema6A* and *PlexA2* mutants, ON SACs are missing portions of their dendritic arbors, and the distal third of ON SAC neurites exhibits a loss of self-avoidance (**Figure 5d**) (Sun et al. 2013). *Sema6A* mutants exhibit impaired DS responses in a subset of posterior motion–preferring ooDSGCs, with tuning of these neurons being broadened such that they fire more robustly in response to nonpreferred motion stimuli than controls. This highlights the importance of SAC morphology for proper DSGC output.

An unrelated family of signaling molecules also plays critical, but distinct, roles in establishing SAC radial dendritic morphology. The protocadherins (Pcdhs), cadherin-related homophilic recognition molecules, are stochastically expressed from a gene cluster of 58 *Pcdh genes* (Zipursky & Sanes 2010). In principle, Pcdhs are quite similar to the DSCAM1 proteins of *Drosophila:* DSCAM1s are represented by many thousands of variants; like Pcdhs, they participate in strict homophilic interactions; and also like Pcdhs, individual protein isoforms are expressed stochastically in neurons. For these reasons, Pcdhs were considered excellent candidates to mediate neurite self-avoidance, similar to *Drosophila Dscam1* (Hattori et al. 2009). The *Pcdh-gamma* (*Pcdhg*) locus produces 22 isoforms that associate to form tetramers. Considering both *trans* and cis interactions, a diversity of over 200,000 protein complexes is predicted from these isoforms (Schreiner & Weiner 2010). A subset of isoforms are expressed stochastically in individual SACs to promote self/nonself discrimination, and in the absence of expression from the *Pcdhg* locus, SAC dendrite self-avoidance fails and secondary and tertiary dendrites emanating from primary branches overlap extensively (Lefebvre et al. 2012).

A subsequent study identified two new roles for Pcdhgs that impact DSGC output directly. First, in the absence of Pcdhg-mediated self-avoidance, SACs form autaptic synapses onto themselves (Kostadinov & Sanes 2015). This is possible because SAC presynaptic release machinery is near sites of postsynaptic specializations. Without self-avoidance, pre- and postsynaptic specializations can appose one another, allowing autapse formation. Second, closely spaced SACs have many intercellular synaptic connections early postnatally that are developmentally pruned later, and Pcdhgs are required for this pruning event (Kostadinov & Sanes 2015). The consequences of increased autaptic and SAC-SAC intercellular connections in *Pcdhg* mutants have interesting implications for the computation of direction selectivity by SACs. Reciprocal lateral inhibition between neighboring antiparallel SAC dendrites sharpens the DS firing response of DSGCs by enhancing centrifugal firing preference within SAC dendrites. If motion occurs along a given SAC (SAC-A) dendrite's null direction (centripetal, or toward its soma), neighboring antiparallel SAC dendrites (SAC-B) will be excited, and they will release GABA onto this SAC-A dendrite to further decrease the SAC-A dendrite's own release of GABA (**Figure 5a**). This lateral inhibition keeps the SAC-A dendrite from inhibiting connected DSGCs in response to motion in their own preferred direction (the SAC-A dendrite's null direction). Impaired developmental synapse pruning and increased autapses in *Pcdhg* mutant SACs likely degrade SAC-intrinsic direction selectivity by increasing parallel inhibitory SAC-SAC synapses. In our example, this would (*a*) impair SAC-B's ability to inhibit the SAC-A dendrite in response to motion that is centripetal to the SACA dendrite, because SAC-B would be more strongly inhibited by parallel synapses, and (*b*) cause SAC-A to be more strongly inhibited by parallel synapses from neighboring SAC dendrites, or autapses, in response to centrifugal motion (Kostadinov & Sanes 2015). This would degrade both SAC-SAC lateral inhibition and DS release of neurotransmitter from SACs in response to centrifugal motion. In *Pcdhg* mutants, the DS output of ventrally preferring ooDSGCs is diminished, underscoring the critical role played by radial SAC dendritic morphology and neurite self-avoidance in DS circuit function.

Molecular machinery that wires starburst amacrine cells to direction-selective ganglion cells. Remarkably, in *Pcdhg*, *Sema6A*, or *PlexA2* mutants, SACs and DSGCs still cofasciculate (Kostadinov & Sanes 2015, Sun et al. 2013), and *Pcdhg* mutant SACs still form synapses onto DSGCs—despite impaired SAC morphology and diminished direction selectivity of DSGC output in these genetic backgrounds. Therefore, SAC self-avoidance and SAC-DSGC cofasciculation are separable events. Which molecules promote SAC-DSGC target recognition? One can imagine an ON-specific pathway to promote target recognition between ON SACs and ON DSGC neurites, whereas an OFF-specific pathway would wire OFF SACs and OFF DSGC neurites. The same cues may wire both ON and OFF-specific SAC/DSGC connections, and still other cues may serve to constrain ON and OFF circuits to their appropriate IPL locations. Additionally, there may be subtypes of SACs that wire with DSGC neurites that have specific motion preferences. In support of this idea, Cdh6 is selectively expressed in a subset of SACs and a subset of ooDSGCs (Kay et al. 2011).

Symmetry is an important theme for SACs, since loss of dendrite symmetry changes the geometry of SAC-SAC and SAC-DSGC interactions, ultimately weakening the DSGC DS response. A single SAC will wire with DSGCs that prefer each of the cardinal directions—upward, down ward, nasal, and temporal (Briggman et al. 2011). What provides the specificity required to wire independent branches of a SAC with antiparallel DSGCs that prefer distinct directions of motion? A recent consideration of SAC symmetry issues suggests a general mechanism for wiring together SACs and DSGCs, and this originates from the observation that SACs express the FERM (4.1/ezrin/radixin/moesin) domain–containing protein 7 (FRMD7) (Yonehara et al. 2016). Mutations at the human *FRMD*7 locus account for more than 70% of idiopathic congenital nystagmus cases (Tarpey et al. 2006). FERM domain proteins mediate signaling events between the cell membrane and the actin cytoskeleton (Moleirinho et al. 2013). Similar to observations in humans, mice without functional FRMD7 lack the horizontal optokinetic reflex (OKR) (Yonehara et al. 2016), which prompts the eye to track the visual scene as it drifts across the retina by generating slipcompensating eye movements (Dhande et al. 2015). However, the vertical OKR is intact in *FRMD7* deficient mice (Yonehara et al. 2016). Because both horizontal and vertical OKR reflexes are driven by oDSGCs [and a subset of ooDSGCs for the horizontal OKR (Dhande et al. 2013)] that respond to motion at slower speeds, further investigation of the DSGCs responsible for the missing horizontal OKR in *FRMD7* mutants was warranted. Nearly all horizontally tuned DSGCs lost their direction selectivity owing to a loss of asymmetric inhibition from nullside SACs (Yonehara et al. 2016). In *FRMD7* mutants, horizontal DSGCs receive symmetric SAC inhibition (**Figure 5d**), which degrades their direction selectivity (Yonehara et al. 2016).

Given the select function of FRMD7 in horizontally tuned asymmetric SAC-DSGC wiring, one might anticipate a restricted localization of FRMD7 to SAC branches oriented along the horizontal axis. Surprisingly, FRMD7 is localized symmetrically throughout the dendrites of wild-type SACs (Yonehara et al. 2016). Therefore, FRMD7 may be part of a molecular machine that sorts molecular cues critical for wiring antiparallel SACs and DSGCs with motion preferences along the same axis, although FRMD7 seems not to provide the specificity required to wire independent SAC dendrite branches with antiparallel DSGCs that prefer one direction and not the other. Nevertheless, this study is the first to suggest a molecular mechanism that in some fashion sorts, or interprets, recognition cues among different SAC branches to align branches of one motion preference with their appropriate antiparallel DSGC targets. Identification of additional molecular components that delineate how SACs wire with antiparallel DSGCs preferring distinct directions of motion will provide significant insight into the central foundations of direction selectivity.

EYE TO BRAIN WIRING

A comprehensive review of the mechanisms that elaborate visual circuitry must consider how RGC axons target their appropriate brain regions. RGC axons navigate out of the retina, extend within the optic tract (OT), project contralaterally or ipsilaterally at the optic chiasm, and then target visual system nuclei appropriate for their tuning properties. Recent reviews on this issue (including Dhande et al. 2015) summarize our general understanding of retinorecipient targeting and its intersection with a broad variety of visual system functions. Here, we consider general principles underlying the development of retinorecipient connectivity in mice and describe recent studies revealing molecular cues that promote connectivity between RGCs—exemplified by DSGCs— and their central targets.

RGCs project to over 40 brain regions (Morin & Studholme 2014). Non–image forming RGCs project to specific nuclei devoted to their functions. For example, ipRGCs, which mediate photoentrainment to light-dark cycles and other physiological responses, project bilaterally to the suprachiasmatic nucleus (SCN), the proposed site of the body's master clock. Most RGCs that participate in image formation arborize within the lateral geniculate nucleus (LGN) of the thalamus, the superior colliculus (SC), or both; furthermore, over 90% of RGCs are thought to have axonal projections terminating in the SC (Dhande et al. 2015). However, there are exceptions to this trend, including the projections of oDSGCs and ooDSGCs that are tuned to slower motion and predominately target nuclei of the accessory optic system (AOS) (Dhande et al. 2013). The central nuclei constituting the AOS are the nucleus of the optic tract (NOT); dorsal terminal nucleus (DTN); medial terminal nucleus (MTN); and, in some vertebrates, the lateral terminal nucleus (LTN). These AOS nuclei are critical for generating the OKR, which, as mentioned above, stabilizes the visual scene on the retina during slow motion by producing compensatory eye movements (Dhande et al. 2013, 2015; Simpson 1984). Ultimately, most scotopic and pho topic visual information, following conveyance to and integration within retinorecipient targets, is processed in the same cortical regions (Barton & Brewer 2015).

Navigating to Central Targets

The first task nascent axons in the developing retina face is to extend toward the optic nerve head (Erskine & Herrera 2007). At the optic disk, the guidance cue netrin-1 attracts RGC axons expressing the netrin-1 receptor DCC, guiding RGC axons out of the optic nerve head to form the optic nerve (**Figure 6**) (de la Torre et al. 1997, Deiner et al. 1997). Beyond the optic nerve head, the next major RGC axon choice point is the optic chiasm, where axons must choose to project through either the ipsilateral or contralateral OT. The majority of RGCs project contralaterally, with only about 3% of RGCs committing to an ipsilateral trajectory (Wilks et al. 2013). Interestingly, RGCs may also choose a retino-retinal (RR) trajectory at the chiasm (Müller & Holländer 1988), and some R-R–projecting RGCs also send collateral axons to the contralateral SC (Avellaneda-Chevrier et al. 2015).

Different classes of guidance cues, morphogens, and extracellular matrix molecules collaborate to steer RGC axons at the optic chiasm (Petros et al. 2008). The secreted guidance cues Slit1 and Slit2 redundantly prevent premature crossing of RGC axons rostral to the normal position of the chiasm, forming a corridor along which axons can navigate to the location of the true chiasm(Plump et al. 2002, Williams et al. 2004). In

Slit1/Slit2 double mutant mice, RGC axons defasciculate and stray into regions anterior to the chiasm, forming an ectopic chiasm (Plump et al. 2002). Repulsive signaling from glial ephrinB2 repels EphB1-expressing RGC axons away from the midline at the chiasm to form the ipsilateral retinal projection (Williams et al. 2003). For a subset of contralateral-projecting RGCs, vascular endothelial growth factor 164 (VEGF164) expressed at the midline acts as a chemoattractant to steer neuropilin-1–expressing axons through the chiasm to the contralateral OT (Erskine et al. 2011). Additionally, Sema6D, expressed in combination with the IgSf cell adhesion molecule NrCAM in radial glia at the chiasm, promotes decussation and midline crossing of RGC axons coexpressing both the PlexA1 receptor and NrCAM (Kuwajima et al. 2012). Because both overlapping and distinct RGC subtypes likely express these guidance receptors, it will be interesting to determine the significance underlying the use of multiple, disparate signaling pathways for RGC axons navigating the optic chiasm. Once RGC axons pass the chiasm, both crossing and noncrossing axons project through the OT toward their final destinations (**Figure 6**).

Choosing a Central Target

After coursing through the OT, RGC axons must identify their appropriate central targets and defasciculate from the OT. The retinorecipient targets and known cues that drive target innervation are depicted in **Figure 6**. The first target RGCs encounter is the SCN, just caudal to the optic chiasm. The cues that direct ipRGC defasciculation and innervation of the SCN remain to be identified. The next major retinorecipient target is the LGN, which consists of the ventral LGN (vLGN) and dorsal LGN (dLGN), separated by the intergeniculate leaflet (IGL). In mice, Cdh6 guides non–image forming RGCs into the vLGN and IGL, which may include subtypes of ipRGCs (Osterhout et al. 2011). ipRGC axons are also guided into the vLGN and IGL by reelin expressed within those targets (Su et al. 2011). Caudal and dorsal to the LGN, Cdh6 also directs RGC axons into the olivary pretectal nucleus, a critical processing center for the pupillary light reflex (Dhande et al. 2015), and the medial division of the posterior pretectal nucleus (Osterhout et al. 2011). The mechanisms underlying image-forming RGC axon target recognition in the mouse are just beginning to be fleshed out. In the dLGN, image-forming RGC axons fail to innervate their target in the absence of a visual cortex (Shanks et al. 2016), indicating that cortical inputs somehow instruct RGC axons to innervate the dLGN (**Figure 6**). The cues affiliated with corticothalamic projections that are responsible for dLGN target recognition remain at large; however, recent studies (discussed below) revealed some of the cues that guide DSGCs to their targets within the AOS.

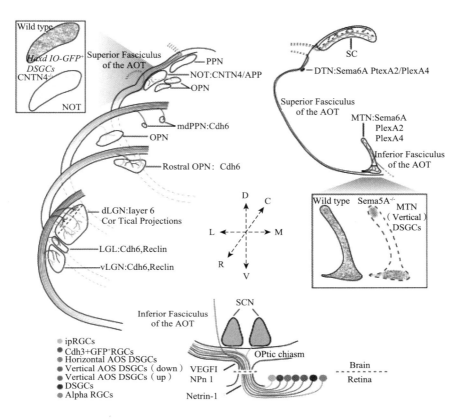

Figure 6

Mechanisms targeting defined retinal ganglion cell (RGC) subtypes to their retinorecipient nuclei. In the retina (*bottom center*), RGC axons are guided through the optic nerve head by netrin-1 signaling through DCC. Contralaterally projecting axons are steered through the optic chiasm by the combined effort of multiple signaling pathways, including vascular endothelial growth factor (VEGF) signaling through neuropilin-1 (Npn1). See the main text for a more comprehensive description of the known signaling pathways that contribute to the RGC axon crossing the optic chiasm. The first retinorecipient target encountered by RGC axons is the suprachiasmatic nucleus (SCN). The inferior fasciculus of the accessory optic tract (AOT) splits off from the main optic tract (OT) just past the SCN, carrying upward vertical motion–preferring direction-selective ganglion cell (DSGC) axons ventrally through the midbrain toward the medial terminal nucleus (MTN). The ensuing targets are displayed as a series of cross-sections through the midbrain, moving from rostral (R) to caudal (C); the molecules and mechanisms driving innervation of these targets are depicted, where known. DSGCs innervate the dorsal lateral geniculate nucleus (dLGN) shell, and alpha RGCs and intrinsically photosensitive RGCs (ipRGCs) innervate the dLGN core. Cortical layer 6 axonal projections are required for proper innervation of the dLGN. Cadherin3-GFP+ (Cdh3-GFP+) RGCs and ipRGCs are cued to enter the inner geniculate leaflet (IGL) and ventral LGN (vLGN) by cadherin 6 (Cdh6) and reelin, respectively. Cdh6 also leads Cdh3-GFP+ RGC axons into the olivary pretectal nucleus (OPN) and medial division of the posterior pretectal nucleus (mdPPN). The OPN is also innervated by ipRGCs. The first nucleus of the accessory optic system (AOS) encountered by RGCs, the nucleus of the optic tract (NOT), is innervated by horizontal motion–preferring DSGCs. NOT targeting depends on contactin 4 and amyloid precursor protein (CNTN4 and APP). In *CNTN4* mutants, the NOT is severely hypoinnervated (*inset, upper left*). Beyond the NOT, the superior fasciculus of the AOT breaks away from the OT and courses ventromedially and caudally. Horizontal DSGC axons within the superior fasciculus terminate in the dorsal terminal nucleus (DTN). Downward-preferring vertical DSGC axons innervate the ventral MTN, whereas upward-preferring vertical DSGC axons within the inferior fasciculus innervate the dorsal MTN. In the absence of semaphorin 6A (Sema6A) or both plexin A2 and plexin A4 (PlexA2 and PlexA4), MTN innervation is nearly abolished, with only minor contributions remaining (*inset, middle right*). The axons of DSGCs, alpha RGCs, and ipRGCs in the main OT continue to the superior colliculus (SC). DSGCs innervate the upper half of the SC, whereas alpha RGCs and ipRGCs innervate the lower half. Other abbreviations: D, dorsal; V, ventral; R, rostral; C, caudal; M, medial; L, lateral.

Projecting Within Central Targets

Following initial target innervation, RGC axons must elaborate synaptic terminals in their appropriate termination zone within that target. Here, we focus on mechanisms that direct targeting of different RGC subtypes, rather than eye-specific segregation of inputs (reviewed elsewhere, including Feller 2009, Guido 2008). RGC axon targeting includes termination along the rostro caudal and mediolateral axes, which can be achieved by topographic mapping (reviewed elsewhere, including Cang & Feldheim 2013) in some nuclei, such as the SC. Topographic maps preserve a complete representation of visual scenery by matching the location of neighboring neurons in the retina to the location of axon termination zones in retinorecipient targets. In addition to mapping in the rostrocaudal and mediolateral axes, RGC axons must terminate within laminar stratifications in the dorsoventral axis. In the SC, the topographic map along the rostrocaudal and mediolateralaxes is repeated throughout the depth of the SC such that the map is composed of a series of superimposed maps segregated into distinct laminar stratifications in the dorsoventral axis (Baier 2013; Huberman et al. 2008, 2010; Robles et al. 2014). The layered complexity of the SC points toward molecular pathways that sort incoming RGC axons into their appropriate stratifications; however, the specific molecules responsible for this sorting step are ill-defined. Whereas target derived ephrins play a critical role in rostrocaudal and mediolateral SC retinotopic targeting (Cang & Feldheim 2013), ephrins As are dispensable for lamination (Sweeney et al. 2015), and so the molecular players that drive mammalian retinorecipient lamination await discovery.

Some retinorecipient targets are laminated and others are not. In the SC, RGCs that encode local object motion and DSGCs project to the upper half of the retinorecipient SC, whereas ipRGCs and alpha RGCs project to the lower half (Dhande et al. 2015, Hattar et al. 2006, Huberman et al. 2008, Kay et al. 2011). The dLGN is another well-studied target with discrete termination zones. In the dLGN, alpha RGCs and ipRGCs project to the core, whereas ooDSGCs project to the dLGN's outer region, or shell (Brown et al. 2010; Ecker et al. 2010; Huberman et al. 2008, 2009; Kay et al. 2011; Kim et al. 2008, 2010; Rivlin-Etzion et al. 2011). Lamination in other image-forming targets has not been reported. The MTN of the AOS, however, does appear to have segregated RGC inputs: The dorsal MTN (dMTN) receives input from oDSGCs that prefer upward motion, and the ventral MTN (vMTN) receives inputs from oDSGCs that prefer downward motion (Yonehara et al. 2009). Additionally, although previous work indicated that lamination was absent from the SCN, a recent study showed that individually traced ipRGC axons project to specific subdomains of the SCN (Fernandez et al. 2016), hinting at functional segregation within this target.

Investigating the mechanisms underlying lamination of retinorecipient targets is now possible, given recent characterizations of transgenic animals expressing fluorescent reporters, or Cre recombinase, in distinct subtypes of RGCs. Some players that wire together DSGCs and their central targets have been described recently, and we discuss them below.

MECHANISMS SHAPING DIRECTIONSELECTIVE CONNECTIVITY IN THE BRAIN

How the axons of a given RGC subtype navigate through each of the various stages of retinorecipient targeting described above is unknown. There are multiple choice points for navigating RGC axons. It is likely that mechanisms defining trajectories taken by RGC axons at critical choice points, such as the optic chiasm, apply more generally to many RGC subtypes, whereas at other choice points, guidance mechanisms are restricted to one or few RGC subtypes. Thorough characterization of guidance strategies employed by RGC axons that convey visual feature–specific information along each phase of their journey through the developing brain provides a framework to understand how other subtypes of RGCs navigate to, and wire with, their postsynaptic targets. Here we discuss the targeting strategies used in the brain by the DS system because recent developments have advanced our knowledge of the mechanisms underlying RGC–central target connectivity.

Wiring Direction Selective Ganglion Cell–Lateral Geniculate Nucleus–Primary Visual Cortex Circuitry

A majority of ooDSGCs project to the SC and the dLGN (Huberman et al. 2009, Kay et al. 2011, Kim et al. 2008, Rivlin-Etzion et al. 2011). The molecular pathways linking together ooDSGCs and the SC or LGN are unknown; however, recent work on ooDSGC connectivity expands our understanding of how the retina and higher processing centers in the cortex wire together. ooDSGCs project to the shell of the dLGN. By injecting adeno-associated virus2–glycoprotein into the dLGN shell, followed by ΔG rabies virus into the superficial layers (I/II) of the primary visual cortex (V1), double infection of sparse dLGN neurons with reconstituted rabies virus was achieved. Subsequent transsynaptic labeling of the RGCs that project to the dually infected dLGN neurons revealed that ooDSGCs wire with thalamocortical neurons that, in turn, project directly to superficial V1 (Cruz-Martin et al. 2014). Thus, the direction selectivity of superficial V1 cortical neurons is likely derived, in part, from the retina by a disynaptic RGC-thalamocortical-V1 circuit.

There are, however, several levels of additional complexity to consider in cortical neuron DS responses. One recent study used a similar monosynaptic rabies virus transsynaptic labeling approach but instead infected single pyramidal neurons in layer II/III of V1 to identify each of the presynaptic inputs to the infected cells and their directional preference (Wertz et al. 2015). Hundreds of neurons across the cortical layers in V1 connect to the singly infected layer II/III neurons. For some of these presynaptic networks, the directional preference of the network is the same as the infected starter cell (Wertz et al. 2015). Where might the direction selectivity of these presynaptic cortical neurons originate? DS responses of deeper cortical neurons may arise from converging LGN inputs with center surround, rather than DS tuning alone (Cruz-Martin et al. 2014, Livingstone 1998).

Is direction selectivity in the dLGN derived entirely from the retina, or are thalamic neurons also DS? In addition to the disynaptic RGCthalamocortical-V1 circuit, dLGN neurons also integrate DS inputs from SC neurons (Bickford et al. 2015). These two sources of direction selectivity do not exclude the possibility that dLGN neurons may themselves be DS. To conserve DS information, one might imagine specificity in wiring between ooDSGCs that respond to a particular motion and thalamic relay neurons that are also tuned to respond to that same direction of motion. In support of this idea, neurons in the superficial SC layer, the stratum griseum superficiale, are highly DS (Inayat et al. 2015), showing that thalamic neurons are indeed capable of DS responses. Whether dLGN neurons are innervated by DSGCs that prefer a single direction of motion, or integrate inputs from DSGCs of varying motion preferences, is unclear. Some thalamocortical neurons receive input from at least 12, and possibly dozens, of distinct RGC types (Hammer et al. 2015, Morgan et al. 2016), masking the origin (or origins) of dLGN direction selectivity.

How Do Direction-Selective Ganglion Cells Wire with Nuclei of the Accessory Optic System?

oDSGCs and ooDSGCs tuned to slow motion project to nuclei of the AOS. Within the AOS, the MTN, which mediates the vertical OKR, and the DTN, which along with the NOT is responsible for the horizontal OKR, are innervated by DSGCs that are guided in part by Sema6A. In *Sema6A* mutants, MTN innervation is almost completely lost, with only a small portion of both the dMTN and vMTN remaining (Sun et al. 2015). Additionally, a substantial portion of oDSGC DTN innervation is also missing in *Sema6A* mutants. PlexA2 and PlexA4 proteins serve as redundant ligands that attract Sema6A-expressing RGC axons via reverse Sema6A signaling, whereby the Sema6A protein serves as a receptor (Sun et al. 2015).

The MTN is innervated by two routes: The inferior fasciculus of the OT branches off of the main OT near the SCN, coursing through the ventral midbrain to innervate the dMTN directly, and the superior fasciculus branches off of the OT near the NOT and DTN, circling back along the lateral edge of the midbrain and then traversing a ventral route to innervate the vMTN (**Figure 6**) (Simpson 1984). Because portions of both the dMTN and the vMTN remain in *Sema6A* mutants, and because Sema6A-expressing oDSGC axons initially target theMTN during embryonic development (Sun et al. 2015), Sema6A appears dispensable for the decision to defasciculate from the OT at either the inferior or superior fasciculi choice points. Similarly, Sema6A is likely not required for initial targeting of the DTN, as some axons still innervate the DTN in the absence of Sema6A. Sema6A is apparently required for a later stage of target recognition in both settings, for instance, to match DSGCs with their specific postsynaptic partners.

In addition to the PlexA2/PlexA4–Sema6A signaling axis, there exists another pathway that targets DSGC axons to an AOS nucleus, the NOT. *CNTN4* is both necessary (**Figure 6,** *left inset*) and sufficient to promote RGC axon targeting and branching within the NOT. CNTN4 is expressed on RGC axons that target the AOS, and a CNTN4 binding partner, amyloid precursor protein (APP), which is expressed on most RGCs, is also required for this retinorecipient targeting (Osterhout et al. 2015). In this study, the authors found that CNTN4 and APP specifically promote DSGC axon branching complexity within the NOT. As more than 90% of RGCs project to the SC, most axonal arbors within other targets represent collateral branches of SC-projecting RGCs. Given the close association of the NOT with the OT and its close proximity to the SC, more than 90% of RGCs are presented with the opportunity to interact with the NOT environment. However, NOT innervation appears to be highly specific (Osterhout et al. 2014). The NOT is directly innervated by DSGCs that mediate the horizontal OKR, with the exception of Drd4-GFP⁺ ooDSGCs that target the NOT transiently during early postnatal development, prior to innervation of the NOT by its appropriate Hoxd10-GFP⁺ DSGC afferents (Osterhout et al. 2014). The timing of Hoxd10-GFP⁺ DSGC elaboration within the NOT corresponds to the timing of Drd4-GFP⁺ axon exit from the NOT. This hints that Drd4-GFP⁺ ooDSGCs may serve as placeholders that exert tight control over NOT innervation, preventing improper innervation of this nucleus by the multitude of RGC subtypes that project to the SC. In line with this idea, researchers observed a substantial increase in the number of axons elaborating collateral branches in the NOT following electroporation of a CNTN4 overexpression construct into early postnatal RGCs (Osterhout et al. 2015), overriding the mechanisms that prevent branching within this structure. So, unlike Sema6A-mediated oDSGC target recognition, which appears to directly wire together specific pre and postsynaptic partners, CNTN and APP acting on RGC axons seem to be a switch that allows axons to enter and elaborate collateral branches within the NOT. This additional layer of complexity underscores the numerous signaling events that coordinate to ensure presynaptic DSGC axons find their appropriate postsynaptic targets.

As mentioned above, Sema6A and PlexA2 are required for SAC lamination and morphological development in the IPL, and PlexA4 is also expressed in the developing IPL, where Sema6A and PlexA4 are essential for dopaminergic AC and M1 ipRGC lamination (Matsuoka et al. 2011b). CNTN4 is expressed in RGCs (Osterhout et al. 2015), as is CNTN2/TAG-1 (Chatzopoulou et al. 2008), and in the chick, CNTN1–5 are each expressed in the developing IPL (Yamagata & Sanes 2012). These expression patterns beg the question, Do these signaling molecules also promote matching of presynaptic ACs and BCs with their postsynaptic RGC targets, in addition to directing target recognition in the brain? It will be interesting to examine whether feature-specific circuitry could be wired at each synapse—from PR to central targets—by the same signaling molecules to maximize efficiency of visual system circuit assembly.

CONCLUDING REMARKS

Wiring visual circuits in both the retina and the brain to achieve appropriate visual coding is a multifaceted and complex process. Although we have taken great strides toward understanding the cellular and molecular mechanisms connecting neurons within retinal circuits, we have only just begun to define the repertoire of molecules that underlie the wiring diagrams of visual circuits in the brain. Beyond defining the basic mechanisms that are involved in specifying synaptic partners, it is now clear that resolving the complexity of subcellular connectivity within visual circuits is the next major challenge. The DS circuit, for example, would be an excellent model for determining how a single presynaptic cell type wires differentially at the subcellular level with a postsynaptic cell to generate distinct and stereotypic responses to visual stimuli. Also, how different presynaptic cell types that provide converging input onto an individual neuron target their synapses to different dendritic domains remains largely unknown. What we have learned so far is that developmental mechanisms may be engaged similarly across circuits in the visual system, but also some mechanisms may be uniquely employed and together create the diverse cellular and subcellular wiring diagrams necessary for dynamically reconstructing the visual scene.

DISCLOSURE STATEMENT

The authors are not aware of any affiliations, memberships, funding, or financial holdings that might be perceived as affecting the objectivity of this review.

ACKNOWLEDGMENTS

We thank Timour Al-Khindi, Onkar Dhande, Andrew Huberman, and Brendan Lilley for helpful comments on the manuscript. Work in the authors' laboratories is supported by the US National Institutes of Health (NIH) [National Institute of Mental Health (P50-MH100023, Project #3, to A.L.K), National Eye Institute (EY10699, EY14358, and EY17101 to R.O.W; EY025114 to R.E.J.)] and the Howard Hughes Medical Institute (HHMI) (A.L.K. and R.E.J.). A.L.K. is an investigator of the HHMI. R.O.W. is an Allen Distinguished Investigator.

LITERATURE CITED

Anderson JR, Jones BW, Watt CB, Shaw MV, Yang JH, et al. 2011. Exploring the retinal connectome. *Mol. Vis.* 17:355–79

Avellaneda-Chevrier VK, Wang X, Hooper ML, Chauhan BC. 2015. The retino-retinal projection: tracing retinal ganglion cells projecting to the contralateral retina. *Neurosci. Lett.* 591:105–9

Baier H. 2013. Synaptic laminae in the visual system: molecular mechanisms forming layers of perception. *Annu. Rev. Cell Dev. Biol.* 29:385–416

Barlow HB, Levick WR. 1965. The mechanism of directionally selective units in rabbit's retina. *J. Physiol.* 178:477–504

Barton B, Brewer AA. 2015. fMRI of the rod scotoma elucidates cortical rod pathways and implications for lesion measurements. *PNAS* 112:5201–6

Bickford ME, Zhou N, Krahe TE, Govindaiah G, Guido W. 2015. Retinal and tectal "driver-like" inputs converge in the shell of the mouse dorsal lateral geniculate nucleus. *J. Neurosci.* 35:10523–34

Bleckert A, Schwartz GW, Turner MH, Rieke F, Wong RO. 2014. Visual space is represented by nonmatching topographies of distinct mouse retinal ganglion cell types. *Curr. Biol.* 24:310–15

Bloomfield SA, Völgyi B. 2009. The diverse functional roles and regulation of neuronal gap junctions in the retina. *Nat. Rev. Neurosci.* 10:495–506

Bolte P, Herrling R, Dorgau B, Schultz K, Feigenspan A, et al. 2016. Expression and localization of connexins in the outer retina of the mouse. *J. Mol. Neurosci.* 58:178–92

Briggman KL, Helmstaedter M, Denk W. 2011. Wiring specificity in the direction-selectivity circuit of the retina. *Nature* 471:183–88

Brown TM, Gias C, Hatori M, Keding SR, Semo M, et al. 2010. Melanopsin contributions to irradiance coding in the thalamocortical visual system. *PLOS Biol.* 8:e1000558

Cang J, Feldheim DA. 2013. Developmental mechanisms of topographic map formation and alignment. *Annu. Rev. Neurosci.* 36:51–77

Cao Y, Sarria I, Fehlhaber KE, Kamasawa N, Orlandi C, et al. 2015. Mechanism for selective synaptic wiring of rod photoreceptors into the retinal circuitry and its role in vision. *Neuron* 87:1248–60

Chalupa LM. 1998. Introduction: development and organization of the retina: cellular, molecular and functional perspectives. *Semin. Cell Dev. Biol.* 9:239–40

Chatzopoulou E, Miguez A, Savvaki M, Levasseur G, Muzerelle A, et al. 2008. Structural requirement of TAG1 for retinal ganglion cell axons and myelin in the mouse optic nerve. *J. Neurosci.* 28:7624–36

Chávez AE, Singer JH, Diamond JS. 2006. Fast neurotransmitter release triggered by Ca influx through AMPAtype glutamate receptors. *Nature* 443:705–8

Cowan CS, Abd-El-Barr M, van der Heijden M, Lo EM, Paul D, et al. 2016. Connexin 36 and rod bipolar cell independent rod pathways drive retinal ganglion cells and optokinetic reflexes. *Vis. Res.* 119:99–109

Cruz-Martin A, El-Danaf RN, Osakada F, Sriram B, Dhande OS, et al. 2014. A dedicated circuit links direction-selective retinal ganglion cells to the primary visual cortex. *Nature* 507:358–61

de la Torre JR, Hopker VH, Ming GL, Poo MM, Tessier-Lavigne M, et al. 1997. Turning of retinal growth cones in a netrin-1 gradient mediated by the netrin receptor DCC. *Neuron* 19:1211–24

Deiner MS, Kennedy TE, Fazeli A, Serafini T, Tessier-Lavigne M, Sretavan DW. 1997. Netrin1 and DCC mediate axon guidance locally at the optic disc: loss of function leads to optic nerve hypoplasia. *Neuron* 19:575–89

Demb JB, Singer JH. 2015. Functional circuitry of the retina. *Annu. Rev. Vis. Sci.* 1:263–89

Dhande OS, Estevez ME, Quattrochi LE, El-DanafRN, Nguyen PL, et al. 2013. Genetic dissection of retinal inputs to brainstem nuclei controlling image stabilization. *J. Neurosci.* 33:17797–813

Dhande OS, Stafford BK, Lim JHA, Huberman AD. 2015. Contributions of retinal ganglion cells to subcortical visual processing and behaviors. *Annu. Rev. Vis. Sci.* 1:291–328

Ding H, Smith RG, Poleg-Polsky A, Diamond JS, Briggman KL. 2016. Species-specific wiring for direction selectivity in the mammalian retina. *Nature* 535:105–10

Duan X, Krishnaswamy A, De la Huerta I, Sanes JR. 2014. Type II cadherins guide assembly of a directionselective retinal circuit. *Cell* 158:793–807

Dunn FA, Della Santina L, Parker ED, Wong RO. 2013. Sensory experience shapes the development of the visual system's first synapse. *Neuron* 80:1159–66

Dunn FA, Wong RO. 2012. Diverse strategies engaged in establishing stereotypic wiring patterns among neurons sharing a common input at the visual system's first synapse. *J. Neurosci.* 32:10306–17

Ecker JL, Dumitrescu ON, Wong KY, Alam NM, Chen SK, et al. 2010. Melanopsin-expressing retinal ganglion-cell photoreceptors: cellular diversity and role in pattern vision. *Neuron* 67:49–60

Eggers ED, Lukasiewicz PD. 2011. Multiple pathways of inhibition shape bipolar cell responses in the retina. *Vis. Neurosci.* 28:95–108

Eggers ED, McCall MA, Lukasiewicz PD. 2007. Presynaptic inhibition differentially shapes transmission in distinct circuits in the mouse retina. *J. Physiol.* 582:569–82

Erskine L, Herrera E. 2007. The retinal ganglion cell axon's journey: insights into molecular mechanisms of axon guidance. *Dev. Biol.* 308:1–14

Erskine L, Reijntjes S, Pratt T, Denti L, Schwarz Q, et al. 2011. VEGF signaling through neuropilin 1 guides commissural axon crossing at the optic chiasm. *Neuron* 70:951–65

Euler T, Detwiler PB, Denk W. 2002. Directionally selective calcium signals in dendrites of starburst amacrine cells. *Nature* 418:845–52

Famiglietti EV. 1991. Synaptic organization of starburst amacrine cells in rabbit retina: analysis of serial thin sections by electron microscopy and graphic reconstruction. *J. Comp. Neurol.* 309:40–70

Farrow K, Teixeira M, Szikra T, Viney TJ, Balint K, et al. 2013. Ambient illumination toggles a neuronal circuit switch in the retina and visual perception at cone threshold. *Neuron* 78:325–38

Feller MB. 2009. Retinal waves are likely to instruct the formation of eye-specific retinogeniculate projections. *Neural Dev.* 4:24

Fernandez DC, Chang YT, Hattar S, Chen SK. 2016. Architecture of retinal projections to the central circadian pacemaker. *PNAS* 113:6047–52

Fletcher EL, Koulen P, Wässle H. 1998. GABAA and GABAC receptors on mammalian rod bipolar cells. *J. Comp. Neurol.* 396:351–65

Fried SI, Munch TA, Werblin FS. 2002. Mechanisms and circuitry underlying directional selectivity in the retina. *Nature* 420:411–14

Fried SI, Munch TA, Werblin FS. 2005. Directional selectivity is formed at multiple levels by laterally offset inhibition in the rabbit retina. *Neuron* 46:117–27

Fuerst PG, Bruce F, Tian M, Wei W, Elstrott J, et al. 2009. DSCAM and DSCAML1 function in self-avoidance in multiple cell types in the developing mouse retina. *Neuron* 64:484–97

Greferath U, Grünert U, Fritschy JM, Stephenson A, Möhler H, Wässle H. 1995. GABAA receptor subunits have differential distributions in the rat retina: in situ hybridization and immunohistochemistry. *J. Comp. Neurol.* 353:553–71

Grimes WN, Zhang J, Tian H, Graydon CW, Hoon M, et al. 2015. Complex inhibitory microcircuitry regulates retinal signaling near visual threshold. *J. Neurophysiol.* 114:341–53

Guido W. 2008. Refinement of the retinogeniculate pathway. *J. Physiol.* 586:4357–62

Günhan E, Choudary PV, Landerholm TE, Chalupa LM. 2002. Depletion of cholinergic amacrine cells by a novel immunotoxin does not perturb the formation of segregated On and Off cone bipolar cell projections. *J. Neurosci.* 22:2265–73

Günhan-Agar E, Kahn D, Chalupa LM. 2000. Segregation of On and Off bipolar cell axonal arbors in the absence of retinal ganglion cells. *J. Neurosci.* 20:306–14

Hammer S, Monavarfeshani A, Lemon T, Su J, Fox MA. 2015. Multiple retinal axons converge onto relay cells in the adult mouse thalamus. *Cell Rep.* 12:1575–83

Hartveit E, Veruki ML. 2012. Electrical synapses between AII amacrine cells in the retina: function and modulation. *Brain Res.* 1487:160–72

Hattar S, Kumar M, Park A, Tong P, Tung J, et al. 2006. Central projections of melanopsin-expressing retinal ganglion cells in the mouse. *J. Comp. Neurol.* 497:326–49

Hattori D, Chen Y, Matthews BJ, Salwinski L, Sabatti C, et al. 2009. Robust discrimination between self and non-self neurites requires thousands of Dscam1 isoforms. *Nature* 461:644–48

Hoon M, Okawa H, Della Santina L, Wong RO. 2014. Functional architecture of the retina: development and disease. *Prog. Retin. Eye Res.* 42:44–84

Hoon M, Sinha R, Okawa H, Suzuki SC, Hirano AA, et al. 2015. Neurotransmission plays contrasting roles in the maturation of inhibitory synapses on axons and dendrites of retinal bipolar cells. *PNAS* 112:12840–45

Huberman AD, Clandinin TR, Baier H. 2010. Molecular and cellular mechanisms of lamina-specific axon targeting. *Cold Spring Harb. Perspect. Biol.* 2:a001743

Huberman AD, Manu M, Koch SM, Susman MW, Lutz AB, et al. 2008. Architecture and activity-mediated refinement of axonal projections from a mosaic of genetically identified retinal ganglion cells. *Neuron* 59:425–38

Huberman AD, Wei W, Elstrott J, Stafford BK, Feller MB, Barres BA. 2009. Genetic identification of an On-Off direction-selective retinal ganglion cell subtype reveals a layer-specific subcortical map of posterior motion. *Neuron* 62:327–34

Inayat S, Barchini J, Chen H, Feng L, Liu X, Cang J. 2015. Neurons in the most superficial lamina of the mouse superior colliculus are highly selective for stimulus direction. *J. Neurosci.* 35:7992–8003

Ivanova E, Müller U, Wässle H. 2006. Characterization of the glycinergic input to bipolar cells of the mouse retina. *Eur. J. Neurosci.* 23:350–64

Joesch M, Meister M. 2016. A neuronal circuit for colour vision based on rod-cone opponency. *Nature* 532:236–39

Kanagawa M, Omori Y, Sato S, Kobayashi K, Miyagoe-Suzuki Y, et al. 2010. Posttranslational maturation of dystroglycan is necessary for pikachurin binding and ribbon synaptic localization. *J. Biol. Chem.* 285:31208–16

Kay JN, De la Huerta I, Kim IJ, Zhang Y, Yamagata M, et al. 2011. Retinal ganglion cells with distinct directional preferences differ in molecular identity, structure, and central projections. *J. Neurosci.* 31:7753–62

Keeley PW, Luna G, Fariss RN, Skyles KA, Madsen NR, et al. 2013. Development and plasticity of outer retinal circuitry following genetic removal of horizontal cells. *J. Neurosci.* 33:17847–62

Keeley PW, Reese BE. 2010. Role of afferents in the differentiation of bipolar cells in the mouse retina. *J. Neurosci.* 30:1677–85

Kerschensteiner D, Morgan JL, Parker ED, Lewis RM, Wong RO. 2009. Neurotransmission selectively regulates synapse formation in parallel circuits in vivo. *Nature* 460:1016–20

Kim IJ, Zhang Y, Meister M, Sanes JR. 2010. Laminar restriction of retinal ganglion cell dendrites and axons: subtype-specific developmental patterns revealed with transgenic markers. *J. Neurosci.* 30:1452–62

Kim IJ, Zhang Y, Yamagata M, Meister M, Sanes JR. 2008. Molecular identification of a retinal cell type that responds to upward motion. *Nature* 452:478–82

Kim JS, Greene MJ, Zlateski A, Lee K, Richardson M, et al. 2014. Spacetime wiring specificity supports direction selectivity in the retina. *Nature* 509:331–36

Kolb H. 1974. The connections between horizontal cells and photoreceptors in the retina of the cat: electron microscopy of Golgi preparations. *J. Comp. Neurol.* 155:1–14

Kolb H. 1979. The inner plexiform layer in the retina of the cat: electron microscopic observations. *J. Neurocytol.* 8:295–329

Kolb H, Famiglietti EV. 1974. Rod and cone pathways in the inner plexiform layer of cat retina. *Science* 186:47–49

Kostadinov D, Sanes JR. 2015. Protocadherin-dependent dendritic self-avoidance regulates neural connectivity and circuit function. *eLife* 4:e08964

Krishnaswamy A, Yamagata M, Duan X, Hong YK, Sanes JR. 2015. Sidekick 2 directs formation of a retinal circuit that detects differential motion. *Nature* 524:466–70

Kuwajima T, Yoshida Y, Takegahara N, Petros TJ, Kumanogoh A, et al. 2012. Optic chiasm presentation of Semaphorin6D in the context of Plexin-A1 and Nr-CAM promotes retinal axon midline crossing. *Neuron* 74:676–90

Lee SC, Meyer A, Schubert T, Huser L, Dedek K, Haverkamp S. 2015. Morphology and connectivity of the small bistratified A8 amacrine cell in the mouse retina. *J. Comp. Neurol.* 523:1529–47

Lefebvre JL, Kostadinov D, Chen WV, Maniatis T, Sanes JR. 2012. Protocadherins mediate dendritic self-avoidance in the mammalian nervous system. *Nature* 488:517–21

Livingstone MS. 1998. Mechanisms of direction selectivity in macaque V1. *Neuron* 20:509–26

Majumdar S, Heinze L, Haverkamp S, Ivanova E, Wässle H. 2007. Glycine receptors of Atype ganglion cells of the mouse retina. *Vis. Neurosci.* 24:471–87

Marc RE, Anderson JR, Jones BW, Sigulinsky CL, Lauritzen JS. 2014. The AII amacrine cell connectome: a dense network hub. *Front. Neural Circuits* 8:104

Matsuoka RL, Chivatakarn O, Badea TC, Samuels IS, Cahill H, et al. 2011a. Class 5 transmembrane semaphorins control selective mammalian retinal lamination and function. *Neuron* 71:460–73

Matsuoka RL, Jiang Z, Samuels IS, NguyenBaCharvet KT, Sun LO, et al. 2012. Guidancecue control of horizontal cell morphology, lamination, and synapse formation in the mammalian outer retina. *J. Neurosci.* 32:6859–68

Matsuoka RL, Nguyen-Ba-Charvet KT, Parray A, Badea TC, Chedotal A, Kolodkin AL. 2011b. Transmembrane semaphorin signalling controls laminar stratification in the mammalian retina. *Nature* 470:259–63

Moleirinho S, TilstonLunel A, Angus L, Gunn-Moore F, Reynolds PA. 2013. The expanding family of FERM proteins. *Biochem. J.* 452:183–93

Morgan JL, Berger DR, Wetzel AW, Lichtman JW. 2016. The fuzzy logic of network connectivity in mouse visual thalamus. *Cell* 165:192–206

Morgan JL, Dhingra A, Vardi N, Wong RO. 2006. Axons and dendrites originate from neuroepithelial-like processes of retinal bipolar cells. *Nat. Neurosci.* 9:85–92

Morgan JL, Soto F, Wong RO, Kerschensteiner D. 2011. Development of cell type-specific connectivity patterns of converging excitatory axons in the retina. *Neuron* 71:1014–21

Morin LP, Studholme KM. 2014. Retinofugal projections in the mouse. *J. Comp. Neurol.* 522:3733–53

Morrie RD, Feller MB. 2015. An asymmetric increase in inhibitory synapse number underlies the development of a direction selective circuit in the retina. *J. Neurosci.* 35:9281–86

Müller M, Holländer H. 1988. A small population of retinal ganglion cells projecting to the retina of the other eye: an experimental study in the rat and the rabbit. *Exp. Brain Res.* 71:611–17

Nakajima Y, Iwakabe H, Akazawa C, Nawa H, Shigemoto R, et al. 1993. Molecular characterization of a novel retinal metabotropic glutamate receptor mGluR6 with a high agonist selectivity for L-2-amino-4 -phosphonobutyrate. *J. Biol. Chem.* 268:11868–73

Okawa H, Della Santina L, Schwartz GW, Rieke F, Wong RO. 2014. Interplay of cell-autonomous and nonautonomous mechanisms tailors synaptic connectivity of converging axons in vivo. *Neuron* 82:125–37

Omori Y, Araki F, Chaya T, Kajimura N, Irie S, et al. 2012. Presynaptic dystroglycan-pikachurin complex regulates the proper synaptic connection between retinal photoreceptor and bipolar cells. *J. Neurosci.* 32:6126–37

Osterhout JA, El-Danaf RN, Nguyen PL, Huberman AD. 2014. Birthdate and outgrowth timing predict cellular mechanisms of axon target matching in the developing visual pathway. *Cell Rep.* 8:1006–17 Osterhout JA, Josten N, Yamada J, Pan F, Wu SW, et al. 2011. Cadherin-6 mediates axontarget matching in a non-image-forming visual circuit. *Neuron* 71:632–39

Osterhout JA, Stafford BK, Nguyen PL, Yoshihara Y, Huberman AD. 2015. Contactin-4 mediates axontarget specificity and functional development of the accessory optic system. *Neuron* 86:985–99

Pang JJ, Gao F, Paul DL, Wu SM. 2012. Rod, M-cone and M/S-cone inputs to hyperpolarizing bipolar cells in the mouse retina. *J. Physiol.* 590:845–54

Park SJ, Kim IJ, Looger LL, Demb JB, Borghuis BG. 2014. Excitatory synaptic inputs to mouse On-Off direction selective retinal ganglion cells lack direction tuning. *J. Neurosci.* 34:3976–81

Perlman I, Kolb H, Nelson R. 2012. *S-potentials and horizontal cells.* Webvision, Moran Eye Cent., Jan. http:// webvision.med.utah.edu/book/part-v-phototransduction-in-rods-and-cones/horizontal-cells/

Petros TJ, Rebsam A, Mason CA. 2008. Retinal axon growth at the optic chiasm: to cross or not to cross. *Annu. Rev. Neurosci.* 31:295–315

Plump AS, Erskine L, Sabatier C, Brose K, Epstein CJ, et al. 2002. Slit1 and Slit2 cooperate to prevent premature midline crossing of retinal axons in the mouse visual system. *Neuron* 33:219–32

Randlett O, MacDonald RB, Yoshimatsu T, Almeida AD, Suzuki SC, et al. 2013. Cellular requirements for building a retinal neuropil. *Cell Rep.* 3:282–90

Raven MA, Oh EC, Swaroop A, Reese BE. 2007. Afferent control of horizontal cell morphology revealed by genetic respecification of rods and cones. *J. Neurosci.* 27:3540–47

Ribic A, Liu X, Crair MC, Biederer T. 2014. Structural organization and function of mouse photoreceptor ribbon synapses involve the immunoglobulin protein synaptic cell adhesion molecule 1. *J. Comp. Neurol.* 522:900–20

Rice DS, Nusinowitz S, Azimi AM, Martinez A, Soriano E, Curran T. 2001. The reelin pathway modulates the structure and function of retinal synaptic circuitry. *Neuron* 31:929–41

Rivlin-Etzion M, Zhou K, Wei W, Elstrott J, Nguyen PL, et al. 2011. Transgenic mice reveal unexpected diversity of On-Off direction-selective retinal ganglion cell subtypes and brain structures involved in motion processing. *J. Neurosci.* 31:8760–69

Robles E, Laurell E, Baier H. 2014. The retinal projectome reveals brain-area-specific visual representations generated by ganglion cell diversity. *Curr. Biol.* 24:2085–96

Sanes JR, Masland RH. 2015. The types of retinal ganglion cells: current status and implications for neuronal classification. *Annu. Rev. Neurosci.* 38:221–46

Sato S, Omori Y, Katoh K, Kondo M, Kanagawa M, et al. 2008. Pikachurin, a dystroglycan ligand, is essential for photoreceptor ribbon synapse formation. *Nat. Neurosci.* 11:923–31

Schreiner D, Weiner JA. 2010. Combinatorial homophilic interaction between γ-protocadherin multimers greatly expands the molecular diversity of cell adhesion. *PNAS* 107:14893–98

Schubert T, Hoon M, Euler T, Lukasiewicz PD, Wong RO. 2013. Developmental regulation and activity-dependent maintenance of GABAergic presynaptic inhibition onto rod bipolar cell axonal terminals. *Neuron* 78:124–37

Schwartz GW, Okawa H, Dunn FA, Morgan JL, Kerschensteiner D, et al. 2012. The spatial structure of a nonlinear receptive field. *Nat. Neurosci.* 15:1572–80

Shanks JA, Ito S, Schaevitz L, Yamada J, Chen B, et al. 2016. Corticothalamic axons are essential for retinal ganglion cell axon targeting to the mouse dorsal lateral geniculate nucleus. *J. Neurosci.* 36:5252–63

Shen N, Qu Y, Yu Y, So KF, Goffinet AM, et al. 2016. *Frizzled3* shapes the development of retinal rod bipolar cells. *Investig. Ophthalmol. Vis. Sci.* 57:2788–96

Simpson JI. 1984. The accessory optic system. *Annu. Rev. Neurosci.* 7:13–41

Soto F, Ma X, Cecil JL, Vo BQ, Culican SM, Kerschensteiner D. 2012. Spontaneous activity promotes synapse formation in a cell-type-dependent manner in the developing retina. *J. Neurosci.* 32:5426–39

Soto F, Watkins KL, Johnson RE, Schottler F, Kerschensteiner D. 2013. NGL2 regulates pathway-specific neurite growth and lamination, synapse formation, and signal transmission in the retina. *J. Neurosci.* 33:11949–59

Stincic T, Smith RG, Taylor WR. 2016. Time course of EPSCs in ON-type starburst amacrine cells is independent of dendritic location. *J. Physiol.* 594:5685–94

Strettoi E, Mears AJ, Swaroop A. 2004. Recruitment of the rod pathway by cones in the absence of rods. *J. Neurosci.* 24:7576–82

Su J, Haner CV, Imbery TE, Brooks JM, Morhardt DR, et al. 2011. Reelin is required for class-specific retinogeniculate targeting. *J. Neurosci.* 31:575–86

Sun LO, Brady CM, Cahill H, AlKhindi T, Sakuta H, et al. 2015. Functional assembly of accessory optic system circuitry critical for compensatory eye movements. *Neuron* 86:971–84

Sun LO, Jiang Z, Rivlin-Etzion M, Hand R, Brady CM, et al. 2013. On and Off retinal circuit assembly by divergent molecular mechanisms. *Science* 342:1241974

Sweeney NT, James KN, Sales EC, Feldheim DA. 2015. Ephrin-As are required for the topographic mapping but not laminar choice of physiologically distinct RGC types. *Dev. Neurobiol.* 75:584–93

Szikra T, Trenholm S, Drinnenberg A, Jüttner J, Raics Z, et al. 2014. Rods in daylight act as relay cells for cone-driven horizontal cell–mediated surround inhibition. *Nat. Neurosci.* 17:1728–35

Tarpey P, Thomas S, Sarvananthan N, Mallya U, Lisgo S, et al. 2006. Mutations in *FRMD7*, a newly identified member of the FERM family, cause X-linked idiopathic congenital nystagmus. *Nat. Genet.* 38:1242–44

Taylor WR, Smith RG. 2012. The role of starburst amacrine cells in visual signal processing. *Vis. Neurosci.*

29:73–81

Thoreson WB, Mangel SC. 2012. Lateral interactions in the outer retina. *Prog. Retin. Eye Res.* 31:407–41

tom Dieck S, Altrock WD, Kessels MM, Qualmann B, Regus H, et al. 2005. Molecular dissection of the photoreceptor ribbon synapse: physical interaction of Bassoon and RIBEYE is essential for the assembly of the ribbon complex. *J. Cell Biol.* 168:825–36

Trotter JH, Klein M, Jinwal UK, Abisambra JF, Dickey CA, et al. 2011. ApoER2 function in the establishment and maintenance of retinal synaptic connectivity. *J. Neurosci.* 31:14413–23

Tsukamoto Y, Morigiwa K, Ishii M, Takao M, Iwatsuki K, et al. 2007. A novel connection between rods and ON cone bipolar cells revealed by ectopic metabotropic glutamate receptor 7 (mGluR7) in mGluR6-deficient mouse retinas. *J. Neurosci.* 27:6261–67

Tsukamoto Y, Omi N. 2013. Functional allocation of synaptic contacts in microcircuits from rods via rod bipolar to AII amacrine cells in the mouse retina. *J. Comp. Neurol.* 521:3541–55

Tsukamoto Y, Omi N. 2014a. Some OFF bipolar cell types make contact with both rods and cones in macaque and mouse retinas. *Front. Neuroanat.* 8:105

Tsukamoto Y, Omi N. 2014b. Effects of mGluR6-deficiency on photoreceptor ribbon synapse formation: comparison of electron microscopic analysis of serial sections with random sections. *Vis. Neurosci.* 31:39–46

Tu HY, Chiao CC. 2016. Cx36 expression in the AII-mediated rod pathway is activity dependent in the developing rabbit retina. *Dev. Neurobiol.* 76:473–86

Tummala SR, Dhingra A, Fina ME, Li JJ, Ramakrishnan H, Vardi N. 2016. Lack of mGluR6-related cascade elements leads to retrograde trans-synaptic effects on rod photoreceptor synapses via matrix-associated proteins. *Eur. J. Neurosci.* 43:1509–22

Vaney DI. 1990. The mosaic of amacrine cells in the mammalian retina. *Prog. Retin. Res.* 9:49–100

Vaney DI, Sivyer B, Taylor WR. 2012. Direction selectivity in the retina: symmetry and asymmetry in structure and function. *Nat. Rev. Neurosci.* 13:194–208

Vaney DI, Taylor WR. 2002. Direction selectivity in the retina. *Curr. Opin. Neurobiol.* 12:405–10

Vlasits AL, Morrie RD, Tran-Van-Minh A, Bleckert A, Gainer CF, et al. 2016. A role for synaptic input distribution in a dendritic computation of motion direction in the retina. *Neuron* 89:1317–30

Völgyi B, Deans MR, Paul DL, Bloomfield SA. 2004. Convergence and segregation of the multiple rod pathways in mammalian retina. *J. Neurosci.* 24:11182–92

Wang Y, Fehlhaber KE, Sarria I, Cao Y, Ingram NT, et al. 2017. The auxiliary calcium channel subunit α2δ4 is required for axonal elaboration, synaptic transmission, and wiring of rod photoreceptors. *Neuron* 93(6):1359–74

Wässle H, Peichl L, Boycott BB. 1981. Dendritic territories of cat retinal ganglion cells. *Nature* 292:344–45

Wertz A, Trenholm S, Yonehara K, Hillier D, Raics Z, et al. 2015. Single-cell–initiated monosynaptic tracing reveals layerspecific cortical network modules. *Science* 349:70–74

Wilks TA, Harvey Alan R, Rodger J. 2013. Seeing with two eyes: integration of binocular retinal projections in the brain. In *Functional Brain Mapping and the Endeavor to Understand the Working Brain*, ed. F Signorelli, D Chirchiglia, pp. 227–50. Rijeka, Croat.: InTech

Williams SE, Mann F, Erskine L, Sakurai T, Wei S, et al. 2003. Ephrin-B2 and EphB1 mediate retinal axon divergence at the optic chiasm. *Neuron* 39:919–35

Williams SE, Mason CA, Herrera E. 2004. The optic chiasm as a midline choice point. *Curr. Opin. Neurobiol.* 14:51–60

Yamagata M, Sanes JR. 2008. Dscam and Sidekick proteins direct lamina-specific synaptic connections in vertebrate retina. *Nature* 451:465–69

Yamagata M, Sanes JR. 2012. Expanding the Ig superfamily code for laminar specificity in retina: expression and role of contactins. *J. Neurosci.* 32:14402–14

Yonehara K, Farrow K, Ghanem A, Hillier D, Balint K, et al. 2013. The first stage of cardinal direction selectivity is localized to the dendrites of retinal ganglion cells. *Neuron* 79:1078–85

Yonehara K, Fiscella M, Drinnenberg A, Esposti F, Trenholm S, et al. 2016. Congenital nystagmus gene FRMD7 is necessary for establishing a neuronal circuit asymmetry for direction selectivity. *Neuron* 89:177–93

Yonehara K, Ishikane H, Sakuta H, Shintani T, Nakamura-Yonehara K, et al. 2009. Identification of retinal ganglion cells and their projections involved in central transmission of information about upward and downward image motion. *PLOS ONE* 4:e4320

Zabouri N, Haverkamp S. 2013. Calcium channeldependent molecular maturation of photoreceptor synapses. *PLOS ONE* 8:e63853

Zhang C, Rompani SB, Roska B, McCall MA. 2014. Adeno-associated virus-RNAi of GlyRα1 and characterization of its synapse-specific inhibition in OFF alpha transient retinal ganglion cells. *J. Neurophysiol.* 112:3125–37

Zipursky SL, Sanes JR. 2010. Chemoaffinity revisited: Dscams, protocadherins, and neural circuit assembly. *Cell* 143:343–53

Neural Mechanisms of Social Cognition in Primates

Marco K. Wittmann,[1,2,*] *Patricia L. Lockwood,*[1,2,*] *and Matthew F.S. Rushworth*[1,2]

[1]Department of Experimental Psychology, University of Oxford, OX1 3UD Oxford,United Kingdom; email: marco.k.wittmann@gmail.com, patricia.lockwood@psy.ox.ac.uk,matthew.rushworth@psy.ox.ac.uk
[2]Wellcome Centre for Integrative Neuroimaging, Oxford Centre for Functional MRI of theBrain, Nuffield Department of Clinical Neurosciences, University of Oxford, OX1 3UD Oxford,United Kingdom

Key Words

social network, dominance, cingulate cortex, superior temporal sulcus, dorsomedial prefrontal cortex

Abstract

Activity in a network of areas spanning the superior temporal sulcus, dorsomedial frontal cortex, and anterior cingulate cortex is concerned with how nonhuman primates negotiate the social worlds in which they live. Central aspects of these circuits are retained in humans. Activity in these areas codes for primates' interactions with one another, their attempts to find out about one another, and their attempts to prevent others from finding out too much about themselves. Moreover, important features of the social world, such as dominance status, cooperation, and competition, modulate activity in these areas. We consider the degree to which activity in these regions is simply encoding an individual's own actions and choices or whether this activity is especially and specifically concerned with social cognition. Recent advances in comparative anatomy and computational modeling may help us to gain deeper insights into the nature and boundaries of primate social cognition.

INTRODUCTION

Why Should We Look for Neural Circuits for Social Cognition?

Perhaps the first question to ask is whether it is reasonable to expect that we might identify neural mechanisms for social cognition in the same way that we can for other cognitive, motor, and perprocesses. Animals possess adaptations that are beneficial for survival in the environments in which they live, and this is also true for primates (Passingham & Wise 2012). Such adaptations occur in the brain just as they do elsewhere in the body. For example, one distinctive primate feature, granular prefrontal cortex, may have evolved as anthropoid primates began to range over ceptual large environments to exploit food sources such as fruiting trees available only intermittently and in restricted areas.

For many primates, an important aspect of the environment is that it contains other individuals. Because primates are well adapted to the physical environments they occupy, it is no surprise that they are adapted to their social environments, too. According to the social brain hypothesis, the complexity of social life is correlated with brain size across primate species (Dunbar & Shultz 2007, Shultz & Dunbar 2010). Advocates of the social brain hypothesis argue that typical social group sizes or the presence of mating strategies that are more cognitively demanding are associated with larger overall brain size.

One possibility is that primate brain mechanisms are best described simply in relation to basic computational processes, any of which might be pressed into service when an animal negotiates its social environment, just as when it negotiates its spatial environment. According to this view, social cognition is the net output of the aggregate activity of these basic processes, none of which are specifically or exclusively concerned with social cognition (Behrens et al. 2009, Rushworth et al. 2013). The other possibility, however, is that some neural mechanisms are especially and specifically concerned with social cognition (Behrens et al. 2009, Rushworth et al. 2013).

Networks of Social Cognition and Cognition of Social Networks

In humans, a network of brain regions contributes to social cognition. They comprise the superior temporal sulcus (STS) and adjacent temporoparietal junction (TPJ), anterior cingulate cortex (ACC), medial and dorsomedial prefrontal cortex (dmPFC), and subcortical regions such as amygdala and striatum (Amodio & Frith 2006, Apps et al. 2016, Ruff & Fehr 2014, Saxe 2006). What brain regions are related to social cognition in nonhuman primates, and are they similar to those in humans?

Like many animals, several nonhuman primate species, such as macaques, live in groups. Macaque groups are organized in dominance hierarchies where every member has a clear rank position relative to all other group members (Maestripieri & Hoffman 2012). Higher rank enables privileged access to food (Boelkins & Wilson 1972). Social bonds within macaque groups attenuate physiological stress responses (Young et al. 2014a) and facilitate increase of social rank through the formation of alliances (Schülke et al. 2010).

One way to pinpoint the neural network supporting primate social cognition is to identify brain regions that change in relation to variation in group lifestyle. Sallet, Mars, Noonan, and colleagues (Mars et al. 2012, Noonan et al. 2014, Sallet et al. 2011) pursued this question in pseudorandomly composed rhesus macaque groups. They compared gray matter in relation to the complexity of each individual's social environment. Just as placing greater demands on the motor system results in measurable changes in specific regions of cortex and white matter (Quallo et al. 2009, Scholz et al. 2009), variation in social environments was associated with variation in brain structure; group size was correlated with gray matter in STS and adjacent temporal lobe cortex and in parts of the anterior prefrontal cortex (aPFC). In addition, during rest, functional connectivity between STS and ACC gyrus (ACCg) increased as social group size increased (**Figure 1a**).

Sliwa & Freiwald (2017) took a different approach in their attempt to map out a primate circuit for social cognition but identified many of the same brain regions. During MRI, macaques watched short movies of interacting conspecifics and a collection of control videos showing macaques interacting with inanimate objects or moving inanimate objects. The authors found responses in STS, aPFC, ACCg, and dmPFC when macaques viewed social interactions (**Figure 1b**).

The STS, aPFC, dmPFC, and ACC are monosynaptically interconnected (Petrides & Pandya 2007, Van Hoesen et al. 1993, Yukie & Shibata 2009), suggesting that they constitute a circuit. Similar or adjacent brain regions also covary with social network size in humans (Bickart et al. 2011, Kanai et al. 2012, Lewis et al. 2011). The fact that certain areas are most affected by an individual's social group size suggests that it may not be correct to think of the whole brain's size as determined by social cognitive demands. Instead, social cognitive demands may fall hardest on a subset of brain systems. Such a view would reconcile aspects of the social brain hypothesis with evidence that other factors, such as diet, have a pronounced effect on overall brain size (DeCasien et al. 2017). Furthermore, these findings suggest broad correspondences between brain regions implicated in social cognition in humans and macaques. In the next sections, we discuss these areas in more detail, outlining human-macaque correspondences where appropriate.

ANATOMY OF SOCIAL COGNITION IN PRIMATES

Superior Temporal Sulcus and Social Cognition

One of the reasons that STS may be an important component of a circuit of social cognition is that it contains patches in which functional MRI (fMRI)-recorded activity increases when macaques look at faces and body parts (Pinsk et al. 2005, 2009; Tsao et al. 2003). Many studies have reported neurons in STS that change their firing rate when macaques look at faces (Perrett et al. 1992, Tsao& Livingstone 2008), and these face-responsive patches in STS identified with fMRI contain high proportions of face-responsive neurons (Bell et al. 2011, Morin et al. 2015, Tsao et al. 2006) (**Figure 2a**).

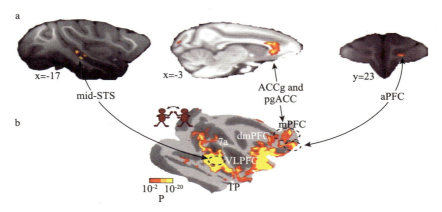

Figure 1

Brain networks of social cognition in macaques. Complementary whole-brain neuroimaging approaches identify mid-STS, ACCg, pgACC, and aPFC. (*a*) These regions are related to the size of the social group a macaque lives in (Sallet et al. 2011), and (*b*) they are amongst the regions that are more active when macaques observe conspecifics interact (Sliwa & Freiwald 2017). Panels *a* and *b* are reprinted and modified from Sallet et al. (2011) and from Sliwa & Freiwald (2017), respectively. Both panels reprinted with permission from AAAS. Abbreviations: ACCg, anterior cingulate gyrus; aPFC, anterior prefrontal cortex; dmPFC, dorsomedial prefrontal cortex; mPFC, medial prefrontal cortex; pgACC, perigenual anterior cingulate cortex; STS, superior temporal sulcus; TP, temporal pole; VLPFC, ventrolateral prefrontal cortex.

Recent investigations of face processing patches in STS have focused on their role in perceptual discrimination of different face identities (Moeller et al. 2017). Animals living in larger social groups must discriminate between more individuals, and as a result, some form of plasticity may occur in STS that results in increased size. There is, however, evidence that STS's role in social cognition is not restricted to discriminating face identities. Sliwa & Freiwald (2017) also emphasize that social interaction–related cortex extends beyond face patches. Both fMRI and neurophysiology experiments demonstrate STS cortex adjacent to face patches responds to the social information conveyed by faces such as emotional expressions and gaze direction (Hadj-Bouziane et al. 2008, Morin et al. 2015, Perrett et al. 1992) (**Figure 2b,c**). STS may also compute predictive information about the face and body movements that another animal is about to make, suggesting a role in determining current attention (Perrett et al. 2009). Some reports have argued STS lesions cause a greater impairment in gaze direction discrimination than in face discrimination and matching (Campbell et al. 1990, Heywood & Cowey 1992). However, it can be difficult to match the difficulty of gaze and face discrimination tasks.

Comparative studies provide further clues to the role of STS in social cognition and suggest a link with human brain areas such as TPJ that are implicated in high-level social cognitive processes such as theory of mind (ToM) (Saxe 2006). Mars and colleagues (2013) recorded restingstate patterns of activity between brain regions in both humans and macaques. Such activity correlations reflect anatomical connections (O'Reilly et al. 2013). Human TPJ has a distinctive connectional fingerprint characterized by strong functional connections to posterior, mid-, and anterior cingulate cortex and the anterior insula—all areas that have clear correspondences in the macaque. Mid-STS in macaques has a partially similar connectional fingerprint to human TPJ, suggesting a comparable functional role. The mid-STS region is similar to the one that increases with social network size (Sallet et al. 2011) and that is activated during observation of social interactions (Sliwa & Freiwald 2017) (**Figure 2d**). In a complementary study, Schwiedrzik and colleagues (2015) have also used functional connectivity measures to estimate connections of face patches in the STS. They also found a functional link with medial frontal cortex areas linked to social cognition.

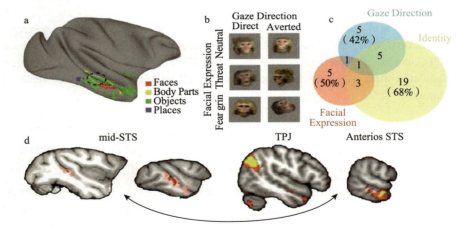

Figure 2

Processing of facial expressions and gaze direction in STS (Morin et al. 2015). (*a*) Using fMRI, patches sensitive to faces (red) are found in macaque STS when monkeys observe stimuli from different visual categories. Neurons recorded from these face-selective regions and adjacent cortex (*arrow*) are sensitive not only to the identity of the observed monkey but also to their gaze direction and facial expression. Example stimuli are shown in panel *b*, and the distribution of neuron selectivities is shown in panel c. (*d*) Macaque mid-STS (*left*) and human TPJ (*right*) resemble one another in the way in which their activity levels are coupled with those in other brain regions; they share aspects of their connectional fingerprints (Mars et al. 2013). Although macaque mid-STS may have the connectional fingerprint that best matches human TPJs, the areas are not identical (Mars et al. 2013). Macaque STS also resembles another human brain area in anterior STS, suggesting all three areas—human anterior STS, human TPJ, and macaque mid-STS—have some relationship. Both human TPJ and anterior STS are active during social cognitive tasks such as ToM (Deen et al. 2015, Mars et al. 2013). Panels *a–c* adapted from Morin et al. (2015) by permission from Oxford University Press. Panel *d* adapted from Mars et al. (2013). Abbreviations: STS, superior temporal sulcus; ToM, theory of mind; TPJ, temporoparietal junction.

Theory of Mind Versus Awareness Relations

The human TPJ has been the focus of considerable interest. Accounts of its function focus on its role in sophisticated aspects of social cognition such as ToM, but it is not clear whether primates other than humans have ToM.

ToM is the ability to represent the beliefs of other agents. If someone has ToM, then they represent not just stimuli and events in the physical world but another person's representations, or beliefs, about those stimuli and events. One common way to assess whether a person or animal has ToM is to examine whether they represent false beliefs. For example, if a person's behavior indicates they simultaneously know about the true disposition of objects in the environment (the banana is in box 1 but not in box 2) and that another agent entertains a false belief about those objects (thinking the banana is in box 2 rather than box 1), then this is strong evidence that they have access both to representations of objects in the world and metarepresentations of what other agents represent, or believe, about those objects. Nonhuman primates have consistently failed false belief tasks (Lurz 2011).

However, nonhuman primates may still be able to track which aspects of the environment another agent is aware of. Martin & Santos (2016) argue that macaques and other nonhuman primates do not track representational relations—others' beliefs that are not true of the real world—but that they do track awareness relations—aspects of the real world another individual is aware of. In one experiment, macaques watched an experimenter place a food item in a green box and then move it to a white box (Marticorena et al. 2011). The monkeys tracked that the experimenter was aware that the food item was in the white as opposed to the green box. They spent more time looking, indicating surprise, when the experimenter returned to the green than the white box. However, the monkeys could not represent false beliefs. They had no expectation as to where the human experimenter would look when the experimenter's vision of the food's transition from green to white box was obscured. Martin and Santos argue an ability to track awareness relations may explain why foraging monkeys behave in a way that reflects knowledge of what competitors are aware of and why they may avoid drawing a competitor's attention as they approach food. At the same time, the ability to track awareness relations, rather than representational relations, would explain why nonhuman primates do not attempt to hide food that another individual has already been aware of, as they do not attempt to inculcate a false belief in another individual.

Gaze-sensitive neurons in STS may be important for tracking awareness relations. However, ToM and tracking others' false beliefs may also depend on the ability to represent counterfactual alternatives. It has even been suggested that internally modeling the behavior of another person is akin to counterfactual inference (Lee & Seo 2016). In simple cases, macaques do take counterfactual rewards into account (Hayden et al. 2009), but the degree to which macaques and humans represent all features of counterfactual choice possibilities may vary. Anatomical differences in anterior prefrontal organization in humans and macaques (Neubert et al. 2014, 2015) might indicate that the contribution anterior prefrontal cortex makes to counterfactual choice in humans (Boorman et al. 2009, 2011) does not generalize to macaques in a straightforward way.

Learning from Others and Preventing Others from Learning About You

Several areas in primate medial frontal cortex carry signals necessary for social cognition. These include the gyral portion of the ACC (ACCg), the sulcus of the ACC (ACCs), and dmPFC, dorsal to ACCs. Activity in dmPFC may be especially important for learning from observing another individual and for preventing other individuals from predicting your behavior.

Yoshida and colleagues (2011, 2012) recorded neurons in dmPFC and ACCs during a rolereversal task. In alternation, two macaques performed a deterministic, binary reversal-learning task with one rewarded and one unrewarded option. Such paradigms are frequently used in studies of reward-guided learning and decision making, and some simple modifications allowed Yoshida et al. to exploit them to study social cognition. One macaque performed two trials while thesecond observed. The actor and observer roles switched for subsequent trials. Observing macaques tracked the other's behavior to learn about rule reversals: Without making an error themselves, they correctly switched to the alternative option after observing an error made by the partner, if that error indicated that the rule had reversed. Some of the neurons recorded in ACCs and dmPFC fired more when one macaque observed the other making a choice compared to when the first made a choice itself, and neurons with this firing profile were more frequent in dmPFC than ACCs (Yoshida et al. 2011). Neurons in both areas encoded when the other macaque made an incorrect choice, but in a substantial subset of cases, activity was unrelated to reward omission per se (Yoshida et al. 2012). The finding that dmPFC activity is related to observation of another individual accords well with findings from human fMRI studies that identified dmPFC activity when expectations about how another person would act were violated (Behrens et al. 2008, Suzuki et al. 2012, Wittmann et al. 2016b).

However, macaque dmPFC is not exclusively concerned with tracking others' actions. It is also concerned with preventing others from predicting one's own behavior. Neurons in dmPFC encode the macaque's own previous choices and whether they were rewarded or not during a matching pennies task (Donahue et al. 2013). In this binary choice task, macaques make choices and receive rewards if a computer opponent makes the same choice (Seo & Lee 2007). The computer is programmed to outplay the macaques based on regularities in their recent choice pattern, so choosing unpredictably is the best strategy. The matching pennies paradigm is similar to paradigms investigating nonsocial reward learning and choice, and accordingly, the analysis of behavioral and neural data is based on computational models, in particular reinforcement learning (RL) models. For example, Seo et al. (2014) find that monkeys' choices are often well predicted by a value expectation computed with an RL model. The value expectation is a longer-term estimate of how good it is to choose an option that is revised each time a new outcome is witnessed. However, to earn rewards in the experiment, the animals should not behave predictably, and thus they also make choices that run counter to what an RL model would predict. Strikingly, the neurons encoding past choices and rewards in dmPFC are especially active when the animal initiates short sequences of choices that systematically deviate from those a reinforcement learner would make (Seo et al. 2014). In such a setting, the animal not only has to prioritize internally generated spontaneous choice over explicitly cued stimulus-response patterns (Nachev et al. 2008), it even has to disregard the type of responses that are directly incentivized by past rewards. Together, the reversal observation (Yoshida et al. 2011, 2012) and matching pennies (Seo et al. 2014) studies suggest dmPFC activity tracks another individual's performance to predict future behavior and an animal's attempts to be unpredictable to others (**Figure 3**).

Moreover, neurons recorded in this or an adjacent dorsomedial area track the degree of insight monkeys have into the correctness of their actions (Middlebrooks & Sommer 2012). An fMRI and inactivation study in macaques suggests a role for this region in judging the accuracy of recently encoded memories and the involvement of more anterior dorsomedial regions in metamemory for more remotely encoded memories (Miyamoto et al. 2017). Metacognitive representations relating to one's own behavior and that of other agents are often considered similar (Carruthers 2009, Frith & Frith 2012).

It is therefore plausible that these dmPFC regions are concerned with metacognitive processes relating to other individuals as well to the monkey's own behavior. In humans, neural activity related to confidence in one's own judgement is also often found relatively anteriorly in rostrolateral prefrontal cortex (De Martino et al. 2013, Fleming et al. 2010).

The macaque studies highlight co-occurrence of self- and other-related choice signals in dmPFC. However, it should be emphasized that the dmPFC recordings in macaques have been conducted in relatively caudal dmPFC areas such as the supplementary eye fields or the posterior supraprincipal dimple, while human social cognition studies have focused on more anterior dmPFC areas such as area 9 (but see Suzuki et al. 2012, figure 3) that are rarely investigated in macaques. Nevertheless, human fMRI studies of dmPFC similarly highlight choice value representations relating to both the self and the other. For example, in one study, participants made choices for themselves on some trials and on behalf of another person, whose preferences were known, on other trials (Nicolle et al. 2012). Activity related to the preference currently not guiding choice (whether one's own or the other's) was found in dmPFC. Human studies further suggest that there are not simply two distinct, overlapping neural representations in dmPFC that are recruited in different circumstances, but that—at least in humans—interactions between self and other representations occur in dmPFC. First, Hampton and colleagues (2008) found a signal in dmPFC reflecting the influence that one individual assumes they have on the actions of another player in an interactive strategic game. Second, this area reflects whether our preferences accord with those held by specific other individuals (Izuma & Adolphs 2013) and how much we are influenced by other opinions (De Martino et al. 2017). Finally, Wittmann and colleagues (2016b) reported that dmPFC activity predicts how our estimates of other people's abilities are biased by how we think about ourselves and vice versa (**Figure 3**).

Anterior Cingulate Cortex and Tracking Motivational Information About Other Individuals

ACC has been implicated in an array of cognitive processes including motivation, decision making, learning, and conflict and error monitoring (Hayden et al. 2011; Holroyd & McClure 2015; Kolling et al. 2014, 2016; Shackman et al. 2011; Ullsperger et al. 2014; Verguts et al. 2015; Wittmann et al. 2016a). ACC's role in social cognition may reflect the contribution it makes to these processes, but some recent results suggest the existence of a representation within ACC that is concerned not with the choices the individual themselves makes but with the choices that another agent might make. One recent study recorded from neurons in ACC as monkeys played a prisoner's dilemma game with another monkey. The activity of many neurons was related to the choices that the monkeys themselves made. However, one class of neurons predicted the other monkeys' choices of whether or not to cooperate. These same neurons largely did not encode features of the monkeys' own choices (Haroush & Williams 2015). Such an activity pattern may suggest ACC codes others' actions as well as one's own.

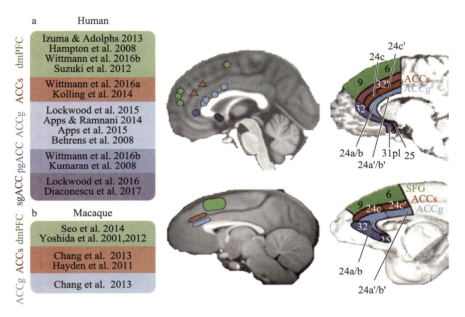

Figure 3

Locations of mPFC responses during social and nonsocial tasks in humans (*top*) and macaques (*bottom*). (*a*)Studies that have identified responses in mPFC during fMRI in humans and the responses' anatomical locations are represented on an MRI scan of the medial surface. Red triangles show anterior cingulate sulcus BOLD response during nonsocial decision making. Circles show locations of BOLD response that tracks social information processing. (*b*) Studies that have recorded from single neurons in mPFC in macaques and the neurons' anatomical locations represented on an MRI scan of the medial surface. Red diamond shows recording site for ACCs that contains neurons that respond during nonsocial reward decision-making tasks. Rectangles show locations of neurons that code for social information. Images on the right taken from Apps and colleagues (2016) and adapted from Rushworth and colleagues (2004). Abbreviations: ACCg, anterior cingulate cortex gyrus; ACCs, anterior cingulate sulcus; BOLD, blood-oxygen-level-dependent; dmPFC, dorsomedial prefrontal cortex; mPFC, medial prefrontal cortex; pgACC, perigenual ACC; sgACC, subgenual ACC; SFG, superior frontal gyrus.

In such a paradigm, it can be difficult to completely disentangle the other individual's actions from the implications that the other's actions have for oneself. For example, it is possible that neurons that encode aspects of the other individuals' choices signal a foregone reward because the actions of the other animal have implications for how much reward the first animal itself will receive.

In some other simple paradigms, however, such ambiguities can be reduced, and it is clear that ACC neural activity can be related to the behavior and motivation of others. Several of these studies have emphasized a sub-region of the ACC—the ACCg. The ACCg plays the most prominent role in social cognition, especially in relation to tracking how motivated another individual is to obtain a goal (Apps et al. 2013b, 2016; Lockwood 2016). In macaques, several lines of evidence also underline ACCg's importance for social cognition. Lesions to ACCg impair normal patterns of social interest in others and cause a reduction in the execution of social behaviors (Rudebeck et al. 2006). Chang and colleagues (2012) recorded from neurons in ACCs, ACCg, and orbitofrontal cortex (OFC) during a social decision-making task. Monkeys were assigned roles of actor (self) and recipient (other). Intriguingly, they found a greater proportion of neurons in ACCg, compared to ACCs and OFC, responded to cues that predicted rewards for other monkeys and during decisions to allocate rewards to other monkeys (**Figure 3**).

The link between ACCg and social processing is also supported by fMRI studies in humans. ACCg signals aspects of others' actions (Behrens et al. 2008), expected rewards for others (Apps & Ramnani 2014, Lockwood et al. 2015), prediction errors when others' expectations are violated (Apps et al. 2015), whether rewarding outcomes are being delivered to others (Apps et al. 2013a, Zhu et al. 2012), or when painful stimulation is predicted for or delivered to another person (Lamm et al. 2011, Lockwood et al. 2013). These patterns are not seen when similar information is processed with reference to oneself (**Figure 3**). Again, many of these studies explain behavior and neural activity in terms of quantities originating from an RL framework, such as decision expectations, decision outcomes, and prediction errors (the positive or negative deviation of outcome from expectation). The RL models employed are structurally very similar to those used when studying nonsocial reward-guided learning in ACCs.

Although most studies in humans have used fMRI, there is now also evidence that single ACC neurons in humans encode social information. Hill and colleagues (2016) recorded from implanted electrodes in the ACCg in patients who were undergoing surgery for intractable epilepsy. Patients performed an RL-based task in which they could learn from the outcomes of their own decisions but also from the outcomes of two other players. Activity in ACCg neurons correlated with both the expected outcome and the actual outcome of trials but only when patients monitored the behavior of another player as opposed to their own behavior. Together, these studies suggest that ACCg plays a crucial role in social cognition in humans and macaques.

Social Prediction Areas in Prefrontal Areas Outside the Social Brain

Several studies have identified prediction errors during social interactions in areas beyond dorsal ACC. Subgenual ACC (sgACC) tracks prosocial prediction errors when a person learns how their own action leads to an outcome for another person (Lockwood et al. 2016) and signals high-level prediction errors regarding uncertainty about an adviser's fidelity or trustworthiness (Diaconescu et al. 2017). Whether this signaling of prediction errors in sgACC is specifically social in nature is unclear because there is initial evidence that this region also tracks prediction errors when learning occurs at a level of abstraction beyond basic stimulus–response association (Iglesias et al. 2013). Moreover, in macaques, this region has been linked to visceromotor control; sgACC lesions disrupt the sustaining of autonomic arousal associated with positive emotional events (Rudebeck et al. 2014). The dorsolateral prefrontal cortex (dlPFC) also tracks social prediction errors in humans (Burke et al. 2010, Suzuki et al. 2012). Although in these studies dlPFC signals were not observed when the players themselves were performing actions, the dlPFC has been linked to a variety of behavioral and cognitive processes beyond social cognition, including strategic decision making, working memory, and attention (Genovesio et al. 2014, Passingham & Wise 2012, Petrides et al. 2012).

Finally, OFC has also been linked to social cognition. Posterior lateral OFC is active during social interaction observation (Sliwa & Freiwald 2017). Macaques have a face-responsive patch in lateral OFC resembling those found in STS (Hadj-Bouziane et al. 2008, Rolls 2007, Tsao et al. 2008b). It has been claimed that a similar region exists in humans (Tsao et al. 2008a), although its location is surprisingly medial. The precise role of OFC face patches remains unclear, but, more generally, OFC plays a central role in learning the reward value of objects and in using these value associations to guide decision making (Rudebeck & Murray 2014). This function of OFC may underlie its recruitment in social situations. Just as OFC learns and infers the values of inanimate objects, other individuals and combinations of individuals are assigned positive or negative values that determine the emotional response they elicit (Rolls 1999). Such values are updated as circumstances change and, in turn, impact on the social interactions that occur between individuals. The lateral OFC is most important for updating such value associations in both humans (Akaishi et al. 2016, Jocham et al. 2016, Noonan et al. 2011) and macaques (Chau et al. 2015, Walton et al. 2010). Lesions of OFC, particularly lateral OFC, disrupt emotional responses (Noonan et al. 2010, Rudebeck et al. 2006), although their impact may depend on adjacent white matter (Rudebeck et al. 2013).

When male monkeys decide between a constant amount of juice or a variable amount of juice and access to a picture of a conspecific, they sometimes pick the variable juice option, particularly if it is accompanied by a picture of a dominant monkey or the perineum of a female monkey (Deaner et al. 2005). This suggests the social stimuli have a value to the male monkeys. OFC neurons respond to such social stimuli more than to other nonsocial stimuli, even if they are associated with reward (Watson & Platt 2012). In some cases, activity levels reflect the varying degree to which male macaques value the opportunity to look at the stimuli, and in other cases, they distinguish between categories of socially significant stimuli, such as faces and female perinea, and the dominance levels of faces (Watson & Platt 2012). Azzi and colleagues (2012) have also identified OFC neurons that track the social rank and identity of other monkeys as well as other OFC neurons tracking the motivational value of rewards obtained in a social context.

Ventral Striatum and Social Behavior

It is well documented that the striatum responds to reward predictions and actual reward outcomes, particularly when outcomes are unexpected. Such a pattern is thought to reflect reinforcement (Schultz 2016). Ventral striatum receives an input from dopamine neurons in the midbrain that carry a prediction error signal, a key feature of RL theory (Schultz et al. 1997). However, the role of this area in social behavior remains incompletely understood (Báez-Mendoza & Schultz 2013).

Like OFC, the striatum may track the value of both social and nonsocial information. Usingthe paradigm devised by Deaner et al. (2005), Klein & Platt (2013) showed that neurons in the caudate division of the striatum responded more to social reward images, whereas neurons in the putamen division of striatum responded more strongly to juice rewards. They suggest that there may be specific signals of social context and fluid rewards in striatum, even though these are used to guide a single action (Klein & Platt 2013).

In a reward-giving task, neurons in striatum responded mostly to the monkey's own reward and infrequently to the other monkeys' reward. However, another class of striatal neurons showed coding of social action without reward, and some of these responses disappeared when a computer replaced the other monkey (Báez-Mendoza et al. 2013). Striatal neurons also signal errors during performances made by the monkeys themselves and by conspecifics that are independent of a negative reward prediction error signal (Báez-Mendoza & Schultz 2016). The striatum may therefore play a role in both identifying social actors and coding one's own reward.

In humans, striatal responses have been found during the processing of social rewards, such as smiling faces, that overlap with similar striatal responses during the processing of monetary rewards (Izuma et al. 2008, Lin et al. 2012, Spreckelmeyer et al. 2009). The response of the striatum to juice, money, and erotic picture rewards has been taken as support for a common currency of reward processing in striatum (Sescousse et al. 2015). It therefore seems likely that the striatum has a domain-general role in learning and reward. This idea is supported by a recent study in humans that showed a blood-oxygen-level-dependent (BOLD) response in ventral striatum to prediction errors when someone learns to benefit themselves, another person, and no one in particular (Lockwood et al. 2016) (**Figure 3**).

COOPERATION AND AFFILIATION

As we have seen, living in groups shapes neural circuits in primates (Noonan et al. 2014, Sallet et al. 2011). Living in groups is beneficial for primates because it helps increase the reward rates of single individuals over the long term (Pulliam & Caraco 1984). However, for this to happen, primates need to work together to some degree. For example, macaques cooperate when they defend against or attack territorially adjacent macaque communities (Boelkins & Wilson 1972) and when they form alliances to win conflicts within the group (Schülke et al. 2010). Such cooperation relies on simple signals conveyed partly through facial expressions and head or gaze direction, which might be encoded in STS. For example, macaques recruit conspecifics during conflicts with other group members by alternating gaze between the potential recruit and the opponent (Maestripieri 1997, Young et al. 2014b). In this sense, cooperation requires conveying and understanding indicators of impending action, but it does not necessarily require deception or ToM (Devaine et al. 2014). There is, however, debate on how similar cooperation really is in humans and other primates (Drayton & Santos 2014, Kaminski et al. 2008, Lakshminarayanan & Santos 2008, Melis et al. 2006, Suchak et al. 2016).

Recent evidence has suggested that neuropeptides such as oxytocin modulate social behaviors such as cooperation. For example, oxytocin affects social behavior in rodents, monkeys, and humans (Tremblay et al. 2017), although there is debate over the size and nature of effects (Leng & Ludwig 2016, Walum et al. 2016). Macaques are fundamentally interested in their conspecifics and willing to trade off reward to see pictures of them (Deaner et al. 2005, Rudebeck et al. 2006). When macaques have the chance to observe each other, inhalation of oxytocin in combination with opioid antagonists (which are supposed to strengthen the oxytocin effects) leads to an increase in social attention toward another macaque (**Figure 4a**) (Dal Monte et al. 2017). Although monkeysrarely sacrifice their own rewards to deliver reward to others (Chang et al. 2012, 2013), they prefer rewarding others compared to no one (Chang et al. 2012, 2013, 2015) as well as rewarding both themselves and another monkey compared to rewarding themselves alone (Ballesta & Duhamel 2015, Chang et al. 2015, Lakshminarayanan & Santos 2008). After intranasal delivery of oxytocin, macaques are more likely to reward conspecifics over no one but also more likely to reward themselves over the conspecific (Chang et al. 2012). Similar effects of increased prosocial choice are observed after direct oxytocin injection in the amygdala (Chang et al. 2015). The amygdala also carries signals related to self and other reward outcomes. One suggestion is that oxytocin might not affect prosociality toward others in general but instead might strengthen the bonds between oneself and one's own group, which could manifest in prosocial behavior toward group members but also in protective aggression toward potential attackers (De Dreu & Kret 2016, De Dreu et al. 2010).

Navigating the social environment also requires representation of one's own place within it (Qu et al. 2017). In humans, the most rostral part of the ACC—the perigenual ACC (pgACC)—shows a characteristic pattern of activity related to learning and knowledge of one's own position within a hierarchy. For example, how much subjects change what they think of their position in a hierarchy on a trial-by-trial basis is correlated with the BOLD signal in pgACC (Kumaran et al. 2016) **(Figure 4b)**. Such signals might reflect the computation of self-values as opposed to the value of specific choices (Murray et al. 2011). Self-related signals are frequently found in pgACC (Denny et al. 2012, Garvert et al. 2015, Kelley et al. 2002, Mitchell et al. 2006, Sul et al. 2015). Computations of self-value in pgACC could be used to estimate the general success rate of one's actions and how likely one is to overcome impending challenges (Wittmann et al. 2016b).

In humans, the amygdala tracks learning of a social as opposed to a nonsocial hierarchy but shows no preference for hierarchies that include oneself versus those that do not include oneself (Kumaran et al. 2012, 2016). Similarly, in macaques, dominance status is correlated with gray matter density in the amygdala, hypothalamus, and raphe nuclei (Noonan et al. 2014). Dominance in macaques is not simply a matter of size or aggression but also a matter of the social coalitions that an individual establishes (Schü lke et al. 2010), and therefore, social dominance likely reflects how successful macaques are in social situations. Perhaps not surprisingly, social dominance is also related to gray matter in STS and aPFC (Noonan et al. 2014, Sallet et al. 2011).

Social interactions often involve decisions to work with others for a common goal. Working together with others can increase our willingness to invest effort (Le Bouc & Pessiglione 2013) and requires alignment of our own actions with those of others (Stolk et al. 2016, Yoshida et al. 2010). A recent fMRI study in humans investigated

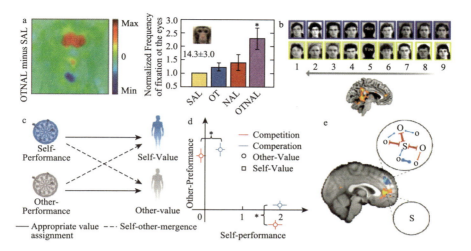

Figure 4

Self-and other-values in social interactions. (*a*) Dal Monte and colleagues (2017) measured social attention using eye tracking when two macaques were seated directly across from each other. Administration of OTNAL led to increased fixation on the other's face and in particular the eyes (*left,* heat map of fixations under OTNAL compared to a SAL control; *right,* frequency of eye fixation per condition; *, P < 0.01). (*b*) Kumaran and colleagues (2016) studied learning about one's own (*yellow*) or a close friend's (*blue*) place in a dominance hierarchy. For hierarchies including oneself only, the updating of hierarchy position during learning was specifically correlated with pgACC activity. (*c*) Self-value and other-value can be learned by tracking how well actions are performed. Self-other-mergence occurs when the performance is attributed to the inappropriate agent and people increase confidence in their own abilities after a partner has performed well (Wittmann et al. 2016b). (*d*) Using logistic GLMs, self-performance and other-performance wereused to predict self-value and other-value as measured by independent ratings. Each data point shows average beta weights (±SEMs) for both predictors. GLMs were performed separately for cooperation and competition conditions. Self-value and other-value were largely governed by appropriate value assignment, but self-other-mergence became apparent when considering the cooperative or competitive context. Self-value was significantly more positively influenced by other-performance in cooperation compared to competition. The analogous effect was observed for other-value (*, $P < 0.005$). (*e*) In the same study, pgACC and dmPFC showed distinct patterns of activity. While pgACC tracked the success of one's own recent actions, dmPFC activity reflected the impact (as shown in panel *d*) that interacting with others had on self- and other-values. Panel *a* adapted from Dal Monte et al. (2017). Panel *b* adapted from Kumaran et al. (2016). Panels *d* and *e* adapted from Wittmann et al. (2016b). Abbreviations: dmPFC, dorsomedial prefrontal cortex; GLM, general linear model; NAL, naloxone; OT, oxytocin; OTNAL, oxytocin combined with naloxone; pgACC, perigenual anterior cingulate cortex; SAL, saline; SEM, standard error of the mean.

how interacting with others affects the representations we hold of self and others (Wittmann et al. 2016b). Participants repeatedly performed games and received performance feedback. Over time, they learned about their own and others' levels of performance. A simple RL model that computed a longer-term average of the performance feedback was used to approximate subjects' trial-by-trial representation of their own and others' performance levels. Importantly, on some trials, they cooperated with each other, and on other trials, they competed. Trial-by-trial performance ratings showed that how subjects rated their own performance was not just a reflection of the feedback they received for their own performance but was also affected by the performance of others. Estimates of one's own self-value increased when cooperating as opposed to competing with a strong partner (**Figure 4c,d**). The same effect was also found when evaluating the partner; during cooperation, strong performances by the players themselves led to higher estimates of the partner's ability. dmPFC signaled the degree to which the performance feedback observed for the other person had an influence on self-value and vice versa. This suggests that self and other related representations in dmPFC (Seo et al. 2014, Yoshida et al. 2011) are used to compute relationships between self and other (Hampton et al. 2008, Izuma & Adolphs 2013) and that dmPFC adjusts the representations we have of ourselves and others as a consequence of interacting together in groups (**Figure 4e**).

CONCLUSION

A network of brain regions carries information for social interactions. In some cases, such as in the OFC, ventral striatum, and amygdala, this social role may be related to a more basic one in fundamental behavioral processes such as learning the reward associations of objects in the environment. In other cases, potentially for ACCg and dmPFC, the activity may be more specifically linked with social cognition. Computational models, as well as studies drawing on comparative anatomy, will be essential for providing precise and quantitative descriptions of the contribution made by such areas to social cognition.

DISCLOSURE STATEMENT

The authors are not aware of any affiliations, memberships, funding, or financial holdings that might be perceived as affecting the objectivity of this review.

ACKNOWLEDGMENTS

Funded by the Wellcome Trust and Medical Research Council (MRC). We are grateful to

Dr. Kentaro Miyamoto for helpful discussions.

LITERATURE CITED

Akaishi R, Kolling N, Brown JW, Rushworth M. 2016. Neural mechanisms of credit assignment in a multicue environment. *J. Neurosci.* 36:1096–112

Amodio DM, Frith CD. 2006. Meeting of minds: the medial frontal cortex and social cognition. *Nat. Rev. Neurosci.* 7:268–77

Apps MAJ, Green R, Ramnani N. 2013a. Reinforcement learning signals in the anterior cingulate cortex code for others' false beliefs. *NeuroImage* 64:1–9

Apps MAJ, Lesage E, Ramnani N. 2015. Vicarious reinforcement learning signals when instructing others. *J. Neurosci.* 35:2904–13

Apps MAJ, Lockwood PL, Balsters JH. 2013b. The role of the midcingulate cortex in monitoring others' decisions. *Front. Neurosci.* 7:251

Apps MAJ, Ramnani N. 2014. The anterior cingulate gyrus signals the net value of others' rewards. *J. Neurosci.* 34:6190–200

Apps MAJ, Rushworth MFS, Chang SWC. 2016. The anterior cingulate gyrus and social cognition: tracking the motivation of others. *Neuron* 90:692–707

Azzi JC, Sirigu A, Duhamel JR. 2012. Modulation of value representation by social context in the primate orbitofrontal cortex. *PNAS* 109:2126–31

Báez-Mendoza R, Harris CJ, Schultz W. 2013. Activity of striatal neurons reflects social action and own reward. *PNAS* 110:16634–39

Báez-Mendoza R, Schultz W. 2013. The role of the striatum in social behavior. *Front. Neurosci.* 7:233

Báez-Mendoza R, Schultz W. 2016. Performance error-related activity in monkey striatum during social interactions. *Sci. Rep.* 6:37199

Ballesta S, Duhamel JR. 2015. Rudimentary empathy in macaques' social decision-making. *PNAS* 112:15516–21

Behrens TEJ, Hunt LT, Rushworth MFS. 2009. The computation of social behavior. *Science* 324:1160–64

Behrens TEJ, Hunt LT, Woolrich MW, Rushworth MFS. 2008. Associative learning of social value. *Nature* 456:245–49

Bell AH, Malecek NJ, Morin EL, Hadj-Bouziane F, Tootell RBH, Ungerleider LG. 2011. Relationship between functional magnetic resonance imaging-identified regions and neuronal category selectivity. *J. Neurosci.* 31:12229–40

Bickart KC, Wright CI, Dautoff RJ, Dickerson BC, Barrett LF. 2011. Amygdala volume and social network size in humans. *Nat. Neurosci.* 14:163–64

Boelkins RC, Wilson AP. 1972. Intergroup social dynamics of the Cayo Santiago rhesus (Macaca mulatta) with special reference to changes in group membership by males. *Primates* 13:125–39

Boorman ED, Behrens TEJ, Rushworth MFS. 2011. Counterfactual choice and learning in a neural network centered on human lateral frontopolar cortex. *PLOS Biol.* 9:e1001093

Boorman ED, Behrens TEJ, Woolrich MW, Rushworth MFS. 2009. How green is the grass on the other side? Frontopolar cortex and the evidence in favor of alternative courses of action. *Neuron* 62:733–43

Burke CJ, Tobler PN, Baddeley M, Schultz W. 2010. Neural mechanisms of observational learning. *PNAS* 107:14431–36

Campbell R, Heywood CA, Cowey A, Regard M, Landis T. 1990. Sensitivity to eye gaze in prosopagnosic patients and monkeys with superior temporal sulcus ablation. *Neuropsychologia* 28:1123–42

Carruthers P. 2009. How we know our own minds: the relationship between mindreading and metacognition. *Behav. Brain Sci.* 32:121–38

Chang SWC, Barter JW, Ebitz RB, Watson KK, Platt ML. 2012. Inhaled oxytocin amplifies both vicarious reinforcement and self reinforcement in rhesus macaques (*Macaca mulatta*). PNAS 109:959–64

Chang SWC, Fagan NA, Toda K, Utevsky AV, Pearson JM, Platt ML. 2015. Neural mechanisms of social decision-making in the primate amygdala. *PNAS* 112:16012–17

Chang SWC, Gariepy JF, Platt ML. 2013. Neuronal reference frames for social decisions in primate frontal cortex. *Nat. Neurosci.* 16:243–50

Chau BK, Sallet J, Papageorgiou GK, Noonan MP, Bell AH, et al. 2015. Contrasting roles for orbitofrontal cortex and amygdala in credit assignment and learning in macaques. *Neuron* 87:1106–18

Dal Monte O, Piva M, Anderson KM, Tringides M, Holmes AJ, Chang SWC. 2017. Oxytocin under opioid antagonism leads to supralinear enhancement of social attention. *PNAS* 114:5247–52

De Dreu CKW, Kret ME. 2016. Oxytocin conditions intergroup relations through upregulated in-group empathy, cooperation, conformity, and defense. *Biol. Psychiatry* 79:165–73

De Dreu CKW, Greer LL, Handgraaf MJJ, Shalvi S, Van Kleef GA, et al. 2010. The neuropeptide oxytocin regulates parochial altruism in intergroup conflict among humans. *Science* 328:1408–11

De Martino B, Bobadilla-Suarez S, Nouguchi T, Sharot T, Love BC. 2017. Social information is integrated into value and confidence judgments according to its reliability. *J. Neurosci.* 37:6066–74

De Martino B, Fleming SM, Garrett N, Dolan RJ. 2013. Confidence in value-based choice. *Nat. Neurosci.* 16:105–10

Deaner RO, Khera AV, Platt ML. 2005. Monkeys pay per view: adaptive valuation of social images by rhesus macaques. *Curr. Biol.* 15:543–48

DeCasien A, Williams S, Higham J. 2017. Primate brain size is predicted by diet but not sociality. *Nat. Ecol.Evol.* 1:0112

Deen B, Koldewyn K, Kanwisher N, Saxe R. 2015. Functional organization of social perception and cognition in the superior temporal sulcus. *Cereb. Cortex* 25:4596–609

Denny BT, Kober H, Wager TD, Ochsner KN. 2012. A meta-analysis of functional neuroimaging studies of self- and other judgments reveals a spatial gradient for mentalizing in medial prefrontal cortex. *J. Cogn. Neurosci.* 24:1742–52

Devaine M, Hollard G, Daunizeau J. 2014. Theory of mind: Did evolution fool us? *PLOS ONE* 9:e 87619

Diaconescu AO, Mathys C, Weber LAE, Kasper L, Mauer J, Stephan KE. 2017. Hierarchical prediction errors in midbrain and septum during social learning. *Soc. Cogn. Affect. Neurosci.* 12:618–34

Donahue CH, Seo H, Lee D. 2013. Cortical signals for rewarded actions and strategic exploration. *Neuron* 80:223–34

Drayton LA, Santos LR. 2014. Capuchins' (Cebus apella) sensitivity to others' goal-directed actions in a helping context. *Anim. Cogn.* 17:689–700

Dunbar RI, Shultz S. 2007. Evolution in the social brain. *Science* 317:1344–47

Fleming SM, Weil RS, Nagy Z, Dolan RJ, Rees G. 2010. Relating introspective accuracy to individual differences in brain structure. *Science* 329:1541–43

Frith CD, Frith U. 2012. Mechanisms of social cognition. *Annu. Rev. Psychol.* 63:287–313

Garvert MM, Moutoussis M, Kurth-Nelson Z, Behrens TEJ, Dolan RJ. 2015. Learning-induced plasticity in medial prefrontal cortex predicts preference malleability. *Neuron* 85:418–28

Genovesio A, Wise SP, Passingham RE. 2014. Prefrontal-parietal function: from foraging to foresight. *Trends Cogn. Sci.* 18:72–81

Hadj-Bouziane F, Bell AH, Knusten TA, Ungerleider LG, Tootell RBH. 2008. Perception of emotional expressions is independent of face selectivity in monkey inferior temporal cortex. *PNAS* 105:5591–96

Hampton AN, Bossaerts P, O'Doherty JP. 2008. Neural correlates of mentalizing-related computations during strategic interactions in humans. *PNAS* 105:6741–46

Haroush K, Williams ZM. 2015. Neuronal prediction of opponent's behavior during cooperative social interchange in primates. *Cell* 160:1233–45

Hayden BY, Pearson JM, Platt ML. 2009. Fictive reward signals in the anterior cingulate cortex. *Science* 324:948–50

Hayden BY, Pearson JM, Platt ML. 2011. Neuronal basis of sequential foraging decisions in a patchy environment. *Nat. Neurosci.* 14:933–39

Heywood CA, Cowey A. 1992. The role of the 'face-cell' area in the discrimination and recognition of faces by monkeys. *Philos. Trans. R. Soc. B* 335:31–38

Hill MR, Boorman ED, Fried I. 2016. Observational learning computations in neurons of the human anterior cingulate cortex. *Nat. Commun.* 7:12722

Holroyd CB, McClure SM. 2015. Hierarchical control over effortful behavior by rodent medial frontal cortex: a computational model. *Psychol. Rev.* 122:54–83

Iglesias S, Mathys C, Brodersen KH, Kasper L, Piccirelli M, et al. 2013. Hierarchical prediction errors in midbrain and basal forebrain during sensory learning. *Neuron* 80:519–30

Izuma K, Adolphs R. 2013. Social manipulation of preference in the human brain. *Neuron* 78:563–73

Izuma K, Saito DN, Sadato N. 2008. Processing of social and monetary rewards in the human striatum. *Neuron* 58:284–94

Jocham G, Brodersen KH, Constantinescu AO, Kahn MC, Ianni AM, et al. 2016. Reward-guided learning with and without causal attribution. *Neuron* 90:177–90

Kaminski J, Call J, Tomasello M. 2008. Chimpanzees know what others know, but not what they believe. *Cognition* 109:224–34

Kanai R, Bahrami B, Roylance R, Rees G. 2012. Online social network size is reflected in human brain structure. *Proc. Biol. Sci.* 279:1327–34

Kelley WM, Macrae CN, Wyland CL, Caglar S, Inati S, Heatherton TF. 2002. Finding the self? An eventrelated fMRI study. *J. Cogn. Neurosci.* 14:785–94

Klein JT, Platt ML. 2013. Social information signaling by neurons in primate striatum. *Curr. Biol.* 23:691–96

Kolling N, Wittmann MK, Behrens TEJ, Boorman ED, Mars RB, Rushworth MFS. 2016. Anterior cingulate cortex and the value of the environment, search, persistence, and model updating. *Nat. Neurosci.* 19:1280–85

Kolling N, Wittmann MK, Rushworth MFS. 2014. Multiple neural mechanisms of decision making and their competition under changing risk pressure. *Neuron* 81:1190–202

Kumaran D, Banino A, Blundell C, Hassabis D, Dayan P. 2016. Computations underlying social hierarchy learning: distinct neural mechanisms for updating and representing self-relevant information. *Neuron* 92:1135–47

Kumaran D, Melo HL, Duzel E. 2012. The emergence and representation of knowledge about social and nonsocial hierarchies. *Neuron* 76:653–66

Lakshminarayanan VR, Santos LR. 2008. Capuchin monkeys are sensitive to others' welfare. *Curr. Biol.* 18:R999–1000

Lamm C, Decety J, Singer T. 2011. Meta-analytic evidence for common and distinct neural networks associated with directly experienced pain and empathy for pain. *NeuroImage* 54:2492–502

Le Bouc R, Pessiglione M. 2013. Imaging social motivation: distinct brain mechanisms drive effort production during collaboration versus competition. *J. Neurosci.* 33:15894–902

Lee D, Seo H. 2016. Neural basis of strategic decision making. *Trends Neurosci.* 39:40–48

Leng G, Ludwig M. 2016. Intranasal oxytocin: myths and delusions. *Biol. Psychiatry* 79:243–50

Lewis PA, Rezaie R, Brown R, Roberts N, Dunbar RI. 2011. Ventromedial prefrontal volume predicts understanding of others and social network size. *NeuroImage* 57:1624–29

Lin A, Adolphs R, Rangel A. 2012. Social and monetary reward learning engage overlapping neural substrates. *Soc. Cogn. Affect. Neurosci.* 7:274–81

Lockwood PL. 2016. The anatomy of empathy: vicarious experience and disorders of social cognition. *Behav. Brain Res.* 311:255–66

Lockwood PL, Apps MAJ, Roiser JP, Viding E. 2015. Encoding of vicarious reward prediction in anterior cingulate cortex and relationship with trait empathy. *J. Neurosci.* 35:13720–27

Lockwood PL, Apps MAJ, Valton V, Viding E, Roiser JP. 2016. Neurocomputational mechanisms of prosocial learning and links to empathy. *PNAS* 113:9763–68

Lockwood PL, Sebastian CL, McCrory EJ, Hyde ZH, Gu X, et al. 2013. Association of callous traits with reduced neural response to others' pain in children with conduct problems. *Curr. Biol.* 23:901–5

Lurz RW. 2011. *Mindreading Animals: The Debate over What Animals Know about Other Minds.* Cambridge, MA: MIT Press

Maestripieri D. 1997. Gestural communication in macaques: usage and meaning of nonvocal signals. *Evol. Commun.* 1:193–222

Maestripieri D, Hoffman CL. 2012. Behavior and social dynamics of rhesus macaques on Cayo Santiago. In *Bones, Genetics, and Behavior of Rhesus Macaques,* ed. Q Wang, pp. 247–62. New York: Springer

Mars RB, Neubert FX, Noonan MP, Sallet J, Toni I, Rushworth MFS. 2012. On the relationship between the "default mode network" and the "social brain." *Front. Hum. Neurosci.* 6:189

Mars RB, Sallet J, Neubert FX, Rushworth MFS. 2013. Connectivity profiles reveal the relationship between brain areas for social cognition in human and monkey temporoparietal cortex. *PNAS* 110:10806–11

Marticorena DC, Ruiz AM, Mukerji C, Goddu A, Santos LR. 2011. Monkeys represent others' knowledge but not their beliefs. *Dev. Sci.* 14:1406–16

Martin A, Santos LR. 2016. What cognitive representations support primate theory of mind? *Trends Cogn. Sci.* 20:375–82

Melis AP, Hare B, Tomasello M. 2006. Chimpanzees recruit the best collaborators. *Science* 311:1297–300

Middlebrooks PG, Sommer MA. 2012. Neuronal correlates of metacognition in primate frontal cortex.

Neuron 75:517–30

Mitchell JP, Macrae CN, Banaji MR. 2006. Dissociable medial prefrontal contributions to judgments of similar and dissimilar others. *Neuron* 50:655–63

Miyamoto K, Osada T, Setsuie R, Takeda M, Tamura K, et al. 2017. Causal neural network of metamemory for retrospection in primates. *Science* 355:188–93

Moeller S, Crapse T, Chang L, Tsao DY. 2017. The effect of face patch microstimulation on perception of faces and objects. *Nat. Neurosci.* 20:743–52

Morin EL, Hadj-Bouziane F, Stokes M, Ungerleider LG, Bell AH. 2015. Hierarchical encoding of social cues in primate inferior temporal cortex. *Cereb. Cortex* 25:3036–45

Murray EA, Wise SP, Drevets WC. 2011. Localization of dysfunction in major depressive disorder: prefrontal cortex and amygdala. *Biol. Psychiatry* 69:e 43–54

Nachev P, Kennard C, Husain M. 2008. Functional role of the supplementary and pre-supplementary motor areas. *Nat. Rev. Neurosci.* 9:856–69

Neubert FX, Mars RB, Sallet J, Rushworth MFS. 2015. Connectivity reveals relationship of brain areas for reward-guided learning and decision making in human and monkey frontal cortex. *PNAS* 112:E 2695–704

Neubert FX, Mars RB, Thomas AG, Sallet J, Rushworth MFS. 2014. Comparison of human ventral frontal cortex areas for cognitive control and language with areas in monkey frontal cortex. *Neuron* 81:700–13

Nicolle A, Klein-Flügge MC, Hunt LT, Vlaev I, Dolan RJ, Behrens TEJ. 2012. An agent independent axis for executed and modeled choice in medial prefrontal cortex. *Neuron* 75:1114–21

Noonan MP, Mars RB, Rushworth MFS. 2011. Distinct roles of three frontal cortical areas in reward-guided behavior. *J. Neurosci.* 31:14399–412

Noonan MP, Sallet J, Mars RB, Neubert FX, O'Reilly JX, et al. 2014. A neural circuit covarying with social hierarchy in macaques. *PLOS Biol.* 12:e1001940

Noonan MP, Walton ME, Behrens TEJ, Sallet J, Buckley MJ, Rushworth MFS. 2010. Separate value comparison and learning mechanisms in macaque medial and lateral orbitofrontal cortex. *PNAS* 107:20547–52

O'Reilly JX, Croxson PL, Jbabdi S, Sallet J, Noonan MP, et al. 2013. Causal effect of disconnection lesions on interhemispheric functional connectivity in rhesus monkeys. *PNAS* 110:13982–87

Passingham RE, Wise SP. 2012. *The Neurobiology of the Prefrontal Cortex: Anatomy, Evolution, and the Origin of Insight.* Oxford, UK: Oxford Univ. Press

Perrett DI, Hietanen JK, Oram MW, Benson PJ. 1992. Organization and functions of cells responsive to faces in the temporal cortex. Philos. *Trans. R. Soc. B* 335:23–30

Perrett DI, Xiao D, Barraclough NE, Keysers C, Oram MW. 2009. Seeing the future: Natural image sequences produce "anticipatory" neuronal activity and bias perceptual report. *Q. J. Exp. Psychol.* 62:2081–104

Petrides M, Pandya D. 2007. Efferent association pathways from the rostral prefrontal cortex in the macaque monkey. *J. Neurosci.* 27:11573–86

Petrides M, Tomaiuolo F, Yeterian EH, Pandya DN. 2012. The prefrontal cortex: comparative architectonic organization in the human and the macaque monkey brains. *Cortex* 48:46–57

Pinsk MA, Arcaro M, Weiner KS, Kalkus JF, Inati SJ, et al. 2009. Neural representations of faces and body parts in macaque and human cortex: a comparative FMRI study. *J. Neurophysiol.* 101:2581–600

Pinsk MA, DeSimone K, Moore T, Gross CG, Kastner S. 2005. Representations of faces and body parts in macaque temporal cortex: a functional MRI study. *PNAS* 102:6996–7001

Pulliam HR, Caraco T. 1984. Living in groups: Is there an optimal group size? In *Behavioral Ecology: An Evolutionary Approach*, ed. JR Krebs, NB Davies, pp. 122–47. Sunderland, MA: Sinauer

Qu C, Ligneul R, Van der Henst JB, Dreher JC. 2017. An integrative interdisciplinary perspective on social dominance hierarchies. *Trends Cogn. Sci.* 21:893–908

Quallo MM, Price CJ, Ueno K, Asamizuya T, Cheng K, et al. 2009. Gray and white matter changes associated with tool-use learning in macaque monkeys. *PNAS* 106:18379–84

Rolls ET. 1999. *The Brain and Emotion*. Oxford, UK: Oxford Univ. Press

Rolls ET. 2007. The representation of information about faces in the temporal and frontal lobes. *Neuropsychologia* 45:124–43

Rudebeck PH, Buckley MJ, Walton ME, Rushworth MFS. 2006. A role for the macaque anterior cingulate gyrus in social valuation. *Science* 313:1310–12

Rudebeck PH, Murray EA. 2014. The orbitofrontal oracle: cortical mechanisms for the prediction and evaluation of specific behavioral outcomes. *Neuron* 84:1143–56

Rudebeck PH, Putnam PT, Daniels TE, Yang T, Mitz AR, et al. 2014. A role for primate subgenual cingulate cortex in sustaining autonomic arousal. *PNAS* 111:5391–96

Rudebeck PH, Saunders RC, Prescott AT, Chau LS, Murray EA. 2013. Prefrontal mechanisms of behavioral flexibility, emotion regulation and value updating. *Nat. Neurosci.* 16:1140–45

Ruff CC, Fehr E. 2014. The neurobiology of rewards and values in social decision making. *Nat. Rev. Neurosci.* 15:549–62

Rushworth MFS, Mars RB, Sallet J. 2013. Are there specialized circuits for social cognition and are they unique to humans? *Curr. Opin. Neurobiol.* 23:436–42

Rushworth MFS, Walton ME, Kennerley SW, Bannerman DM. 2004. Action sets and decisions in the medial frontal cortex. *Trends Cogn. Sci.* 8:410–17

Sallet J, Mars RB, Noonan MP, Andersson JL, O'Reilly JX, et al. 2011. Social network size affects neural circuits in macaques. *Science* 334:697–700

Saxe R. 2006. Uniquely human social cognition. *Curr. Opin. Neurobiol.* 16:235–39

Scholz J, Klein MC, Behrens TEJ, Johansen-Berg H. 2009. Training induces changes in white-matter architecture. *Nat. Neurosci.* 12:1370–71

Schülke O, Bhagavatula J, Vigilant L, Ostner J. 2010. Social bonds enhance reproductive success in male macaques. *Curr. Biol.* 20:2207–10

Schultz W. 2016. Dopamine reward prediction-error signalling: a two-component response. *Nat. Rev. Neurosci.* 17:183–95

Schultz W, Dayan P, Montague PR. 1997. A neural substrate of prediction and reward. *Science* 275:1593–99

Schwiedrzik CM, Zarco W, Everling S, Freiwald WA. 2015. Face patch resting state networks link face processing to social cognition. *PLOS Biol.* 13:e1002245

Seo H, Cai X, Donahue CH, Lee D. 2014. Neural correlates of strategic reasoning during competitive games. *Science* 346:340–43

Seo H, Lee D. 2007. Temporal filtering of reward signals in the dorsal anterior cingulate cortex during a mixed-strategy game. *J. Neurosci.* 27:8366–77

Sescousse G, Li Y, Dreher JC. 2015. A common currency for the computation of motivational values in the human striatum. *Soc. Cogn. Affect. Neurosci.* 10:467–73

Shackman AJ, Salomons TV, Slagter HA, Fox AS, Winter JJ, Davidson RJ. 2011. The integration of negative affect, pain and cognitive control in the cingulate cortex. *Nat. Rev. Neurosci.* 12:154–67

Shultz S, Dunbar R. 2010. Encephalization is not a universal macroevolutionary phenomenon in mammals but is associated with sociality. *PNAS* 107:21582–86

Sliwa J, Freiwald WA. 2017. A dedicated network for social interaction processing in the primate brain. *Science* 356:745–49

Spreckelmeyer KN, Krach S, Kohls G, Rademacher L, Irmak A, et al. 2009. Anticipation of monetary and social reward differently activates mesolimbic brain structures in men and women. *Soc. Cogn. Affect. Neurosci.* 4:158–65

Stolk A, Verhagen L, Toni I. 2016. Conceptual alignment: how brains achieve mutual understanding. *Trends Cogn. Sci.* 20:180–91

Suchak M, Eppley TM, Campbell MW, Feldman RA, Quarles LF, de Waal FBM. 2016. How chimpanzees cooperate in a competitive world. *PNAS* 113:10215–20

Sul S, Tobler PN, Hein G, Leiberg S, Jung D, et al. 2015. Spatial gradient in value representation along the medial prefrontal cortex reflects individual differences in prosociality. *PNAS* 112:7851–56

Suzuki S, Harasawa N, Ueno K, Gardner JL, Ichinohe N, et al. 2012. Learning to simulate others' decisions. *Neuron* 74:1125–37

Tremblay S, Sharika KM, Platt ML. 2017. Social decision-making and the brain: a comparative perspective. *Trends Cogn. Sci.* 21:265–76

Tsao DY, Freiwald WA, Knutsen TA, Mandeville JB, Tootell RBH. 2003. Faces and objects in macaque cerebral cortex. *Nat. Neurosci.* 6:989–95

Tsao DY, Freiwald WA, Tootell RBH, Livingstone MS. 2006. A cortical region consisting entirely of face-selective cells. *Science* 311:670–74

Tsao DY, Livingstone MS. 2008. Mechanisms of face perception. *Annu. Rev. Neurosci.* 31:411–37

Tsao DY, Moeller S, Freiwald WA. 2008a. Comparing face patch systems in macaques and humans. *PNAS* 105:19514–19

Tsao DY, Schweers N, Moeller S, Freiwald WA. 2008b. Patches of face-selective cortex in the macaque frontal lobe. *Nat. Neurosci.* 11:877–79

Ullsperger M, Fischer AG, Nigbur R, Endrass T. 2014. Neural mechanisms and temporal dynamics of performance monitoring. *Trends Cogn. Sci.* 18:259–67

Van Hoesen GW, Morecraft RJ, Vogt BA. 1993. Connections of the monkey cingulate cortex. In *Neurobiology of Cingulate Cortex and Limbic Thalamus,* ed. BA Vogt, M Gabriel, pp. 249–84. Boston: Birkhauser

Verguts T, Vassena E, Silvetti M. 2015. Adaptive effort investment in cognitive and physical tasks: a neuro-computational model. *Front. Behav. Neurosci.* 9:57

Walton ME, Behrens TEJ, Buckley MJ, Rudebeck PH, Rushworth MFS. 2010. Separable learning systems in the macaque brain and the role of orbitofrontal cortex in contingent learning. *Neuron* 65:927–39

Walum H, Waldman ID, Young LJ. 2016. Statistical and methodological considerations for the interpretation of intranasal oxytocin studies. *Biol. Psychiatry* 79:251–57

Watson KK, Platt ML. 2012. Social signals in primate orbitofrontal cortex. *Curr. Biol.* 22:2268–73

Wittmann MK, Kolling N, Akaishi R, Chau BK, Brown JW, et al. 2016a. Predictive decision making driven by multiple time-linked reward representations in the anterior cingulate cortex. *Nat. Commun.* 7:12327

Wittmann MK, Kolling N, Faber NS, Scholl J, Nelissen N, Rushworth MFS. 2016b. Self-other mergence in the frontal cortex during cooperation and competition. *Neuron* 91:482–93

Yoshida K, Saito N, Iriki A, Isoda M. 2011. Representation of others' action by neurons in monkey medial frontal cortex. *Curr. Biol.* 21:249–53

Yoshida K, Saito N, Iriki A, Isoda M. 2012. Social error monitoring in macaque frontal cortex. *Nat. Neurosci.* 15:1307–12

Yoshida W, Seymour B, Friston KJ, Dolan RJ. 2010. Neural mechanisms of belief inference during cooperative games. *J. Neurosci.* 30:10744–51

Young C, Majolo B, Heistermann M, Schü lke O, Ostner J. 2014a. Responses to social and environmental stress are attenuated by strong male bonds in wild macaques. *PNAS* 111:18195–200

Young C, Majolo B, Schü lke O, Ostner J. 2014b. Male social bonds and rank predict supporter selection in cooperative aggression in wild Barbary macaques. *Anim. Behav.* 95:23–32

Yukie M, Shibata H. 2009. Temporocingulate interactions in the monkey. In *Cingulate Neurobiology and Disease,* ed. B Vogt, pp. 95–162. New York: Oxford Univ. Press

Zhu L, Mathewson KE, Hsu M. 2012. Dissociable neural representations of reinforcement and belief prediction errors underlie strategic learning. *PNAS* 109:1419–24

图书在版编目(CIP)数据

从心灵到神经：认知神经科学文献选编 = From
Mind to Nervous System： Selected Articles on
Cognitive Neuroscience / 燕燕，李福华，白玉主编
. -- 北京：社会科学文献出版社，2019.12
　　ISBN 978 - 7 - 5201 - 5919 - 7

　　Ⅰ.①从… 　Ⅱ.①燕… ②李… ③白… 　Ⅲ.①认知科
学 - 文集 　Ⅳ. ①B842.1

中国版本图书馆 CIP 数据核字(2019)第 291531 号

从心灵到神经：认知神经科学文献选编

主　　编 / 燕　燕　李福华　白　玉

出 版 人 / 谢寿光
责任编辑 / 周　琼
文稿编辑 / 李家莲　杜红春

出　　版 / 社会科学文献出版社·社会政法分社 (010) 59367156
　　　　　地址：北京市北三环中路甲 29 号院华龙大厦　邮编：100029
　　　　　网址：www. ssap. com. cn
发　　行 / 市场营销中心 (010) 59367081　59367083
印　　装 / 北京盛通印刷股份有限公司

规　　格 / 开　本：787mm × 1092mm　1/16
　　　　　印　张：35　字　数：743 千字
版　　次 / 2019 年 12 月第 1 版　2019 年 12 月第 1 次印刷
书　　号 / ISBN 978 - 7 - 5201 - 5919 - 7
定　　价 / 168.00 元